# Nonlinear Physical Science

# Nonlinear Physical Science

*Nonlinear Physical Science* focuses on recent advances of fundamental theories and principles, analytical and symbolic approaches, as well as computational techniques in nonlinear physical science and nonlinear mathematics with engineering applications.

Topics of interest in *Nonlinear Physical Science* include but are not limited to:

- New findings and discoveries in nonlinear physics and mathematics
- Nonlinearity, complexity and mathematical structures in nonlinear physics
- Nonlinear phenomena and observations in nature and engineering
- Computational methods and theories in complex systems
- Lie group analysis, new theories and principles in mathematical modeling
- Stability, bifurcation, chaos and fractals in physical science and engineering
- Nonlinear chemical and biological physics
- Discontinuity, synchronization and natural complexity in the physical sciences

## Series editors

Albert C.J. Luo
Department of Mechanical and Industrial Engineering
Southern Illinois University Edwardsville
Edwardsville, IL 62026-1805, USA
e-mail: aluo@siue.edu

Nail H. Ibragimov
Department of Mathematics and Science
Blekinge Institute of Technology
S-371 79 Karlskrona, Sweden
e-mail: nib@bth.se

## International Advisory Board

More information about this series at http://www.springer.com/series/8389

Albert C.J. Luo

# Discretization and Implicit Mapping Dynamics

Higher
Education
Press

 Springer

Albert C.J. Luo
Department of Mechanical and Industrial
  Engineering
Southern Illinois University Edwardsville
Edwardsville, IL
USA

ISSN 1867-8440             ISSN 1867-8459   (electronic)
Nonlinear Physical Science
ISBN 978-3-662-47274-3       ISBN 978-3-662-47275-0   (eBook)
DOI 10.1007/978-3-662-47275-0

Jointly published with Higher Education Press, Beijing
ISBN: 978-7-04-042835-3 Higher Education Press, Beijing

Library of Congress Control Number: 2015939425

Springer Heidelberg New York Dordrecht London

Printed on acid-free paper

Springer-Verlag GmbH Berlin Heidelberg is part of Springer Science+Business Media
(www.springer.com)

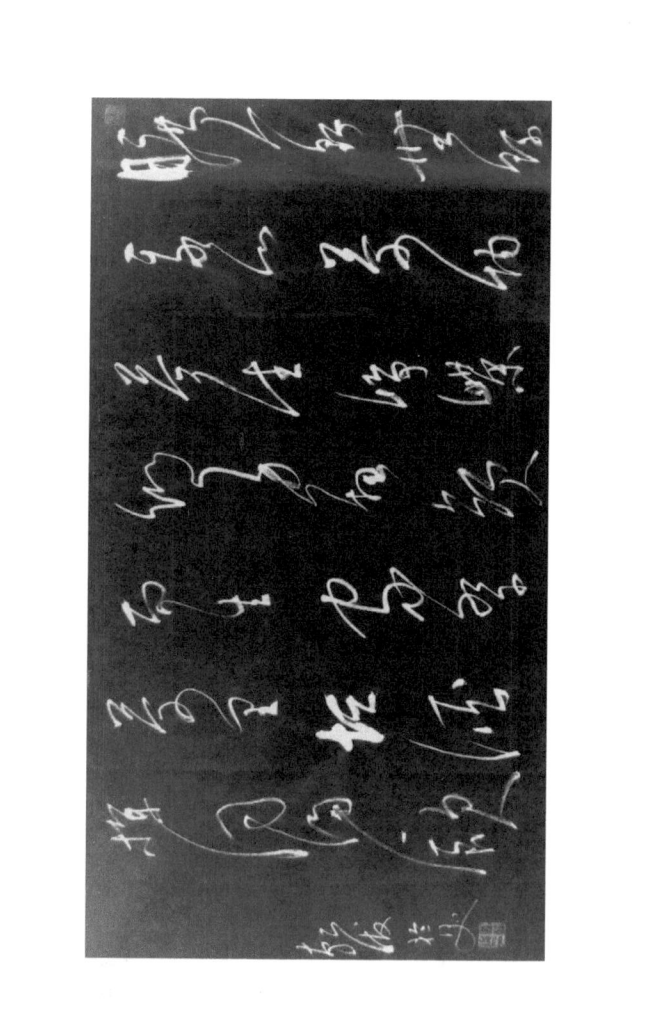

# Preface

This book discusses discretization of differential equations of continuous nonlinear systems and implicit mapping dynamics of periodic flows to chaos. In recent years, approximate analytical solutions for periodic motions to chaos in continuous nonlinear systems were developed by the author through finite Fourier series. However, for many nonlinear dynamical systems, it is difficult to achieve such approximate analytical solutions of periodic motions to chaos. With computer extensive applications in numerical computations, one has used the discrete forms of differential equations of nonlinear systems to obtain numerical solutions via recurrent iterations. The discrete forms in recurrent iterations will cause accumulated computational errors for numerical results. Once the iteration number increases, numerical results given by the discrete forms cannot approximately represent true solutions of nonlinear dynamical systems. To improve the computational accuracy, one has tried to adopt implicit maps as discrete forms to achieve numerical results. However, such implicit mapping forms cannot be iterated directly, which cause the difficulty to extensive applications of discrete implicit maps in continuous nonlinear systems. In this book, the author would like to systematically discuss implicit mapping dynamics of periodic motions to chaos in continuous dynamical systems, and discrete Fourier series based on the discrete nodes of periodic motions will be used to obtain the harmonic responses in frequency space, which can be measured from experiments.

This book includes six chapters. In Chap. 1, a brief literature survey is completed. Chapter 2 reviewed the nonlinear theory for stability and bifurcation of fixed points in discrete nonlinear systems. In Chap. 3, discretization of differential equations is discussed comprehensively. The explicit and implicit discrete schemes in nonlinear dynamical system are discussed through one-step and multi-step discretization of differential equations, and the corresponding stability and convergence of the explicit and implicit discrete maps are discussed. In Chap. 4, implicit mapping dynamics of period-m fixed points in discrete dynamical systems are discussed with positive and negative discrete maps, and the complete solutions of Ying-Yang states of period-m fixed points are presented. In Chap. 5, the methodology for the solutions of periodic motions in continuous dynamical systems with/without time delay is presented through the mapping dynamics of discrete implicit mappings under specific truncated errors. The discrete Fourier series of periodic motions are discussed from discrete nodes of periodic motions, and the corresponding approximate analytical expression can be obtained. Harmonic amplitude quantity levels can be analyzed for periodic motions in continuous nonlinear systems. Chapter 6 discusses the bifurcation trees of periodic motions to chaos in the Duffing oscillator to demonstrate the implicit mapping dynamics of the discretized Duffing oscillator. Such semi-analytical results of periodic motions in the Duffing oscillator are compared with the approximate analytical solutions of periodic motions based on the finite Fourier series solutions.

Finally, I would like to appreciate my former student, Dr. Yu Guo, for completing all numerical computations. Herein, I thank my wife (Sherry X. Huang) and my children (Yanyi Luo, Robin Ruo-Bing Luo, and Robert Zong-Yuan Luo) for their understanding and infinite support.

Albert C.J. Luo

# Contents

# Chapter 1
# Introduction

For solutions of periodic motions in nonlinear dynamical systems, analytical and numerical techniques have been adopted. The analytical methods include the method of averaging, perturbation methods, harmonic balance method, and generalized harmonic balance method. Through the analytical methods, one can obtain the analytical expressions of approximate solutions of periodic motions in dynamical systems. The numerical methods are based on discrete maps obtained by discretization of differential equations for dynamical systems. The discrete maps include explicit and implicit maps. The explicit maps can be directly used to obtain numerical solutions of differential equations for dynamical systems, but the computational errors for the recurrence iteration of explicit maps will be accumulated in numerical results. Once the recurrence iteration times become large, the numerical results may not be adequate for numerical solutions of dynamical systems. Herein, implicit maps will be used to develop mapping structures for periodic motions. The implicit maps cannot be simply used by the recurrence iteration. For periodic flows in nonlinear dynamics, mapping structures based on implicit maps can be developed. Of course, an explicit mapping can be expressed by an implicit map as a special case. Based on the mapping structures, analytical prediction of periodic flows in nonlinear dynamical systems can be completed. The mapping structure gives a set of nonlinear algebraic equations, which can be solved. Without the recurrence iteration, the solution errors of node points of periodic flows are fixed without computational errors caused by iterations. The purpose of this book is to develop a semi-analytical method for periodic flows to chaos in nonlinear dynamical systems with/without time delay through implicit mapping structures.

## 1.1 A Brief History

To determine periodic flows in nonlinear dynamical systems, existing techniques for periodic motions in nonlinear systems are reviewed briefly. The analytical methods for periodic motions are discussed first. Lagrange (1788) developed the method of averaging for periodic motions in the three-body problem as a perturbation of the two-body problem. The idea is based on the solutions of linear

© Higher Education Press, Beijing and Springer-Verlag Berlin Heidelberg 2015
A.C.J. Luo, *Discretization and Implicit Mapping Dynamics*,
Nonlinear Physical Science, DOI 10.1007/978-3-662-47275-0_1

systems. Such an idea was further extended by Poincare in the end of the nineteenth century. Thus, Poincare (1899) developed the perturbation theory for motions of celestial bodies. van der Pol (1920) used the method of averaging for the periodic solutions of oscillation systems in circuits. Such an application caused great interest in the perturbation theory for the approximate analytical solution of periodic motions in nonlinear systems. Until 1928, the asymptotic validity of the method of averaging was not proved. Fatou (1928) gave the proof of the asymptotic validity through the solution existence theorems of differential equations. Krylov and Bogoliubov (1935) further developed the method of averaging, and the detailed presentation was given in Bogoliubov and Mitropolsky (1961). Hayashi (1964) presented the perturbation methods including averaging method and principle of harmonic balance. Barkham and Soudack (1969) extended the Krylov–Bogoliubov method for the approximate solutions of nonlinear autonomous second-order differential equations [also see, Barkham and Soudack (1970)]. Nayfeh (1973) employed the multiple-scale perturbation method to develop approximate solutions of periodic motions in the Duffing oscillators. Holmes and Rand (1976) discussed the stability and bifurcation of periodic motions in the Duffing oscillator. Nayfeh and Mook (1979) used the perturbation method to investigate nonlinear structural vibrations, and Holmes (1979) demonstrated chaotic motions in nonlinear oscillators through the Duffing oscillator with a twin-well potential. Ueda (1980) numerically simulated chaos by period-doubling of periodic motions of Duffing oscillators. A generalized harmonic balance approach was used by Garcia-Margallo and Bejarano (1987) to determine approximate solutions of nonlinear oscillations with strong nonlinearity. Rand and Armbruster (1987) determined the stability of periodic solutions by the perturbation method and bifurcation theory. Yuste and Bejarano (1989) employed the elliptic functions instead of trigonometric functions to extend the Krylov–Bogoliubov method. Coppola and Rand (1990) used the averaging method with elliptic functions for approximation of limit cycle. Wang et al. (1992) used the harmonic balance method and the Floquet theory to investigate the nonlinear behaviors of the Duffing oscillator with a bounded potential well [also see, Kuo et al. (1992)]. Luo and Han (1997) determined the stability and bifurcation conditions of periodic motions of the Duffing oscillator. However, only symmetric periodic motions of the Duffing oscillators were investigated. Luo and Han (1999) investigated the analytical prediction of chaos in nonlinear rods through the Duffing oscillator. Peng et al. (2008) presented the approximate symmetric solution of period-1 motions in the Duffing oscillator by the harmonic balance method with three harmonic terms. Luo (2012a) developed a generalized harmonic balance method for the approximate analytical solutions of periodic motions and chaos in nonlinear dynamical systems. This method used the finite-term Fourier series to approximately express periodic motions, and the coefficients are time-varying. With averaging, a dynamical system of coefficients is obtained, and from such a dynamical system, the approximate solutions of periodic motions are achieved and the corresponding stability and bifurcation analysis are completed. Luo and Huang (2012a) used such a generalized harmonic balance method with finite terms for the analytical solutions of period-1 motions in the Duffing oscillator with a twin-well potential. Luo and Huang (2012b) also employed a generalized harmonic balance method to find analytical solutions of period-$m$ motions in such a Duffing oscillator. The analytical bifurcation trees of periodic motions in the Duffing oscillator to chaos were obtained [also see, (Luo and Huang 2012c, d, 2013a, b, c, d)]. Such analytical bifurcation trees show the connection from periodic solution to chaos analytically. For a better understanding of nonlinear behaviors in nonlinear dynamical systems, analytical bifurcation trees of period-1 motions to chaos in a periodically forced oscillator with quadratic nonlinearity were presented in Luo and Yu (2013a, b, 2015), and period-m motions in the periodically forced van der Pol equation were presented in Luo and Laken (2013). The analytical solutions of periodic oscillations in the van der Pol oscillator can be used to verify the conclusions in Cartwright and Littlewood (1947) and Levinson (1948). The results for the parametric quadratic nonlinear oscillator in Luo and Yu (2014) analytically show the complicated period-1 motions and the corresponding bifurcation structures. The detailed presentation for analytical methods for periodic flows in nonlinear dynamical systems can be found in Luo (2014a, b).

In recent years, time-delayed systems are of great interest since such systems extensively exist in engineering (e.g., Tlusty 2000; Hu and Wang 2002). The infinite dimensional state space causes the significant difficulty to solve such time-delayed problems. Thus, one used numerical methods for the corresponding complicated behaviors. On the other hand, one is interested in the stability and bifurcation of equilibriums of the time-delayed systems (e.g., Stepan 1989; Sun 2009; Insperger and Stepan 2011). In addition, one is also interested in analytical solutions of periodic motions in time-delayed dynamical systems. Perturbation methods have been used for such periodic motions in delayed dynamical systems. For instance, the approximate solutions of the time-delayed nonlinear oscillator were investigated by the method of multiple scales (e.g., Hu et al. 1998; Wang and Hu 2006). The harmonic balance method was also used to determine approximate solutions of periodic motions in delayed nonlinear oscillators [e.g., MacDonald (1995); Liu and Kalmar-Nagy (2010); Lueng and Guo (2014)]. However, such approximate solutions of periodic motions in the time-delayed oscillators are based on one or two harmonic terms, which are not accurate enough. In addition, the corresponding stability and bifurcation analysis of such approximate solutions of periodic motions may not be adequate. Luo (2013) presented an alternative way for the accurate analytical solutions of periodic flows in time-delayed dynamical systems (see also, Luo 2014c). This method is without any small-parameter requirement. In addition, this approach can also be applied to the coefficient varying with time. Luo and Jin (2014a) analytically presented the bifurcation tree of period-1 motions to chaos in a periodically forced, time-delayed quadratic nonlinear oscillator. Luo and Jin (2014b, c, d) discussed the bifurcation trees of period-m motions to chaos in the periodically forced Duffing oscillator with a linear time-delayed displacement.

From the literature survey, for some simple nonlinear systems, the approximate analytical solutions of periodic motions can be obtained. However, for most of the nonlinear dynamical systems, it is very difficult to obtain analytical solutions of periodic motions. Thus, numerical results of periodic motions in complicated nonlinear dynamical systems become very significant in engineering. In fact, human being has a long history as old as human civilization to use numerical algorithms to get approximate numerical results instead of exact results. For instance, the Rhind Papyrus of ancient Egypt describes a root-finding method for solving a simple equation in about 1650 BC, and Archimedes of Syracuse (287–212 BC) used numerical algorithm to approximately compute lengths, areas, and volumes of geometric figures. Based on the ideas and spirits of numerical approximations, Isaac Newton and Gottfried Leibnitz developed the calculus by infinitesimal elements to linear approximation and infinitesimal summarization to integration. Because of calculus development, one can describe more complicated mathematical models for real physical problems, but it is very difficult to solve such accurate mathematical models explicitly. This is an important impetus for one to develop numerical methods to get approximate solutions of the accurate mathematical models. Thus, Newton developed several numerical methods to find approximate solutions. For instance, numerical methods for root-finding and polynomial interpolation were developed by Newton. Since then, Euler (1707–1783), Lagrange (1736–1813), and Gauss (1777–1855) further developed numerical methods for approximate results, such as Euler method for differential equations, Lagrange interpolation method, and Gauss interpolation. The more detailed information about numerical methods can be found in Goldestine (1977).

This book will focus on numerical methods for nonlinear dynamical systems. For this issue, Euler developed an explicit method to achieve approximate solutions numerically. Such Euler method is a one-step discrete method. This method is still used in numerical computation, but its computational accuracy is very low, and numerical solutions are not accurate. Bashforth and Adams (1883) presented a multistep discrete method for numerical solutions of differential equations. Moulton (1926) extended such a method to the Adams–Moulton method. The Adams–Bashforth method is the explicit method as a predictor, and the Adams–Moulton method is the implicit method as a corrector. In addition, the Adams–Bashforth method can be extended for the practical application of the Taylor series method as presented in Nordsieck (1962). Milne (1949) used the entire interval for integration based on Newton–Cotes quadrature formulas. The recent theory of linear multi-step method was systematically discussed by Dahlquist (1956, 1959). The general formulas were presented, and the corresponding consistency, stability, and convergence were discussed by the linear stability theory. Runge (1895) started modern one-step methods with the order of two and three for numerical solutions of differential equations. Heun (1900) raised the order of the method from two and three to four. Kutta (1901) gave the formulation of the method with the order conditions. Nystrom (1925) made the correction of the fifth-order method of Kutta and showed how to apply the Runge–Kutta method to the second-order differential equations. Butcher (1963) discussed the coefficients of Runge–Kutta method, and the implicit Runge–Kutta methods were presented in Butcher (1964, 1975).

With extensive applications of computers, numerical computations become very popular to obtain numerical results for differential equations through discretization.

Once the discrete maps are obtained for dynamical systems, discrete dynamical systems can be used to investigate nonlinear dynamics of dynamical systems. Based on nonlinear maps, one discovered the existence of chaotic motions in nonlinear dynamical systems through iteration of discrete maps.

In 2005, Luo (2005a, b) presented a mapping dynamics of discrete dynamical systems which is a more generalized symbolic dynamics. The systematical description of mapping dynamics in discontinuous dynamical systems was presented in Luo (2009). The discrete maps can be any implicit and/or explicit functions rather than explicit maps in numerical iterative methods only. From discrete mapping structures, periodic motions in discrete dynamical systems can be predicted analytically, and the stability and bifurcation analysis of periodic motions in nonlinear dynamical systems can be completed. Such an idea was applied to discontinuous dynamical systems in Luo (2009, 2012b, c).

## 1.2  Book Layout

The main body in this book will discuss discretization of differential equations of nonlinear continuous dynamical systems to obtain implicit maps for periodic flows. The mapping structures will be employed to analytically predict the periodic flows in nonlinear continuous systems, and the corresponding stability and bifurcation can be discussed.

In Chap. 2, a theory for nonlinear discrete systems is reviewed. The local and global theories of stability and bifurcation for nonlinear discrete systems are discussed. The stability switching and bifurcation on specific eigenvectors of the linearized system at fixed points under a specific period are presented. The higher order singularity and stability for nonlinear discrete systems on the specific eigenvectors are discussed.

In Chap. 3, the discretization of continuous systems is presented. The explicit and implicit discrete maps are discussed for numerical predictions of continuous systems. Basic discrete schemes are presented which include forward and backward Euler methods, and midpoint and trapezoidal rule methods. An introduction to Runge–Kutta methods is presented, and the Taylor series method and second-order Runge–Kutta method are introduced. The explicit Runge–Kutta methods for third and fourth order are systematically presented. The implicit Runge–Kutta methods are discussed based on the polynomial interpolation, which include a generalized implicit Runge–Kutta method, Guass method, Radau method, and Lotta methods. In addition to one-step methods, implicit and explicit multi-step methods are discussed, including Adams–Bashforth method, Adams–Moulton methods, and explicit and implicit Adams methods.

In Chap. 4 presented is a Ying–Yang theory for implicit, discrete, nonlinear systems with consideration of positive and negative iterations of discrete iterative maps. In existing analysis, the solutions relative to "Yang" in nonlinear dynamical systems are extensively investigated. However, the solutions pertaining to "Ying" in nonlinear dynamical systems are not discussed too much. A set of concepts on "Ying" and "Yang" in implicit, nonlinear, discrete dynamical systems are introduced. Based on the Ying–Yang theory, the complete dynamics of implicit discrete systems can be discussed. A discrete dynamical system with the Henon map is investigated as an example. Period-m solutions, stability, and bifurcations for multi-step, implicit discrete systems are discussed.

In Chap. 5, periodic flows in continuous nonlinear systems are discussed through discrete implicit mappings. The period-1 flows in nonlinear systems are discussed by the one-step discrete maps, and then, the period-m flows in nonlinear dynamical systems are also discussed through the one-step discrete maps. Multi-step, implicit discrete maps are employed to discuss the period-1 and period-m motions in nonlinear dynamical systems. Periodic flows in nonlinear time-delayed dynamical systems are discussed with time-delay discrete nodes interpolated by two non-delay discrete nodes. In addition, periodic flows in time-delayed nonlinear dynamical systems are also discussed through the delay nodes determined by integration. Through the discrete nodes in periodic flows, the periodic flows are approximated by the discrete Fourier series and the frequency space of the periodic flows can be determined through amplitude spectrums.

In Chap. 6, periodic motions in the Duffing oscillator are discussed through the mapping structures of discrete implicit maps. The discrete implicit maps are obtained from the differential equation of the Duffing oscillator. From mapping structures, bifurcation trees of periodic motions are predicted analytically through nonlinear algebraic equations of implicit maps, and the corresponding stability and bifurcation analysis of periodic motions in the bifurcation trees are presented. The bifurcation trees of periodic motions are also presented through the harmonic amplitudes of the discrete Fourier series. Finally, from the analytical prediction, numerical simulation results of periodic motions are performed to verify the analytical prediction. The harmonic amplitude spectrums are also presented, and the corresponding analytical expression of periodic motions can be obtained approximately.

# References

Barkham, P. G. D., & Soudack, A. C. (1969). An extension to the method of Krylov and Bogoliubov. *International Journal of Control, 10*, 377–392.

Barkham, P. G. D., & Soudack, A. C. (1970). Approximate solutions of nonlinear, non-autonomous second-order differential equations. *International Journal of Control, 11*, 763–767.

Bashforth, F., & Adams, J. C. (1883). *Theories of capillary action*. Cambridge University Press: London.

Bogolyubov, N. N., & Mitropolsky, Y. A. (1961). *Asymptotic methods in the theory of non-linear oscillations*. New York: Gordon and Breach.

Butcher, J. C. (1963). Coefficients for the study of Rung-Kutta integration processes. *Journal of Australian Mathematical Society, 3*, 185–201.

Butcher, J. C. (1964). Implicit Runge-Kutta processes. *Mathematics of Computation, 19*, 408–417.

Butcher, J. C. (1975). A stability property of implicit Runge-Kutta methods. *BIT Numerical Mathematics, 15*, 358–361.

Cartwright, M. L., & Littlewood, J. E. (1947). On nonlinear differential equations of the second order II. The equation $\ddot{y} + kf(y)\dot{y} + g(y,k) = p(t) = p_1(t) + kp_2(t)$, $k > 0$, $f(y) \geq 1$. *Annals of Mathematics, 48*, 472–494.

Coppola, V. T., & Rand, R. H. (1990). Averaging using elliptic functions: Approximation of limit cycle. *Acta Mechanica, 81*, 125–142.

Dahlqist, G. (1956). Convergence and stability in the numerical integration of ordinary differential equations. *Mathematica Scandinavica, 4*, 33–53.

Dahlqist, G. (1959). Stability and error bounds in the numerical integration of ordinary differential equations. *Trans Royal Inst Technology, 130*, 1–87. Stockholm.

Fatou, P. (1928). Sur le mouvement d'un systeme soumis `a des forces a courte periode. *Bull Soc Math, 56*, 98–139.

Garcia-Margallo, J. D., & Bejarano, J. D. (1987). A generalization of the method of harmonic balance. *Journal of Sound and Vibration, 116*, 591–595.

Goldestine, H. (1977). *A history of numerical analysis: From the 16th through the 19th century.* New York: Springer.

Hayashi, C. (1964). *Nonlinear oscillations in physical systems.* New York: McGraw-Hill Book Company.

Holmes, P. J. (1979). A nonlinear oscillator with strange attractor. *Philosophical Transactions of the Royal Society, A292*, 419–448.

Holmes, P. J., & Rand, D. A. (1976). Bifurcations of duffing equation; An application of catastrophe theory. *Quarterly Applied Mathematics, 35*, 495–509.

Hu, H. Y., & Wang, Z. H. (2002). *Dynmaics of controlled mechanical systems with delayed feedback.* Berlin: Springer.

Hu, H. Y., Dowell, E. H., & Virgin, L. N. (1998). Resonance of harmonically forced duffing oscillator with time-delay state feedback. *Nonlinear Dynamics, 15*(4), 311–327.

Huen, K. (1900). Neue methoden zur approximativen inegration de differentialgleichungen einer unabhangigen veranderlichen. *Z Math Phys, 45*, 23–38.

Insperger, T., & Stepan, G. (2011). *Semi-discretization for time-delay systems: Stability and engineering applications.* New York: Springer.

Kao, Y. H., Wang, C. S., & Yang, T. H. (1992). Influences of harmonic coupling on bifurcations in duffing oscillator with bounded potential wells. *Journal of Sound and Vibration, 159*, 13–21.

Krylov, N. M., & Bogolyubov, N. N. (1935). *Methodes approchees de la mecanique non-lineaire dans leurs application a l'Aeetude de la perturbation des mouvements periodiques de divers phenomenes de resonance s'y rapportant*, Academie des Sciences d'Ukraine:Kiev (in French).

Kutta, W. (1901). Beitrag zur naherungsweisen intergration totaler differentialgleichungen. *Z Math Phys, 46*, 435–453.

Lagrange, J. L. (1788). *Mecanique Analytique* (Vol. 2), (edition Albert Balnchard: Paris, 1965).

Leung, A. Y. T., & Guo, Z. (2014). Bifurcation of the periodic motions in nonlinear delayed oscillators. *Journal of Vibration and Control, 20*, 501–517.

Levinson, N. (1948). A simple second order differential equation with singular motions. In: *Proceedings of the National Academy of Science of the United States of America* (Vol. *34*, Issue No. 1, pp. 13–15).

Liu, L., & Kalmar-Nagy, T. (2010). High-dimensional harmonic balance analysis for second-order delay-differential equations. *Journal of Vibration and Control, 16*(7–8), 1189–1208.

Luo, A. C. J. (2005a). The mapping dynamics of periodic motions for a three-piecewise linear system under a periodic excitation. *Journal of Sound and Vibration, 283*, 723–748.

Luo, A. C. J. (2005b). A theory for non-smooth dynamic systems on the connectable domains. *Communications in Nonlinear Science and Numerical Simulation, 10*, 1–55.

Luo, A. C. J. (2009). *Discontinuous dynamical systems on time-varying domains.* Higher Education Press/Springer: Beijing/Heidelberg.

Luo, A. C. J. (2012a). *Continuous Dynamical Systems.* Beijing/Glen Carbon: Higher Education Press/L and H Scientific.

Luo, A. C. J. (2012b). *Regularity and complexity in dynamical systems*. New York: Springer.

Luo, A. C. J. (2012c). *Discrete and switching dynamical systems*. Beijing/Glen Carbon: Higher Education Press/L and H Scientific.

Luo, A. C. J. (2013). Analytical solutions for periodic motions to chaos in nonlinear systems with/without time-delay. *International Journal of Dynamics and Control, 1*, 330–359.

Luo, A. C. J. (2014a). *Toward analytical chaos in nonlinear systems*. New York: Wiley.

Luo, A. C. J. (2014b). *Analytical routes to chaos in nonlinear engineering*. New York: Wiley.

Luo, A. C. J. (2014c). On analytical routes to chaos in nonlinear systems. *International Journal of Bifurcation and Chaos, 24*, Article no.: 1430013 (28 pages).

Luo, A. C. J., & Han, R. P. S. (1997). A quantitative stability and bifurcation analyses of a generalized duffing oscillator with strong nonlinearity. *Journal of Franklin Institute, 334B*, 447–459.

Luo, A. C. J., & Han, R. P. S. (1999). Analytical predictions of chaos in a nonlinear rod. *Journal of Sound and Vibration, 227*(3), 523–544.

Luo, A. C. J., & Huang, J. Z. (2012a). Approximate solutions of periodic motions in nonlinear systems via a generalized harmonic balance. *Journal of Vibration and Control, 18*, 1661–1671.

Luo, A. C. J., & Huang, J. Z. (2012b). Analytical dynamics of period-m flows and chaos in nonlinear systems. *International Journal of Bifurcation and Chaos, 22*(4), Article No. 1250093 (29 pages).

Luo, A. C. J., & Huang, J. Z. (2012c). Analytical routes of period-1 motions to chaos in a periodically forced Duffing oscillator with a twin-well potential. *Journal of Applied Nonlinear Dynamics, 1*, 73–108.

Luo, A. C. J., & Huang, J. Z. (2012d). Unstable and stable period-*m* motions in a twin-well potential duffing oscillator. *Discontinuity, Nonlinearity, and Complexity, 1*, 113–145.

Luo, A. C. J., & Huang, J. Z. (2013a). Analytical solutions for asymmetric periodic motions to chaos in a hardening duffing oscillator. *Nonlinear Dynamics, 72*, 417–438.

Luo, A. C. J., & Huang, J. Z. (2013b). Analytical period-3 motions to chaos in a hardening duffing oscillator. *Nonlinear Dynamics, 73*, 1905–1932.

Luo, A. C. J., & Huang, J. Z. (2013c). An analytical prediction of period-1 motions to chaos in a softening Duffing oscillator. *International Journal of Bifurcation and Chaos, 23*, Article No: 1350086 (31 pages).

Luo, A. C. J., & Huang, J. Z. (2013d). Period-3 motions to chaos in a softening Duffing oscillator. *International Journal of Bifurcation and Chaos, 24*(3), Article no.: 1430010 (26 pages).

Luo, A. C. J., & Jin, H. X. (2014a). Bifurcation trees of period-m motions in a periodically forced, time-delayed, quadratic nonlinear oscillator. *Discontinuity, Nonlinearity, and Complexity, 3*, 87–107.

Luo, A. C. J., & Jin, H. X. (2014b). Complex period-1 motions of a periodically forced duffing oscillator with a time-delay feedback. *International Journal of Dynamics and Control*. doi:10.1007/s40435-014-0091-8.

Luo, A. C. J., & Jin, H. X. (2014c). Period-m motions to chaos in a periodically forced duffing oscillator with a time-delayed displacement. *International Journal of Bifurcation and Chaos*. doi:10.1142/S0218127414501260.

Luo, A. C. J., & Jin, H. X. (2014d). Period-3 motions to chaos in a periodically forced duffing oscillator with a linear time-delay. *International Journal of Dynamics and Control*. doi:10.1007/s40435-014-0116-3.

Luo, A. C. J., & Laken, A. B. (2013). Analytical solutions for period-m motions in a periodically forced van der Pol oscillator. *International Journal of Dynamics and Control, 1*(2), 99–115.

Luo, A. C. J., & Yu, B. (2013a). Analytical solutions for stable and unstable period-1 motion in a periodically forced oscillator with quadratic nonlinearity. *ASME Journal of Vibration and Acoustics, 135*, Article No: 034503 (5 pages).

Luo, A. C. J., & Yu, B. (2013b). Period-*m* motions and bifurcation trees in a periodically excited, quadratic nonlinear oscillator. *Discontinuity, Nonlinearity, and Complexity, 2*, 263–288.

Luo, A. C. J., & Yu, B. (2014). Bifurcation tree of periodic motions to chaos in a parametric, quadratic nonlinear oscillator. *International Journal of Bifurications and Chaos, 24*, Article no.: 1450075 (28 pages).

Luo, A. C. J., & Yu, B. (2015). Complex period-1 motions in a periodically forced, quadratic nonlinear oscillator. *Journal of Vibration and Control. 21*, 896–906.

MacDonald, N. (1995). Harmonic balance in delay-differential equations. *Journal of Sounds and Vibration, 186*(4), 649–656.

Milne, W. E. (1949). A note on the numerical integration of differential equations. *Journal of Research of the National Bureau of Standards, 43*, 537–542.

Moulton, F. R. (1926). *New methods in exterier balistics*. University of Chicago Press: Chicago.

Nayfeh, A. H. (1973). *Perturbation methods*. New York: John Wiley.

Nayfeh, A. H., & Mook, D. T. (1979). *Nonlinear oscillation*. New York: John Wiley.

Norsieck, A. (1962). On numerical integration of ordinary differential equation. *Mathematics of Computation, 16*, 22–49.

Nystrom, E. J. (1925). Uber die numerische integration von differentialgleichungen. *Acta Soc Sci Fennicae, 50*(13), 1–55.

Peng, Z. K., Lang, Z. Q., Billings, S. A., & Tomlinson, G. R. (2008). Comparisons between harmonic balance and nonlinear output frequency response function in nonlinear system analysis. *Journal of Sound and Vibration, 311*, 56–73.

Poincare, H. (1899). *Methodes Nouvelles de la Mecanique Celeste* (Vol. 3). Paris: Gauthier-Villars.

Rand, R.H. and Armbruster, D. (1987). *Perturbation Methods, Bifurcation Theory, and Computer Algebra* (Applied Mathematical Sciences, No. **65**, Springer-Verlag, New York).

Runge, C. (1895). Uberdie numerische auflosung von differentialgleichungen. *Mathematische Annalen, 46*, 167–178.

Stepan, G. (1989). *Retarded dynamical systems*. Harlow: Longman.

Sun, J. Q. (2009). A method of continuous time approximation of delayed dynamical systems. *Communications in Nonlinear Science and Numerical Simulation, 14*(4), 998–1007.

Tlusty, J. (2000). *Manufacturing processes and equipment*. New Jersey: Prentice Hall.

Ueda, Y. (1980). Explosion of strange attractors exhibited by the Duffing equations. *Annuals of the New York Academy of Science, 357*, 422–434.

van der Pol, B. (1920). A theory of the amplitude of free and forced triode vibrations. *Radio Review, 1*(701–710), 754–762.

Wang, H., & Hu, H. Y. (2006). Remarks on the perturbation methods in solving the second order delay differential equations. *Nonlinear Dynamics, 33*, 379–398.

Wang, C. S., Kao, Y. H., Huang, J. C., & Gou, Y. H. (1992). Potential dependence of the bifurcation structure in generalized Duffing oscillators. *Physical Review A, 45*, 3471–3485.

Yuste, S. B., & Bejarano, J. D. (1989). Extension and improvement to the Krylov-Bogoliubov method that use elliptic functions. *International Journal of Control, 49*, 1127–1141.

# Chapter 2
# Nonlinear Discrete Systems

In this chapter, a theory for nonlinear discrete systems is reviewed. The local and global theory of stability and bifurcation for nonlinear discrete systems is presented. The stability switching and bifurcation on specific eigenvectors of the linearized system at fixed points under a specific period are discussed. The higher-order singularity and stability for nonlinear discrete systems on the specific eigenvectors are also presented.

## 2.1 Definitions

**Definition 2.1** For $\Omega_\alpha \subseteq \mathscr{R}^n$ and $\Lambda \subseteq \mathscr{R}^m$ with $\alpha \in \mathbb{Z}$, consider a vector function $\mathbf{f}_\alpha : \Omega_\alpha \times \Lambda \to \Omega_\alpha$ which is $C^r$ $(r \geq 1)$-continuous, and there is a discrete (or difference) equation in a form of

$$\mathbf{x}_{k+1} = \mathbf{f}_\alpha(\mathbf{x}_k, \mathbf{p}_\alpha) \text{ for } \mathbf{x}_k, \ \mathbf{x}_{k+1} \in \Omega_\alpha, \quad k \in \mathbb{Z} \text{ and } \mathbf{p}_\alpha \in \Lambda \tag{2.1}$$

with an initial condition of $\mathbf{x}_k = \mathbf{x}_0$, the solution of Eq. (2.1) is given by

$$\mathbf{x}_k = \underbrace{\mathbf{f}_\alpha(\mathbf{f}_\alpha(\ldots(\mathbf{f}_\alpha(\mathbf{x}_0, \mathbf{p}_\alpha))))}_{k} \tag{2.2}$$

$$\text{for } \mathbf{x}_k \in \Omega_\alpha, \quad k \in \mathbb{Z} \text{ and } \mathbf{p} \in \Lambda.$$

(i)   The difference equation with the initial condition is called a *discrete dynamical system*.
(ii)  The vector function $\mathbf{f}_\alpha(\mathbf{x}_k, \mathbf{p}_\alpha)$ is called a *discrete vector field* on $\Omega_\alpha$.
(iii) The solution $\mathbf{x}_k$ for each $k \in \mathbb{Z}$ is called a *flow* of discrete system.
(iv)  The solution $\mathbf{x}_k$ for all $k \in \mathbb{Z}$ on domain $\Omega_\alpha$ is called the trajectory, phase curve, or orbit of the discrete dynamical system, which is defined as

$$\Gamma = \{\mathbf{x}_k | \mathbf{x}_{k+1} = \mathbf{f}_\alpha(\mathbf{x}_k, \mathbf{p}_\alpha) \text{ for } k \in \mathbb{Z} \text{ and } \mathbf{p}_\alpha \in \Lambda\} \subseteq \cup_\alpha \Omega_\alpha. \tag{2.3}$$

© Higher Education Press, Beijing and Springer-Verlag Berlin Heidelberg 2015
A.C.J. Luo, *Discretization and Implicit Mapping Dynamics*,
Nonlinear Physical Science, DOI 10.1007/978-3-662-47275-0_2

(v)  The discrete dynamical system is called *a uniform discrete system* if

$$\mathbf{x}_{k+1} = \mathbf{f}_\alpha(\mathbf{x}_k, \mathbf{p}_\alpha) = \mathbf{f}(\mathbf{x}_k, \mathbf{p}) \quad \text{for } k \in \mathbb{Z} \text{ and } \mathbf{x}_k \in \Omega_\alpha. \tag{2.4}$$

Otherwise, this discrete dynamical system is called a *non-uniform discrete system*.

**Definition 2.2** For the discrete dynamical system in Eq. (2.1), the relation between state $\mathbf{x}_k$ and state $\mathbf{x}_{k+1}$ ($k \in \mathbb{Z}$) is called a discrete map if

$$P_\alpha : \mathbf{x}_k \xrightarrow{\ \mathbf{f}_\alpha\ } \mathbf{x}_{k+1} \quad \text{and} \quad \mathbf{x}_{k+1} = P_\alpha \mathbf{x}_k \tag{2.5}$$

with the following properties:

$$P_{(k;l)} : \mathbf{x}_k \xrightarrow{\ \mathbf{f}_{\alpha_1},\mathbf{f}_{\alpha_2},\dots,\mathbf{f}_{\alpha_l}\ } \mathbf{x}_{k+l} \quad \text{and} \quad \mathbf{x}_{k+l} = P_{\alpha_l} \circ P_{\alpha_{l-1}} \circ \cdots \circ P_{\alpha_1} \mathbf{x}_k \tag{2.6}$$

where

$$P_{(k;l)} = P_{\alpha_l} \circ P_{\alpha_{l-1}} \circ \cdots \circ P_{\alpha_1}. \tag{2.7}$$

If $P_{\alpha_l} = P_{\alpha_{l-1}} = \cdots = P_{\alpha_1} = P_\alpha$, then

$$P_{(\alpha;l)} \equiv P_\alpha^{(l)} = P_\alpha \circ P_\alpha \circ \cdots \circ P_\alpha \tag{2.8}$$

with

$$P_\alpha^{(n)} = P_\alpha \circ P_\alpha^{(n-1)} \quad \text{and} \quad P_\alpha^{(0)} = \mathbf{I}. \tag{2.9}$$

The total map with *l*-different sub-maps is shown in Fig. 2.1. The map $P_{\alpha_k}$ with the relation function $\mathbf{f}_{\alpha_k}(\alpha_k \in \mathbb{Z})$ is given by Eq. (2.5). The total map $P_{(k;l)}$ is given in Eq. (2.7). The domains $\Omega_{\alpha_k}(\alpha_k \in \mathbb{Z})$ can fully overlap each other or can be completely separated without any intersection.

**Definition 2.3** For a vector function in $\mathbf{f}_\alpha \in \mathscr{R}^n$, $\mathbf{f}_\alpha : \mathscr{R}^n \to \mathscr{R}^n$. The operator norm of $\mathbf{f}_\alpha$ is defined by

$$\|\mathbf{f}_\alpha\| = \sum_{i=1}^{n} \max_{\|\mathbf{x}_k\| \le 1, \mathbf{p}_\alpha} |f_{\alpha(i)}(\mathbf{x}_k, \mathbf{p}_\alpha)|. \tag{2.10}$$

For an $n \times n$ matrix $\mathbf{f}_\alpha(\mathbf{x}_k, \mathbf{p}_\alpha) = \mathbf{A}_\alpha \mathbf{x}_k$ and $\mathbf{A}_\alpha = (a_{ij})_{n \times n}$, the corresponding norm is defined by

$$\|\mathbf{A}_\alpha\| = \sum_{i,j=1}^{n} |a_{ij}|. \tag{2.11}$$

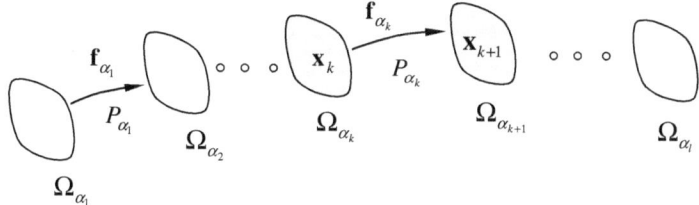

**Fig. 2.1** Maps and vector functions on each sub-domain for discrete dynamical system

**Definition 2.4** For $\Omega_\alpha \subseteq \mathscr{R}^n$ and $\Lambda \subseteq \mathscr{R}^m$ with $\alpha \in \mathbb{Z}$, the vector function $\mathbf{f}_\alpha(\mathbf{x}_k, \mathbf{p}_\alpha)$ with $\mathbf{f}_\alpha : \Omega_\alpha \times \Lambda \to \mathscr{R}^n$ is differentiable at $\mathbf{x}_k \in \Omega_\alpha$ if

$$\frac{\partial \mathbf{f}_\alpha(\mathbf{x}_k, \mathbf{p}_\alpha)}{\partial \mathbf{x}_k}\bigg|_{(\mathbf{x}_k, \mathbf{p})} = \lim_{\Delta \mathbf{x}_k \to \mathbf{0}} \frac{\mathbf{f}_\alpha(\mathbf{x}_k + \Delta \mathbf{x}_k, \mathbf{p}_\alpha) - \mathbf{f}_\alpha(\mathbf{x}_k, \mathbf{p}_\alpha)}{\Delta \mathbf{x}_k}. \tag{2.12}$$

$\partial \mathbf{f}_\alpha / \partial \mathbf{x}_k$ is called the spatial derivative of $\mathbf{f}_\alpha(\mathbf{x}_k, \mathbf{p}_\alpha)$ at $\mathbf{x}_k$, and the derivative is given by the Jacobian matrix

$$\frac{\partial \mathbf{f}_\alpha(\mathbf{x}_k, \mathbf{p}_\alpha)}{\partial \mathbf{x}_k} = \left[\frac{\partial f_{\alpha(i)}}{\partial x_{k(j)}}\right]_{n \times n}. \tag{2.13}$$

**Definition 2.5** For $\Omega_\alpha \subseteq \mathscr{R}^n$ and $\Lambda \subseteq \mathscr{R}^m$, consider a vector function $\mathbf{f}(\mathbf{x}_k, \mathbf{p})$ with $\mathbf{f} : \Omega_\alpha \times \Lambda \to \mathscr{R}^n$ where $\mathbf{x}_k \in \Omega_\alpha$ and $\mathbf{p} \in \Lambda$ with $k \in \mathbb{Z}$. The vector function $\mathbf{f}(\mathbf{x}_k, \mathbf{p})$ is said to satisfy the Lipschitz condition if

$$\|\mathbf{f}(\mathbf{y}_k, \mathbf{p}) - \mathbf{f}(\mathbf{x}_k, \mathbf{p})\| \le L\|\mathbf{y}_k - \mathbf{x}_k\| \tag{2.14}$$

with $\mathbf{x}_k, \mathbf{y}_k \in \Omega_\alpha$ and $L$ a constant. The constant $L$ is called the Lipschitz constant.

## 2.2 Fixed Points and Stability

**Definition 2.6** Consider a discrete, dynamical system $\mathbf{x}_{k+1} = \mathbf{f}_\alpha(\mathbf{x}_k, \mathbf{p}_\alpha)$ in Eq. (2.4).

(i)  A point $\mathbf{x}_k^* \in \Omega_\alpha$ is called a fixed point or period-1 solution of a discrete nonlinear system $\mathbf{x}_{k+1} = \mathbf{f}_\alpha(\mathbf{x}_k, \mathbf{p}_\alpha)$ under a map $P_\alpha$ if for $\mathbf{x}_{k+1} = \mathbf{x}_k = \mathbf{x}_k^*$

$$\mathbf{x}_k^* = \mathbf{f}_\alpha(\mathbf{x}_k^*, \mathbf{p}). \tag{2.15}$$

The linearized system of the nonlinear discrete system $\mathbf{x}_{k+1} = \mathbf{f}_\alpha(\mathbf{x}_k, \mathbf{p}_\alpha)$ in Eq. (2.4) at the fixed point $\mathbf{x}_k^*$ is given by

$$\mathbf{y}_{k+1} = DP_\alpha(\mathbf{x}_k^*, \mathbf{p})\mathbf{y}_k = D\mathbf{f}_\alpha(\mathbf{x}_k^*, \mathbf{p})\mathbf{y}_k \tag{2.16}$$

where

$$\mathbf{y}_k = \mathbf{x}_k - \mathbf{x}_k^* \quad \text{and} \quad \mathbf{y}_{k+1} = \mathbf{x}_{k+1} - \mathbf{x}_{k+1}^*. \tag{2.17}$$

(ii)  A set of points $\mathbf{x}_j^* \in \Omega_{\alpha_j}$ $(\alpha_j \in \mathbb{Z})$ is called the fixed point set or period-1 point set of the total map $P_{(k;l)}$ with $l$-different sub-maps in nonlinear discrete system of Eq. (2.5) if

$$\mathbf{x}_{k+j+1}^* = \mathbf{f}_{\alpha_{j'}}(\mathbf{x}_{k+j}^*, \mathbf{p}_{\alpha_{j'}}) \quad \text{for } j \in \mathbb{Z}_+ \text{ and } j' = \mathrm{mod}(j, l) + 1;$$
$$\mathbf{x}_{k+\mathrm{mod}(j,l)}^* = \mathbf{x}_k^*. \tag{2.18}$$

The linearized equation of the total map $P_{(k;l)}$ gives

$$\mathbf{y}_{k+j+1} = DP_{\alpha_{j'}}(\mathbf{x}_{k+j}^*, \mathbf{p}_{\alpha_{j'}})\mathbf{y}_{k+j} = D\mathbf{f}_{\alpha_{j'}}(\mathbf{x}_{k+j}^*, \mathbf{p}_{\alpha_{j'}})\mathbf{y}_{k+j} \text{ with}$$
$$\mathbf{y}_{k+j+1} = \mathbf{x}_{k+j+1} - \mathbf{x}_{k+j+1}^* \quad \text{and} \quad \mathbf{y}_{k+j} = \mathbf{x}_{k+j} - \mathbf{x}_{k+j}^* \text{ for} \tag{2.19}$$
$$j \in \mathbb{Z}_+ \quad \text{and} \quad j' = \mathrm{mod}(j, l) + 1.$$

The resultant equation for each individual map is

$$\mathbf{y}_{k+j+1} = DP_{(k,l)}(\mathbf{x}_k^*, \mathbf{p})\mathbf{y}_{k+j} \quad \text{for } j \in \mathbb{Z}_+ \tag{2.20}$$

where

$$DP_{(k,n)}(\mathbf{x}_k^*, \mathbf{p}) = \prod_{j=1}^1 DP_{\alpha_j}(\mathbf{x}_{k+j-1}^*, \mathbf{p})$$
$$= DP_{\alpha_l}(\mathbf{x}_{k+l-1}^*, \mathbf{p}_{\alpha_n}) \cdot \cdots \cdot DP_{\alpha_2}(\mathbf{x}_{k+1}^*, \mathbf{p}_{\alpha_2}) \cdot DP_{\alpha_1}(\mathbf{x}_k^*, \mathbf{p}_{\alpha_1})$$
$$= D\mathbf{f}_{(\alpha_l)}(\mathbf{x}_{k+l-1}^*, \mathbf{p}_{\alpha_n}) \cdot \cdots \cdot D\mathbf{f}_{(\alpha_2)}(\mathbf{x}_{k+1}^*, \mathbf{p}_{\alpha_2}) \cdot D\mathbf{f}_{(\alpha_1)}(\mathbf{x}_k^*, \mathbf{p}_{\alpha_1}). \tag{2.21}$$

The fixed point $\mathbf{x}_k^*$ lies in the intersected set of two domains $\Omega_k$ and $\Omega_{k+1}$, as shown in Fig. 2.2. In the vicinity of the fixed point $\mathbf{x}_k^*$, the incremental relations in the two domains $\Omega_k$ and $\Omega_{k+1}$ are different. In other words, setting $\mathbf{y}_k = \mathbf{x}_k - \mathbf{x}_k^*$ and $\mathbf{y}_{k+1} = \mathbf{x}_{k+1} - \mathbf{x}_{k+1}^*$, the corresponding linearization is generated as in Eq. (2.16). Similarly, the fixed point of the total map with $n$-different sub-maps requires the intersection set of two domains $\Omega_k$ and $\Omega_{k+n}$, and there are a set of equations to obtain the fixed points from Eq. (2.18). The other values of fixed points lie in different domains, i.e., $\mathbf{x}_j^* \in \Omega_j$ $(j = k+1, k+2, \ldots, k+n-1)$, as shown in Fig. 2.3.

The corresponding linearized equations are given in Eq. (2.19). From Eq. (2.20), the local characteristics of the total map can be discussed as a single map. Thus, the

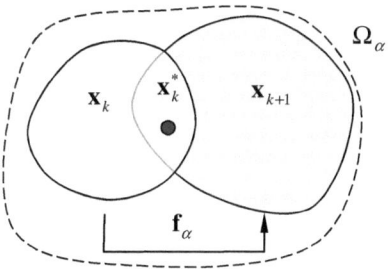

**Fig. 2.2** A fixed point between domains $\Omega_k$ and $\Omega_{k+1}$ for a discrete dynamical system

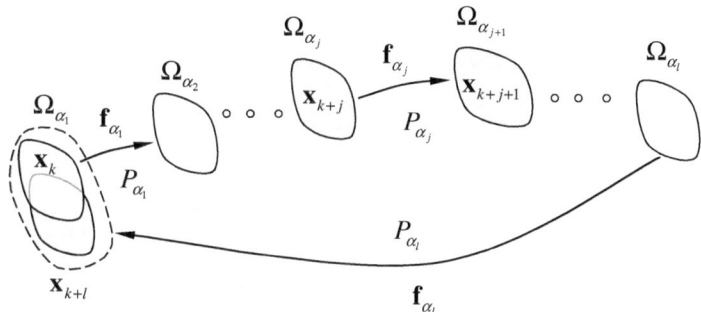

**Fig. 2.3** Fixed points with $l$-maps for discrete dynamical system

dynamical characteristics for the fixed point of the single map will be discussed comprehensively, and the fixed points for resultant map are applicable. The results can be extended to any period-$m$ flows with $P^{(m)}$.

**Definition 2.7** Consider a discrete, nonlinear dynamical system $\mathbf{x}_{k+1} = \mathbf{f}(\mathbf{x}_k, \mathbf{p})$ in Eq. (2.4) with a fixed point $\mathbf{x}_k^*$. The linearized system of the discrete nonlinear system in the neighborhood of $\mathbf{x}_k^*$ is $\mathbf{y}_{k+1} = D\mathbf{f}(\mathbf{x}_k^*, \mathbf{p})\mathbf{y}_k$ ($\mathbf{y}_l = \mathbf{x}_l - \mathbf{x}_k^*$ and $l = k, k+1$) in Eq. (2.16). The matrix $D\mathbf{f}(\mathbf{x}_k^*, \mathbf{p})$ possesses $n_1$ real eigenvalues $|\lambda_j| < 1$ ($j \in N_1$), $n_2$ real eigenvalues $|\lambda_j| > 1$ ($j \in N_2$), $n_3$ real eigenvalues $\lambda_j = 1$ ($j \in N_3$), and $n_4$ real eigenvalues $\lambda_j = -1$ ($j \in N_4$). $N = \{1, 2, \ldots, n\}$ and $N_i = \{i_1, i_2, \ldots, i_{n_i}\} \cup \emptyset$ ($i = 1, 2, 3, 4$) with $i_m \in N$ ($m = 1, 2, \ldots, n_i$) and $\Sigma_{i=1}^4 n_i = n$. $N_i \subseteq N \cup \emptyset$, $\cup_{i=1}^4 N_i = N$, $N_i \cap N_p = \emptyset$ ($p \neq i$). $N_i = \emptyset$ if $n_i = 0$. The corresponding eigenvectors for contraction, expansion, invariance, and flip oscillation are $\{\mathbf{v}_j\}$ ($j \in N_i$) ($i = 1, 2, 3, 4$), respectively. The stable, unstable, invariant, and flip subspaces of $\mathbf{y}_{k+1} = D\mathbf{f}(\mathbf{x}_k^*, \mathbf{p})\mathbf{y}_k$ in Eq. (2.16) are linear subspace spanned by $\{\mathbf{v}_j\}$ ($j \in N_i$) ($i = 1, 2, 3, 4$), respectively, i.e.,

$$\mathscr{E}^{\mathrm{s}} = \mathrm{span}\left\{\mathbf{v}_j \,\middle|\, \begin{array}{l} (D\mathbf{f}(\mathbf{x}_k^*, \mathbf{p}) - \lambda_j \mathbf{I})\mathbf{v}_j = \mathbf{0}, \\ |\lambda_j| < 1, j \in N_1 \subseteq N \cup \varnothing \end{array}\right\};$$

$$\mathscr{E}^{\mathrm{u}} = \mathrm{span}\left\{\mathbf{v}_j \,\middle|\, \begin{array}{l} (D\mathbf{f}(\mathbf{x}_k^*, \mathbf{p}) - \lambda_j \mathbf{I})\mathbf{v}_j = \mathbf{0}, \\ |\lambda_j| > 1, j \in N_2 \subseteq N \cup \varnothing \end{array}\right\};$$

$$\mathscr{E}^{\mathrm{i}} = \mathrm{span}\left\{\mathbf{v}_j \,\middle|\, \begin{array}{l} (D\mathbf{f}(\mathbf{x}_k^*, \mathbf{p}) - \lambda_j \mathbf{I})\mathbf{v}_j = \mathbf{0}, \\ \lambda_j = 1, j \in N_3 \subseteq N \cup \varnothing \end{array}\right\};$$   (2.22)

$$\mathscr{E}^{\mathrm{f}} = \mathrm{span}\left\{\mathbf{v}_j \,\middle|\, \begin{array}{l} (D\mathbf{f}(\mathbf{x}_k^*, \mathbf{p}) - \lambda_j \mathbf{I})\mathbf{v}_j = \mathbf{0}, \\ \lambda_j = -1, j \in N_4 \subseteq N \cup \varnothing \end{array}\right\}$$

where

$$\mathscr{E}^{\mathrm{s}} = \mathscr{E}_{\mathrm{m}}^{\mathrm{s}} \cup \mathscr{E}_{\mathrm{o}}^{\mathrm{s}} \cup \mathscr{E}_{\mathrm{z}}^{\mathrm{s}} \text{ with}$$

$$\mathscr{E}_{\mathrm{m}}^{\mathrm{s}} = \mathrm{span}\left\{\mathbf{v}_j \,\middle|\, \begin{array}{l} (D\mathbf{f}(\mathbf{x}_k^*, \mathbf{p}) - \lambda_j \mathbf{I})\mathbf{v}_j = \mathbf{0}, \\ 0 < \lambda_j < 1, j \in N_1^{\mathrm{m}} \subseteq N \cup \varnothing \end{array}\right\};$$

$$\mathscr{E}_{\mathrm{o}}^{\mathrm{s}} = \mathrm{span}\left\{\mathbf{v}_j \,\middle|\, \begin{array}{l} (D\mathbf{f}(\mathbf{x}_k^*, \mathbf{p}) - \lambda_j \mathbf{I})\mathbf{v}_j = \mathbf{0}, \\ -1 < \lambda_j < 0, j \in N_1^{\mathrm{o}} \subseteq N \cup \varnothing \end{array}\right\};$$   (2.23)

$$\mathscr{E}_{\mathrm{z}}^{\mathrm{s}} = \mathrm{span}\left\{\mathbf{v}_j \,\middle|\, \begin{array}{l} (D\mathbf{f}(\mathbf{x}_k^*, \mathbf{p}) - \lambda_j \mathbf{I})\mathbf{v}_j = \mathbf{0}, \\ \lambda_j = 0, j \in N_1^{\mathrm{z}} \subseteq N \cup \varnothing \end{array}\right\};$$

$$\mathscr{E}^{\mathrm{u}} = \mathscr{E}_{\mathrm{m}}^{\mathrm{u}} \cup \mathscr{E}_{\mathrm{o}}^{\mathrm{u}} \text{ with}$$

$$\mathscr{E}_{\mathrm{m}}^{\mathrm{u}} = \mathrm{span}\left\{\mathbf{v}_j \,\middle|\, \begin{array}{l} (D\mathbf{f}(\mathbf{x}_k^*, \mathbf{p}) - \lambda_j \mathbf{I})\mathbf{v}_j = \mathbf{0}, \\ \lambda_j > 1, j \in N_2^{\mathrm{m}} \subseteq N \cup \varnothing \end{array}\right\};$$   (2.24)

$$\mathscr{E}_{\mathrm{o}}^{\mathrm{u}} = \mathrm{span}\left\{\mathbf{v}_j \,\middle|\, \begin{array}{l} (D\mathbf{f}(\mathbf{x}_k^*, \mathbf{p}) - \lambda_j \mathbf{I})\mathbf{v}_j = \mathbf{0}, \\ -1 > \lambda_j, j \in N_2^{\mathrm{o}} \subseteq N \cup \varnothing \end{array}\right\}.$$

Herein, subscripts "m" and "o" represent the monotonic and oscillatory evolutions.

**Definition 2.8** Consider a discrete, nonlinear dynamical system $\mathbf{x}_{k+1} = \mathbf{f}(\mathbf{x}_k, \mathbf{p})$ in Eq. (2.4) with a fixed point $\mathbf{x}_k^*$. The linearized system of the discrete nonlinear system in the neighborhood of $\mathbf{x}_k^*$ is $\mathbf{y}_{k+1} = D\mathbf{f}(\mathbf{x}_k^*, \mathbf{p})\mathbf{y}_k$ ($\mathbf{y}_l = \mathbf{x}_l - \mathbf{x}_k^*$ and $l = k, k+1$) in Eq. (2.16). The matrix $D\mathbf{f}(\mathbf{x}_k^*, \mathbf{p})$ has complex eigenvalues $\alpha_j \pm i\beta_j$ with eigenvectors $\mathbf{u}_j \pm i\mathbf{v}_j$ ($j \in \{1, 2, \ldots, n\}$), and the base of vector is

$$\mathbf{B} = \left\{\mathbf{u}_1, \mathbf{v}_1, \ldots, \mathbf{u}_j, \mathbf{v}_j, \ldots, \mathbf{u}_n, \mathbf{v}_n\right\}.$$   (2.25)

The stable, unstable, center subspaces of $\mathbf{y}_{k+1} = D\mathbf{f}_k(\mathbf{x}_k^*, \mathbf{p})\mathbf{y}_k$ in Eq. (2.16) are linear subspaces spanned by $\{\mathbf{u}_j, \mathbf{v}_j\}$ ($j \in N_i$, $i = 1, 2, 3$), respectively. Set $N = \{1, 2, \ldots, n\}$ plus $N_i = \{i_1, i_2, \ldots, i_{n_i}\} \cup \varnothing \subseteq N \cup \varnothing$ with $i_m \in N$ ($m = 1, 2, \ldots, n_i$) and $\Sigma_{i=1}^4 n_i = n$. $\cup_{i=1}^4 N_i = N$ with $N_i \cap N_p = \varnothing (p \neq i)$. $N_i = \varnothing$ if $n_i = 0$. The stable, unstable, center subspaces of $\mathbf{y}_{k+1} = D\mathbf{f}(\mathbf{x}_k^*, \mathbf{p})\mathbf{y}_k$ in Eq. (2.16) are defined by

$$
\mathscr{E}^{\mathrm{s}} = \mathrm{span} \left\{ (\mathbf{u}_j, \mathbf{v}_j) \left|
\begin{array}{l}
r_j = \sqrt{\alpha_j^2 + \beta_j^2} < 1, \\
(D\mathbf{f}(\mathbf{x}_k^*, \mathbf{p}) - (\alpha_j \pm i\beta_j)\mathbf{I})(\mathbf{u}_j \pm i\mathbf{v}_j) = \mathbf{0}, \\
j \in N_1 \subseteq \{1, 2, \ldots, n\} \cup \varnothing
\end{array}
\right. \right\};
$$

$$
\mathscr{E}^{\mathrm{u}} = \mathrm{span} \left\{ (\mathbf{u}_j, \mathbf{v}_j) \left|
\begin{array}{l}
r_j = \sqrt{\alpha_j^2 + \beta_j^2} > 1, \\
(D\mathbf{f}(\mathbf{x}_k^*, \mathbf{p}) - (\alpha_j \pm i\beta_j)\mathbf{I})(\mathbf{u}_j \pm i\mathbf{v}_j) = \mathbf{0}, \\
j \in N_2 \subseteq \{1, 2, \ldots, n\} \cup \varnothing
\end{array}
\right. \right\}; \tag{2.26}
$$

$$
\mathscr{E}^{\mathrm{c}} = \mathrm{span} \left\{ (\mathbf{u}_j, \mathbf{v}_j) \left|
\begin{array}{l}
r_j = \sqrt{\alpha_j^2 + \beta_j^2} = 1, \\
(D\mathbf{f}(\mathbf{x}_k^*, \mathbf{p}) - (\alpha_j \pm i\beta_j)\mathbf{I})(\mathbf{u}_j \pm i\mathbf{v}_j) = \mathbf{0}, \\
j \in N_3 \subseteq \{1, 2, \ldots, n\} \cup \varnothing
\end{array}
\right. \right\}.
$$

**Definition 2.9** Consider a discrete, nonlinear dynamical system $\mathbf{x}_{k+1} = \mathbf{f}(\mathbf{x}_k, \mathbf{p})$ in Eq. (2.4) with a fixed point $\mathbf{x}_k^*$. The linearized system of the discrete nonlinear system in the neighborhood of $\mathbf{x}_k^*$ is $\mathbf{y}_{k+1} = D\mathbf{f}(\mathbf{x}_k^*, \mathbf{p})\mathbf{y}_k$ ($\mathbf{y}_l = \mathbf{x}_l - \mathbf{x}_k^*$ and $l = k, k+1$) in Eq. (2.16). The fixed point or period-1 point is *hyperbolic* if no any eigenvalues of $D\mathbf{f}(\mathbf{x}_k^*, \mathbf{p})$ are on the unit circle (i.e., $|\lambda_i| \neq 1$ for $i = 1, 2, \ldots, n$).

**Theorem 2.1** *Consider a discrete, nonlinear dynamical system $\mathbf{x}_{k+1} = \mathbf{f}(\mathbf{x}_k, \mathbf{p})$ in Eq. (2.4) with a fixed point $\mathbf{x}_k^*$. The linearized system of the discrete nonlinear system in the neighborhood of $\mathbf{x}_k^*$ is $\mathbf{y}_{k+1} = D\mathbf{f}(\mathbf{x}_k^*, \mathbf{p})\mathbf{y}_k$ ($\mathbf{y}_j = \mathbf{x}_j - \mathbf{x}_k^*$ and $j = k, k+1$) in Eq. (2.16). The eigenspace of $D\mathbf{f}(\mathbf{x}_k^*, \mathbf{p})$ (i.e., $\mathscr{E} \subseteq \mathscr{R}^n$) in the linearized dynamical system is expressed by direct sum of three subspaces*

$$
\mathscr{E} = \mathscr{E}^{\mathrm{s}} \oplus \mathscr{E}^{\mathrm{u}} \oplus \mathscr{E}^{\mathrm{c}} \tag{2.27}
$$

*where $\mathscr{E}^{\mathrm{s}}, \mathscr{E}^{\mathrm{u}}$ and $\mathscr{E}^{\mathrm{c}}$ are the stable, unstable, and center subspaces, respectively.*

*Proof* The proof can be referred to Luo (2011). $\qquad\square$

**Definition 2.10** Consider a discrete, nonlinear dynamical system $\mathbf{x}_{k+1} = \mathbf{f}(\mathbf{x}_k, \mathbf{p})$ in Eq. (2.4) with a fixed point $\mathbf{x}_k^*$. Suppose there is a neighborhood of the equilibrium $\mathbf{x}_k^*$ as $U_k(\mathbf{x}_k^*) \subset \Omega_k$, and in the neighborhood,

$$
\lim_{\|\mathbf{y}_k\| \to 0} \frac{\|\mathbf{f}(\mathbf{x}_k^* + \mathbf{y}_k, \mathbf{p}) - D\mathbf{f}(\mathbf{x}_k^*, \mathbf{p})\mathbf{y}_k\|}{\|\mathbf{y}_k\|} = 0, \tag{2.28}
$$

and

$$
\mathbf{y}_{k+1} = D\mathbf{f}(\mathbf{x}_k^*, \mathbf{p})\mathbf{y}_k. \tag{2.29}
$$

(i)   A $C^r$ invariant manifold

$$\mathscr{S}_{\mathrm{loc}}(\mathbf{x}_k, \mathbf{x}_k^*) = \{\mathbf{x}_k \in U_k(\mathbf{x}_k^*) |\ \lim_{j \to +\infty} \mathbf{x}_{k+j} = \mathbf{x}_k^* \quad \text{and}$$

$$\mathbf{x}_{k+j} \in U_k(\mathbf{x}_k^*) \text{ with } j \in \mathbb{Z}_+\} \tag{2.30}$$

is called the local stable manifold of $\mathbf{x}_k^*$, and the corresponding global stable manifold is defined as

$$\mathscr{S}(\mathbf{x}_k, \mathbf{x}_k^*) = \cup_{j \in \mathbb{Z}_-} \mathbf{f}(\mathscr{S}_{\mathrm{loc}}(\mathbf{x}_{k+j}, \mathbf{x}_{k+j}^*)) = \cup_{j \in \mathbb{Z}_-} \mathbf{f}^{(j)}(\mathscr{S}_{\mathrm{loc}}(\mathbf{x}_k, \mathbf{x}_k^*)). \tag{2.31}$$

(ii)  A $C^r$ invariant manifold $\mathscr{U}_{\mathrm{loc}}(\mathbf{x}_k, \mathbf{x}_k^*)$

$$\mathscr{U}_{\mathrm{loc}}(\mathbf{x}_k, \mathbf{x}_k^*) = \{\mathbf{x}_k \in U_k(\mathbf{x}_k^*) |\ \lim_{j \to -\infty} \mathbf{x}_{k+j} = \mathbf{x}_k^* \text{ and}$$

$$\mathbf{x}_{k+j} \in U_k(\mathbf{x}_k^*) \text{ with } j \in \mathbb{Z}_-\} \tag{2.32}$$

is called the local unstable manifold of $\mathbf{x}^*$, and the corresponding global unstable manifold is defined as

$$\mathscr{U}(\mathbf{x}_k, \mathbf{x}_k^*) = \cup_{j \in \mathbb{Z}_+} \mathbf{f}(\mathscr{U}_{\mathrm{loc}}(\mathbf{x}_{k+j}, \mathbf{x}_{k+j}^*)) = \cup_{j \in \mathbb{Z}_+} \mathbf{f}^{(j)}(\mathscr{U}_{\mathrm{loc}}(\mathbf{x}_k, \mathbf{x}_k^*)). \tag{2.33}$$

(iii) A $C^{r-1}$ invariant manifold $\mathscr{C}_{\mathrm{loc}}(\mathbf{x}, \mathbf{x}^*)$ is called the center manifold of $\mathbf{x}^*$ if $\mathscr{C}_{\mathrm{loc}}(\mathbf{x}, \mathbf{x}^*)$ possesses the same dimension of $\mathscr{E}^c$ for $\mathbf{x}^* \in \mathscr{C}_{\mathrm{loc}}(\mathbf{x}, \mathbf{x}^*)$, and the tangential space of $\mathscr{C}_{\mathrm{loc}}(\mathbf{x}, \mathbf{x}^*)$ is identical to $\mathscr{E}^c$.

As in continuous dynamical systems, the stable and unstable manifolds are unique, but the center manifold is not unique. If the nonlinear vector field $\mathbf{f}$ is $C^\infty$-continuous, then a $C^r$ center manifold can be found for any $r < \infty$.

**Theorem 2.2** *Consider a discrete, nonlinear dynamical system $\mathbf{x}_{k+1} = \mathbf{f}(\mathbf{x}_k, \mathbf{p})$ in Eq. (2.4) with a hyperbolic fixed point $\mathbf{x}_k^*$. The corresponding solution is $\mathbf{x}_{k+j} = \mathbf{f}(\mathbf{x}_{k+j-1}, \mathbf{p})$ with $j \in \mathbb{Z}$. Suppose there is a neighborhood of the hyperbolic fixed point $\mathbf{x}_k^*$ (i.e., $U_k(\mathbf{x}_k^*) \subset \Omega_\alpha$), and $\mathbf{f}(\mathbf{x}_k, \mathbf{p})$ is $C^r$ ($r \geq 1$)-continuous in $U_k(\mathbf{x}_k^*)$. The linearized system is $\mathbf{y}_{k+j+1} = D\mathbf{f}(\mathbf{x}_k^*, \mathbf{p})\mathbf{y}_{k+j}$ ($\mathbf{y}_{k+j} = \mathbf{x}_{k+j} - \mathbf{x}_k^*$) in $U_k(\mathbf{x}_k^*)$. If the homeomorphism between the local invariant subspace $E(\mathbf{x}_k^*) \subset U_k(\mathbf{x}_k^*)$ and the eigenspace $\mathscr{E}$ of the linearized system exists with the condition in Eq. (2.28), the local invariant subspace is decomposed by*

$$E(\mathbf{x}_k, \mathbf{x}_k^*) = \mathscr{S}_{\mathrm{loc}}(\mathbf{x}_k, \mathbf{x}_k^*) \oplus \mathscr{U}_{\mathrm{loc}}(\mathbf{x}_k, \mathbf{x}_k^*). \tag{2.34}$$

(a)  *The local stable invariant manifold $\mathscr{S}_{\mathrm{loc}}(\mathbf{x}, \mathbf{x}^*)$ possesses the following properties:*

   (i)  *for $\mathbf{x}_k^* \in \mathscr{S}_{\mathrm{loc}}(\mathbf{x}_k, \mathbf{x}_k^*)$, $\mathscr{S}_{\mathrm{loc}}(\mathbf{x}_k, \mathbf{x}_k^*)$ possesses the same dimension of $\mathscr{E}^s$ and the tangential space of $\mathscr{S}_{\mathrm{loc}}(\mathbf{x}_k, \mathbf{x}_k^*)$ is identical to $\mathscr{E}^s$;*

(ii) *for* $\mathbf{x}_k \in \mathscr{S}_{\mathrm{loc}}(\mathbf{x}_k, \mathbf{x}_k^*)$, $\mathbf{x}_{k+j} \in \mathscr{S}_{\mathrm{loc}}(\mathbf{x}_k, \mathbf{x}_k^*)$ *and* $\lim_{j \to \infty} \mathbf{x}_{k+j} = \mathbf{x}_k^*$ *for all* $j \in \mathbb{Z}_+$;

(iii) *For* $\mathbf{x}_k \notin \mathscr{S}_{\mathrm{loc}}(\mathbf{x}_k, \mathbf{x}_k^*)$, $\|\mathbf{x}_{k+j} - \mathbf{x}_k^*\| \geq \delta$ *for* $\delta > 0$ *with* $j, j_1 \in \mathbb{Z}_+$ *and* $j \geq j_1 \geq 0$.

(b) *The local unstable invariant manifold* $\mathscr{U}_{\mathrm{loc}}(\mathbf{x}_k, \mathbf{x}_k^*)$ *possesses the following properties*:

(i) *for* $\mathbf{x}_k^* \in \mathscr{U}_{\mathrm{loc}}(\mathbf{x}_k, \mathbf{x}_k^*)$, $\mathscr{U}_{\mathrm{loc}}(\mathbf{x}_k, \mathbf{x}_k^*)$ *possesses the same dimension of* $\mathscr{E}^{\mathrm{u}}$ *and the tangential space of* $\mathscr{U}_{\mathrm{loc}}(\mathbf{x}_k, \mathbf{x}_k^*)$ *is identical to* $\mathscr{E}^{\mathrm{u}}$;

(ii) *for* $\mathbf{x}_k \in \mathscr{U}_{\mathrm{loc}}(\mathbf{x}_k, \mathbf{x}_k^*)$, $\mathbf{x}_{k+j} \in \mathscr{U}_{\mathrm{loc}}(\mathbf{x}_k, \mathbf{x}_k^*)$ *and* $\lim_{j \to -\infty} \mathbf{x}_{k+j} = \mathbf{x}_k^*$ *for all* $j \in \mathbb{Z}_-$;

(iii) *for* $\mathbf{x}_k \notin \mathscr{U}_{\mathrm{loc}}(\mathbf{x}_k, \mathbf{x}_k^*)$, $\|\mathbf{x}_{k+j} - \mathbf{x}_k^*\| \geq \delta$ *for* $\delta > 0$ *with* $j_1, j \in \mathbb{Z}_-$ *and* $j \leq j_1 \leq 0$.

*Proof* See Nitecki (1971). □

**Theorem 2.3** *Consider a discrete, nonlinear dynamical system* $\mathbf{x}_{k+1} = \mathbf{f}(\mathbf{x}_k, \mathbf{p})$ *in Eq. (2.4) with a fixed point* $\mathbf{x}_k^*$. *The corresponding solution is* $\mathbf{x}_{k+j} = \mathbf{f}(\mathbf{x}_{k+j-1}, \mathbf{p})$ *with* $j \in \mathbb{Z}$. *Suppose there is a neighborhood of the fixed point* $\mathbf{x}_k^*$ *(i.e.,* $U_k(\mathbf{x}_k^*) \subset \Omega_\alpha$*), and* $\mathbf{f}(\mathbf{x}_k, \mathbf{p})$ *is* $C^r$ *(*$r \geq 1$*)-continuous in* $U_k(\mathbf{x}_k^*)$. *The linearized system is* $\mathbf{y}_{k+j+1} = D\mathbf{f}(\mathbf{x}_k^*, \mathbf{p})\mathbf{y}_{k+j}$ *(*$\mathbf{y}_{k+j} = \mathbf{x}_{k+j} - \mathbf{x}_k^*$*) in* $U_k(\mathbf{x}_k^*)$. *If the homeomorphism between the local invariant subspace* $E(\mathbf{x}_k^*) \subset U_k(\mathbf{x}_k^*)$ *and the eigenspace* $\mathscr{E}$ *of the linearized system exists with the condition in* Eq. (2.28), *in addition to the local stable and unstable invariant manifolds, there is a* $C^{r-1}$ *center manifold* $\mathscr{C}_{\mathrm{loc}}(\mathbf{x}_k, \mathbf{x}_k^*)$. *The center manifold possesses the same dimension of* $\mathscr{E}^{\mathrm{c}}$ *for* $\mathbf{x}^* \in \mathscr{C}_{\mathrm{loc}}(\mathbf{x}_k, \mathbf{x}_k^*)$, *and the tangential space of* $\mathscr{C}_{\mathrm{loc}}(\mathbf{x}, \mathbf{x}^*)$ *is identical to* $\mathscr{E}^{\mathrm{c}}$. *Thus, the local invariant subspace is decomposed by*

$$E(\mathbf{x}_k, \mathbf{x}_k^*) = \mathscr{S}_{\mathrm{loc}}(\mathbf{x}_k, \mathbf{x}_k^*) \oplus \mathscr{U}_{\mathrm{loc}}(\mathbf{x}_k, \mathbf{x}_k^*) \oplus \mathscr{C}_{\mathrm{loc}}(\mathbf{x}_k, \mathbf{x}_k^*). \qquad (2.35)$$

*Proof* See Guckenhiemer and Holmes (1990). □

**Definition 2.11** Consider a discrete, nonlinear dynamical system $\mathbf{x}_{k+1} = \mathbf{f}(\mathbf{x}_k, \mathbf{p})$ in Eq. (2.4) on domain $\Omega_\alpha \in \mathscr{R}^n$. Suppose there is a metric space $(\Omega_\alpha, \rho)$, then the map $P$ under the vector function $\mathbf{f}(\mathbf{x}_k, \mathbf{p})$ is called the contraction map if

$$\rho(\mathbf{x}_{k+1}^{(1)}, \mathbf{x}_{k+1}^{(2)}) = \rho(\mathbf{f}(\mathbf{x}_k^{(1)}, \mathbf{p}), \mathbf{f}(\mathbf{x}_k^{(2)}, \mathbf{p})) \leq \lambda \rho(\mathbf{x}_k^{(1)}, \mathbf{x}_k^{(2)}) \qquad (2.36)$$

for $\lambda \in (0, 1)$ and $\mathbf{x}_k^{(1)}, \mathbf{x}_k^{(2)} \in \Omega_\alpha$ with $\rho(\mathbf{x}_k^{(1)}, \mathbf{x}_k^{(2)}) = \|\mathbf{x}_k^{(1)} - \mathbf{x}_k^{(2)}\|$.

**Theorem 2.4** *Consider a discrete, nonlinear dynamical system* $\mathbf{x}_{k+1} = \mathbf{f}(\mathbf{x}_k, \mathbf{p})$ *in Eq. (2.4) on domain* $\Omega_\alpha \in \mathscr{R}^n$. *Suppose there is a metric space* $(\Omega_\alpha, \rho)$, *if the map* $P$ *under the vector function* $\mathbf{f}(\mathbf{x}_k, \mathbf{p})$ *is the contraction map, then there is a unique fixed point* $\mathbf{x}_k^*$ *which is globally stable.*

*Proof* The proof can be referred to Luo (2011). □

**Definition 2.12** Consider a discrete, nonlinear dynamical system $\mathbf{x}_{k+1} = \mathbf{f}(\mathbf{x}_k, \mathbf{p})$ in Eq. (2.4) with a fixed point $\mathbf{x}_k^*$. The corresponding solution is given by $\mathbf{x}_{k+j} = \mathbf{f}(\mathbf{x}_{k+j-1}, \mathbf{p})$ with $j \in \mathbb{Z}$. Suppose there is a neighborhood of the fixed point $\mathbf{x}_k^*$ (i.e., $U_k(\mathbf{x}_k^*) \subset \Omega_\alpha$), and $\mathbf{f}(\mathbf{x}_k, \mathbf{p})$ is $C^r$ $(r \geq 1)$-continuous in $U_k(\mathbf{x}_k^*)$. The linearized system is $\mathbf{y}_{k+j+1} = D\mathbf{f}(\mathbf{x}_k^*, \mathbf{p})\mathbf{y}_{k+j}$ $(\mathbf{y}_{k+j} = \mathbf{x}_{k+j} - \mathbf{x}_k^*)$ in $U_k(\mathbf{x}_k^*)$. Consider a real eigenvalue $\lambda_i$ of matrix $D\mathbf{f}(\mathbf{x}_k^*, \mathbf{p})$ $(i \in N = \{1, 2, \ldots, n\})$ and there is a corresponding eigenvector $\mathbf{v}_i$. On the invariant eigenvector $\mathbf{v}_k^{(i)} = \mathbf{v}_i$, consider $\mathbf{y}_k^{(i)} = c_k^{(i)} \mathbf{v}_i$ and $\mathbf{y}_{k+1}^{(i)} = c_{k+1}^{(i)} \mathbf{v}_i = \lambda_i c_k^{(i)} \mathbf{v}_i$, and thus, $c_{k+1}^{(i)} = \lambda_i c_k^{(i)}$.

(i)   $\mathbf{x}_k^{(i)}$ on the direction $\mathbf{v}_i$ is stable if

$$\lim_{k \to \infty} |c_k^{(i)}| = \lim_{k \to \infty} |(\lambda_i)^k| \times |c_0^{(i)}| = 0 \quad \text{for } |\lambda_i| < 1. \tag{2.37}$$

(ii)   $\mathbf{x}_k^{(i)}$ on the direction $\mathbf{v}_i$ is unstable if

$$\lim_{k \to \infty} |c_k^{(i)}| = \lim_{k \to \infty} |(\lambda_i)^k| \times |c_0^{(i)}| = \infty \quad \text{for } |\lambda_i| > 1. \tag{2.38}$$

(iii)   $\mathbf{x}_k^{(i)}$ on the direction $\mathbf{v}_i$ is invariant if

$$\lim_{k \to \infty} c_k^{(i)} = \lim_{k \to \infty} (\lambda_i)^k c_0^{(i)} = c_0^{(i)} \quad \text{for } \lambda_i = 1. \tag{2.39}$$

(iv)   $\mathbf{x}_k^{(i)}$ on the direction $\mathbf{v}_i$ is flipped if

$$\left.\begin{aligned}
\lim_{2k \to \infty} c_k^{(i)} &= \lim_{2k \to \infty} (\lambda_i)^{2k} \times c_0^{(i)} = c_0^{(i)} \\
\lim_{2k+1 \to \infty} c_k^{(i)} &= \lim_{2k+1 \to \infty} (\lambda_i)^{2k+1} \times c_0^{(i)} = -c_0^{(i)}
\end{aligned}\right\} \text{for } \lambda_i = -1. \tag{2.40}$$

(v)   $\mathbf{x}_k^{(i)}$ on the direction $\mathbf{v}_i$ is degenerate if

$$c_k^{(i)} = (\lambda_i)^k c_0^{(i)} = 0 \quad \text{for } \lambda_i = 0. \tag{2.41}$$

**Definition 2.13** Consider a discrete, nonlinear dynamical system $\mathbf{x}_{k+1} = \mathbf{f}(\mathbf{x}_k, \mathbf{p})$ in Eq. (2.4) with a fixed point $\mathbf{x}_k^*$. The corresponding solution is given by $\mathbf{x}_{k+j} = \mathbf{f}(\mathbf{x}_{k+j-1}, \mathbf{p})$ with $j \in \mathbb{Z}$. Suppose there is a neighborhood of the fixed point $\mathbf{x}_k^*$ (i.e., $U_k(\mathbf{x}_k^*) \subset \Omega_\alpha$), and $\mathbf{f}(\mathbf{x}_k, \mathbf{p})$ is $C^r$ $(r \geq 1)$-continuous in $U_k(\mathbf{x}_k^*)$. Consider a pair of complex eigenvalues $\alpha_i \pm i\beta_i$ of matrix $D\mathbf{f}(\mathbf{x}_k^*, \mathbf{p})$ $(i \in N = \{1, 2, \ldots, n\}, \mathbf{i} = \sqrt{-1})$ and there is a corresponding eigenvector $\mathbf{u}_i \pm i\mathbf{v}_i$. On the invariant plane of $(\mathbf{u}_k^{(i)}, \mathbf{v}_k^{(i)}) = (\mathbf{u}_i, \mathbf{v}_i)$, consider $\mathbf{x}_k^{(i)} = \mathbf{x}_{k+}^{(i)} + \mathbf{x}_{k-}^{(i)}$ with

$$\mathbf{x}_k^{(i)} = c_k^{(i)}\mathbf{u}_i + d_k^{(i)}\mathbf{v}_i, \mathbf{x}_{k+1}^{(i)} = c_{k+1}^{(i)}\mathbf{u}_i + d_{k+1}^{(i)}\mathbf{v}_i. \tag{2.42}$$

Thus, $\mathbf{c}_k^{(i)} = (c_k^{(i)}, d_k^{(i)})^{\mathrm{T}}$ with

$$\mathbf{c}_{k+1}^{(i)} = \mathbf{E}_i\mathbf{c}_k^{(i)} = r_i\mathbf{R}_i\mathbf{c}_k^{(i)} \tag{2.43}$$

where

$$\mathbf{E}_i = \begin{bmatrix} \alpha_i & \beta_i \\ -\beta_i & \alpha_i \end{bmatrix} \quad \text{and} \quad \mathbf{R}_i = \begin{bmatrix} \cos\theta_i & \sin\theta_i \\ -\sin\theta_i & \cos\theta_i \end{bmatrix},$$

$$r_i = \sqrt{\alpha_i^2 + \beta_i^2}, \quad \cos\theta_i = \alpha_i/r_i \quad \text{and} \quad \sin\theta_i = \beta_i/r_i; \tag{2.44}$$

and

$$\mathbf{E}_i^k = \begin{bmatrix} \alpha_i & \beta_i \\ -\beta_i & \alpha_i \end{bmatrix}^k \quad \text{and} \quad \mathbf{R}_i^k = \begin{bmatrix} \cos k\theta_i & \sin k\theta_i \\ -\sin k\theta_i & \cos k\theta_i \end{bmatrix}. \tag{2.45}$$

(i) $\mathbf{x}_k^{(i)}$ on the plane of $(\mathbf{u}_i, \mathbf{v}_i)$ is spirally stable if

$$\lim_{k\to\infty} ||\mathbf{c}_k^{(i)}|| = \lim_{k\to\infty} r_i^k ||\mathbf{R}_i^k|| \times ||\mathbf{c}_0^{(i)}|| = 0 \quad \text{for } r_i = |\lambda_i| < 1. \tag{2.46}$$

(ii) $\mathbf{x}_k^{(i)}$ on the plane of $(\mathbf{u}_i, \mathbf{v}_i)$ is spirally unstable if

$$\lim_{k\to\infty} ||\mathbf{c}_k^{(i)}|| = \lim_{k\to\infty} r_i^k ||\mathbf{R}_i^k|| \times ||\mathbf{c}_0^{(i)}|| = \infty \quad \text{for } r_i = |\lambda_i| > 1. \tag{2.47}$$

(iii) $\mathbf{x}_k^{(i)}$ on the plane of $(\mathbf{u}_i, \mathbf{v}_i)$ is on the invariant circles if

$$||\mathbf{c}_k^{(i)}|| = r_i^k ||\mathbf{R}_i^k|| \times ||\mathbf{c}_0^{(i)}|| = ||\mathbf{c}_0^{(i)}|| \quad \text{for } r_i = |\lambda_i| = 1. \tag{2.48}$$

(iv) $\mathbf{x}_k^{(i)}$ on the plane of $(\mathbf{u}_i, \mathbf{v}_i)$ is degenerate in the direction of $\mathbf{u}_i$ if $\beta_i = 0$.

**Definition 2.14** Consider a discrete, nonlinear dynamical system $\mathbf{x}_{k+1} = \mathbf{f}(\mathbf{x}_k, \mathbf{p})$ in Eq. (2.4) with a fixed point $\mathbf{x}_k^*$. The corresponding solution is given by $\mathbf{x}_{k+j} = \mathbf{f}(\mathbf{x}_{k+j-1}, \mathbf{p})$ with $j \in \mathbb{Z}$. Suppose there is a neighborhood of the fixed point $\mathbf{x}_k^*$ (i.e., $U_k(\mathbf{x}_k^*) \subset \Omega_\alpha$), and $\mathbf{f}(\mathbf{x}_k, \mathbf{p})$ is $C^r$ ($r \geq 1$)-continuous in $U_k(\mathbf{x}_k^*)$ with Eq. (2.28). The linearized system is $\mathbf{y}_{k+j+1} = D\mathbf{f}(\mathbf{x}_k^*, \mathbf{p})\mathbf{y}_{k+j}$ ($\mathbf{y}_{k+j} = \mathbf{x}_{k+j} - \mathbf{x}_k^*$) in $U_k(\mathbf{x}_k^*)$. The matrix $D\mathbf{f}(\mathbf{x}_k^*, \mathbf{p})$ possesses $n$ eigenvalues $\lambda_i$ ($i = 1, 2, \ldots, n$).

(i) The fixed point $\mathbf{x}_k^*$ is called a hyperbolic point if $|\lambda_i| \neq 1$ ($i = 1, 2, \ldots, n$).

(ii) The fixed point $\mathbf{x}_k^*$ is called a sink if $|\lambda_i| < 1$ ($i = 1, 2, \ldots, n$).

(iii)  The fixed point $\mathbf{x}_k^*$ is called a source if $|\lambda_i| > 1$ $(i = 1, 2, \ldots, n)$.
(iv)  The fixed point $\mathbf{x}_k^*$ is called a center if $|\lambda_i| = 1$ $(i = 1, 2, \ldots, n)$ with distinct eigenvalues.

**Definition 2.15** Consider a discrete, nonlinear dynamical system $\mathbf{x}_{k+1} = \mathbf{f}(\mathbf{x}_k, \mathbf{p})$ in Eq. (2.4) with a fixed point $\mathbf{x}_k^*$. The corresponding solution is given by $\mathbf{x}_{k+j} = \mathbf{f}(\mathbf{x}_{k+j-1}, \mathbf{p})$ with $j \in \mathbb{Z}$. Suppose there is a neighborhood of the fixed point $\mathbf{x}_k^*$ (i.e., $U_k(\mathbf{x}_k^*) \subset \Omega_\alpha$), and $\mathbf{f}(\mathbf{x}_k, \mathbf{p})$ is $C^r$ $(r \geq 1)$-continuous in $U_k(\mathbf{x}_k^*)$ with Eq. (2.28). The linearized system is $\mathbf{y}_{k+j+1} = D\mathbf{f}(\mathbf{x}_k^*, \mathbf{p})\mathbf{y}_{k+j}$ $(\mathbf{y}_{k+j} = \mathbf{x}_{k+j} - \mathbf{x}_k^*)$ in $U_k(\mathbf{x}_k^*)$. The matrix $D\mathbf{f}(\mathbf{x}_k^*, \mathbf{p})$ possesses $n$ eigenvalues $\lambda_i$ $(i = 1, 2, \ldots, n)$.

(i)  The fixed point $\mathbf{x}_k^*$ is called a stable node if $|\lambda_i| < 1 (i = 1, 2, \ldots, n)$.
(ii)  The fixed point $\mathbf{x}_k^*$ is called an unstable node if $|\lambda_i| > 1$ $(i = 1, 2, \ldots, n)$.
(iii)  The fixed point $\mathbf{x}_k^*$ is called an $(l_1 : l_2)$-saddle if at least one $|\lambda_i| > 1$ $(i \in L_1 \subset \{1, 2, \ldots, n\})$ and the other $|\lambda_j| < 1$ $(j \in L_2 \subset \{1, 2, \ldots, n\})$ with $L_1 \cup L_2 = \{1, 2, \ldots, n\}$ and $L_1 \cap L_2 = \emptyset$. $l_1 = span(L_1)$ and $l_2 = span(L_2)$.
(iv)  The fixed point $\mathbf{x}_k^*$ is called an $l$th-order degenerate case if $\lambda_i = 0$ $(i \in L \subseteq \{1, 2, \ldots, n\})$.

**Definition 2.16** Consider a discrete, nonlinear dynamical system $\mathbf{x}_{k+1} = \mathbf{f}(\mathbf{x}_k, \mathbf{p})$ in Eq. (2.4) with a fixed point $\mathbf{x}_k^*$. The corresponding solution is given by $\mathbf{x}_{k+j} = \mathbf{f}(\mathbf{x}_{k+j-1}, \mathbf{p})$ with $j \in \mathbb{Z}$. Suppose there is a neighborhood of the fixed point $\mathbf{x}_k^*$ (i.e., $U_k(\mathbf{x}_k^*) \subset \Omega_\alpha$), and $\mathbf{f}(\mathbf{x}_k, \mathbf{p})$ is $C^r(r \geq 1)$-continuous in $U_k(\mathbf{x}_k^*)$ with Eq. (2.28). The linearized system is $\mathbf{y}_{k+j+1} = D\mathbf{f}(\mathbf{x}_k^*, \mathbf{p})\mathbf{y}_{k+j}$ $(\mathbf{y}_{k+j} = \mathbf{x}_{k+j} - \mathbf{x}_k^*)$ in $U_k(\mathbf{x}_k^*)$. The matrix $D\mathbf{f}(\mathbf{x}_k^*, \mathbf{p})$ possesses $n$-pairs of complex eigenvalues $\lambda_i$ $(i = 1, 2, \ldots, n)$.

(i)  The fixed point $\mathbf{x}_k^*$ is called a spiral sink if $|\lambda_i| < 1$ $(i = 1, 2, \ldots, n)$ and $\mathrm{Im}\lambda_j \neq 0$ $(j \in \{1, 2, \ldots, n\})$.
(ii)  The fixed point $\mathbf{x}_k^*$ is called a spiral source if $|\lambda_i| > 1$ $(i = 1, 2, \ldots, n)$ with $\mathrm{Im}\lambda_j \neq 0$ $(j \in \{1, 2, \ldots, n\})$.
(iii)  The fixed point $\mathbf{x}_k^*$ is called a center if $|\lambda_i| = 1$ with distinct $\mathrm{Im}\lambda_i \neq 0$ $(i = 1, 2, \ldots, n)$.

The generalized stability and bifurcation of flows in linearized, nonlinear dynamical systems in Eq. (2.4) will be discussed as follows.

**Definition 2.17** Consider a discrete, nonlinear dynamical system $\mathbf{x}_{k+1} = \mathbf{f}(\mathbf{x}_k, \mathbf{p})$ in Eq. (2.4) with a fixed point $\mathbf{x}_k^*$. The corresponding solution is given by $\mathbf{x}_{k+s} = \mathbf{f}(\mathbf{x}_{k+s-1}, \mathbf{p})$ with $s \in \mathbb{Z}$. Suppose there is a neighborhood of the fixed point $\mathbf{x}_k^*$ (i.e., $U_k(\mathbf{x}_k^*) \subset \Omega_\alpha$), and $\mathbf{f}(\mathbf{x}_k, \mathbf{p})$ is $C^r$ $(r \geq 1)$-continuous in $U_k(\mathbf{x}_k^*)$ with Eq. (2.28). The linearized system is $\mathbf{y}_{k+s+1} = D\mathbf{f}(\mathbf{x}_k^*, \mathbf{p})\mathbf{y}_{k+s}$ $(\mathbf{y}_{k+s} = \mathbf{x}_{k+s} - \mathbf{x}_k^*)$ in $U_k(\mathbf{x}_k^*)$. The matrix $D\mathbf{f}(\mathbf{x}_k^*, \mathbf{p})$ possesses $n$ eigenvalues $\lambda_i$ $(i = 1, 2, \ldots, n)$. Set $N = \{1, 2, \ldots, m, m + 1, \ldots, (n + m)/2\}$, $N_j = \{j_1, j_2, \ldots, j_{n_j}\} \cup \emptyset$ with $j_p \in N$ $(p = 1, 2, \ldots, n_j; j = 1, 2, \ldots, 7)$, $\Sigma_{j=1}^{4} n_j = m$ and $2\Sigma_{j=5}^{7} n_j = n - m$. $\cup_{j=1}^{7} N_j = N$ with $N_j \cap N_l = \emptyset (l \neq j)$. $N_j = \emptyset$ if $n_j = 0$. $N_\alpha = N_\alpha^m \cup N_\alpha^o$ $(\alpha = 1, 2)$ and $N_\alpha^m \cap N_\alpha^o = \emptyset$ with $n_\alpha^m + n_\alpha^o = n_\alpha$

where superscripts "m" and "o" represent monotonic and oscillatory evolutions. The matrix $D\mathbf{f}(\mathbf{x}_k^*, \mathbf{p})$ possesses $n_1$-stable, $n_2$-unstable, $n_3$-invariant, and $n_4$-flip real eigenvectors plus $n_5$-stable, $n_6$-unstable, and $n_7$-center pairs of complex eigenvectors. Without repeated complex eigenvalues of $|\lambda_i| = 1$ ($i \in N_3 \cup N_4 \cup N_7$), an iterative response of $\mathbf{x}_{k+1} = \mathbf{f}(\mathbf{x}_k, \mathbf{p})$ is an $([n_1^m, n_1^o] : [n_2^m, n_2^o] : [n_3; \kappa_3] : [n_4; \kappa_4] | n_5 : n_6 : n_7)$ flow in the neighborhood of the fixed point $\mathbf{x}_k^*$. With repeated complex eigenvalues of $|\lambda_i| = 1$ ($i \in N_3 \cup N_4 \cup N_7$), an iterative response of $\mathbf{x}_{k+1} = \mathbf{f}(\mathbf{x}_k, \mathbf{p})$ is an $([n_1^m, n_1^o] : [n_2^m, n_2^o] : [n_3; \kappa_3] : [n_4; \kappa_4] | n_5 : n_6 : [n_7, l; \kappa_7])$ flow in the neighborhood of the fixed point $\mathbf{x}_k^*$, where $\kappa_j \in \{\emptyset, m_j\}$ ($j = 3, 4, 7$). The meanings of notations in the afore-mentioned structures are defined as follows:

(i) $[n_1^m, n_1^o]$ represents $n_1$-sinks with $n_1^m$-monotonic convergence and $n_1^o$-oscillatory convergence among $n_1$-directions of $\mathbf{v}_i$ if $|\lambda_i| < 1$ ($i \in N_1$ and $1 \le n_1 \le n$) with distinct or repeated eigenvalues.

(ii) $[n_2^m, n_2^o]$ represents $n_2$-sources with $n_2^m$-monotonic divergence and $n_2^o$-oscillatory divergence among $n_2$-directions of $\mathbf{v}_i$ if $|\lambda_i| > 1$ ($i \in N_2$ and $1 \le n_2 \le n$) with distinct or repeated eigenvalues.

(iii) $n_3 = 1$ represents an invariant center on 1-direction of $\mathbf{v}_i$ if $\lambda_i = 1$ ($i \in N_3$ and $n_3 = 1$).

(iv) $n_4 = 1$ represents a flip center on 1-direction of $\mathbf{v}_i$ if $\lambda_i = -1$ ($i \in N_4$ and $n_4 = 1$).

(v) $n_5$ represents $n_5$-spiral sinks on $n_5$-pairs of $(\mathbf{u}_i, \mathbf{v}_i)$ if $|\lambda_i| < 1$ and $\mathrm{Im}\lambda_i \ne 0$ ($i \in N_5$ and $1 \le n_5 \le n$) with distinct or repeated eigenvalues.

(vi) $n_6$ represents $n_6$-spiral sources on $n_6$-directions of $(\mathbf{u}_i, \mathbf{v}_i)$ if $|\lambda_i| > 1$ and $\mathrm{Im}\lambda_i \ne 0$ ($i \in N_6$ and $1 \le n_6 \le n$) with distinct or repeated eigenvalues.

(vii) $n_7$ represents $n_7$-invariant centers on $n_7$-pairs of $(\mathbf{u}_i, \mathbf{v}_i)$ if $|\lambda_i| = 1$ and $\mathrm{Im}\lambda_i \ne 0$ ($i \in N_7$ and $1 \le n_7 \le n$) with distinct eigenvalues.

(viii) $\emptyset$ represents none if $n_j = 0$ ($j \in \{1, 2, \ldots, 7\}$).

(ix) $[n_3; \kappa_3]$ represents $(n_3 - \kappa_3)$-invariant centers on $(n_3 - \kappa_3)$-directions of $\mathbf{v}_{i_3}$ ($i_3 \in N_3$) and $\kappa_3$-sources in $\kappa_3$-directions of $\mathbf{v}_{j_3}$ ($j_3 \in N_3$ and $j_3 \ne i_3$) if $\lambda_i = 1$ ($i \in N_3$ and $n_3 \le n$) with the $(\kappa_3 + 1)$th-order nilpotent matrix $\mathbf{N}_3^{\kappa_3+1} = \mathbf{0}$ ($0 < \kappa_3 \le n_3 - 1$).

(x) $[n_3; \emptyset]$ represents $n_3$-invariant centers on $n_3$-directions of $\mathbf{v}_i$ if $\lambda_i = 1$ ($i \in N_3$ and $1 < n_3 \le n$) with a nilpotent matrix $\mathbf{N}_3 = \mathbf{0}$.

(xi) $[n_4; \kappa_4]$ represents $(n_4 - \kappa_4)$-flip oscillatory centers on $(n_4 - \kappa_4)$-directions of $\mathbf{v}_{i_4}$ ($i_4 \in N_4$) and $\kappa_4$-sources in $\kappa_4$-directions of $\mathbf{v}_{j_4}$ ($j_4 \in N_4$ and $j_4 \ne i_4$) if $\lambda_i = -1$ ($i \in N_4$ and $n_4 \le n$) with the $(\kappa_4 + 1)$th-order nilpotent matrix $\mathbf{N}_4^{\kappa_4+1} = \mathbf{0}$ ($0 < \kappa_4 \le n_4 - 1$).

(xii) $[n_4; \emptyset]$ represents $n_4$ flip oscillatory centers on $n_4$-directions of $\mathbf{v}_i$ if $\lambda_i = -1$ ($i \in N_4$ and $1 < n_4 \le n$) with a nilpotent matrix $\mathbf{N}_4 = \mathbf{0}$.

(xiii) $[n_7, l; \kappa_7]$ represents $(n_7 - \kappa_7)$-invariant centers on $(n_7 - \kappa_7)$-pairs of $(\mathbf{u}_{i_7}, \mathbf{v}_{i_7})$ ($i_7 \in N_7$) and $\kappa_7$-sources on $\kappa_7$-pairs of $(\mathbf{u}_{j_7}, \mathbf{v}_{j_7})$ ($j_7 \in N_7$ and $j_7 \ne i_7$) if

$|\lambda_i| = 1$ and $\text{Im}\lambda_i \neq 0$ $(i \in N_7$ and $n_7 \leq n)$ for $(l+1)$-pairs of repeated eigenvalues with the $(\kappa_7 + 1)$th-order nilpotent matrix $\mathbf{N}_7^{\kappa_7+1} = \mathbf{0}$ $(0 < \kappa_7 \leq l)$.

(xiv)  $[n_7, l; \varnothing]$ represents $n_7$-invariant centers on $n_7$-pairs of $(\mathbf{u}_i, \mathbf{v}_i)$ if $|\lambda_i| = 1$ and $\text{Im}\lambda_i \neq 0$ $(i \in N_7$ and $1 \leq n_7 \leq n)$ for $(l+1)$-pairs of repeated eigenvalues with a nilpotent matrix $\mathbf{N}_7 = \mathbf{0}$.

## 2.3 Stability Switching Theory

To extend the idea of Definitions 2.11 and 2.12, a new function will be defined to determine the stability and the stability state switching.

**Definition 2.18** Consider a discrete, nonlinear dynamical system $\mathbf{x}_{k+1} = \mathbf{f}(\mathbf{x}_k, \mathbf{p})$ $\in \mathscr{R}^n$ in Eq. (2.4) with a fixed point $\mathbf{x}_k^*$. The corresponding solution is given by $\mathbf{x}_{k+j} = \mathbf{f}(\mathbf{x}_{k+j-1}, \mathbf{p})$ with $j \in \mathbb{Z}$. Suppose there is a neighborhood of the fixed point $\mathbf{x}_k^*$ (i.e., $U_k(\mathbf{x}_k^*) \subset \Omega$), and $\mathbf{f}(\mathbf{x}_k, \mathbf{p})$ is $C^r (r \geq 1)$-continuous in $U_k(\mathbf{x}_k^*)$ with Eq. (2.28). The linearized system is $\mathbf{y}_{k+j+1} = D\mathbf{f}(\mathbf{x}_k^*, \mathbf{p})\mathbf{y}_{k+j}$ $(\mathbf{y}_{k+j} = \mathbf{x}_{k+j} - \mathbf{x}_k^*)$ in $U_k(\mathbf{x}_k^*)$ and there are $n$ linearly independent vectors $\mathbf{v}_i$ $(i = 1, 2, \ldots, n)$. For a perturbation of fixed point $\mathbf{y}_k = \mathbf{x}_k - \mathbf{x}_k^*$, let $\mathbf{y}_k^{(i)} = c_k^{(i)}\mathbf{v}_i$ and $\mathbf{y}_{k+1}^{(i)} = c_{k+1}^{(i)}\mathbf{v}_i$,

$$s_k^{(i)} = \mathbf{v}_i^{\mathrm{T}} \cdot \mathbf{y}_k = \mathbf{v}_i^{\mathrm{T}} \cdot (\mathbf{x}_k - \mathbf{x}_k^*) \tag{2.49}$$

where $s_k^{(i)} = c_k^{(i)} ||\mathbf{v}_i||^2$. Define the following functions

$$G_i(\mathbf{x}_k, \mathbf{p}) = \mathbf{v}_i^{\mathrm{T}} \cdot [\mathbf{f}(\mathbf{x}_k, \mathbf{p}) - \mathbf{x}_k^*] \tag{2.50}$$

and

$$\begin{aligned}
G_{s_k^{(i)}}^{(1)}(\mathbf{x}_k, \mathbf{p}) &= \mathbf{v}_i^{\mathrm{T}} \cdot D_{s_k^{(i)}}\mathbf{f}(\mathbf{x}_k(s_k^{(i)}), \mathbf{p}) = \mathbf{v}_i^{\mathrm{T}} \cdot D_{\mathbf{x}_k}\mathbf{f}(\mathbf{x}_k(s_k^{(i)}), \mathbf{p})\partial_{c_k^{(i)}}\mathbf{x}_k\partial_{s_k^{(i)}}c_k^{(i)} \\
&= \mathbf{v}_i^{\mathrm{T}} \cdot D_{\mathbf{x}}\mathbf{f}(\mathbf{x}_k(s_k^{(i)}), \mathbf{p})\mathbf{v}_i ||\mathbf{v}_i||^{-2}
\end{aligned} \tag{2.51}$$

$$G_{s_k^{(i)}}^{(m)}(\mathbf{x}_k, \mathbf{p}) = \mathbf{v}_i^{\mathrm{T}} \cdot D_{s_k^{(i)}}^{(m)}\mathbf{f}(\mathbf{x}_k(s_k^{(i)}), \mathbf{p}) = \mathbf{v}_i^{\mathrm{T}} \cdot D_{s_k^{(i)}}(D_{s_k^{(i)}}^{(m-1)}\mathbf{f}(\mathbf{x}_k(s_k^{(i)}), \mathbf{p})) \tag{2.52}$$

where $D_{s_k^{(i)}}(\cdot) = \partial(\cdot)/\partial s_k^{(i)}$ and $D_{s_k^{(i)}}^{(m)}(\cdot) = D_{s_k^{(i)}}(D_{s_k^{(i)}}^{(m-1)}(\cdot))$.

**Definition 2.19** Consider a discrete, nonlinear dynamical system $\mathbf{x}_{k+1} = \mathbf{f}(\mathbf{x}_k, \mathbf{p})$ $\in \mathscr{R}^n$ in Eq. (2.4) with a fixed point $\mathbf{x}_k^*$. The corresponding solution is given by $\mathbf{x}_{k+j} = \mathbf{f}(\mathbf{x}_{k+j-1}, \mathbf{p})$ with $j \in \mathbb{Z}$. Suppose there is a neighborhood of the fixed point $\mathbf{x}_k^*$ (i.e., $U_k(\mathbf{x}_k^*) \subset \Omega$), and $\mathbf{f}(\mathbf{x}_k, \mathbf{p})$ is $C^r$ $(r \geq 1)$-continuous in $U_k(\mathbf{x}_k^*)$ with Eq. (2.28). The linearized system is $\mathbf{y}_{k+j+1} = D\mathbf{f}(\mathbf{x}_k^*, \mathbf{p})\mathbf{y}_{k+j}$ $(\mathbf{y}_{k+j} = \mathbf{x}_{k+j} - \mathbf{x}_k^*)$ in

$U_k(\mathbf{x}_k^*)$ and there are $n$ linearly independent vectors $\mathbf{v}_i$ $(i = 1, 2, \ldots, n)$. For a perturbation of fixed point $\mathbf{y}_k = \mathbf{x}_k - \mathbf{x}_k^*$, let $\mathbf{y}_k^{(i)} = c_k^{(i)}\mathbf{v}_i$ and $\mathbf{y}_{k+1}^{(i)} = c_{k+1}^{(i)}\mathbf{v}_i$.

(i)  $\mathbf{x}_{k+j}$ $(j \in \mathbb{Z})$ at fixed point $\mathbf{x}_k^*$ on the direction $\mathbf{v}_i$ is stable if

$$|\mathbf{v}_i^{\mathrm{T}} \cdot (\mathbf{x}_{k+1} - \mathbf{x}_k^*)| < |\mathbf{v}_i^{\mathrm{T}} \cdot (\mathbf{x}_k - \mathbf{x}_k^*)| \tag{2.53}$$

for $\mathbf{x}_k \in U(\mathbf{x}_k^*) \subset \Omega_\alpha$. The fixed point $\mathbf{x}_k^*$ is called a sink (or stable node) on the direction $\mathbf{v}_i$.

(ii)  $\mathbf{x}_{k+j}$ $(j \in \mathbb{Z})$ at fixed point $\mathbf{x}_k^*$ on the direction $\mathbf{v}_i$ is unstable if

$$|\mathbf{v}_i^{\mathrm{T}} \cdot (\mathbf{x}_{k+1} - \mathbf{x}_k^*)| > |\mathbf{v}_i^{\mathrm{T}} \cdot (\mathbf{x}_k - \mathbf{x}_k^*)| \tag{2.54}$$

for $\mathbf{x}_k \in U(\mathbf{x}_k^*) \subset \Omega_\alpha$. The fixed point $\mathbf{x}_k^*$ is called a source (or unstable node) on the direction $\mathbf{v}_i$.

(iii)  $\mathbf{x}_{k+j}$ $(j \in \mathbb{Z})$ at fixed point $\mathbf{x}_k^*$ on the direction $\mathbf{v}_i$ is invariant if

$$\mathbf{v}_i^{\mathrm{T}} \cdot (\mathbf{x}_{k+1} - \mathbf{x}_k^*) = \mathbf{v}_i^{\mathrm{T}} \cdot (\mathbf{x}_k - \mathbf{x}_k^*) \tag{2.55}$$

for $\mathbf{x}_k \in U(\mathbf{x}_k^*) \subset \Omega_\alpha$. The fixed point $\mathbf{x}_k^*$ is called to be degenerate on the direction $\mathbf{v}_i$.

(iv)  $\mathbf{x}_{k+j}^{(i)}$ $(j \in \mathbb{Z})$ at fixed point $\mathbf{x}_k^*$ on the direction $\mathbf{v}_i$ is symmetrically flipped if

$$(\mathrm{v})\ \mathbf{v}_i^{\mathrm{T}} \cdot (\mathbf{x}_{k+1} - \mathbf{x}_k^*) = -\mathbf{v}_i^{\mathrm{T}} \cdot (\mathbf{x}_k - \mathbf{x}_k^*) \tag{2.56}$$

for $\mathbf{x}_k \in U(\mathbf{x}_k^*) \subset \Omega_\alpha$. The fixed point $\mathbf{x}_k^*$ is called to be degenerate on the direction $\mathbf{v}_i$.

The stability of fixed points for a specific eigenvector is presented in Fig. 2.4. The solid curve is $\mathbf{v}_i^{\mathrm{T}} \cdot \mathbf{x}_{k+1} = \mathbf{v}_i^{\mathrm{T}} \cdot \mathbf{f}(\mathbf{x}_k, \mathbf{p})$. The circular symbol is fixed point. The shaded regions are stable. The horizontal solid line is for a degenerate case. The vertical solid line is for a line with infinite slope. The monotonically stable node (sink) is presented in Fig. 2.4a. From the fixed point $\mathbf{x}_k^*$, let $\mathbf{y}_k = \mathbf{x}_k - \mathbf{x}_k^*$ and $\mathbf{y}_{k+1} = \mathbf{x}_{k+1} - \mathbf{x}_k^*$. $\mathbf{v}_i^{\mathrm{T}} \cdot \mathbf{x}_k = \mathbf{v}_i^{\mathrm{T}} \cdot \mathbf{x}_{k+1}$ and $\mathbf{v}_i^{\mathrm{T}} \cdot \mathbf{y}_{k+1} = -\mathbf{v}_i^{\mathrm{T}} \cdot \mathbf{y}_k$ are represented by dashed and dotted lines, respectively. The iterative responses approach the fixed point. However, the monotonically unstable (source) is presented in Fig. 2.4b. The iterative responses go away from the fixed point. Similarly, the oscillatory stable node (sink) after iteration with a flip $\mathbf{v}_i^{\mathrm{T}} \cdot \mathbf{y}_k = -\mathbf{v}_i^{\mathrm{T}} \cdot \mathbf{y}_{k+1}$ is presented in Fig. 2.4c. The dashed and dotted lines are used for two lines $\mathbf{v}_i^{\mathrm{T}} \cdot \mathbf{y}_{k+1} = -\mathbf{v}_i^{\mathrm{T}} \cdot \mathbf{y}_k$ and $\mathbf{v}_i^{\mathrm{T}} \cdot \mathbf{x}_k = \mathbf{v}_i^{\mathrm{T}} \cdot \mathbf{x}_{k+1}$, respectively. In a similar fashion, the oscillatory unstable node (source) is presented in Fig. 2.4d. This illustration can be easily observed from the stability of fixed points. In Fig. 2.4e, f, the oscillatory stable and unstable nodes are presented as usual through the two-time iterations.

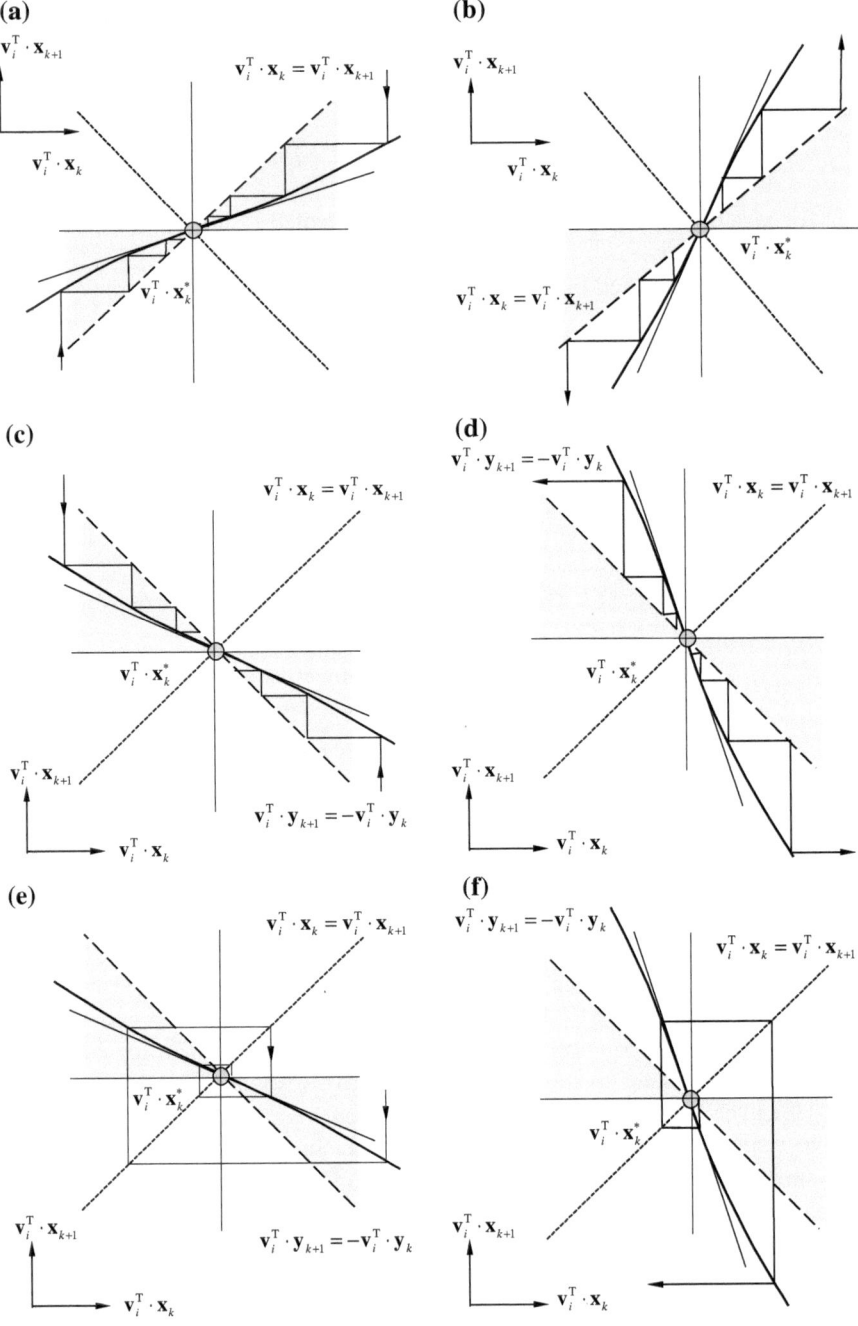

**Fig. 2.4** Stability of fixed points: **a** monotonically stable node (sink); **b** monotonically unstable node (source); **c** oscillatory stable node (sink) and **d** oscillatory unstable node (source); **e** oscillatory stable node (sink) and **f** oscillatory unstable node (sink). Shaded areas are stable zones. ($\mathbf{y}_k = \mathbf{x}_k - \mathbf{x}_k^*$ and $\mathbf{y}_{k+1} = \mathbf{x}_{k+1} - \mathbf{x}_k^*$)

**Theorem 2.5** *Consider a discrete, nonlinear dynamical system $\mathbf{x}_{k+1} = \mathbf{f}(\mathbf{x}_k, \mathbf{p})$ $\in \mathcal{R}^n$ in Eq. (2.4) with a fixed point $\mathbf{x}_k^*$. The corresponding solution is given by $\mathbf{x}_{k+j} = \mathbf{f}(\mathbf{x}_{k+j-1}, \mathbf{p})$ with $j \in \mathbb{Z}$. Suppose there is a neighborhood of the fixed point $\mathbf{x}_k^*$ (i.e., $U_k(\mathbf{x}_k^*) \subset \Omega$), and $\mathbf{f}(\mathbf{x}_k, \mathbf{p})$ is $C^r$ $(r \geq 1)$-continuous in $U_k(\mathbf{x}_k^*)$ with Eq. (2.28). The linearized system is $\mathbf{y}_{k+j+1} = D\mathbf{f}(\mathbf{x}_k^*, \mathbf{p})\mathbf{y}_{k+j}$ $(\mathbf{y}_{k+j} = \mathbf{x}_{k+j} - \mathbf{x}_k^*)$ in $U_k(\mathbf{x}_k^*)$ and there are n linearly independent vectors $\mathbf{v}_i$ $(i = 1, 2, \ldots, n)$. For a perturbation of fixed point $\mathbf{y}_k = \mathbf{x}_k - \mathbf{x}_k^*$, let $\mathbf{y}_k^{(i)} = c_k^{(i)} \mathbf{v}_i$ and $\mathbf{y}_{k+1}^{(i)} = c_{k+1}^{(i)} \mathbf{v}_i$.*

(i) $\mathbf{x}_{k+j}$ $(j \in \mathbb{Z})$ *at fixed point $\mathbf{x}_k^*$ on the direction $\mathbf{v}_i$ is stable if and only if*

$$G_{s_k^{(i)}}^{(1)}(\mathbf{x}_k^*, \mathbf{p}) = \lambda_i \in (-1, 1) \tag{2.57}$$

    *for $\mathbf{x}_k \in U(\mathbf{x}_k^*) \subset \Omega_\alpha$.*

(ii) $\mathbf{x}_{k+j}$ $(j \in \mathbb{Z})$ *at fixed point $\mathbf{x}_k^*$ on the direction $\mathbf{v}_i$ is unstable if and only if*

$$G_{s_k^{(i)}}^{(1)}(\mathbf{x}_k^*, \mathbf{p}) = \lambda_i \in (1, \infty) \, and \, (-\infty, -1) \tag{2.58}$$

    *for $\mathbf{x}_k \in U(\mathbf{x}_k^*) \subset \Omega_\alpha$.*

(iii) $\mathbf{x}_{k+j}$ $(j \in \mathbb{Z})$ *at fixed point $\mathbf{x}_k^*$ on the direction $\mathbf{v}_i$ is invariant if and only if*

$$G_{s_k^{(i)}}^{(1)}(\mathbf{x}_k^*, \mathbf{p}) = \lambda_i = 1 \quad and \quad G_{s_k^{(i)}}^{(m_i)}(\mathbf{x}_k^*, \mathbf{p}) = 0 \, m_i = 2, 3, \ldots \tag{2.59}$$

    *for $\mathbf{x}_k \in U(\mathbf{x}_k^*) \subset \Omega_\alpha$.*

(iv) $\mathbf{x}_{k+j}^{(i)}$ $(j \in \mathbb{Z})$ *at fixed point $\mathbf{x}_k^*$ on the direction $\mathbf{v}_k$ is symmetrically flip if and only if*

$$G_{s_k^{(i)}}^{(1)}(\mathbf{x}_k^*, \mathbf{p}) = \lambda_i = -1 \quad and \quad G_{s_k^{(i)}}^{(m_i)}(\mathbf{x}_k^*, \mathbf{p}) = 0 \, m_i = 2, 3, \ldots \tag{2.60}$$

    *for $\mathbf{x}_k \in U(\mathbf{x}_k^*) \subset \Omega_\alpha$.*

*Proof* The proof can be referred to Luo (2012). ◻

The monotonic stability of fixed points with higher-order singularity for a specific eigenvector is presented in Fig. 2.5. The solid curve is $\mathbf{v}_i^T \cdot \mathbf{x}_{k+1} = \mathbf{v}_i^T \cdot \mathbf{f}(\mathbf{x}_k, \mathbf{p})$. The circular symbol is fixed pointed. The shaded regions are stable. The horizontal solid line is also for the degenerate case. The vertical solid line is for a line with infinite slope. The monotonically stable node (sink) of the $(2m_i + 1)$th-order is sketched in Fig. 2.5a. The dashed and dotted lines are for $\mathbf{v}_i^T \cdot \mathbf{x}_k = \mathbf{v}_i^T \cdot \mathbf{x}_{k+1}$ and $\mathbf{v}_i^T \cdot \mathbf{y}_{k+1} = -\mathbf{v}_i^T \cdot \mathbf{y}_k$, respectively. The nonlinear curve lies in the stable zone, and the iterative responses approach the fixed point. However, the monotonically

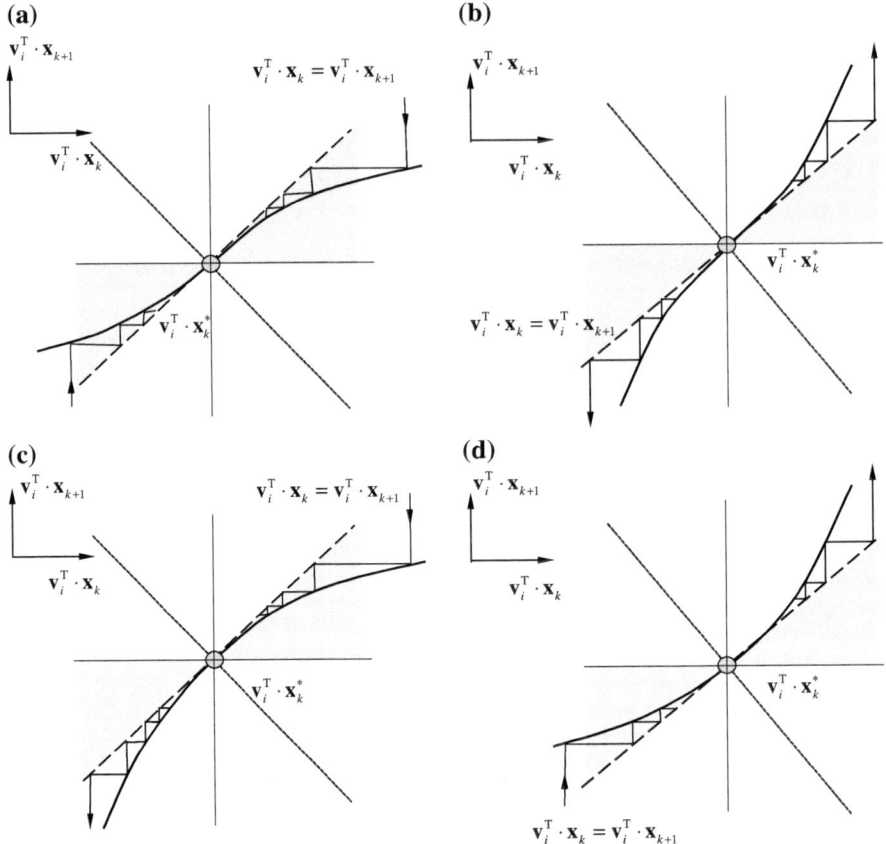

**Fig. 2.5** Monotonic stability of fixed points with higher-order singularity: **a** monotonically stable node (sink) of $(2m_i + 1)$th-order, **b** monotonically unstable node (source) of $(2m_i + 1)$th-order, **c** monotonically lower saddle of $(2m_i)$th-order, and **d** monotonically upper saddle of $(2m_i)$th-order. *Shaded areas* are stable zones. ($\mathbf{y}_k = \mathbf{x}_k - \mathbf{x}_k^*$ and $\mathbf{y}_{k+1} = \mathbf{x}_{k+1} - \mathbf{x}_k^*$)

unstable (source) of the $(2m_i + 1)$th-order is presented in Fig. 2.5b. The nonlinear curve lies in the unstable zone, and the iterative responses go away from the fixed point. The monotonically lower saddle of the $(2m_i)$th-order is presented in Fig. 2.5c. The nonlinear curve is tangential to the line of $\mathbf{v}_i^{\mathrm{T}} \cdot \mathbf{x}_k = \mathbf{v}_i^{\mathrm{T}} \cdot \mathbf{x}_{k+1}$ with the $(2m_i)$th-order, and the upper branch is in the stable zone and the lower branch is in the unstable zone. Similarly, the monotonically upper saddle of the $(2m_i)$th-order is presented in Fig. 2.5d. The oscillatory stability of fixed points with higher-order singularity for a specific eigenvector after iteration with a flip $\mathbf{v}_i^{\mathrm{T}} \cdot \mathbf{y}_k = -\mathbf{v}_i^{\mathrm{T}} \cdot \mathbf{y}_{k+1}$ is presented in Fig. 2.6. The oscillatory stable node (sink) of the $(2m_i + 1)$th-order is sketched in Fig. 2.6a. The dashed and dotted lines are for $\mathbf{v}_i^{\mathrm{T}} \cdot \mathbf{y}_{k+1} = -\mathbf{v}_i^{\mathrm{T}} \cdot \mathbf{y}_k$

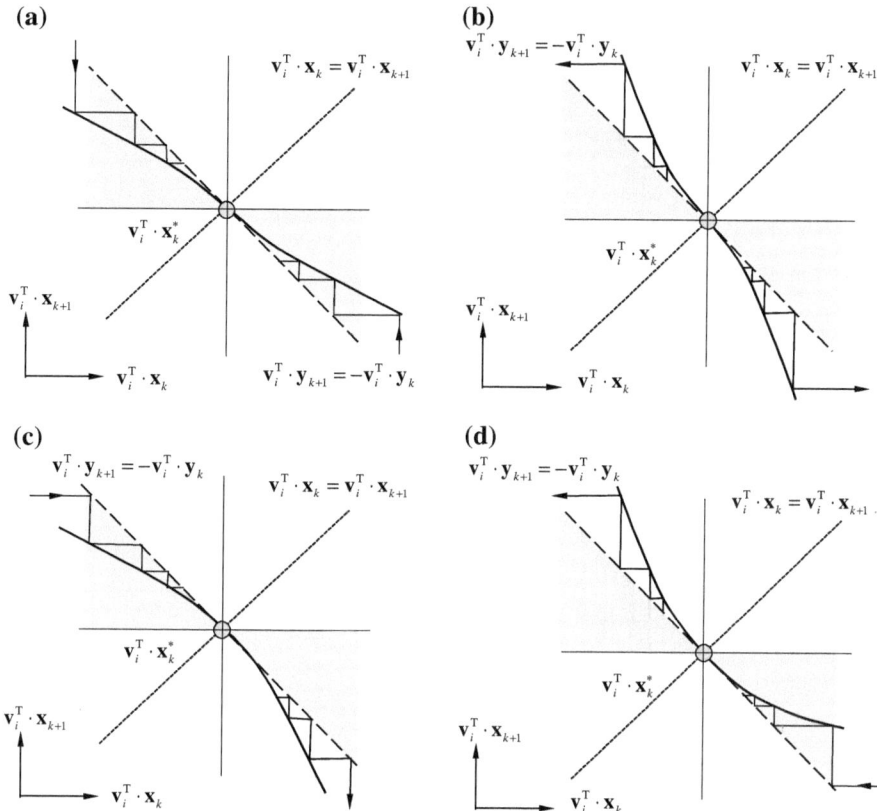

**Fig. 2.6** Oscillatory stability of fixed points with higher-order singularity after iteration with a flip $\mathbf{v}_i^T \cdot \mathbf{y}_k = -\mathbf{v}_i^T \cdot \mathbf{y}_{k+1}$: **a** oscillatory stable node (sink) of $(2m_i + 1)$th-order, **b** oscillatory unstable node (source) of $(2m_i + 1)$th-order, **c** oscillatory lower saddle of $(2m_i)$th-order, and **d** oscillatory upper saddle of $(2m_i)$th-order. *Shaded areas* are stable zones. ($\mathbf{y}_k = \mathbf{x}_k - \mathbf{x}_k^*$ and $\mathbf{y}_{k+1} = \mathbf{x}_{k+1} - \mathbf{x}_k^*$)

and $\mathbf{v}_i^T \cdot \mathbf{x}_k = \mathbf{v}_i^T \cdot \mathbf{x}_{k+1}$, respectively. The nonlinear curve lies in the stable zone, and the iterative responses approach the fixed point. However, the oscillatory unstable (source) of the $(2m_i + 1)$th-order is presented in Fig. 2.6b. The nonlinear curve lies in the unstable zone, and the iterative responses go away from the fixed point. The oscillatory lower saddle of the $(2m_i)$th-order is presented in Fig. 2.6c. The nonlinear curve is tangential to and below the line of $\mathbf{v}_i^T \cdot \mathbf{y}_{k+1} = -\mathbf{v}_i^T \cdot \mathbf{y}_k$ with the $(2m_i)$th-order, and the upper branch is in the stable zone and the lower branch is in the unstable zone. Finally, the oscillatory upper saddle of the $(2m_i)$th-order is presented in Fig. 2.6d. For clear illustrations, the oscillatory stability of fixed points with higher-order singularity for the two-time iterations is presented in Fig. 2.7.

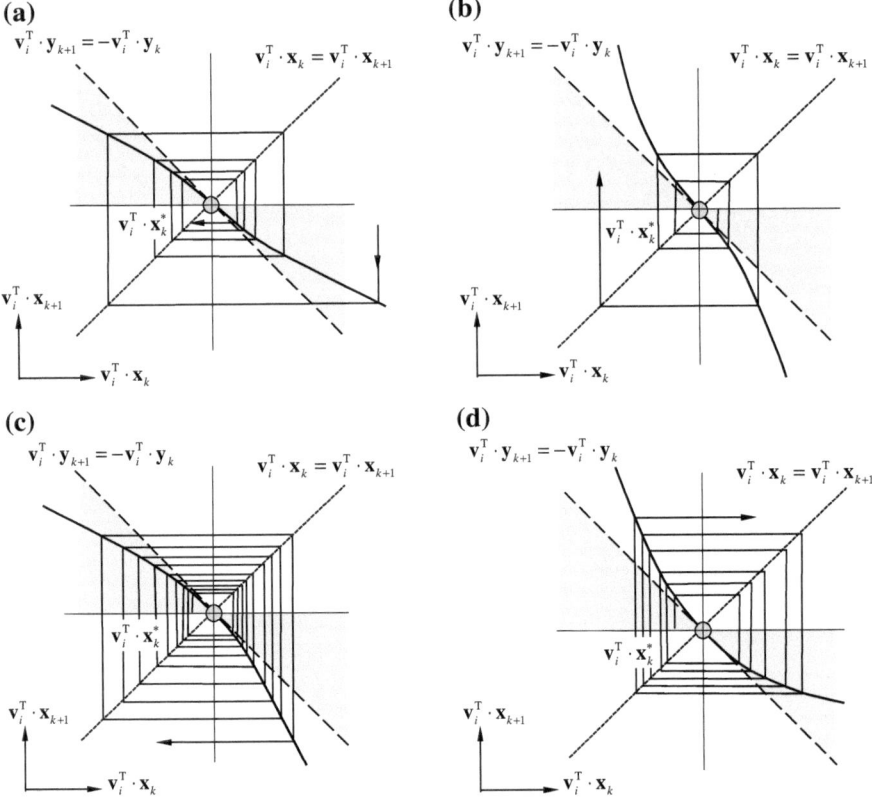

**Fig. 2.7** Oscillatory stability of fixed points with higher-order singularity for the two-time iterations: **a** oscillatory stable node (sink) of $(2m_i + 1)$th-order, **b** oscillatory unstable node (source) of $(2m_i + 1)$th-order, **c** oscillatory lower saddle of $(2m_i)$th-order, and **d** oscillatory upper saddle of $(2m_i)$th-order. *Shaded areas* are stable zones. ($\mathbf{y}_k = \mathbf{x}_k - \mathbf{x}_k^*$ and $\mathbf{y}_{k+1} = \mathbf{x}_{k+1} - \mathbf{x}_k^*$)

**Definition 2.20** Consider a discrete, nonlinear dynamical system $\mathbf{x}_{k+1} = \mathbf{f}(\mathbf{x}_k, \mathbf{p})$ $\in \mathcal{R}^{2n}$ in Eq. (2.4) with a fixed point $\mathbf{x}_k^*$. The corresponding solution is given by $\mathbf{x}_{k+j} = \mathbf{f}(\mathbf{x}_{k+j-1}, \mathbf{p})$ with $j \in \mathbb{Z}$. Suppose there is a neighborhood of the fixed point $\mathbf{x}_k^*$ (i.e., $U_k(\mathbf{x}_k^*) \subset \Omega$), and $\mathbf{f}(\mathbf{x}_k, \mathbf{p})$ is $C^r$ $(r \geq 1)$-continuous in $U_k(\mathbf{x}_k^*)$ with Eq. (2.28). The linearized system is $\mathbf{y}_{k+j+1} = D\mathbf{f}(\mathbf{x}_k^*, \mathbf{p})\mathbf{y}_{k+j}$ ($\mathbf{y}_{k+j} = \mathbf{x}_{k+j} - \mathbf{x}_k^*$) in $U_k(\mathbf{x}_k^*)$ and there are $n$ linearly independent vectors $\mathbf{v}_i$ $(i = 1, 2, \ldots, n)$. For a perturbation of fixed point $\mathbf{y}_k = \mathbf{x}_k - \mathbf{x}_k^*$, let $\mathbf{y}_k^{(i)} = c_k^{(i)}\mathbf{v}_i$ and $\mathbf{y}_{k+1}^{(i)} = c_{k+1}^{(i)}\mathbf{v}_i$.

(i)  $\mathbf{x}_{k+j}$ $(j \in \mathbb{Z})$ at fixed point $\mathbf{x}_k^*$ on the direction $\mathbf{v}_i$ is monotonically stable of the $(2m_i + 1)$th-order if

$$G^{(1)}_{s^{(i)}_k}(\mathbf{x}^*_k, \mathbf{p}) = \lambda_i = 1,$$

$$G^{(r_i)}_{s^{(i)}_k}(\mathbf{x}^*_k, \mathbf{p}) = 0 \quad \text{for } r_i = 2, 3, \dots, 2m_i,$$

$$G^{(2m_i+1)}_{s^{(i)}_k}(\mathbf{x}^*_k, \mathbf{p}) \neq 0,$$

$$|\mathbf{v}^T_i \cdot (\mathbf{x}_{k+1} - \mathbf{x}^*_k)| < |\mathbf{v}^T_i \cdot (\mathbf{x}_k - \mathbf{x}^*_k)|$$

(2.61)

for $\mathbf{x}_k \in U_k(\mathbf{x}^*_k) \subset \Omega_\alpha$. The fixed point $\mathbf{x}^*_k$ is called a monotonic sink (or stable node) of the $(2m_i + 1)$th-order on the direction $\mathbf{v}_i$.

(ii)  $\mathbf{x}_{k+j}$ ($j \in \mathbb{Z}$) at fixed point $\mathbf{x}^*_k$ on the direction $\mathbf{v}_i$ is monotonically unstable of the $(2m_i + 1)$th-order if

$$G^{(1)}_{s^{(i)}_k}(\mathbf{x}^*_k, \mathbf{p}) = \lambda_i = 1,$$

$$G^{(r_i)}_{s^{(i)}_k}(\mathbf{x}^*_k, \mathbf{p}) = 0 \quad \text{for } r_i = 2, 3, \dots, 2m_i;$$

$$G^{(2m_i+1)}_{s^{(i)}_k}(\mathbf{x}^*_k, \mathbf{p}) \neq 0;$$

$$|\mathbf{v}^T_i \cdot (\mathbf{x}_{k+1} - \mathbf{x}^*_k)| > |\mathbf{v}^T_i \cdot (\mathbf{x}_k - \mathbf{x}^*_k)|$$

(2.62)

for $\mathbf{x}_k \in U_k(\mathbf{x}^*_k) \subset \Omega_\alpha$. The fixed point $\mathbf{x}^*_k$ is called a monotonic source (or unstable node) of the $(2m_i + 1)$th-order on the direction $\mathbf{v}_i$.

(iii)  $\mathbf{x}_{k+j}$ ($j \in \mathbb{Z}$) at fixed point $\mathbf{x}^*_k$ on the direction $\mathbf{v}_i$ is monotonically unstable of the $(2m_i)$th-order, lower saddle if

$$G^{(1)}_{s^{(i)}_k}(\mathbf{x}^*_k, \mathbf{p}) = \lambda_i = 1,$$

$$G^{(r_i)}_{s^{(i)}_k}(\mathbf{x}^*_k, \mathbf{p}) = 0 \quad \text{for } r_i = 2, 3, \dots, 2m_i - 1;$$

$$G^{(2m_i)}_{s^{(i)}_k}(\mathbf{x}^*_k, \mathbf{p}) \neq 0,$$

$$|\mathbf{v}^T_i \cdot (\mathbf{x}_{k+1} - \mathbf{x}^*_k)| < |\mathbf{v}^T_i \cdot (\mathbf{x}_k - \mathbf{x}^*_k)| \quad \text{for } s^{(i)}_k > 0,$$

$$|\mathbf{v}^T_i \cdot (\mathbf{x}_{k+1} - \mathbf{x}^*_k)| > |\mathbf{v}^T_i \cdot (\mathbf{x}_k - \mathbf{x}^*_k)| \quad \text{for } s^{(i)}_k < 0$$

(2.63)

for $\mathbf{x}_k \in U_k(\mathbf{x}^*_k) \subset \Omega_\alpha$. The fixed point $\mathbf{x}^*_k$ is called a monotonic, lower saddle of the $(2m_i)$th-order on the direction $\mathbf{v}_i$.

(iv)  $\mathbf{x}_{k+j}$ ($j \in \mathbb{Z}$) at fixed point $\mathbf{x}^*_k$ on the direction $\mathbf{v}_i$ is monotonically unstable of the $(2m_i)$th-order, upper saddle if

$$G^{(1)}_{s^{(i)}_k}(\mathbf{x}^*_k, \mathbf{p}) = \lambda_i = 1,$$

$$G^{(r_i)}_{s^{(i)}_k}(\mathbf{x}^*_k, \mathbf{p}) = 0 \quad \text{for } r_i = 2, 3, \dots, 2m_i - 1;$$

$$G^{(2m_i)}_{s^{(i)}_k}(\mathbf{x}^*_k, \mathbf{p}) \neq 0,$$

$$|\mathbf{v}_i^{\mathrm{T}} \cdot (\mathbf{x}_{k+1} - \mathbf{x}_k^*)| > |\mathbf{v}_i^{\mathrm{T}} \cdot (\mathbf{x}_k - \mathbf{x}_k^*)| \quad \text{for } s_k^{(i)} > 0,$$
$$|\mathbf{v}_i^{\mathrm{T}} \cdot (\mathbf{x}_{k+1} - \mathbf{x}_k^*)| < |\mathbf{v}_i^{\mathrm{T}} \cdot (\mathbf{x}_k - \mathbf{x}_k^*)| \quad \text{for } s_k^{(i)} < 0 \tag{2.64}$$

for $\mathbf{x}_k \in U_k(\mathbf{x}_k^*) \subset \Omega_\alpha$. The fixed point $\mathbf{x}_k^*$ is called a monotonic, upper saddle of the $(2m_i)$th-order on the direction $\mathbf{v}_i$.

(v)  $\mathbf{x}_{k+j}$ $(j \in \mathbb{Z})$ at fixed point $\mathbf{x}_k^*$ on the direction $\mathbf{v}_i$ is oscillatory stable of the $(2m_i + 1)$th-order if

$$G_{s_k^{(i)}}^{(1)}(\mathbf{x}_k^*, \mathbf{p}) = \lambda_i = -1,$$
$$G_{s_k^{(i)}}^{(r_i)}(\mathbf{x}_k^*, \mathbf{p}) = 0 \quad \text{for } r_i = 2, 3, \ldots, 2m_i;$$
$$G_{s_k^{(i)}}^{(2m_i+1)}(\mathbf{x}_k^*, \mathbf{p}) \neq 0;$$
$$|\mathbf{v}_i^{\mathrm{T}} \cdot (\mathbf{x}_{k+1} - \mathbf{x}_k^*)| < |\mathbf{v}_i^{\mathrm{T}} \cdot (\mathbf{x}_k - \mathbf{x}_k^*)| \tag{2.65}$$

for $\mathbf{x}_k \in U_k(\mathbf{x}_k^*) \subset \Omega_\alpha$. The fixed point $\mathbf{x}_k^*$ is called an oscillatory sink (or stable node) of the $(2m_i + 1)$th-order on the direction $\mathbf{v}_i$.

(vi)  $\mathbf{x}_{k+j}$ $(j \in \mathbb{Z})$ at fixed point $\mathbf{x}_k^*$ on the direction $\mathbf{v}_i$ is oscillatory unstable of the $(2m_i + 1)$th-order if

$$G_{s_k^{(i)}}^{(1)}(\mathbf{x}_k^*, \mathbf{p}) = \lambda_i = -1;$$
$$G_{s_k^{(i)}}^{(r_i)}(\mathbf{x}_k^*, \mathbf{p}) = 0 \quad \text{for } r_i = 2, 3, \ldots, 2m_i;$$
$$G_{s_k^{(i)}}^{(2m_i+1)}(\mathbf{x}_k^*, \mathbf{p}) \neq 0,$$
$$|\mathbf{v}_i^{\mathrm{T}} \cdot (\mathbf{x}_{k+1} - \mathbf{x}_k^*)| > |\mathbf{v}_i^{\mathrm{T}} \cdot (\mathbf{x}_k - \mathbf{x}_k^*)| \tag{2.66}$$

for $\mathbf{x}_k \in U_k(\mathbf{x}_k^*) \subset \Omega_\alpha$. The fixed point $\mathbf{x}_k^*$ is called an oscillatory source (or unstable node) of the $(2m_i + 1)$th-order on the direction $\mathbf{v}_i$.

(vii)  $\mathbf{x}_{k+j}$ $(j \in \mathbb{Z})$ at fixed point $\mathbf{x}_k^*$ on the direction $\mathbf{v}_i$ is oscillatory unstable of the $(2m_i)$th-order, lower saddle if

$$G_{s_k^{(i)}}^{(1)}(\mathbf{x}_k^*, \mathbf{p}) = \lambda_i = -1,$$
$$G_{s_k^{(i)}}^{(r_i)}(\mathbf{x}_k^*, \mathbf{p}) = 0 \quad \text{for } r_i = 2, 3, \ldots, 2m_i - 1;$$
$$G_{s_k^{(i)}}^{(2m_i)}(\mathbf{x}_k^*, \mathbf{p}) \neq 0,$$
$$|\mathbf{v}_i^{\mathrm{T}} \cdot (\mathbf{x}_{k+1} - \mathbf{x}_k^*)| > |\mathbf{v}_i^{\mathrm{T}} \cdot (\mathbf{x}_k - \mathbf{x}_k^*)| \quad \text{for } s_k^{(i)} > 0,$$
$$|\mathbf{v}_i^{\mathrm{T}} \cdot (\mathbf{x}_{k+1} - \mathbf{x}_k^*)| < |\mathbf{v}_i^{\mathrm{T}} \cdot (\mathbf{x}_k - \mathbf{x}_k^*)| \quad \text{for } s_k^{(i)} < 0 \tag{2.67}$$

for $\mathbf{x}_k \in U_k(\mathbf{x}_k^*) \subset \Omega_\alpha$. The fixed point $\mathbf{x}_k^*$ is called an oscillatory lower saddle of the $(2m_i)$th-order on the direction $\mathbf{v}_i$.

(viii)   $\mathbf{x}_{k+j}$ ($j \in \mathbb{Z}$) at fixed point $\mathbf{x}_k^*$ on the direction $\mathbf{v}_i$ is oscillatory unstable of the $(2m_i)$th-order, upper saddle if

$$G_{s_k^{(i)}}^{(1)}(\mathbf{x}_k^*, \mathbf{p}) = \lambda_i = -1,$$

$$G_{s_k^{(i)}}^{(r_i)}(\mathbf{x}_k^*, \mathbf{p}) = 0 \quad \text{for } r_i = 2, 3, \ldots, 2m_i - 1;$$

$$G_{s_k^{(i)}}^{(2m_i)}(\mathbf{x}_k^*, \mathbf{p}) \neq 0, \tag{2.68}$$

$$|\mathbf{v}_i^{\mathrm{T}} \cdot (\mathbf{x}_{k+1} - \mathbf{x}_k^*)| < |\mathbf{v}_i^{\mathrm{T}} \cdot (\mathbf{x}_k - \mathbf{x}_k^*)| \quad \text{for } s_k^{(i)} > 0,$$

$$|\mathbf{v}_i^{\mathrm{T}} \cdot (\mathbf{x}_{k+1} - \mathbf{x}_k^*)| > |\mathbf{v}_i^{\mathrm{T}} \cdot (\mathbf{x}_k - \mathbf{x}_k^*)| \quad \text{for } s_k^{(i)} < 0$$

for $\mathbf{x}_k \in U_k(\mathbf{x}_k^*) \subset \Omega_\alpha$. The fixed point $\mathbf{x}_k^*$ is called an oscillatory, upper saddle of the $(2m_i)$th-order on the direction $\mathbf{v}_i$.

**Theorem 2.6** *Consider a discrete, nonlinear dynamical system* $\mathbf{x}_{k+1} = \mathbf{f}(\mathbf{x}_k, \mathbf{p}) \in \mathscr{R}^n$ *in Eq. (2.4) with a fixed point* $\mathbf{x}_k^*$. *The corresponding solution is given by* $\mathbf{x}_{k+j} = \mathbf{f}(\mathbf{x}_{k+j-1}, \mathbf{p})$ *with* $j \in \mathbb{Z}$. *Suppose there is a neighborhood of the fixed point* $\mathbf{x}_k^*$ *(i.e.,* $U_k(\mathbf{x}_k^*) \subset \Omega$*), and* $\mathbf{f}(\mathbf{x}_k, \mathbf{p})$ *is* $C^r$ *(*$r \geq 1$*)-continuous in* $U_k(\mathbf{x}_k^*)$ *with Eq. (2.28). The linearized system is* $\mathbf{y}_{k+j+1} = D\mathbf{f}(\mathbf{x}_k^*, \mathbf{p})\mathbf{y}_{k+j}$ *(*$\mathbf{y}_{k+j} = \mathbf{x}_{k+j} - \mathbf{x}_k^*$*) in* $U_k(\mathbf{x}_k^*)$ *and there are n linearly independent vectors* $\mathbf{v}_i$ *(*$i = 1, 2, \ldots, n$*). For a perturbation of fixed point* $\mathbf{y}_k = \mathbf{x}_k - \mathbf{x}_k^*$, *let* $\mathbf{y}_k^{(i)} = c_k^{(i)} \mathbf{v}_i$ *and* $\mathbf{y}_{k+1}^{(i)} = c_{k+1}^{(i)} \mathbf{v}_i$.

(i)   $\mathbf{x}_{k+j}$ ($j \in \mathbb{Z}$) *at fixed point* $\mathbf{x}_k^*$ *on the direction* $\mathbf{v}_i$ *is monotonically stable of the* $(2m_i + 1)$*th-order if and only if*

$$G_{s_k^{(i)}}^{(1)}(\mathbf{x}_k^*, \mathbf{p}) = \lambda_i = 1,$$

$$G_{s_k^{(i)}}^{(r_i)}(\mathbf{x}_k^*, \mathbf{p}) = 0 \ \text{for } r_i = 2, 3, \ldots, 2m_i, \tag{2.69}$$

$$G_{s_k^{(i)}}^{(2m_i+1)}(\mathbf{x}_k^*, \mathbf{p}) < 0$$

*for* $\mathbf{x}_k \in U(\mathbf{x}_k^*) \subset \Omega_\alpha$.

(ii)   $\mathbf{x}_{k+j}$ ($j \in \mathbb{Z}$) *at fixed point* $\mathbf{x}_k^*$ *on the direction* $\mathbf{v}_i$ *is monotonically unstable of the* $(2m_i + 1)$*th-order if and only if*

$$G_{s_k^{(i)}}^{(1)}(\mathbf{x}_k^*, \mathbf{p}) = \lambda_i = 1,$$

$$G_{s_k^{(i)}}^{(r_i)}(\mathbf{x}_k^*, \mathbf{p}) = 0 \ \text{for } r_i = 2, 3, \ldots, 2m_i, \tag{2.70}$$

$$G_{s_k^{(i)}}^{(2m_i+1)}(\mathbf{x}_k^*, \mathbf{p}) > 0$$

*for* $\mathbf{x}_k \in U(\mathbf{x}_k^*) \subset \Omega_\alpha$.

(iii)   $\mathbf{x}_{k+j}$ ($j \in \mathbb{Z}$) *at fixed point* $\mathbf{x}_k^*$ *on the direction* $\mathbf{v}_i$ *is monotonically unstable of the* $(2m_i)$*th-order, lower saddle if and only if*

$$G^{(1)}_{s^{(i)}_k}(\mathbf{x}^*_k, \mathbf{p}) = \lambda_i = 1,$$

$$G^{(r_i)}_{s^{(i)}_k}(\mathbf{x}^*_k, \mathbf{p}) = 0 \quad for \ r_i = 2, 3, \ldots, 2m_i - 1,$$

$$G^{(2m_i)}_{s^{(i)}_k}(\mathbf{x}^*_k, \mathbf{p}) < 0 \quad stable \ for \ s^{(i)}_k > 0;$$

$$G^{(2m_i)}_{s^{(i)}_k}(\mathbf{x}^*_k, \mathbf{p}) < 0 \quad unstable \ for \ s^{(i)}_k < 0$$

$$(2.71)$$

*for* $\mathbf{x}_k \in U(\mathbf{x}^*_k) \subset \Omega_\alpha.$

(iv) $\mathbf{x}_{k+j}$ ($j \in \mathbb{Z}$) *at fixed point* $\mathbf{x}^*_k$ *on the direction* $\mathbf{v}_i$ *is monotonically unstable of the* $(2m_i)$th-*order if and only if*

$$G^{(1)}_{s^{(i)}_k}(\mathbf{x}^*_k, \mathbf{p}) = \lambda_i = 1,$$

$$G^{(r_i)}_{s^{(i)}_k}(\mathbf{x}^*_k, \mathbf{p}) = 0 \quad for \ r_i = 2, 3, \ldots, 2m_i - 1,$$

$$G^{(2m_i)}_{s^{(i)}_k}(\mathbf{x}^*_k, \mathbf{p}) > 0 \quad unstable \ for \ s^{(i)}_k > 0;$$

$$G^{(2m_i)}_{s^{(i)}_k}(\mathbf{x}^*_k, \mathbf{p}) > 0 \quad stable \ for \ s^{(i)}_k < 0$$

$$(2.72)$$

*for* $\mathbf{x}_k \in U(\mathbf{x}^*_k) \subset \Omega_\alpha.$

(v) $\mathbf{x}_{k+j}$ ($j \in \mathbb{Z}$) *at fixed point* $\mathbf{x}^*_k$ *on the direction* $\mathbf{v}_i$ *is oscillatory stable of the* $(2m_i + 1)$th-*order if and only if*

$$G^{(1)}_{s^{(i)}_k}(\mathbf{x}^*_k, \mathbf{p}) = \lambda_i = -1,$$

$$G^{(r_i)}_{s^{(i)}_k}(\mathbf{x}^*_k, \mathbf{p}) = 0 \quad for \ r_i = 2, 3, \ldots, 2m_i,$$

$$G^{(2m_i+1)}_{s^{(i)}_k}(\mathbf{x}^*_k, \mathbf{p}) > 0$$

$$(2.73)$$

*for* $\mathbf{x}_k \in U(\mathbf{x}^*_k) \subset \Omega_\alpha.$

(vi) $\mathbf{x}_{k+j}$ ($j \in \mathbb{Z}$) *at fixed point* $\mathbf{x}^*_k$ *on the direction* $\mathbf{v}_i$ *is oscillatory unstable of the* $(2m_i + 1)$th-*order if and only if*

$$G^{(1)}_{s^{(i)}_k}(\mathbf{x}^*_k, \mathbf{p}) = \lambda_i = -1,$$

$$G^{(r_i)}_{s^{(i)}_k}(\mathbf{x}^*_k, \mathbf{p}) = 0 \quad for \ r_i = 2, 3, \ldots, 2m_i,$$

$$G^{(2m_i+1)}_{s^{(i)}_k}(\mathbf{x}^*_k, \mathbf{p}) < 0$$

$$(2.74)$$

*for* $\mathbf{x}_k \in U(\mathbf{x}^*_k) \subset \Omega_\alpha.$

(vii) $\mathbf{x}_{k+j}$ ($j \in \mathbb{Z}$) *at fixed point* $\mathbf{x}^*_k$ *on the direction* $\mathbf{v}_i$ *is oscillatory unstable of the* $(2m_i)$th-*order, upper saddle if and only if*

$$G^{(1)}_{s^{(i)}_k}(\mathbf{x}^*_k, \mathbf{p}) = \lambda_i = -1,$$

$$G^{(r_i)}_{s^{(i)}_k}(\mathbf{x}^*_k, \mathbf{p}) = 0 \ \ for \ r_i = 2, 3, \ldots, 2m_i - 1,$$

$$G^{(2m_i)}_{s^{(i)}_k}(\mathbf{x}^*_k, \mathbf{p}) > 0 \ \ stable \ for \ s^{(i)}_k > 0;$$

$$G^{(2m_i)}_{s^{(i)}_k}(\mathbf{x}^*_k, \mathbf{p}) > 0 \ \ unstable \ for \ s^{(i)}_k < 0$$

(2.75)

*for* $\mathbf{x}_k \in U(\mathbf{x}^*_k) \subset \Omega_\alpha$.

(viii) $\mathbf{x}_{k+j}$ ($j \in \mathbb{Z}$) *at fixed point* $\mathbf{x}^*_k$ *on the direction* $\mathbf{v}_i$ *is oscillatory unstable of the* $(2m_i)$th-*order, lower saddle if and only if*

$$G^{(1)}_{s^{(i)}_k}(\mathbf{x}^*_k, \mathbf{p}) = \lambda_i = -1,$$

$$G^{(r_i)}_{s^{(i)}_k}(\mathbf{x}^*_k, \mathbf{p}) = 0 \ \ for \ r_i = 2, 3, \ldots, 2m_i - 1,$$

$$G^{(2m_i)}_{s^{(i)}_k}(\mathbf{x}^*_k, \mathbf{p}) < 0 \ \ stable \ for \ s^{(i)}_k < 0;$$

$$G^{(2m_i)}_{s^{(i)}_k}(\mathbf{x}^*_k, \mathbf{p}) < 0 \ \ unstable \ for \ s^{(i)}_k > 0$$

(2.76)

*for* $\mathbf{x}_k \in U(\mathbf{x}^*_k) \subset \Omega_\alpha$.

*Proof* The proof can be referred to Luo (2012). □

**Definition 2.21** Consider a discrete, nonlinear dynamical system $\mathbf{x}_{k+1} = \mathbf{f}(\mathbf{x}_k, \mathbf{p})$ $\in \mathcal{R}^n$ in Eq. (2.4) with a fixed point $\mathbf{x}^*_k$. The corresponding solution is given by $\mathbf{x}_{k+j} = \mathbf{f}(\mathbf{x}_{k+j-1}, \mathbf{p})$ with $j \in \mathbb{Z}$. Suppose there is a neighborhood of the fixed point $\mathbf{x}^*_k$ (i.e., $U_k(\mathbf{x}^*_k) \subset \Omega$), and $\mathbf{f}(\mathbf{x}_k, \mathbf{p})$ is $C^r(r \geq 1)$-continuous in $U_k(\mathbf{x}^*_k)$ with Eq. (2.28). The linearized system is $\mathbf{y}_{k+j+1} = D\mathbf{f}(\mathbf{x}^*_k, \mathbf{p})\mathbf{y}_{k+j}$ ($\mathbf{y}_{k+j} = \mathbf{x}_{k+j} - \mathbf{x}^*_k$) in $U_k(\mathbf{x}^*_k)$. Consider a pair of complex eigenvalues $\alpha_i \pm i\beta_i$ ($i \in N = \{1, 2, \ldots, n\}$, $\mathbf{i} = \sqrt{-1}$) of matrix $D\mathbf{f}(\mathbf{x}^*, \mathbf{p})$ with a pair of eigenvectors $\mathbf{u}_i \pm i\mathbf{v}_i$. On the invariant plane of $(\mathbf{u}_i, \mathbf{v}_i)$, consider $\mathbf{r}^{(i)}_k = \mathbf{y}^{(i)}_k = \mathbf{y}^{(i)}_{k+} + \mathbf{y}^{(i)}_{k-}$ with

$$\mathbf{r}^{(i)}_k = c^{(i)}_k \mathbf{u}_i + d^{(i)}_k \mathbf{v}_i,$$

$$\mathbf{r}^{(i)}_{k+1} = c^{(i)}_{k+1} \mathbf{u}_i + d^{(i)}_{k+1} \mathbf{v}_i$$

(2.77)

and

$$c^{(i)}_k = \frac{1}{\Delta}[\Delta_2(\mathbf{u}^T_i \cdot \mathbf{y}_k) - \Delta_{12}(\mathbf{v}^T_i \cdot \mathbf{y}_k)],$$

$$d^{(i)}_k = \frac{1}{\Delta}[\Delta_1(\mathbf{v}^T_i \cdot \mathbf{y}_k) - \Delta_{12}(\mathbf{u}^T_i \cdot \mathbf{y}_k)];$$

$$\Delta_1 = ||\mathbf{u}_i||^2, \Delta_2 = ||\mathbf{v}_i||^2, \Delta_{12} = \mathbf{u}^T_i \cdot \mathbf{v}_i;$$

$$\Delta = \Delta_1 \Delta_2 - \Delta^2_{12}.$$

(2.78)

Consider a polar coordinate of $(r_k, \theta_k)$ defined by

$$c_k^{(i)} = r_k^{(i)} \cos \theta_k^{(i)}, \quad \text{and} \quad d_k^{(i)} = r_k^{(i)} \sin \theta_k^{(i)};$$
$$r_k^{(i)} = \sqrt{(c_k^{(i)})^2 + (d_k^{(i)})^2}, \quad \text{and} \quad \theta_k^{(i)} = \arctan(d_k^{(i)}/c_k^{(i)}). \tag{2.79}$$

Thus,

$$c_{k+1}^{(i)} = \frac{1}{\Delta}[\Delta_2 G_{c_k^{(i)}}(\mathbf{x}_k, \mathbf{p}) - \Delta_{12} G_{d_k^{(i)}}(\mathbf{x}_k, \mathbf{p})]$$
$$d_{k+1}^{(i)} = \frac{1}{\Delta}[\Delta_1 G_{d_k^{(i)}}(\mathbf{x}_k, \mathbf{p}) - \Delta_{12} G_{c_k^{(i)}}(\mathbf{x}_k, \mathbf{p})] \tag{2.80}$$

where

$$G_{c_k^{(i)}}(\mathbf{x}_k, \mathbf{p}) = \mathbf{u}_i^{\mathrm{T}} \cdot [\mathbf{f}(\mathbf{x}_k, \mathbf{p}) - \mathbf{x}_k^*] = \sum_{m_i=1}^{\infty} \frac{1}{m_i!} G_{c_k^{(i)}}^{(m_i)}(\theta_k^{(i)})(r_k^{(i)})^{m_i},$$
$$G_{d_k^{(i)}}(\mathbf{x}_k, \mathbf{p}) = \mathbf{v}_i^{\mathrm{T}} \cdot [\mathbf{f}(\mathbf{x}_k, \mathbf{p}) - \mathbf{x}_k^*] = \sum_{m_i=1}^{\infty} \frac{1}{m_i!} G_{d_k^{(i)}}^{(m_i)}(\theta_k^{(i)})(r_k^{(i)})^{m_i}; \tag{2.81}$$

$$G_{c_k^{(i)}}^{(m_i)}(\theta_k^{(i)}) = \mathbf{u}_i^{\mathrm{T}} \cdot \partial_{\mathbf{x}_k}^{(m_i)} \mathbf{f}(\mathbf{x}_k, \mathbf{p})[\mathbf{u}_i \cos \theta_k^{(i)} + \mathbf{v}_i \sin \theta_k^{(i)}]^{m_i}\Big|_{(\mathbf{x}_k^*, \mathbf{p})},$$
$$G_{d_k^{(i)}}^{(m_i)}(\theta_k^{(i)}) = \mathbf{v}_i^{\mathrm{T}} \cdot \partial_{\mathbf{x}_k}^{(m_i)} \mathbf{f}(\mathbf{x}_k, \mathbf{p})[\mathbf{u}_i \cos \theta_k^{(i)} + \mathbf{v}_i \sin \theta_k^{(i)}]^{m_i}\Big|_{(\mathbf{x}_k^*, \mathbf{p})}. \tag{2.82}$$

Thus,

$$r_{k+1}^{(i)} = \sqrt{(c_{k+1}^{(i)})^2 + (d_{k+1}^{(i)})^2} = \sqrt{\sum_{m=2}^{\infty} (r_k^{(i)})^{m_i} G_{r_{k+1}^{(i)}}^{(m_i)}(\theta_k^{(i)})}$$
$$= \sqrt{G_{r_{k+1}^{(i)}}^{(2)} r_k^{(i)} \sqrt{1 + (G_{r_{k+1}^{(i)}}^{(2)})^{-1} \sum_{m=3}^{\infty} (r_k^{(i)})^{m_i-2} G_{r_{k+1}^{(i)}}^{(m_i)}(\theta_k^{(i)})}} \tag{2.83}$$
$$\theta_{k+1}^{(i)} = \arctan(d_{k+1}^{(i)}/c_{k+1}^{(i)})$$

where

$$G_{r_{k+1}^{(i)}}^{(m_i)}(\theta_k^{(i)})$$
$$= \sum_{r_i=1}^{\infty} \sum_{s_i=1}^{\infty} \frac{1}{r_i! s_i!}[G_{c_{k+1}^{(i)}}^{(r_i)}(\theta_k^{(i)}) G_{c_{k+1}^{(i)}}^{(s_i)}(\theta_k^{(i)}) + G_{d_{k+1}^{(i)}}^{(r_i)}(\theta_k^{(i)}) G_{d_{k+1}^{(i)}}^{(s_i)}(\theta_k^{(i)})]\delta_{m_i}^{(r_i+s_i)} \tag{2.84}$$
$$= \frac{1}{m_i!} \sum_{r_i=1}^{m_i} C_{m_i}^{r_i} G_{c_{k+1}^{(i)}}^{(r_i)}(\theta_k^{(i)}) G_{c_{k+1}^{(i)}}^{(m_i-r_i)}(\theta_k^{(i)}) + G_{d_{k+1}^{(i)}}^{(r_i)}(\theta_k^{(i)}) G_{d_{k+1}^{(i)}}^{(m_i-r_i)}(\theta_k^{(i)})$$

and

$$G^{(m_i)}_{c^{(i)}_{k+1}}(\theta^{(i)}_k) = \frac{1}{\Delta}[\Delta_2 G^{(m_i)}_{c^{(i)}_k}(\theta^{(i)}_k) - \Delta_{12} G^{(m_i)}_{d^{(i)}_k}(\theta^{(i)}_k)],$$

$$G^{(m_i)}_{d^{(i)}_{k+1}}(\theta^{(i)}_k) = \frac{1}{\Delta}[\Delta_1 G^{(m_i)}_{d^{(i)}_k}(\theta^{(i)}_k) - \Delta_{12} G^{(m_i)}_{c^{(i)}_k}(\theta^{(i)}_k)].$$

(2.85)

From the foregoing definition, consider the first-order terms of G-function

$$G^{(1)}_{c^{(i)}_k}(\mathbf{x}_k, \mathbf{p}) = G^{(1)}_{c^{(i)}_k 1}(\mathbf{x}_k, \mathbf{p}) + G^{(1)}_{c^{(i)}_k 2}(\mathbf{x}_k, \mathbf{p}),$$

$$G^{(1)}_{d^{(i)}_k}(\mathbf{x}_k, \mathbf{p}) = G^{(1)}_{d^{(i)}_k 1}(\mathbf{x}_k, \mathbf{p}) + G^{(1)}_{d^{(i)}_k 2}(\mathbf{x}_k, \mathbf{p})$$

(2.86)

where

$$
\begin{aligned}
G^{(1)}_{c^{(i)}_k 1}(\mathbf{x}_k, \mathbf{p}) &= \mathbf{u}^{\mathrm{T}}_i \cdot D_{\mathbf{x}_k}\mathbf{f}(\mathbf{x}_k, \mathbf{p})\partial_{c^{(i)}_k}\mathbf{x}_k = \mathbf{u}^{\mathrm{T}}_i \cdot D_{\mathbf{x}_k}\mathbf{f}(\mathbf{x}_k, \mathbf{p})\mathbf{u}_i \\
&= \mathbf{u}^{\mathrm{T}}_i \cdot (-\beta_i \mathbf{v}_i + \alpha_i \mathbf{u}_i) = \alpha_i \Delta_1 - \beta_i \Delta_{12}, \\
G^{(1)}_{c^{(i)}_k 2}(\mathbf{x}_k, \mathbf{p}) &= \mathbf{u}^{\mathrm{T}}_i \cdot D_{\mathbf{x}_k}\mathbf{f}(\mathbf{x}_k, \mathbf{p})\partial_{d^{(i)}_k}\mathbf{x}_k = \mathbf{u}^{\mathrm{T}}_i \cdot D_{\mathbf{x}_k}\mathbf{f}(\mathbf{x}_k, \mathbf{p})\mathbf{v}_i \\
&= \mathbf{u}^{\mathrm{T}}_i \cdot (\beta_i \mathbf{u}_i + \alpha_i \mathbf{v}_i) = \alpha_i \Delta_{12} + \beta_i \Delta_1;
\end{aligned}
$$

(2.87)

and

$$
\begin{aligned}
G^{(1)}_{d^{(i)}_k 1}(\mathbf{x}_k, \mathbf{p}) &= \mathbf{v}^{\mathrm{T}}_i \cdot D_{\mathbf{x}_k}\mathbf{f}(\mathbf{x}_k, \mathbf{p})\partial_{c^{(i)}_k}\mathbf{x}_k = \mathbf{v}^{\mathrm{T}}_i \cdot D_{\mathbf{x}_k}\mathbf{f}(\mathbf{x}_k, \mathbf{p})\mathbf{u}_i \\
&= \mathbf{v}^{\mathrm{T}}_i \cdot (-\beta_i \mathbf{v}_i + \alpha_i \mathbf{u}_i) = -\beta_i \Delta_2 + \alpha_i \Delta_{12}, \\
G^{(1)}_{d^{(i)}_k 2}(\mathbf{x}, \mathbf{p}) &= \mathbf{v}^{\mathrm{T}}_i \cdot D_{\mathbf{x}_k}\mathbf{f}(\mathbf{x}_k, \mathbf{p})\partial_{d^{(i)}_k}\mathbf{x}_k = \mathbf{v}^{\mathrm{T}}_i \cdot D_{\mathbf{x}_k}\mathbf{f}(\mathbf{x}_k, \mathbf{p})\mathbf{v}_i \\
&= \mathbf{v}^{\mathrm{T}}_i \cdot (\beta_i \mathbf{u}_i + \alpha_i \mathbf{v}_i) = \alpha_i \Delta_2 + \beta_i \Delta_{12}.
\end{aligned}
$$

(2.88)

Substitution of Eqs. (2.86)–(2.88) into Eq. (2.82) gives

$$
\begin{aligned}
G^{(1)}_{c^{(i)}_k}(\theta^{(i)}_k) &= G^{(1)}_{c^{(i)}_k 1}(\mathbf{x}_k, \mathbf{p})\cos\theta^{(i)}_k + G^{(1)}_{c^{(i)}_k 2}(\mathbf{x}_k, \mathbf{p})\sin\theta^{(i)}_k \\
&= (\alpha_i \Delta_1 - \beta_i \Delta_{12})\cos\theta^{(i)}_k + (\alpha_i \Delta_{12} + \beta_i \Delta_1)\sin\theta^{(i)}_k, \\
G^{(1)}_{d^{(i)}_k}(\theta^{(i)}_k) &= G^{(1)}_{d^{(i)}_k 1}(\mathbf{x}_k, \mathbf{p})\cos\theta^{(i)}_k + G^{(1)}_{d^{(i)}_k 2}(\mathbf{x}_k, \mathbf{p})\sin\theta^{(i)}_k \\
&= (-\beta_i \Delta_2 + \alpha_i \Delta_{12})\cos\theta^{(i)}_k + (\alpha_i \Delta_2 + \beta_i \Delta_{12})\sin\theta^{(i)}_k.
\end{aligned}
$$

(2.89)

From Eq. (2.85), we have

$$G^{(1)}_{c^{(i)}_{k+1}}(\theta^{(i)}_k) = \frac{1}{\Delta}[\Delta_2 G^{(1)}_{c^{(i)}_k}(\theta^{(i)}_k) - \Delta_{12} G^{(1)}_{d^{(i)}_k}(\theta^{(i)}_k)]$$
$$= \alpha_i \cos\theta^{(i)}_k + \beta_i \sin\theta^{(i)}_k,$$
$$G^{(1)}_{d^{(i)}_{k+1}}(\theta^{(i)}_k) = \frac{1}{\Delta}[\Delta_1 G^{(1)}_{d^{(i)}_k}(\theta^{(i)}_k) - \Delta_{12} G^{(1)}_{c^{(i)}_k}(\theta^{(i)}_k)]$$
$$= \alpha_i \sin\theta^{(i)}_k - \beta_i \cos\theta^{(i)}_k. \tag{2.90}$$

Thus,

$$G^{(2)}_{r^{(i)}_{k+1}}(\theta^{(i)}_k) = [G^{(1)}_{c^{(i)}_{k+1}}(\theta^{(i)}_k)G^{(1)}_{c^{(i)}_{k+1}}(\theta^{(i)}_k) + G^{(1)}_{d^{(i)}_{k+1}}(\theta^{(i)}_k)G^{(1)}_{d^{(i)}_{k+1}}(\theta^{(i)}_k)]$$
$$= \alpha_i^2 + \beta_i^2. \tag{2.91}$$

Furthermore, Eq. (2.83) gives

$$r^{(i)}_{k+1} = \rho_i r^{(i)}_k + o(r^{(i)}_k) \quad\text{and}\quad \theta^{(i)}_{k+1} = \theta^{(i)}_k - \vartheta_i + o(r^{(i)}_k). \tag{2.92}$$

where

$$\vartheta_i = \arctan(\beta_i/\alpha_i) \quad\text{and}\quad \rho_i = \sqrt{\alpha_i^2 + \beta_i^2}. \tag{2.93}$$

As $r^{(i)}_k \ll 1$ and $r^{(i)}_k \to 0$, we have

$$r^{(i)}_{k+1} = \rho_i r^{(i)}_k \quad\text{and}\quad \theta^{(i)}_{k+1} = \vartheta_i - \theta^{(i)}_k. \tag{2.94}$$

With an initial condition of $r^{(i)}_k = r^0_k$ and $\theta^{(i)}_k = \theta^{(i)}_k$, the corresponding solution of Eq. (2.94) is

$$r^{(i)}_{k+j} = (\rho_i)^j r^0_k \quad\text{and}\quad \theta^{(i)}_{k+j} = j\vartheta_i - \theta^{(i)}_k. \tag{2.95}$$

From Eqs. (2.80), (2.81), and (2.90), we have

$$c^{(i)}_{k+1} = \alpha_i r^{(i)}_k \cos\theta^{(i)}_k + \beta_i r^{(i)}_k \sin\theta^{(i)}_k = \alpha_i c^{(i)}_k + \beta_i d^{(i)}_k,$$
$$d^{(i)}_{k+1} = \alpha_i r^{(i)}_k \sin\theta^{(i)}_k - \beta_i r^{(i)}_k \cos\theta^{(i)}_k = -\beta_i c^{(i)}_k + \alpha_i d^{(i)}_k. \tag{2.96}$$

That is,

$$\left\{\begin{matrix} c^{(i)}_{k+1} \\ d^{(i)}_{k+1} \end{matrix}\right\} = \begin{bmatrix} \alpha_i & \beta_i \\ -\beta_i & \alpha_i \end{bmatrix}\left\{\begin{matrix} c^{(i)}_k \\ d^{(i)}_k \end{matrix}\right\} = \rho_i\begin{bmatrix} \cos\vartheta_i & \sin\vartheta_i \\ -\sin\vartheta_i & \cos\vartheta_i \end{bmatrix}\left\{\begin{matrix} c^{(i)}_k \\ d^{(i)}_k \end{matrix}\right\}. \tag{2.97}$$

From the foregoing equation, we have

$$\left\{ \begin{matrix} c_{k+j}^{(i)} \\ d_{k+j}^{(i)} \end{matrix} \right\} = \left[ \begin{matrix} \alpha_i & \beta_i \\ -\beta_i & \alpha_i \end{matrix} \right]^j \left\{ \begin{matrix} c_k^{(i)} \\ d_k^{(i)} \end{matrix} \right\} = (\rho_i)^j \left[ \begin{matrix} \cos j\vartheta_i & \sin j\vartheta_i \\ -\sin j\vartheta_i & \cos j\vartheta_i \end{matrix} \right] \left\{ \begin{matrix} c_k^{(i)} \\ d_k^{(i)} \end{matrix} \right\}. \tag{2.98}$$

**Definition 2.22** Consider a discrete, nonlinear dynamical system $\mathbf{x}_{k+1} = \mathbf{f}(\mathbf{x}_k, \mathbf{p}) \in \mathscr{R}^n$ in Eq. (2.4) with a fixed point $\mathbf{x}_k^*$. The corresponding solution is given by $\mathbf{x}_{k+j} = \mathbf{f}(\mathbf{x}_{k+j-1}, \mathbf{p})$ with $j \in \mathbb{Z}$. Suppose there is a neighborhood of the fixed point $\mathbf{x}_k^*$ (i.e., $U_k(\mathbf{x}_k^*) \subset \Omega$), and $\mathbf{f}(\mathbf{x}_k, \mathbf{p})$ is $C^r$ $(r \geq 1)$-continuous in $U_k(\mathbf{x}_k^*)$ with Eq. (2.28). The linearized system is $\mathbf{y}_{k+j+1} = D\mathbf{f}(\mathbf{x}_k^*, \mathbf{p})\mathbf{y}_{k+j}$ $(\mathbf{y}_{k+j} = \mathbf{x}_{k+j} - \mathbf{x}_k^*)$ in $U_k(\mathbf{x}_k^*)$. Consider a pair of complex eigenvalues $\alpha_i \pm i\beta_i$ $(i \in N = \{1, 2, \ldots, n\}$, $\mathbf{i} = \sqrt{-1})$ of matrix $D\mathbf{f}(\mathbf{x}^*, \mathbf{p})$ with a pair of eigenvectors $\mathbf{u}_i \pm i\mathbf{v}_i$. On the invariant plane of $(\mathbf{u}_i, \mathbf{v}_i)$, consider $\mathbf{r}_k^{(i)} = \mathbf{y}_k^{(i)} = \mathbf{y}_{k+}^{(i)} + \mathbf{y}_{k-}^{(i)}$ with Eqs. (2.73) and (2.75). For any arbitrarily small $\varepsilon > 0$, the stability of the fixed point $\mathbf{x}_k^*$ on the invariant plane of $(\mathbf{u}_i, \mathbf{v}_i)$ can be determined.

(i)  $\mathbf{x}_k^{(i)}$ at the fixed point $\mathbf{x}_k^*$ on the plane of $(\mathbf{u}_i, \mathbf{v}_i)$ is spirally stable if

$$r_{k+1}^{(i)} - r_k^{(i)} < 0. \tag{2.99}$$

(ii)  $\mathbf{x}_k^{(i)}$ at the fixed point $\mathbf{x}_k^*$ on the plane of $(\mathbf{u}_i, \mathbf{v}_i)$ is spirally unstable if

$$r_{k+1}^{(i)} - r_k^{(i)} > 0. \tag{2.100}$$

(iii)  $\mathbf{x}_k^{(i)}$ at the fixed point $\mathbf{x}_k^*$ on the plane of $(\mathbf{u}_i, \mathbf{v}_i)$ is spirally stable with the $m_i$th-order singularity if for $\theta_k^{(i)} \in [0, 2\pi]$

$$\begin{aligned} &\rho_i = \sqrt{\alpha_i^2 + \beta_i^2} = 1, \\ &G_{r_{k+1}^{(i)}}^{(s_k^{(i)})}(\theta_k) = 0 \quad \text{for } s_k^{(i)} = 1, 2, \ldots, m_i - 1, \\ &r_{k+1}^{(i)} - r_k^{(i)} < 0. \end{aligned} \tag{2.101}$$

(iv)  $\mathbf{x}_k^{(i)}$ at the fixed point $\mathbf{x}_k^*$ on the plane of $(\mathbf{u}_i, \mathbf{v}_i)$ is spirally unstable with the $m_i$th-order singularity if for $\theta_k^{(i)} \in [0, 2\pi]$

$$\begin{aligned} &\rho_i = \sqrt{\alpha_i^2 + \beta_i^2} = 1, \\ &G_{r_{k+1}^{(i)}}^{(s_k^{(i)})}(\theta_k) = 0 \quad \text{for } s_k^{(i)} = 1, 2, \ldots, m_i - 1, \\ &r_{k+1}^{(i)} - r_k^{(i)} > 0. \end{aligned} \tag{2.102}$$

(v)  $\mathbf{x}_k^{(i)}$ at the fixed point $\mathbf{x}_k^*$ on the plane of $(\mathbf{u}_i, \mathbf{v}_i)$ is circular if for $\theta_k^{(i)} \in [0, 2\pi]$

$$r_{k+1}^{(i)} - r_k^{(i)} = 0. \tag{2.103}$$

(vi)  $\mathbf{x}_k^{(i)}$ at the fixed point $\mathbf{x}_k^*$ on the plane of $(\mathbf{u}_i, \mathbf{v}_i)$ is degenerate in the direction of $\mathbf{u}_i$ if

$$\beta_i = 0 \quad \text{and} \quad \theta_{k+1}^{(i)} - \theta_k^{(i)} = 0. \tag{2.104}$$

**Theorem 2.7** *Consider a discrete, nonlinear dynamical system* $\mathbf{x}_{k+1} = \mathbf{f}(\mathbf{x}_k, \mathbf{p}) \in \mathscr{R}^n$ *in Eq. (2.4) with a fixed point* $\mathbf{x}_k^*$. *The corresponding solution is given by* $\mathbf{x}_{k+j} = \mathbf{f}(\mathbf{x}_{k+j-1}, \mathbf{p})$ *with* $j \in \mathbb{Z}$. *Suppose there is a neighborhood of the fixed point* $\mathbf{x}_k^*$ *(i.e.,* $U_k(\mathbf{x}_k^*) \subset \Omega$), *and* $\mathbf{f}(\mathbf{x}_k, \mathbf{p})$ *is* $C^r$ *(*$r \geq 1$*)-continuous in* $U_k(\mathbf{x}_k^*)$ *with Eq. (2.28). The linearized system is* $\mathbf{y}_{k+j+1} = D\mathbf{f}(\mathbf{x}_k^*, \mathbf{p})\mathbf{y}_{k+j}$ *(*$\mathbf{y}_{k+j} = \mathbf{x}_{k+j} - \mathbf{x}_k^*$*) in* $U_k(\mathbf{x}_k^*)$. *Consider a pair of complex eigenvalues* $\alpha_i \pm i\beta_i$ *(*$i \in N = \{1, 2, \ldots, n\}$, $\mathbf{i} = \sqrt{-1}$) *of matrix* $D\mathbf{f}(\mathbf{x}^*, \mathbf{p})$ *with a pair of eigenvectors* $\mathbf{u}_i \pm i\mathbf{v}_i$. *On the invariant plane of* $(\mathbf{u}_i, \mathbf{v}_i)$, *consider* $\mathbf{r}_k^{(i)} = \mathbf{y}_k^{(i)} = \mathbf{y}_{k+}^{(i)} + \mathbf{y}_{k-}^{(i)}$ *with Eqs. (2.73) and (2.75). For any arbitrarily small* $\varepsilon > 0$, *the stability of the equilibrium* $\mathbf{x}_k^*$ *on the invariant plane of* $(\mathbf{u}_i, \mathbf{v}_i)$ *can be determined.*

(i)  $\mathbf{x}_k^{(i)}$ *at the fixed point* $\mathbf{x}_k^*$ *on the plane of* $(\mathbf{u}_k, \mathbf{v}_k)$ *is spirally stable if and only if*

$$\rho_i < 1. \tag{2.105}$$

(ii)  $\mathbf{x}_k^{(i)}$ *at the fixed point* $\mathbf{x}_k^*$ *on the plane of* $(\mathbf{u}_i, \mathbf{v}_i)$ *is spirally unstable if and only if*

$$\rho_i > 1. \tag{2.106}$$

(iii)  $\mathbf{x}_k^{(i)}$ *at the fixed point* $\mathbf{x}_k^*$ *on the plane of* $(\mathbf{u}_i, \mathbf{v}_i)$ *is stable with the* $m_i$th-order *singularity if and only if for* $\theta_k^{(i)} \in [0, 2\pi]$

$$
\begin{aligned}
&\rho_i = \sqrt{\alpha_i^2 + \beta_i^2} = 1, \\
&G_{r_k^{(i)}}^{(s_k^{(i)})}(\theta_k^{(i)}) = 0 \quad \text{for } s_k = 1, 2, \ldots, m_i - 1, \\
&G_{r_k^{(i)}}^{(m_i)}(\theta_k^{(i)}) < 0.
\end{aligned}
\tag{2.107}
$$

(iv)  $\mathbf{x}_k^{(i)}$ *at the fixed point* $\mathbf{x}_k^*$ *on the plane of* $(\mathbf{u}_i, \mathbf{v}_i)$ *is spirally unstable with the* $m_i$th-order *singularity if and only if for* $\theta_k^{(i)} \in [0, 2\pi]$

$$\rho_i = \sqrt{\alpha_i^2 + \beta_i^2} = 1,$$

$$G_{r_k^{(i)}}^{(s_k^{(i)})}(\theta_k^{(i)}) = 0 \ \ for \ s_k^{(i)} = 0, 1, 2, \ldots, m_i - 1, \tag{2.108}$$

$$G_{r_k^{(i)}}^{(m_i)}(\theta_k^{(i)}) > 0.$$

(v)   $x_k^{(i)}$ at the fixed point $\mathbf{x}_k^*$ on the plane of $(\mathbf{u}_i, \mathbf{v}_i)$ is circular if and only if for $\theta_k^{(i)} \in [0, 2\pi]$

$$\rho_i = \sqrt{\alpha_i^2 + \beta_i^2} = 1,$$

$$G_{r_k^{(i)}}^{(s_k^{(i)})}(\theta_k^{(i)}) = 0 \ \ for \ s_k^{(i)} = 0, 1, 2, \ldots. \tag{2.109}$$

*Proof* The proof can be referred to Luo (2011).                    □

## 2.4  Bifurcation Theory

**Definition 2.23** Consider a discrete, nonlinear dynamical system $\mathbf{x}_{k+1} = \mathbf{f}(\mathbf{x}_k, \mathbf{p}) \in \mathscr{R}^n$ in Eq. (2.4) with a fixed point $\mathbf{x}_k^*$. The corresponding solution is given by $\mathbf{x}_{k+j} = \mathbf{f}(\mathbf{x}_{k+j-1}, \mathbf{p})$ with $j \in \mathbb{Z}$. Suppose there is a neighborhood of the fixed point $\mathbf{x}_k^*$ (i.e., $U_k(\mathbf{x}_k^*) \subset \Omega$), and $\mathbf{f}(\mathbf{x}_k, \mathbf{p})$ is $C^r$ ($r \geq 1$)-continuous in $U_k(\mathbf{x}_k^*)$ with Eq. (2.28). The linearized system is $\mathbf{y}_{k+j+1} = D\mathbf{f}(\mathbf{x}_k^*, \mathbf{p})\mathbf{y}_{k+j}$ ($\mathbf{y}_{k+j} = \mathbf{x}_{k+j} - \mathbf{x}_k^*$) in $U_k(\mathbf{x}_k^*)$ and there are $n$ linearly independent vectors $\mathbf{v}_i$ ($i = 1, 2, \ldots, n$). For a perturbation of fixed point $\mathbf{y}_k = \mathbf{x}_k - \mathbf{x}_k^*$, let $\mathbf{y}_k^{(i)} = c_k^{(i)}\mathbf{v}_i$ and $\mathbf{y}_{k+1}^{(i)} = c_{k+1}^{(i)}\mathbf{v}_i$.

$$s_k^{(i)} = \mathbf{v}_i^{\mathrm{T}} \cdot \mathbf{y}_k = \mathbf{v}_i^{\mathrm{T}} \cdot (\mathbf{x}_k - \mathbf{x}_k^*) \tag{2.110}$$

where $s_k^{(i)} = c_k^{(i)}\|\mathbf{v}_i\|^2$.

$$s_{k+1}^{(i)} = \mathbf{v}_i^{\mathrm{T}} \cdot \mathbf{y}_{k+1} = \mathbf{v}_i^{\mathrm{T}} \cdot [\mathbf{f}(\mathbf{x}_k, \mathbf{p}) - \mathbf{x}_k^*]. \tag{2.111}$$

In the vicinity of point $(\mathbf{x}_{k(0)}^*, \mathbf{p}_0)$, $\mathbf{v}_i^{\mathrm{T}} \cdot \mathbf{f}(\mathbf{x}_k, \mathbf{p})$ can be expanded for $(0 < \theta < 1)$ as

$$\begin{aligned}
\mathbf{v}_i^{\mathrm{T}} \cdot [\mathbf{f}(\mathbf{x}_k, \mathbf{p}) - \mathbf{x}_{k(0)}^*] = {}& a_i(s_k^{(i)} - s_{k(0)}^{(i)*}) + \mathbf{b}_i^{\mathrm{T}} \cdot (\mathbf{p} - \mathbf{p}_0) \\
& + \sum_{q=2}^{m} \frac{1}{q!} \sum_{r=0}^{q} C_q^r \mathbf{a}_i^{(q-r,r)}(s_k^{(i)} - s_{k(0)}^{(i)*})^{q-r}(\mathbf{p} - \mathbf{p}_0)^r \\
& + \frac{1}{(m+1)!}[(s_k^{(i)} - s_{k(0)}^{(i)*})\partial_{s_k^{(i)}} + (\mathbf{p} - \mathbf{p}_0)\partial_{\mathbf{p}}]^{m+1} \\
& \times (\mathbf{v}_i^{\mathrm{T}} \cdot \mathbf{f}(\mathbf{x}_{k(0)}^* + \theta\Delta\mathbf{x}_k, \mathbf{p}_0 + \theta\Delta\mathbf{p}))
\end{aligned} \tag{2.112}$$

where

$$a_i = \mathbf{v}_i^{\mathrm{T}} \cdot \partial_{s_k^{(i)}} \mathbf{f}(\mathbf{x}_k, \mathbf{p})\Big|_{(\mathbf{x}_{k(0)}^*, \mathbf{p}_0)}, \quad \mathbf{b}_i^{\mathrm{T}} = \mathbf{v}_i^{\mathrm{T}} \cdot \partial_{\mathbf{p}} \mathbf{f}(\mathbf{x}_k, \mathbf{p})\Big|_{(\mathbf{x}_{k(0)}^*, \mathbf{p}_0)},$$

$$\mathbf{a}_i^{(r,s)} = \mathbf{v}_i^{\mathrm{T}} \cdot \partial_{s_k^{(i)}}^{(r)} \partial_{\mathbf{p}}^{(s)} \mathbf{f}(\mathbf{x}_k, \mathbf{p})\Big|_{(\mathbf{x}_{k(0)}^*, \mathbf{p}_0)}. \tag{2.113}$$

If $a_i = 1$ and $\mathbf{p} = \mathbf{p}_0$, the stability of the fixed point $\mathbf{x}_k^*$ on an eigenvector $\mathbf{v}_i$ changes from stable to unstable state (or from unstable to stable state). The bifurcation manifold on the direction of $\mathbf{v}_i$ is determined by

$$\mathbf{b}_i^{\mathrm{T}} \cdot (\mathbf{p} - \mathbf{p}_0) + \sum_{q=2}^{m} \frac{1}{q!} \sum_{r=0}^{q} C_q^r \mathbf{a}_i^{(q-r,r)} (s_k^{(i)} - s_{k(0)}^{(i)*})^{q-r} (\mathbf{p} - \mathbf{p}_0)^r = 0. \tag{2.114}$$

In the neighborhood of $(\mathbf{x}_{k(0)}^*, \mathbf{p}_0)$, when other components of fixed point $\mathbf{x}_k^*$ on the eigenvector of $\mathbf{v}_j$ for all $j \neq i, (i, j \in N)$ do not change their stability states, Eq. (2.114) possesses $l$-branch solutions of equilibrium $s_k^{(i)*}$ $(0 < l \leq m)$ with $l_1$-stable and $l_2$-unstable solutions $(l_1, l_2 \in \{0, 1, 2, \ldots, l\})$. Such $l$-branch solutions are called the bifurcation solutions of fixed point $\mathbf{x}_k^*$ on the eigenvector of $\mathbf{v}_i$ in the neighborhood of $(\mathbf{x}_{k(0)}^*, \mathbf{p}_0)$. Such a bifurcation at point $(\mathbf{x}_{k(0)}^*, \mathbf{p}_0)$ is called the hyperbolic bifurcation of $m$th-order on the eigenvector of $\mathbf{v}_i$. Consider two special cases herein.

(i) If

$$\mathbf{a}_i^{(1,1)} = \mathbf{0} \quad \text{and} \quad \mathbf{b}_i^{\mathrm{T}} \cdot (\mathbf{p} - \mathbf{p}_0) + \frac{1}{2!} a_i^{(2,0)} (s_k^{(i)*} - s_{k0}^{(i)*})^2 = 0 \tag{2.115}$$

where

$$a_i^{(2,0)} = \mathbf{v}_i^{\mathrm{T}} \cdot \partial_{s_k^{(i)}}^{(2)} \partial_{\mathbf{p}}^{(0)} \mathbf{f}(\mathbf{x}_k, \mathbf{p})\Big|_{(\mathbf{x}_{k(0)}^*, \mathbf{p}_0)} = \mathbf{v}_i^{\mathrm{T}} \cdot \partial_{s_k^{(i)}}^{(2)} \mathbf{f}(\mathbf{x}_k, \mathbf{p})\Big|_{(\mathbf{x}_{k(0)}^*, \mathbf{p}_0)}$$

$$= \mathbf{v}_i^{\mathrm{T}} \cdot \partial_{\mathbf{x}}^{(2)} \mathbf{f}(\mathbf{x}_k, \mathbf{p})(\mathbf{v}_k \mathbf{v}_k)\Big|_{(\mathbf{x}_{k(0)}^*, \mathbf{p}_0)} = G_{s_k^{(i)}}^{(2)}(\mathbf{x}_{k(0)}^*, \mathbf{p}_0) \neq 0, \tag{2.116}$$

$$\mathbf{b}_i^{\mathrm{T}} = \mathbf{v}_i^{\mathrm{T}} \cdot \partial_{\mathbf{p}} \mathbf{f}(\mathbf{x}_k, \mathbf{p})\Big|_{(\mathbf{x}_{k(0)}^*, \mathbf{p}_0)} \neq \mathbf{0},$$

$$a_i^{(2,0)} \times [\mathbf{b}_i^{\mathrm{T}} \cdot (\mathbf{p} - \mathbf{p}_0)] < 0, \tag{2.117}$$

such a bifurcation at point $(\mathbf{x}_0^*, \mathbf{p}_0)$ is called the *saddle–node* bifurcation on the eigenvector of $\mathbf{v}_i$.

(ii) If

$$\mathbf{b}_i^{\mathrm{T}} \cdot (\mathbf{p} - \mathbf{p}_0) = 0 \quad \text{and}$$

$$\mathbf{a}_i^{(1,1)} \cdot (\mathbf{p} - \mathbf{p}_0)(s_k^{(i)*} - s_{k(0)}^{(i)*}) + \frac{1}{2!} a_i^{(2,0)} (s_k^{(i)*} - s_{k(0)}^{(i)*})^2 = 0 \tag{2.118}$$

where

$$a_i^{(2,0)} = \mathbf{v}_i^{\mathrm{T}} \cdot \partial_{s_k^{(i)}}^{(2)} \partial_{\mathbf{p}}^{(0)} \mathbf{f}(\mathbf{x}_k, \mathbf{p}) \Big|_{(\mathbf{x}_{k(0)}^*, \mathbf{p}_0)} = \mathbf{v}_i^{\mathrm{T}} \cdot \partial_{s_k^{(i)}}^{(2)} \mathbf{f}(\mathbf{x}_k, \mathbf{p}) \Big|_{(\mathbf{x}_0^*, \mathbf{p}_0)}$$

$$= \mathbf{v}_i^{\mathrm{T}} \cdot \partial_{\mathbf{x}_k}^{(2)} \mathbf{f}(\mathbf{x}_k, \mathbf{p})(\mathbf{v}_i \mathbf{v}_i) \Big|_{(\mathbf{x}_{k(0)}^*, \mathbf{p}_0)} = G_{s_k^{(i)}}^{(2)}(\mathbf{x}_{k(0)}^*, \mathbf{p}_0) \neq 0,$$

$$\mathbf{a}_i^{(1,1)} = \mathbf{v}_i^{\mathrm{T}} \cdot \partial_{s_k^{(i)}}^{(1)} \partial_{\mathbf{p}}^{(1)} \mathbf{f}(\mathbf{x}_k, \mathbf{p}) \Big|_{(\mathbf{x}_{k(0)}^*, \mathbf{p}_0)} = \mathbf{v}_i^{\mathrm{T}} \cdot \partial_{s_k^{(i)}} \partial_{\mathbf{p}} \mathbf{f}(\mathbf{x}_k, \mathbf{p}) \Big|_{(\mathbf{x}_{k(0)}^*, \mathbf{p}_0)} \tag{2.119}$$

$$= \mathbf{v}_i^{\mathrm{T}} \cdot \partial_{\mathbf{x}_k} \partial_{\mathbf{p}} \mathbf{f}(\mathbf{x}_k, \mathbf{p}) \mathbf{v}_i \Big|_{(\mathbf{x}_{k(0)}^*, \mathbf{p}_0)} \neq \mathbf{0},$$

$$a_i^{(2,0)} \times [\mathbf{a}_i^{(1,1)} \cdot (\mathbf{p} - \mathbf{p}_0)] < 0, \tag{2.120}$$

such a bifurcation at point $(\mathbf{x}_{k(0)}^*, \mathbf{p}_0)$ is called the *transcritical* bifurcation on the eigenvector of $\mathbf{v}_i$.

**Definition 2.24** Consider a discrete, nonlinear dynamical system $\mathbf{x}_{k+1} = \mathbf{f}(\mathbf{x}_k, \mathbf{p}) \in \mathcal{R}^n$ in Eq. (2.4) with a fixed point $\mathbf{x}_k^*$. The corresponding solution is given by $\mathbf{x}_{k+j} = \mathbf{f}(\mathbf{x}_{k+j-1}, \mathbf{p})$ with $j \in \mathbb{Z}$. Suppose there is a neighborhood of the fixed point $\mathbf{x}_k^*$ (i.e., $U_k(\mathbf{x}_k^*) \subset \Omega$), and $\mathbf{f}(\mathbf{x}_k, \mathbf{p})$ is $C^r$ ($r \geq 1$)-continuous in $U_k(\mathbf{x}_k^*)$ with Eq. (2.28). The linearized system is $\mathbf{y}_{k+j+1} = D\mathbf{f}(\mathbf{x}_k^*, \mathbf{p})\mathbf{y}_{k+j}$ ($\mathbf{y}_{k+j} = \mathbf{x}_{k+j} - \mathbf{x}_k^*$) in $U_k(\mathbf{x}_k^*)$ and there are $n$ linearly independent vectors $\mathbf{v}_i$ ($i = 1, 2, \ldots, n$). For a perturbation of fixed point $\mathbf{y}_k = \mathbf{x}_k - \mathbf{x}_k^*$, let $\mathbf{y}_k^{(i)} = c_k^{(i)} \mathbf{v}_i$ and $\mathbf{y}_{k+1}^{(i)} = c_{k+1}^{(i)} \mathbf{v}_i$. Equations (2.110), (2.111), and (2.113) hold. In the vicinity of point $(\mathbf{x}_{k0}^*, \mathbf{p}_0)$, $\mathbf{v}_i^{\mathrm{T}} \cdot \mathbf{f}(\mathbf{x}_k, \mathbf{p})$ can be expended for $(0 < \theta < 1)$ as

$$\mathbf{v}_i^{\mathrm{T}} \cdot [\mathbf{f}(\mathbf{x}_k, \mathbf{p}) - \mathbf{x}_{k+1(0)}^*] = a_i(s_k^{(i)} - s_{k(0)}^{(i)*}) + \mathbf{b}_i^{\mathrm{T}} \cdot (\mathbf{p} - \mathbf{p}_0)$$

$$+ \sum_{q=2}^{m} \frac{1}{q!} \sum_{r=0}^{q} C_q^r \mathbf{a}_i^{(q-r,r)} (s_k^{(i)} - s_{k(0)}^{(i)*})^{q-r} (\mathbf{p} - \mathbf{p}_0)^r$$

$$+ \frac{1}{(m+1)!} [(s_k^{(i)} - s_{k(0)}^{(i)*}) \partial_{s_k^{(i)}} + (\mathbf{p} - \mathbf{p}_0) \partial_{\mathbf{p}}]^{m+1}$$

$$\times (\mathbf{v}_k^{\mathrm{T}} \cdot \mathbf{f}(\mathbf{x}_{k0}^* + \theta \Delta \mathbf{x}_k, \mathbf{p}_0 + \theta \Delta \mathbf{p})) \tag{2.121}$$

and

$$\mathbf{v}_i^{\mathrm{T}} \cdot [\mathbf{f}(\mathbf{x}_{k+1}, \mathbf{p}) - \mathbf{x}_{k(0)}^*] = a_i(s_{k+1}^{(i)} - s_{k+1(0)}^*) + \mathbf{b}_i^{\mathrm{T}} \cdot (\mathbf{p} - \mathbf{p}_0)$$

$$+ \sum_{q=2}^{m} \frac{1}{q!} \sum_{r=0}^{q} C_q^r \mathbf{a}_i^{(q-r,r)} (s_{k+1}^{(i)} - s_{k+1(0)}^{(i)*})^{q-r} (\mathbf{p} - \mathbf{p}_0)^r$$

$$+ \frac{1}{(m+1)!} [(s_{k+1}^{(i)} - s_{k+1(0)}^*) \partial_{s_{k+1}^{(i)}} + (\mathbf{p} - \mathbf{p}_0) \partial_{\mathbf{p}}]^{m+1}$$

$$\times (\mathbf{v}_i^{\mathrm{T}} \cdot \mathbf{f}(\mathbf{x}_{k+1(0)}^* + \theta \Delta \mathbf{x}_{k+1}, \mathbf{p}_0 + \theta \Delta \mathbf{p})). \tag{2.122}$$

If $a_i = -1$ and $\mathbf{p} = \mathbf{p}_0$, the stability of current equilibrium $\mathbf{x}_k^*$ on an eigenvector $\mathbf{v}_i$ changes from stable to unstable state (or from unstable to stable state). The bifurcation manifold in the direction of $\mathbf{v}_i$ is determined by

$$
\begin{aligned}
&\mathbf{b}_i^{\mathrm{T}} \cdot (\mathbf{p} - \mathbf{p}_0) + a_i(s_k^{(i)*} - s_{k(0)}^{(i)*}) \\
&+ \sum_{q=2}^m \frac{1}{q!} \sum_{r=0}^q C_q^r a_i^{(q-r,r)} (s_k^{(i)} - s_{k(0)}^{(i)*})^{q-r} (\mathbf{p} - \mathbf{p}_0)^r = (s_{k+1}^{(i)*} - s_{k+1(0)}^{(i)*}); \\
&\mathbf{b}_i^{\mathrm{T}} \cdot (\mathbf{p} - \mathbf{p}_0) + a_i(s_{k+1}^{(i)*} - s_{k+1(0)}^{(i)*}) \\
&+ \sum_{q=2}^m \frac{1}{q!} \sum_{r=0}^q C_q^r a_i^{(q-r,r)} (s_{k+1}^{(i)} - s_{k+1(0)}^{(i)*})^{q-r} (\mathbf{p} - \mathbf{p}_0)^r = (s_k^{(i)*} - s_{k(0)}^{(i)*}).
\end{aligned}
\tag{2.123}
$$

In the neighborhood of $(\mathbf{x}_{k(0)}^*, \mathbf{p}_0)$, when other components of fixed point $\mathbf{x}_{k(0)}^*$ on the eigenvector of $\mathbf{v}_j$ for all $j \neq i$, $(j, i \in N)$ do not change their stability states, Eq. (2.123) possesses $l$-branch solutions of equilibrium $s_k^{(i)*}$ $(0 < l \leq m)$ with $l_1$-stable and $l_2$-unstable solutions $(l_1, l_2 \in \{0, 1, 2, \ldots, l\})$. Such $l$-branch solutions are called the bifurcation solutions of fixed point $\mathbf{x}_k^*$ on the eigenvector of $\mathbf{v}_i$ in the neighborhood of $(\mathbf{x}_{k(0)}^*, \mathbf{p}_0)$. Such a bifurcation at point $(\mathbf{x}_{k(0)}^*, \mathbf{p}_0)$ is called the *hyperbolic bifurcation* of $m$th-order with doubling iterations on the eigenvector of $\mathbf{v}_i$. Consider a special case. If

$$
\begin{aligned}
&\mathbf{b}_i^{\mathrm{T}} \cdot (\mathbf{p} - \mathbf{p}_0) = 0, \ a_i = -1, \ a_i^{(2,0)} = 0, \quad a_i^{(2,1)} = 0, \quad a_i^{(1,2)} = 0, \\
&[\mathbf{a}^{(1,1)} \cdot (\mathbf{p} - \mathbf{p}_0) + a_i](s_k^{(i)*} - s_{k(0)}^{(i)*}) + \frac{1}{3!} a_i^{(3,0)} (s_k^* - s_{k(0)}^*)^3 = (s_{k+1}^{(i)*} - s_{k+1(0)}^{(i)*}), \\
&[\mathbf{a}^{(1,1)} \cdot (\mathbf{p} - \mathbf{p}_0) + a_i](s_{k+1}^{(i)*} - s_{k+1(0)}^{(i)*}) + \frac{1}{3!} a_i^{(3,0)} (s_{k+1}^* - s_{k+1(0)}^*)^3 = s_k^{(i)*} - s_{k(0)}^{(i)*})
\end{aligned}
\tag{2.124}
$$

where

$$
\begin{aligned}
a_i^{(3,0)} &= \mathbf{v}_i^{\mathrm{T}} \cdot \partial_{s_k^{(i)}}^{(3)} \partial_{\mathbf{p}}^{(0)} \mathbf{f}(\mathbf{x}_k, \mathbf{p}) \Big|_{(\mathbf{x}_{k(0)}^*, \mathbf{p}_0)} = \mathbf{v}_i^{\mathrm{T}} \cdot \partial_{s_k^{(i)}}^{(3)} \mathbf{f}(\mathbf{x}_k, \mathbf{p}) \Big|_{(\mathbf{x}_{k(0)}^*, \mathbf{p}_0)} \\
&= \mathbf{v}_i^{\mathrm{T}} \cdot \partial_{\mathbf{x}_k}^{(3)} \mathbf{f}(\mathbf{x}_k, \mathbf{p})(\mathbf{v}_i \mathbf{v}_i \mathbf{v}_i) \Big|_{(\mathbf{x}_{k(0)}^*, \mathbf{p}_0)} = G_i^{(3)}(\mathbf{x}_{k(0)}^*, \mathbf{p}_0) \neq 0, \\
\mathbf{a}_i^{(1,1)} &= \mathbf{v}_i^{\mathrm{T}} \cdot \partial_{s_k^{(1)}}^{(1)} \partial_{\mathbf{p}}^{(1)} \mathbf{f}(\mathbf{x}_k, \mathbf{p}) \Big|_{(\mathbf{x}_{k(0)}^*, \mathbf{p}_0)} = \mathbf{v}_i^{\mathrm{T}} \cdot \partial_{s_k^{(i)}} \partial_{\mathbf{p}} \mathbf{f}(\mathbf{x}_k, \mathbf{p}) \Big|_{(\mathbf{x}_{k(0)}^*, \mathbf{p}_0)} \\
&= \mathbf{v}_i^{\mathrm{T}} \cdot \partial_{\mathbf{x}_k} \partial_{\mathbf{p}} \mathbf{f}(\mathbf{x}_k, \mathbf{p}) \mathbf{v}_i \Big|_{(\mathbf{x}_{k(0)}^*, \mathbf{p}_0)} \neq \mathbf{0},
\end{aligned}
\tag{2.125}
$$

$$
a_i^{(3,0)} \times [\mathbf{a}_i^{(1,1)} \cdot (\mathbf{p} - \mathbf{p}_0)] < 0,
\tag{2.126}
$$

such a bifurcation at point $(\mathbf{x}_{k(0)}^*, \mathbf{p}_0)$ is called the *pitchfork* bifurcation (or period-doubling bifurcation) on the eigenvector of $\mathbf{v}_i$.

For the saddle–node bifurcation of the first kind, the $(2m)$th-order singularity of the fixed point at the bifurcation point exists as a saddle of the $(2m)$th-order. For the transcritical bifurcation, the $(2m)$th-order singularity of the fixed point at the bifurcation point exists as a saddle of the $(2m)$th-order. However, for the stable pitchfork bifurcation (or saddle–node bifurcation of the second kind, or period-doubling bifurcation), the $(2m + 1)$th-order singularity of the fixed point at the bifurcation point exists as an oscillatory sink of the $(2m + 1)$th-order. For the unstable pitchfork bifurcation (or the unstable saddle–node bifurcation of the second kind, or unstable period-doubling bifurcation), the $(2m + 1)$th-order singularity of the fixed point at the bifurcation point exists as an oscillatory source of the $(2m + 1)$th-order.

**Definition 2.25** Consider a discrete, nonlinear dynamical system $\mathbf{x}_{k+1} = \mathbf{f}(\mathbf{x}_k, \mathbf{p}) \in \mathscr{R}^n$ in Eq. (2.4) with a fixed point $\mathbf{x}_k^*$. The corresponding solution is given by $\mathbf{x}_{k+j} = \mathbf{f}(\mathbf{x}_{k+j-1}, \mathbf{p})$ with $j \in \mathbb{Z}$. Suppose there is a neighborhood of the fixed point $\mathbf{x}_k^*$ (i.e., $U_k(\mathbf{x}_k^*) \subset \Omega$), and $\mathbf{f}(\mathbf{x}_k, \mathbf{p})$ is $C^r$ ($r \geq 1$)-continuous in $U_k(\mathbf{x}_k^*)$ with Eq. (2.28). The linearized system is $\mathbf{y}_{k+j+1} = D\mathbf{f}(\mathbf{x}_k^*, \mathbf{p})\mathbf{y}_{k+j}$ ($\mathbf{y}_{k+j} = \mathbf{x}_{k+j} - \mathbf{x}_k^*$) in $U_k(\mathbf{x}_k^*)$. Consider a pair of complex eigenvalues $\alpha_i \pm i\beta_i$ ($i \in N = \{1, 2, \ldots, n\}$, $i = \sqrt{-1}$) of matrix $D\mathbf{f}(\mathbf{x}^*, \mathbf{p})$ with a pair of eigenvectors $\mathbf{u}_i \pm i\mathbf{v}_i$. On the invariant plane of $(\mathbf{u}_i, \mathbf{v}_i)$, consider $\mathbf{r}_k^{(i)} = \mathbf{y}_k^{(i)} = \mathbf{y}_{k+}^{(i)} + \mathbf{y}_{k-}^{(i)}$ with

$$\mathbf{r}_k^{(i)} = c_k^{(i)}\mathbf{u}_i + d_k^{(i)}\mathbf{v}_i \quad \text{and} \quad \mathbf{r}_{k+1}^{(i)} = c_{k+1}^{(i)}\mathbf{u}_i + d_{k+1}^{(i)}\mathbf{v}_i. \tag{2.127}$$

and

$$c_k^{(i)} = \frac{1}{\Delta}[\Delta_2(\mathbf{u}_i^{\mathrm{T}} \cdot \mathbf{y}_k) - \Delta_{12}(\mathbf{v}_i^{\mathrm{T}} \cdot \mathbf{y}_k)],$$

$$d_k^{(i)} = \frac{1}{\Delta}[\Delta_1(\mathbf{v}_i^{\mathrm{T}} \cdot \mathbf{y}_k) - \Delta_{12}(\mathbf{u}_i^{\mathrm{T}} \cdot \mathbf{y}_k)]; \tag{2.128}$$

$$\Delta_1 = \|\mathbf{u}_i\|^2, \ \Delta_2 = \|\mathbf{v}_i\|^2, \ \Delta_{12} = \mathbf{u}_i^{\mathrm{T}} \cdot \mathbf{v}_i;$$

$$\Delta = \Delta_1\Delta_2 - \Delta_{12}^2.$$

Consider a polar coordinate of $(r_k, \theta_k)$ defined by

$$c_k^{(i)} = r_k^{(i)}\cos\theta_k^{(i)}, \quad \text{and} \quad d_k^{(i)} = r_k^{(i)}\sin\theta_k^{(i)};$$

$$r_k^{(i)} = \sqrt{(c_k^{(i)})^2 + (d_k^{(i)})^2}, \quad \text{and} \quad \theta_k^{(i)} = \arctan(d_k^{(i)}/c_k^{(i)}). \tag{2.129}$$

Thus,

$$c_{k+1}^{(i)} = \frac{1}{\Delta}[\Delta_2 G_{c_k^{(i)}}(\mathbf{x}_k, \mathbf{p}) - \Delta_{12} G_{d_k^{(i)}}(\mathbf{x}_k, \mathbf{p})],$$

$$d_{k+1}^{(i)} = \frac{1}{\Delta}[\Delta_1 G_{d_k^{(i)}}(\mathbf{x}_k, \mathbf{p}) - \Delta_{12} G_{c_k^{(i)}}(\mathbf{x}_k, \mathbf{p})] \tag{2.130}$$

where

$$G_{c_k^{(i)}}(\mathbf{x}_k, \mathbf{p}) = \mathbf{u}_i^{\mathrm{T}} \cdot [\mathbf{f}(\mathbf{x}_k, \mathbf{p}) - \mathbf{x}_{k(0)}^*]$$

$$= \mathbf{a}_i^{\mathrm{T}} \cdot (\mathbf{p} - \mathbf{p}_0) + a_{i11}(c_k^{(i)} - c_{k(0)}^{(i)*}) + a_{i12}(d_k^{(i)} - d_k^{(i)*})$$

$$+ \sum_{q=2}^{m_i} \frac{1}{q!} \sum_{r=0}^{q} C_{m_i}^{r_i} \mathbf{G}_{c_k^{(i)}}^{(m_i - r_i, r_i)}(\mathbf{x}_k^*, \mathbf{p}_0)(\mathbf{p} - \mathbf{p}_0)^{r_i}(r_k^{(i)})^{m_i - r_i}$$

$$+ \frac{1}{(m_i + 1)!}[(c_k^{(i)} - c_{k(0)}^{(i)*})\partial_{c_k^{(i)}} + (d_k^{(i)} - d_{k(0)}^{(i)*})\partial_{d_k^{(i)}} + (\mathbf{p} - \mathbf{p}_0)\partial_{\mathbf{p}}]^{m_i + 1}$$

$$\times (\mathbf{u}_i^{\mathrm{T}} \cdot \mathbf{f}(\mathbf{x}_{k0}^* + \theta \Delta \mathbf{x}_k, \mathbf{p}_0 + \theta \Delta \mathbf{p})),$$

$$G_{d_k^{(i)}}(\mathbf{x}_k, \mathbf{p}) = \mathbf{v}_i^{\mathrm{T}} \cdot [\mathbf{f}(\mathbf{x}_k, \mathbf{p}) - \mathbf{x}_{k(0)}^*]$$

$$= \mathbf{b}_i^{\mathrm{T}} \cdot (\mathbf{p} - \mathbf{p}_0) + a_{i21}(c_k^{(i)} - c_{k(0)}^{(i)*}) + a_{i22}(d_k^{(i)} - d_{k(0)}^{(i)*})$$

$$+ \sum_{q=2}^{m_i} \frac{1}{q!} \sum_{r=0}^{q} C_{m_i}^{r_i} \mathbf{G}_{d_k^{(i)}}^{(m_i - r_i, r_i)}(\mathbf{x}_k^*, \mathbf{p}_0)(\mathbf{p} - \mathbf{p}_0)^{r_i} r_k^{m_i - r_i}$$

$$+ \frac{1}{(m_i + 1)!}[(c_k^{(i)} - c_{k(0)}^{(i)*})\partial_{c_k^{(i)}} + (d_k^{(i)} - d_{k(0)}^{(i)*})\partial_{d_k^{(i)}} + (\mathbf{p} - \mathbf{p}_0)\partial_{\mathbf{p}}]^{m_i + 1}$$

$$\times (\mathbf{v}_i^{\mathrm{T}} \cdot \mathbf{f}(\mathbf{x}_{k(0)}^* + \theta \Delta \mathbf{x}, \mathbf{p}_0 + \theta \Delta \mathbf{p}));$$

$$(2.131)$$

and

$$\mathbf{G}_{c_k^{(i)}}^{(s,r)}(\mathbf{x}_{k(0)}^*, \mathbf{p}_0)$$

$$= \mathbf{u}_i^{\mathrm{T}} \cdot [\partial_{\mathbf{x}_k}()\mathbf{u}_i \cos \theta_k^{(i)} + \partial_{\mathbf{x}_k}()\mathbf{v}_i \sin \theta_k^{(i)}]^s \partial_{\mathbf{p}}^{(r)} \mathbf{f}(\mathbf{x}_k, \mathbf{p})\Big|_{(\mathbf{x}_{k(0)}^*, \mathbf{p}_0)},$$

$$\mathbf{G}_{d_k^{(i)}}^{(s,r)}(\mathbf{x}_{k(0)}^*, \mathbf{p}_0)$$

$$= \mathbf{v}_i^{\mathrm{T}} \cdot [\partial_{\mathbf{x}_k}()\mathbf{u}_i \cos \theta_k^{(i)} + \partial_{\mathbf{x}_k}()\mathbf{v}_i \sin \theta_k^{(i)}]^s \partial_{\mathbf{p}}^{(r)} \mathbf{f}(\mathbf{x}_k, \mathbf{p})\Big|_{(\mathbf{x}_{k(0)}^*, \mathbf{p}_0)};$$

$$(2.132)$$

$$\mathbf{a}_i^{\mathrm{T}} = \mathbf{u}_i^{\mathrm{T}} \cdot \partial_{\mathbf{p}} \mathbf{f}(\mathbf{x}_k, \mathbf{p}), \quad \mathbf{b}_i^{\mathrm{T}} = \mathbf{v}_i^{\mathrm{T}} \cdot \partial_{\mathbf{p}} \mathbf{f}(\mathbf{x}_k, \mathbf{p});$$

$$a_{i11} = \mathbf{u}_i^{\mathrm{T}} \cdot \partial_{\mathbf{x}_k} \mathbf{f}(\mathbf{x}_k, \mathbf{p})\mathbf{u}_i, \quad a_{i12} = \mathbf{u}_i^{\mathrm{T}} \cdot \partial_{\mathbf{x}_k} \mathbf{f}(\mathbf{x}_k, \mathbf{p})\mathbf{u}_i; \quad (2.133)$$

$$a_{i21} = \mathbf{v}_i^{\mathrm{T}} \cdot \partial_{\mathbf{x}_k} \mathbf{f}(\mathbf{x}_k, \mathbf{p})\mathbf{u}_i, \quad a_{i22} = \mathbf{v}_i^{\mathrm{T}} \cdot \partial_{\mathbf{x}_k} \mathbf{f}(\mathbf{x}_k, \mathbf{p})\mathbf{v}_i.$$

Suppose

$$\mathbf{a}_i = \mathbf{0} \quad \text{and} \quad \mathbf{b}_i = \mathbf{0} \qquad (2.134)$$

then

$$
\begin{aligned}
r_{k+1}^{(i)} &= \sqrt{(c_{k+1}^{(i)})^2 + (d_{k+1}^{(i)})^2} = \sqrt{\sum_{m=2}^{\infty} (r_k^{(i)})^m G_{r_{k+1}^{(i)}}^{(m)}} \\
&= \sqrt{G_{r_{k+1}^{(i)}}^{(2,0)}}\, r_k^{(i)} \sqrt{1 + \lambda^{(i)} + \sum_{m=3}^{\infty} \lambda_m^{(i)} (r_k^{(i)})^{m-2}} \\
\theta_{k+1}^{(i)} &= \arctan(d_{k+1}^{(i)}/c_{k+1}^{(i)})
\end{aligned}
\tag{2.135}
$$

where

$$
\begin{aligned}
G_{r_{k+1}^{(i)}}^{(2)} &= G_{r_{k+1}^{(i)}}^{(2,0)} + G_{r_{k+1}^{(i)}}^{(1,1)} \quad \text{and} \quad \lambda^{(i)} = G_{r_{k+1}^{(i)}}^{(1,1)} / G_{r_{k+1}^{(i)}}^{(2,0)} \quad \text{with} \\
G_{r_{k+1}^{(i)}}^{(2,0)} &= [G_{c_{k+1}^{(i)}}^{(1,0)}(\theta_k^{(i)}, \mathbf{p}_0)]^2 + [G_{d_{k+1}^{(i)}}^{(1,0)}(\theta_k^{(i)}, \mathbf{p}_0)]^2, \\
G_{r_{k+1}^{(i)}}^{(1,1)} &= \sum_{r=1}^{M} \sum_{s=1}^{M} [\mathbf{G}_{c_{k+1}^{(i)}}^{(1,r)}(\theta_k^{(i)}, \mathbf{p}_0) \cdot \mathbf{G}_{c_{k+1}^{(i)}}^{(1,s)}(\theta_k^{(i)}, \mathbf{p}_0)(\mathbf{p} - \mathbf{p}_0)^{r+s} \\
&\quad + [\mathbf{G}_{d_{k+1}^{(i)}}^{(1,r)}(\theta_k^{(i)}, \mathbf{p}_0) \cdot \mathbf{G}_{d_{k+1}^{(i)}}^{(1,s)}(\theta_k^{(i)}, \mathbf{p}_0) \cdot (\mathbf{p} - \mathbf{p}_0)^{r+s}];
\end{aligned}
\tag{2.136}
$$

and

$$
\begin{aligned}
\lambda_m^{(i)} &= G_{r_{k+1}^{(i)}}^{(m)} / G_{r_{k+1}^{(i)}}^{(2,0)} \quad \text{with} \\
G_{r_{k+1}^{(i)}}^{(m)} &= \sum_{m_i=0}^{M} \sum_{m_j=0}^{M} \frac{1}{m_i!} \frac{1}{m_j!} [\mathbf{G}_{c_{k+1}^{(i)}}^{(m_i-r_i, r_i)}(\theta_k^{(i)}, \mathbf{p}_0) \cdot (\mathbf{p} - \mathbf{p}_0)^{m_i - r_i}] \\
&\quad \times \mathbf{G}_{c_{k+1}^{(i)}}^{(m_j-s_j, s_j)}(\theta_k^{(i)}, \mathbf{p}_0) \cdot (\mathbf{p} - \mathbf{p}_0)^{m_j - s_j} \\
&\quad + \mathbf{G}_{d_{k+1}^{(i)}}^{(m_i-r_i, r_i)}(\theta_k^{(i)}, \mathbf{p}_0) \cdot (\mathbf{p} - \mathbf{p}_0)^{m_i - r_i}] \\
&\quad \times \mathbf{G}_{d_{k+1}^{(i)}}^{(m_j-s_j, s_j)}(\theta_k^{(i)}, \mathbf{p}_0) \cdot (\mathbf{p} - \mathbf{p}_0)^{m_j - s_j}] \delta_m^{(r_i + s_j)} \\
&= \frac{1}{m!} \sum_{r=1}^{m-1} C_m^r \sum_{s=1}^{M} \frac{1}{s!} \frac{1}{(2M - m)!} [\mathbf{G}_{c_{k+1}^{(i)}}^{(r,s)}(\theta_k^{(i)}, \mathbf{p}_0) \cdot \mathbf{G}_{c_{k+1}^{(i)}}^{(m-r, 2M-m-r)}(\theta_k^{(i)}, \mathbf{p}_0) \\
&\quad + \mathbf{G}_{d_{k+1}^{(i)}}^{(r,s)}(\theta_k^{(i)}, \mathbf{p}_0) \cdot \mathbf{G}_{d_{k+1}^{(i)}}^{(m-r, 2M-m-r)}(\theta_k^{(i)}, \mathbf{p}_0)] \cdot (\mathbf{p} - \mathbf{p}_0)^{M-m},
\end{aligned}
\tag{2.137}
$$

$$
\begin{aligned}
\mathbf{G}_{c_{k+1}^{(i)}}^{(m-r,r)}(\theta_k, \mathbf{p}_0) &= \frac{1}{\Delta}[\Delta_2 \mathbf{G}_{c_k^{(i)}}^{(m-r,r)}(\mathbf{x}_{k(0)}^*, \mathbf{p}_0) - \Delta_{12} \mathbf{G}_{d_k^{(i)}}^{(m-r,r)}(\mathbf{x}_{k(0)}^*, \mathbf{p}_0)], \\
\mathbf{G}_{d_{k+1}^{(i)}}^{(m-r,r)}(\theta_k, \mathbf{p}_0) &= \frac{1}{\Delta}[\Delta_1 \mathbf{G}_{d_k^{(i)}}^{(m-r,r)}(\mathbf{x}_{k(0)}^*, \mathbf{p}_0) - \Delta_{12} \mathbf{G}_{c_k^{(i)}}^{(m-r,r)}(\mathbf{x}_{k(0)}^*, \mathbf{p}_0)].
\end{aligned}
\tag{2.138}
$$

If $G^{(2,0)}_{r^{(i)}_{k+1}} = 1$ and $\mathbf{p} = \mathbf{p}_0$, the stability of current fixed point $\mathbf{x}^*_k$ on an eigenvector plane of $(\mathbf{u}_i, \mathbf{v}_i)$ changes from stable to unstable state (or from unstable to stable state). The bifurcation manifold in the direction of $\mathbf{v}_i$ is determined by

$$\lambda^{(i)} + \sum_{m=3}^{\infty} \lambda^{(i)}_m (r^{(i)}_k)^{m-2} = 0. \tag{2.139}$$

Such a bifurcation at the fixed point $(\mathbf{x}^*_{k(0)}, \mathbf{p}_0)$ is called the generalized Neimark bifurcation on the eigenvector plane of $(\mathbf{u}_i, \mathbf{v}_i)$.

For a special case, if

$$\lambda^{(i)} + \lambda^{(i)}_4 (r^{(i)}_k)^2 = 0, \quad \text{for} \quad \lambda^{(i)} \times \lambda^{(i)}_4 < 0 \quad \text{and} \quad \lambda^{(i)}_3 = 0 \tag{2.140}$$

such a bifurcation at point $(\mathbf{x}^*_0, \mathbf{p}_0)$ is called the Neimark bifurcation on the eigenvector plane of $(\mathbf{u}_i, \mathbf{v}_i)$.

For the repeating eigenvalues of $DP(\mathbf{x}^*_k, \mathbf{p})$, the bifurcation of fixed point $\mathbf{x}^*_k$ can be similarly discussed in the foregoing Theorems 2.5 and 2.6. Herein, such a procedure will not be repeated. From the foregoing analysis of the Neimark bifurcation, the Neimark bifurcation points possess the higher-order singularity of the flow in discrete dynamical system in the radial direction. For the stable Neimark bifurcation, the $m$th-order singularity of the flow at the bifurcation point exists as a sink of the $m$th-order in the radial direction. For the unstable Neimark bifurcation, the $m$th-order singularity of the flow at the bifurcation point exists as a source of the $m$th-order in the radial direction.

Consider a 2D map

$$P : \mathbf{x}_k \rightarrow \mathbf{x}_{k+1} \quad \text{with } \mathbf{x}_{k+1} = \mathbf{f}(\mathbf{x}_k, \mathbf{p}) \tag{2.141}$$

where $\mathbf{x}_k = (x_k, y_k)^{\mathrm{T}}$ and $\mathbf{f} = (f_1, f_2)^{\mathrm{T}}$ with a parameter vector $\mathbf{p}$. The period-$n$ fixed point for Eq. (2.141) is $(\mathbf{x}^*_k, \mathbf{p})$, i.e., $P^{(n)}\mathbf{x}^*_k = \mathbf{x}^*_{k+n}$ where $P^{(n)} = P \circ P^{(n-1)}$ and $P^{(0)} = 1$, and its stability and bifurcation conditions are given as follows.

(i) period-doubling (flip or pitchfork) bifurcation

$$\mathrm{tr}(DP^{(n)}) + \det(DP^{(n)}) + 1 = 0, \tag{2.142}$$

(ii) saddle–node bifurcation

$$\det(DP^{(n)}) + 1 = \mathrm{tr}(DP^{(n)}), \tag{2.143}$$

(iii) Neimark bifurcation

$$\det(DP^{(n)}) = 1. \tag{2.144}$$

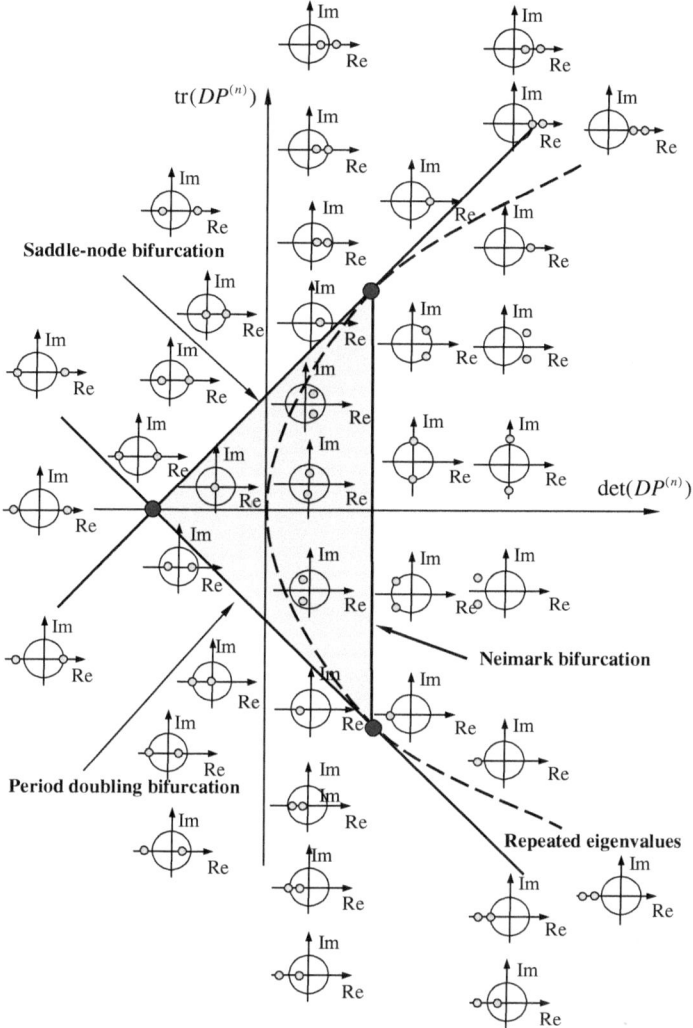

**Fig. 2.8** Stability and bifurcation diagrams through the complex plane of eigenvalues for 2D discrete dynamical systems

The bifurcation and stability conditions for the solution of period-$n$ for Eq. (2.141) are summarized in Fig. 2.8 with $\det(DP^{(n)}) = \det(DP^{(n)}(\mathbf{x}^*_{k(0)}, \mathbf{p}_0))$ and $\mathrm{tr}(DP^{(n)}) = \mathrm{tr}(DP^{(n)}(\mathbf{x}^*_{k(0)}, \mathbf{p}_0))$. The thick dashed lines are bifurcation lines. The stability of the fixed point is given by the eigenvalues in complex plane. The stability of the fixed point for higher-dimensional systems can be identified by using a naming of stability for linear dynamical systems in Luo (2011, 2012). The saddle–node bifurcation possesses stable saddle–node bifurcation (critical) and unstable saddle–node bifurcation (degenerate).

# References

Guckenhiemer, J., & Holmes, P. (1990). *Nonlinear oscillations, dynamical systems, and bifurcations of vector fields*. New-York: Springer.

Luo, A. C. J. (2011). *Regularity and complexity in dynamical systems*. New York: Springer.

Luo, A. C. J. (2012). *Discrete and switching dynamical systems*. Glen Carbon: HEP-L&H Scientific.

Nitecki, Z. (1971). *Differentiable dynamics: An introduction to the orbit structures of diffeomorphisms*. Cambridge, MA: MIT Press.

# Chapter 3
# Discretization of Continuous Systems

In this chapter, the discretization of continuous systems is presented. The explicit and implicit discrete maps are discussed for numerical predictions of continuous systems. Basic discrete schemes are presented which include forward and backward Euler methods, midpoint, and trapezoidal rule method. An introduction to Runge–Kutta methods is presented, and the Taylor series method and second-order Runge–Kutta method are introduced. The explicit Runge–Kutta methods for third and fourth order are systematically presented. The implicit Runge–Kutta methods are discussed based on the polynomial interpolation, which include a generalized implicit Runge–Kutta method, Gauss method, Radau method, and Lotta methods. In addition to one-step methods, implicit and explicit multi-step methods are discussed, including Adams–Bashforth method, Adams–Moulton methods, and explicit and implicit Adams methods.

## 3.1 Continuous Systems

**Definition 3.1** For $I \subseteq \mathcal{R}$, $\Omega \subseteq \mathcal{R}^n$, and $\Lambda \subseteq \mathcal{R}^m$, consider a vector function $\mathbf{f} \colon \Omega \times I \times \Lambda \to \mathcal{R}^n$ which is $C^r(r \geq 1)$-continuous, and there is an ordinary differential equation in a form of

$$\dot{\mathbf{x}} = \mathbf{f}(\mathbf{x}, t, \mathbf{p}) \quad \text{for} \quad t \in I, \, \mathbf{x} \in \Omega \text{ and } \mathbf{p} \in \Lambda \tag{3.1}$$

where $\dot{\mathbf{x}} = d\mathbf{x}/dt$ is differentiation with respect to time $t$, which is simply called the velocity vector of the state variables $\mathbf{x}$. With an initial condition of $\mathbf{x}(t_0) = \mathbf{x}_0$, the solution of Eq. (3.1) is given by

$$\mathbf{x}(t) = \mathbf{\Phi}(\mathbf{x}_0, t - t_0, \mathbf{p}). \tag{3.2}$$

(i) The ordinary differential equation with the initial condition is called a *dynamical system*.
(ii) The vector function $\mathbf{f}(\mathbf{x}, t, \mathbf{p})$ is called a *vector field* on domain $\Omega$.

© Higher Education Press, Beijing and Springer-Verlag Berlin Heidelberg 2015
A.C.J. Luo, *Discretization and Implicit Mapping Dynamics*,
Nonlinear Physical Science, DOI 10.1007/978-3-662-47275-0_3

(iii)  The solution $\Phi(\mathbf{x}_0, t - t_0, \mathbf{p})$ is called the *flow* of dynamical systems.

(iv)  The projection of the solution $\Phi(\mathbf{x}_0, t - t_0, \mathbf{p})$ on domain $\Omega$ is called the trajectory, phase curve, or orbit of dynamical system, which is defined as follows:

$$\Gamma = \{\mathbf{x}(t) \in \Omega | \mathbf{x}(t) = \Phi(\mathbf{x}_0, t - t_0, \mathbf{p}) \text{ for } t \in I\} \subset \Omega. \qquad (3.3)$$

**Definition 3.2** If the vector field of the dynamical system in Eq. (3.1) is independent of time, then such a system is called an autonomous dynamical system. Thus, Eq. (3.1) becomes

$$\dot{\mathbf{x}} = \mathbf{f}(\mathbf{x}, \mathbf{p}) \text{ for } t \in I \subseteq \mathscr{R}, \mathbf{x} \in \Omega \subseteq \mathscr{R}^n \text{ and } \mathbf{p} \in \Lambda \subseteq \mathscr{R}^m \qquad (3.4)$$

Otherwise, such a system is called non-autonomous dynamical systems if the vector field of the dynamical system in Eq. (3.1) is dependent on time and state variables.

**Definition 3.3** For a vector function $\mathbf{f} \in \mathscr{R}^n$ with $\mathbf{x} \in \mathscr{R}^n$, the operator norm of $\mathbf{f}$ is defined by

$$\|\mathbf{f}\| = \sum_{i=1}^{n} \max_{\|\mathbf{x}\| \le 1, \mathbf{p} \in \Lambda} |f_i(\mathbf{x}, \mathbf{p})|. \qquad (3.5)$$

For an $n \times n$ matrix $\mathbf{f}(\mathbf{x}, \mathbf{p}) = \mathbf{A}\mathbf{x}$ with $\mathbf{A} = (a_{ij})_{n \times n}$, the corresponding norm is defined by

$$\|\mathbf{A}\| = \sum_{i,j=1}^{n} |a_{ij}|. \qquad (3.6)$$

**Definition 3.4** For a vector function $\mathbf{x}(t) = (x_1, x_2, \ldots, x_n)^{\mathrm{T}} \in \mathscr{R}^n$, the derivative and integral of $\mathbf{x}(t)$ are defined by

$$\begin{aligned} \frac{d\mathbf{x}(t)}{dt} &= \left( \frac{dx_1(t)}{dt}, \frac{dx_2(t)}{dt}, \ldots, \frac{dx_n(t)}{dt} \right)^{\mathrm{T}}, \\ \int \mathbf{x}(t) dt &= \left( \int x_1(t) dt, \int x_2(t) dt, \ldots, \int x_n(t) dt \right)^{\mathrm{T}}. \end{aligned} \qquad (3.7)$$

For an $n \times n$ matrix $\mathbf{A} = (a_{ij})_{n \times n}$, the corresponding derivative and integral are defined by

$$\frac{d\mathbf{A}(t)}{dt} = \left( \frac{da_{ij}(t)}{dt} \right)_{n \times n} \quad \text{and} \quad \int \mathbf{A}(t) dt = \left( \int a_{ij}(t) dt \right)_{n \times n}. \qquad (3.8)$$

**Definition 3.5** For $I \subseteq \mathcal{R}$, $\Omega \subseteq \mathcal{R}^n$, and $\Lambda \subseteq \mathcal{R}^m$, the vector function $\mathbf{f}(\mathbf{x}, t, \mathbf{p})$ with $\mathbf{f} \colon \Omega \times I \times \Lambda \to \mathcal{R}^n$ is differentiable at $\mathbf{x}_0 \in \Omega$ if

$$\left. \frac{\partial \mathbf{f}(\mathbf{x}, t, \mathbf{p})}{\partial \mathbf{x}} \right|_{(\mathbf{x}_0, t, \mathbf{p})} = \lim_{\Delta \mathbf{x} \to 0} \frac{\mathbf{f}(\mathbf{x}_0 + \Delta \mathbf{x}, t, \mathbf{p}) - \mathbf{f}(\mathbf{x}_0, t, \mathbf{p})}{\Delta \mathbf{x}}. \tag{3.9}$$

$\partial \mathbf{f} / \partial \mathbf{x}$ is called the spatial derivative of $\mathbf{f}(\mathbf{x}, t, \mathbf{p})$ at $\mathbf{x}_0$, and the derivative is given by the Jacobian matrix

$$\frac{\partial \mathbf{f}(\mathbf{x}, t, \mathbf{p})}{\partial \mathbf{x}} = (\partial f_i / \partial x_j)_{n \times n}. \tag{3.10}$$

**Definition 3.6** For $I \subseteq \mathcal{R}$, $\Omega \subseteq \mathcal{R}^n$, and $\Lambda \subseteq \mathcal{R}^m$, consider a vector function $\mathbf{f}(\mathbf{x}, t, \mathbf{p})$ with $\mathbf{f} \colon \Omega \times I \times \Lambda \to \mathcal{R}^n$, $t \in I$ and $\mathbf{x} \in \Omega$ and $\mathbf{p} \in \Lambda$. The vector function $\mathbf{f}(\mathbf{x}, t, \mathbf{p})$ is said to satisfy the Lipschitz condition with respect to $\mathbf{x}$ for $I \times \Omega \times \Lambda$, if

$$\|\mathbf{f}(\mathbf{x}_2, t, \mathbf{p}) - \mathbf{f}(\mathbf{x}_1, t, \mathbf{p})\| \leq L \|\mathbf{x}_2 - \mathbf{x}_1\| \tag{3.11}$$

with $\mathbf{x}_1, \mathbf{x}_2 \in \Omega$ and $L$ a constant. The constant $L$ is called the Lipschitz constant.

**Theorem 3.1** *Consider a dynamical system as*

$$\dot{\mathbf{x}} = \mathbf{f}(\mathbf{x}, t, \mathbf{p}) \quad \text{with } \mathbf{x}(t_0) = \mathbf{x}_0 \tag{3.12}$$

*with $t_0, t \in I = [t_1, t_2]$, $\mathbf{x} \in \Omega = \{\mathbf{x} | \|\mathbf{x} - \mathbf{x}_0\| \leq d\}$ and $\mathbf{p} \in \Lambda$. If the vector function $\mathbf{f}(\mathbf{x}, t, \mathbf{p})$ is $C^r$-continuous ($r \geq 1$) in $G = \Omega \times I \times \Lambda$, then the dynamical system in Eq. (3.12) has one and only one solution $\mathbf{\Phi}(\mathbf{x}_0, t - t_0, \mathbf{p})$ for*

$$|t - t_0| \leq \min(t_2 - t_1, d/M) \quad \text{with } M = \max_G \|\mathbf{f}\|. \tag{3.13}$$

*Proof* The proof of this theorem can be referred to the book by Coddington and Levinson (1955). ∎

**Theorem 3.2** (Gronwall) *Suppose there is a continuous real-valued function $g(t) \geq 0$ to satisfy*

$$g(t) \leq \delta_1 \int_{t_0}^{t} g(\tau) \mathrm{d}\tau + \delta_2 \tag{3.14}$$

*for all $t \in [t_0, t_1]$ and $\delta_1$ and $\delta_2$ are positive constants. For $t \in [t_0, t_1]$, one obtains*

$$g(t) \leq \delta_2 \mathrm{e}^{\delta_1(t - t_0)}. \tag{3.15}$$

*Proof* The proof can be referred to Luo (2012). ∎

**Theorem 3.3** *Consider a dynamical system as* $\dot{\mathbf{x}} = \mathbf{f}(\mathbf{x}, t, \mathbf{p})$ *with* $\mathbf{x}(t_0) = \mathbf{x}_0$ *in Eq. (3.12) with* $t_0, t \in I = [t_1, t_2]$, $\mathbf{x} \in \Omega = \{\mathbf{x} | \|\mathbf{x} - \mathbf{x}_0\| \leq d\}$ *and* $\mathbf{p} \in \Lambda$. *The vector function* $\mathbf{f}(\mathbf{x}, t, \mathbf{p})$ *is* $C^r$-*continuous* $(r \geq 1)$ *in* $G = \Omega \times I \times \Lambda$. *If the solution of* $\dot{\mathbf{x}} = \mathbf{f}(\mathbf{x}, t, \mathbf{p})$ *with* $\mathbf{x}(t_0) = \mathbf{x}_0$ *is* $\mathbf{x}(t)$ *on* $G$ *and the solution of* $\dot{\mathbf{y}} = \mathbf{f}(\mathbf{y}, t, \mathbf{p})$ *with* $\mathbf{y}(t_0) = \mathbf{y}_0$ *is* $\mathbf{y}(t)$ *on* $G$. *For a given* $\varepsilon > 0$, *if* $\|\mathbf{x}_0 - \mathbf{y}_0\| \leq \varepsilon$, *then*

$$\|\mathbf{x}(t) - \mathbf{y}(t)\| \leq \varepsilon e^{L(t-t_0)} \quad \text{on } I \times \Lambda. \tag{3.16}$$

*Proof* The proof can be referred to Luo (2012).                                    $\square$

## 3.2  Basic Discretization

**Definition 3.7** Consider the integration of Eq. (3.1) during the time interval $t \in [t_0, t_M]$ as

$$\mathbf{x}(t) = \mathbf{x}(t_0) + \int_{t_0}^{t} \mathbf{f}(\mathbf{x}, t, \mathbf{p}) dt \quad \text{with } \mathbf{x}(t_0) = \mathbf{x}_0. \tag{3.17}$$

Subdivision of the interval $t \in [t_0, t_M]$ into $M$ subintervals gives

$$t_{k+1} = t_k + h_k \quad \text{for } k = 0, 1, 2, \ldots, M - 1 \tag{3.18}$$

where $h_k$ is called the $k$th step size. Thus, the total interval $[t_0, t_M] = \cup_{k=0}^{M-1}[t_k, t_{k+1}]$ $[t_0, t_M] = \cup_{k=0}^{M-1}[t_k, t_{k+1}]$.

From the time subintervals, the integration during the entire interval can be written as

$$\mathbf{x}(t_M) = x(t_0) + \sum_{k=0}^{M-1} \int_{t_k}^{t_{k+1}} \mathbf{f}(\mathbf{x}, t, \mathbf{p}) dt = x(t_0) + \sum_{k=0}^{M-1} [\mathbf{x}(t_{k+1}) - \mathbf{x}(t_k)] \tag{3.19}$$

with

$$\mathbf{x}(t_{k+1}) = \mathbf{x}(t_k) + \int_{t_k}^{t_{k+1}} \mathbf{f}(\mathbf{x}, t, \mathbf{p}) dt \quad \text{for } k = 1, 2, \ldots, M. \tag{3.20}$$

From the calculus, we can find a point $t_k^c \in [t_k, t_{k+1}]$ with solutions $\mathbf{x}(t_k^c)$ which give $\mathbf{f}_c = \mathbf{f}(\mathbf{x}_c, t_c, \mathbf{p})$ and the mean value theorem generates

$$\mathbf{x}(t_{k+1}) = \mathbf{x}(t_k) + h_k \mathbf{f}(\mathbf{x}_k^c, t_k^c, \mathbf{p}) \quad \text{for } k = 1, 2, \ldots, M \tag{3.21}$$

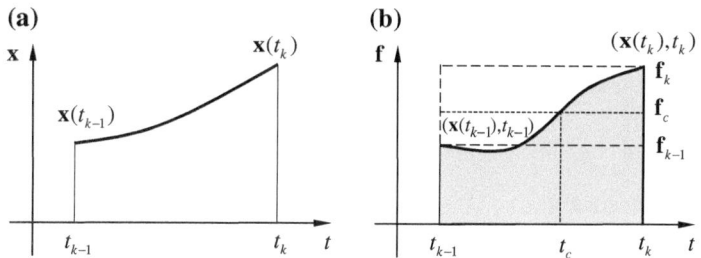

**Fig. 3.1  a** Solution and **b** integration during the time subinterval $t \in [t_{k-1}, t_k]$

where

$$t_k^c = t_k + \theta_k h_k \quad \text{for } k = 1, 2, \ldots, M. \tag{3.22}$$

The solution and integration during the time interval $t \in [t_{k-1}, t_k]$ is geometrically illustrated in Fig. 3.1.

Consider  $\mathbf{f}(\mathbf{x}_k^c, t_k^c, \mathbf{p}) \approx \boldsymbol{\Phi}(\mathbf{x}_k, t_k, \mathbf{p})$  with  $\mathbf{x}(t_{k+1}) \approx \mathbf{x}_{k+1}$  and  $\mathbf{x}(t_k) \approx \mathbf{x}_k$. Equation (3.21) becomes

$$\mathbf{x}_{k+1} = \mathbf{x}_k + h_k \boldsymbol{\Phi}(\mathbf{x}_k, t_k, \mathbf{p}) \quad \text{for } k = 1, 2, \ldots, M. \tag{3.23}$$

The function $\boldsymbol{\Phi}(\mathbf{x}_k, t_k, \mathbf{p})$ is called the increment function. The above equation is obtained from the single-step discrete method.

### 3.2.1  Forward Euler's Method

If $h_{k+1} \to 0$, then $\mathbf{f}(\mathbf{x}_k^c, t_k^c, \mathbf{p}) \approx \mathbf{f}(\mathbf{x}_k, t_k, \mathbf{p})$ and Eq. (3.21) becomes

$$\mathbf{x}_{k+1} = \mathbf{x}_k + h_{k+1}\mathbf{f}(\mathbf{x}_k, t_k, \mathbf{p}) \quad \text{for } k = 0, 1, 2, \ldots, M - 1. \tag{3.24}$$

**Definition 3.8** The discretization of Eq. (3.1) during the time interval $[t_k, t_{k+1}]$ as

$$t_{k+1} = t_k + h_{k+1}, \quad \mathbf{x}_{k+1} = \mathbf{x}_k + h_{k+1}\mathbf{f}(\mathbf{x}_k, t_k, \mathbf{p}) \quad \text{for } k = 0, 1, 2, \ldots, M - 1 \tag{3.25}$$

is called the *Forward Euler's* discrete approximation (or Euler's discrete approximation).

**Definition 3.9** Suppose that the points $(t_k, \mathbf{x}_k)$ for $k = 0, 1, 2, \ldots, M - 1$ is the set of discrete approximation of Eq. (3.1), and the solution $\mathbf{x} = \mathbf{x}(t)$ is the unique solution of Eq. (3.1). The global discretization error $\mathbf{e}_k$ is defined by

$$\mathbf{e}_k = \mathbf{x}(t_k) - \mathbf{x}_k \quad \text{for } k = 0, 1, 2, \ldots, M. \tag{3.26}$$

The local discretization error $\boldsymbol{\varepsilon}_{k+1} = \mathbf{x}(t_{k+1}) - \mathbf{x}_{k+1}$ is defined by

$$\boldsymbol{\varepsilon}_{k+1} = \mathbf{x}(t_{k+1}) - \mathbf{x}_k - h_{k+1}\boldsymbol{\Phi}(\mathbf{x}_k, t_k, \mathbf{p}) \quad \text{for } k = 0, 1, 2, \ldots, M-1. \tag{3.27}$$

The final global error is

$$E(\mathbf{x}(t_M), h) = \|\mathbf{x}(t_M) - \mathbf{x}_M\|. \tag{3.28}$$

**Theorem 3.4** *Suppose the solution* $\mathbf{x} = \mathbf{x}(t)$ *is the unique solution of* Eq. (3.1). *If* $\mathbf{x}(t) \in C^2[t_0, t_M]$ *and the points* $(t_k, \mathbf{x}_k)$ *for* $k = 0, 1, 2, \ldots, M-1$ *is the sequence of Euler's approximation of* Eq. (3.1), *then the local error with* $\mathbf{x}(t_k) = \mathbf{x}_k$ *is*

$$\|\boldsymbol{\varepsilon}_{k+1}\| = \|\mathbf{x}(t_{k+1}) - \mathbf{x}_k - h_{k+1}\mathbf{f}(\mathbf{x}_k, t_k, \mathbf{p})\| = O(h_{k+1}^2) \tag{3.29}$$

*and the global error for* $h_k = h$ *(*$k = 1, 2, 3, \ldots$*) and* $t_k = kh$ *is*

$$\|\mathbf{e}_k\| \le \frac{hL}{2K}(e^{t_k K} - 1) \tag{3.30}$$

*with*

$$L = \max_{k \in \{1,2,\ldots,M\}} \left\| D\mathbf{f}(\mathbf{x}(t_k^c), t_k^c, \mathbf{p}) \right\| \quad \text{and} \quad K = \max_{k \in \{1,2,\ldots,M\}} \|\mathbf{A}_k\|. \tag{3.31}$$

*For a fixed* $t_k = kh$, *as* $h$ *goes to zero,*

$$\|\mathbf{e}_k\| = \|\mathbf{x}(t_k) - \mathbf{x}_k\| = O(h). \tag{3.32}$$

*Proof* Consider the Taylor series of $\mathbf{x}(t_{k+1})$ at point $(t_k, \mathbf{x}(t_k))$ as

$$\begin{aligned}
\mathbf{x}(t_{k+1}) &= \mathbf{x}(t_k + h_{k+1}) = \mathbf{x}(t_k) + h_{k+1}\dot{\mathbf{x}}(t_k) + \frac{1}{2}h_{k+1}^2\ddot{\mathbf{x}}(t_k^c) \\
&= \mathbf{x}(t_k) + h_{k+1}\mathbf{f}(\mathbf{x}(t_k), t_k, \mathbf{p}) + \frac{1}{2}h_{k+1}^2 D\mathbf{f}(\mathbf{x}(t_k^c), t_k^c, \mathbf{p})
\end{aligned}$$

and the Euler's approximation gives

$$\mathbf{x}_{k+1} = \mathbf{x}_k + h_{k+1}\mathbf{f}(\mathbf{x}_k, t_k, \mathbf{p}).$$

If $\mathbf{x}(t_k) = \mathbf{x}_k$, then $\mathbf{f}(\mathbf{x}(t_k), t_k, \mathbf{p}) = \mathbf{f}(\mathbf{x}_k, t_k, \mathbf{p})$. Therefore, the local error is

$$\boldsymbol{\varepsilon}_{k+1} = \mathbf{x}(t_{k+1}) - \mathbf{x}_k - h_{k+1}\mathbf{f}(\mathbf{x}_k, t_k, \mathbf{p}) = \frac{1}{2}h_{k+1}^2 D\mathbf{f}(\mathbf{x}(t_k^c), t_k^c, \mathbf{p}).$$

Thus, the local error at the $(k+1)$th step is

$$\|\boldsymbol{\varepsilon}_{k+1}\| = \|\mathbf{x}(t_{k+1}) - \mathbf{x}_k - h_{k+1}\mathbf{f}(\mathbf{x}_k, t_k, \mathbf{p})\| = O(h_{k+1}^2).$$

If $\mathbf{x}(t_k) \neq \mathbf{x}_k$, then $\mathbf{f}(\mathbf{x}(t_k), t_k, \mathbf{p}) \neq \mathbf{f}(\mathbf{x}_k, t_k, \mathbf{p})$. Thus,

$$\begin{aligned}
\mathbf{x}(t_{k+1}) - \mathbf{x}_{k+1} &= \mathbf{x}(t_k) - \mathbf{x}_k + h_{k+1}[\mathbf{f}(\mathbf{x}(t_k), t_k, \mathbf{p}) - \mathbf{f}(\mathbf{x}_k, t_k, \mathbf{p})] \\
&\quad + \frac{1}{2}h_{k+1}^2 D\mathbf{f}(\mathbf{x}(t_k^c), t_k^c, \mathbf{p})
\end{aligned}$$

and the Lipschitz condition gives

$$\mathbf{x}(t_{k+1}) - \mathbf{x}_{k+1} = (\mathbf{I} + h_{k+1}\mathbf{A}_k)(\mathbf{x}(t_k) - \mathbf{x}_k) + \boldsymbol{\varepsilon}_{k+1},$$

where $\mathbf{A}_k = D_\mathbf{x}\mathbf{f}(\boldsymbol{\xi}_k, t_k, \mathbf{p})$ and $\|\boldsymbol{\xi}_k\| \in (\|\mathbf{x}(t_k)\|, \|\mathbf{x}_k\|)$. So,

$$\mathbf{e}_{k+1} = (\mathbf{I} + h_{k+1}\mathbf{A}_k)\mathbf{e}_k + \boldsymbol{\varepsilon}_{k+1}$$

gives

$$\|\mathbf{e}_{k+1}\| \leq \|(\mathbf{I} + h_{k+1}\mathbf{A}_k)\| \times \|\mathbf{e}_k\| + \|\boldsymbol{\varepsilon}_{k+1}\|.$$

For $h_k = h$ $(k = 1, 2, 3, \ldots)$ and $t_k = kh$, letting

$$L = \max_{k \in \{1,2,\ldots,M\}} \|D\mathbf{f}(\mathbf{x}(t_k^c), t_k^c, \mathbf{p})\| \quad \text{and} \quad K = \max_{k \in \{1,2,\ldots,M\}} \|\mathbf{A}_k\|,$$

consider a simple discrete equation as

$$z_{k+1} = (1 + hK)z_k + \frac{1}{2}hL \quad \text{with} \quad z_0 = 0$$

and

$$z_k = \frac{1}{2}h^2L\left(\sum_{j=0}^{k-1}(1 + hK)^j\right) = \frac{hL}{2K}\left[(1 + hK)^k - 1\right] \quad \text{for } k = 1, 2, \ldots, M.$$

For $k > 0$, we have $1 + hK > 0$. If $z_k \geq \|\mathbf{e}_k\|$, then $z_{k+1} \geq \|\mathbf{e}_{k+1}\|$. Since $t_k = kh$, for $k \to \infty$, we have

$$\lim_{k \to \infty}\left(1 + \frac{t_k}{k}K\right)^k = e^{t_k K}.$$

so

$$\|\mathbf{e}_k\| \leq \frac{hL}{2K}(e^{t_k K} - 1).$$

For a fixed $t_k = kh$, as $h$ goes to zero, we have

$$\|\mathbf{e}_k\| = \|\mathbf{x}(t_k) - \mathbf{x}_k\| = O(h) \quad \text{for } k = 1, 2, \ldots, M.$$

This theorem is proved.                                                            □

For each step, if the global error is enlarged, then such a discrete approximation is unstable. Consider the global error for the $(k + 1)$th step as

$$\mathbf{e}_{k+1} = (\mathbf{I} + h_{k+1}\mathbf{A}_k)\mathbf{e}_k. \tag{3.33}$$

If $\mathbf{e}_{k+1} = \lambda\mathbf{e}_k$, then the foregoing equation becomes

$$(\mathbf{I} + h_{k+1}\mathbf{A}_k - \lambda\mathbf{I})\mathbf{e}_k = \mathbf{0}. \tag{3.34}$$

The corresponding eigenvalues are determined by

$$|\mathbf{I} + h_{k+1}\mathbf{A}_k - \lambda\mathbf{I}| = \mathbf{0}. \tag{3.35}$$

If all eigenvalues $|\lambda_j| < 1$ $(j = 1, 2, \ldots, n)$, then

$$\|\mathbf{e}_{k+1}\| < \|\mathbf{e}_k\|. \tag{3.36}$$

In other words, the global error will not be enlarged, which implies that the forward Euler's method gives a stable approximation. For a one-dimensional system, we have $\mathbf{A}_k = L_k$ and $\mathbf{I} = 1$. Equation (3.35) gives the stability interval for the forward Euler's method as

$$|1 + h_{k+1}L_k| < 1 \quad \Leftrightarrow \quad -2 < h_{k+1}L_k < 0. \tag{3.37}$$

Since $h_{k+1} > 0$, we have $L_k < 0$. Thus, $0 < h_{k+1} < -2/L_k$ or $h_{k+1} < |2/L_k|$. If we consider a local error is controlled with a small positive $\varepsilon$, letting

$$\|\boldsymbol{\varepsilon}_{k+1}\| = \left\|\frac{1}{2}h_{k+1}^2 D\mathbf{f}(\mathbf{x}(t_k^c), t_k^c, \mathbf{p})\right\| = \frac{1}{2}h_{k+1}^2\|D\mathbf{f}(\mathbf{x}(t_k^c), t_k^c, \mathbf{p})\| \leq \varepsilon \tag{3.38}$$

then

$$h_{k+1} \leq \sqrt{\frac{2\varepsilon}{\|D\mathbf{f}(\mathbf{x}(t_k^c), t_k^c, \mathbf{p})\|}}. \tag{3.39}$$

The step size can be controlled through Eq. (3.39). This is based on the absolute error control. For the relative error control, setting $\|\boldsymbol{\varepsilon}_{k+1}\| \leq \varepsilon \|\mathbf{x}_k - \mathbf{x}_{k-1}\|$, we have

$$h_{k+1} \leq \sqrt{\frac{2\varepsilon \|\mathbf{x}_k - \mathbf{x}_{k-1}\|}{\|D\mathbf{f}(\mathbf{x}(t_k^c), t_k^c, \mathbf{p})\|}}. \tag{3.40}$$

In practical computation, $D\mathbf{f}(\mathbf{x}(t_k^c), t_k^c, \mathbf{p})$ is first estimated by $D\mathbf{f}(\mathbf{x}(t_{k-1}^c), t_{k-1}^c, \mathbf{p})$ to select the step size $h_{k+1}$. That is,

$$D\mathbf{f}(\mathbf{x}(t_{k-1}^c), t_{k-1}^c, \mathbf{p}) \approx \frac{\mathbf{f}(\mathbf{x}(t_k), t_k, \mathbf{p}) - \mathbf{f}(\mathbf{x}(t_{k-1}), t_{k-1}, \mathbf{p})}{t_k - t_{k-1}}. \tag{3.41}$$

Once this step size $h_{k+1}$ is determined, we have

$$D\mathbf{f}(\mathbf{x}(t_k^c), t_k^c, \mathbf{p}) \approx \frac{\mathbf{f}(\mathbf{x}(t_{k+1}), t_{k+1}, \mathbf{p}) - \mathbf{f}(\mathbf{x}(t_k), t_k, \mathbf{p})}{t_{k+1} - t_k}. \tag{3.42}$$

Using Eq. (3.40), we finally select the step size $h_{k+1}$ for next step. For other discussion on the step size control, readers can refer Kahaner et al. (1989).

The forward Euler's method is a simple, lower order, explicit method that is not recommended in practice. The explicit discrete method can be computed from the values of the previous step. For the Euler's method, only the value of one previous step is needed to compute values of the next step. This is why the Euler's method is a forwarded discrete method. The significance of the problem with the computer round-off error for the Euler's method is caused by its low accuracy. For higher order discrete methods, the simple accumulation of the round-off error will be less significant.

### 3.2.2  Backward Euler's Method

If $h_{k+1} \to 0$, then $\mathbf{f}(\mathbf{x}_k^c, t_k^c, \mathbf{p}) \approx \mathbf{f}(\mathbf{x}_{k+1}, t_{k+1}, \mathbf{p})$ and Eq. (3.21) becomes

$$\mathbf{x}_{k+1} = \mathbf{x}_k + h_{k+1} \mathbf{f}(\mathbf{x}_{k+1}, t_{k+1}, \mathbf{p}) \quad \text{for } k = 0, 1, 2, \ldots, M - 1. \tag{3.43}$$

**Definition 3.10** The discretization of Eq. (3.1) during the time interval $[t_k, t_{k+1}]$ as

$$t_{k+1} = t_k + h_{k+1}, \quad \mathbf{x}_{k+1} = \mathbf{x}_k + h_{k+1} \mathbf{f}(\mathbf{x}_{k+1}, t_{k+1}, \mathbf{p}) \quad \text{for } k = 0, 1, 2, \ldots, M - 1 \tag{3.44}$$

is called the backward Euler's discrete approximation.

**Theorem 3.5** *Suppose that the solution* $\mathbf{x} = \mathbf{x}(t)$ *is the unique solution of* Eq. (3.1).
*If* $\mathbf{x}(t) \in C^2[t_0, t_M]$ *and the points* $(t_k, \mathbf{x}_k)$ *for* $k = 0, 1, 2, \ldots, M - 1$ *is the sequence
of the backward Euler's approximation of* Eq. (3.1), *then*

$$\|\boldsymbol{\varepsilon}_{k+1}\| = \|\mathbf{x}(t_{k+1}) - \mathbf{x}_k - h_{k+1}\mathbf{f}(\mathbf{x}_{k+1}, t_{k+1}, \mathbf{p})\| = O(h_{k+1}^2) \qquad (3.45)$$

*and the global error for* $h_k = h$ $(k = 1, 2, 3, \ldots)$ *and* $t_k = kh$ *is*

$$\|\mathbf{e}_k\| \le \frac{hL}{2K}(e^{t_k K} - 1) \qquad (3.46)$$

*with*

$$L = \max_{k \in \{1,2,\ldots,M\}} \|D\mathbf{f}(\mathbf{x}(t_k^c), t_k^c, \mathbf{p})\| \quad \text{and} \quad K = \max_{k \in \{1,2,\ldots,M\}} \|\mathbf{A}_{k+1}\|. \qquad (3.47)$$

*For a fixed* $t_k = kh$, *as* $h$ *goes to zero,*

$$\|\mathbf{e}_k\| = \|\mathbf{x}(t_k) - \mathbf{x}_k\| = O(h). \qquad (3.48)$$

*Proof* Consider the Taylor series of $\mathbf{x}(t_k)$ at point $(t_{k+1}, \mathbf{x}(t_{k+1}))$ as

$$\mathbf{x}(t_k) = \mathbf{x}(t_{k+1} - h_{k+1}) = \mathbf{x}(t_{k+1}) - h_{k+1}\dot{\mathbf{x}}(t_{k+1}) + \frac{1}{2}h_{k+1}^2\ddot{\mathbf{x}}(t_{k+1}^c)$$

$$= \mathbf{x}(t_{k+1}) - h_{k+1}\mathbf{f}(\mathbf{x}(t_{k+1}), t_{k+1}, \mathbf{p}) + \frac{1}{2}h_{k+1}^2 D\mathbf{f}(\mathbf{x}(t_{k+1}^c), t_{k+1}^c, \mathbf{p})$$

and

$$\mathbf{x}(t_{k+1}) = \mathbf{x}(t_k) + h_{k+1}\mathbf{f}(\mathbf{x}(t_{k+1}), t_{k+1}, \mathbf{p}) - \frac{1}{2}h_{k+1}^2 D\mathbf{f}(\mathbf{x}(t_{k+1}^c), t_{k+1}^c, \mathbf{p}).$$

The backward Euler's approximation gives

$$\mathbf{x}_{k+1} = \mathbf{x}_k + h_{k+1}\mathbf{f}(\mathbf{x}_{k+1}, t_{k+1}, \mathbf{p}).$$

(i)  If $\mathbf{x}(t_k) = \mathbf{x}_k$, then the local error is

$$\boldsymbol{\varepsilon}_{k+1} = \mathbf{x}(t_{k+1}) - \mathbf{x}_k - h_{k+1}\mathbf{f}(\mathbf{x}_{k+1}, t_{k+1}, \mathbf{p})$$

$$= h_{k+1}[\mathbf{f}(\mathbf{x}(t_{k+1}), t_{k+1}, \mathbf{p}) - \mathbf{f}(\mathbf{x}_{k+1}, t_{k+1}, \mathbf{p})] - \frac{1}{2}h_{k+1}^2 D\mathbf{f}(\mathbf{x}(t_k^c), t_k^c, \mathbf{p})$$

with

$$\mathbf{f}(\mathbf{x}(t_{k+1}), t_{k+1}, \mathbf{p}) = \mathbf{f}(\mathbf{x}_{k+1}, t_{k+1}, \mathbf{p}) + D_{\mathbf{x}}\mathbf{f}(\mathbf{x}_{k+1}^c, t_{k+1}, \mathbf{p})(\mathbf{x}(t_{k+1}) - \mathbf{x}_{k+1})$$

$$= \mathbf{f}(\mathbf{x}_{k+1}, t_{k+1}, \mathbf{p}) + \mathbf{A}_k\boldsymbol{\varepsilon}_{k+1}$$

where $\mathbf{A}_k = D_{\mathbf{x}}\mathbf{f}(\xi_{k+1}, t_{k+1}, \mathbf{p})$ and $\|\xi_{k+1}\| \in (\|\mathbf{x}(t_{k+1})\|, \|\mathbf{x}_{k+1}\|)$. So

$$\varepsilon_{k+1} = -\frac{1}{2}h_{k+1}^2(\mathbf{I} - h_{k+1}\mathbf{A}_k)^{-1}D\mathbf{f}(\mathbf{x}(t_k^c), t_k^c, \mathbf{p}) + O(h_{k+1}^3).$$

Thus, the local error is

$$\|\varepsilon_{k+1}\| = \|\mathbf{x}(t_{k+1}) - \mathbf{x}_k - h_{k+1}\mathbf{f}(\mathbf{x}_{k+1}, t_{k+1}, \mathbf{p})\| = O(h_{k+1}^2).$$

(ii)  If $\mathbf{x}(t_{k+1}) \neq \mathbf{x}_{k+1}$, then $\mathbf{f}(\mathbf{x}(t_{k+1}), t_{k+1}, \mathbf{p}) \neq \mathbf{f}(\mathbf{x}_{k+1}, t_{k+1}, \mathbf{p})$, thus

$$\mathbf{x}(t_{k+1}) - \mathbf{x}_{k+1} = \mathbf{x}(t_k) - \mathbf{x}_k + h_{k+1}[\mathbf{f}(\mathbf{x}(t_{k+1}), t_{k+1}, \mathbf{p}) - \mathbf{f}(\mathbf{x}_{k+1}, t_{k+1}, \mathbf{p})]$$
$$-\frac{1}{2}h_{k+1}^2 D\mathbf{f}(\mathbf{x}(t_{k+1}^c), t_{k+1}^c, \mathbf{p})$$

and the Lipschitz condition yields

$$\mathbf{x}(t_{k+1}) - \mathbf{x}_{k+1} = \mathbf{x}(t_k) - \mathbf{x}_k + h_{k+1}\mathbf{A}_k(\mathbf{x}(t_{k+1}) - \mathbf{x}_{k+1}) + \delta_{k+1}$$

where $\delta_{k+1} = -\frac{1}{2}h_{k+1}D\mathbf{f}(\mathbf{x}(t_k^c), t_k^c, \mathbf{p})$. So

$$\mathbf{e}_{k+1} = \mathbf{e}_k + h_{k+1}\mathbf{A}_k\mathbf{e}_{k+1} + \delta_{k+1}$$

gives

$$\|\mathbf{e}_{k+1}\| \leq \|\mathbf{e}_k\| + h_{k+1}\|\mathbf{A}_k\| \times \|\mathbf{e}_{k+1}\| + \|\delta_{k+1}\|.$$

For $h_k = h$ ($k = 1, 2, 3, \ldots$) and $t_k = kh$, letting

$$L = \max_{k \in \{1,2,\ldots,M\}} \|D\mathbf{f}(\mathbf{x}(t_k^c), t_k^c, \mathbf{p})\| \quad \text{and} \quad K = \max_{k \in \{1,2,\ldots,M\}} \|\mathbf{A}_k\|,$$

consider a simple discrete equation as

$$z_{k+1} = z_k + (hK)z_{k+1} + \frac{1}{2}hL \quad \text{with } z_0 = 0$$

and

$$z_k = \frac{1}{2}h^2 L \sum_{j=0}^{k-1}(1 + hK)^j = \frac{hL}{2K}[(1 + hK)^k - 1] \quad \text{for } k = 1, 2, \ldots, M.$$

For $k > 0$, we have $1 + hK > 0$. If $z_k \geq \|\mathbf{e}_k\|$, then $z_{k+1} \geq \|\mathbf{e}_{k+1}\|$. Since $t_k = kh$, for $k \to \infty$, we have

$$\lim_{k \to \infty}(1 - \frac{t_k}{k}K)^{-k} = e^{t_k K}.$$

so

$$\|\mathbf{e}_k\| \leq \frac{hL}{2K}(e^{t_kK} - 1).$$

For a fixed $t_k = kh$, as $h$ goes to zero, we have

$$\|\mathbf{e}_k\| = \|\mathbf{x}(t_k) - \mathbf{x}_k\| = O(h) \quad \text{for } k = 1, 2, \ldots, M.$$

This theorem is proved.                                                    □

Consider the global error for the $(k + 1)$th step with the following relation

$$\mathbf{e}_{k+1} = (\mathbf{I} - h_{k+1}\mathbf{A}_k)^{-1}\mathbf{e}_k. \tag{3.49}$$

If $\mathbf{e}_{k+1} = \lambda\mathbf{e}_k$, then the foregoing equation becomes

$$\left[(\mathbf{I} - h_{k+1}\mathbf{A}_k)^{-1} - \lambda\mathbf{I}\right]\mathbf{e}_k = \mathbf{0}. \tag{3.50}$$

The corresponding eigenvalues are produced by

$$|(\mathbf{I} - h_{k+1}\mathbf{A}_k)^{-1} - \lambda\mathbf{I}| = \mathbf{0}. \tag{3.51}$$

If all eigenvalues $|\lambda_j| < 1$ ($j = 1, 2, \ldots, n$), then

$$\|\mathbf{e}_{k+1}\| < \|\mathbf{e}_k\|. \tag{3.52}$$

In other words, the global error will not be enlarged. The backward Euler's method gives a stable approximation. For one-dimensional systems, we have $\mathbf{A}_k = L_k$ and $\mathbf{I} = 1$. Equation (3.51) gives the stability interval for the backward Euler's method as

$$|(1 - h_{k+1}L_k)^{-1}| < 1. \tag{3.53}$$

Since $h_{k+1} > 0$, if $L_k < 0$, then the foregoing equation always exists.

The backward Euler's method is also a lower order method that is not recommended in practice. However, this method is an implicit discrete method because the backward method uses the value at next step to evaluate the vector field (right-hand side in a differential equation). For this method, the iteration and Newton–Raphson method should be used to determine values for each step, which will be discussed later. An implicit method is much more stable than an explicit method. The backward Euler's method possesses a large stability range. For any positive step size, the amplification factor is less than one, and the errors will not be magnified. Such a method is called to be absolutely stable. The significance of the problem with the computer round-off error for the backward Euler's method is also caused by its low accuracy but it is better than the forward Euler's method.

### 3.2.3 Trapezoidal Rule Discretization

If $h_{k+1} \to 0$, then $\mathbf{f}(\mathbf{x}_k^c, t_k^c, \mathbf{p}) \approx \frac{1}{2}[\mathbf{f}(\mathbf{x}_k, t_k, \mathbf{p}) + \mathbf{f}(\mathbf{x}_{k+1}, t_{k+1}, \mathbf{p})]$ and Eq. (3.21) becomes

$$\mathbf{x}_{k+1} = \mathbf{x}_k + \frac{1}{2}h_{k+1}[\mathbf{f}(\mathbf{x}_k, t_k, \mathbf{p}) + \mathbf{f}(\mathbf{x}_{k+1}, t_{k+1}, \mathbf{p})] \quad \text{for } k = 0, 1, 2, \ldots, M - 1.$$
$$(3.54)$$

**Definition 3.11** The discretization of Eq. (3.1) during the time interval $[t_k, t_{k+1}]$ as

$$t_{k+1} = t_k + h_k, \quad \mathbf{x}_{k+1} = \mathbf{x}_k + \frac{1}{2}h_{k+1}[\mathbf{f}(\mathbf{x}_k, t_k, \mathbf{p}) + \mathbf{f}(\mathbf{x}_{k+1}, t_{k+1}, \mathbf{p})]$$
$$\text{for } k = 0, 1, 2, \ldots, M - 1$$
$$(3.55)$$

is called the *trapezoidal rule* discrete approximation (or Heun's method, or trapezoidal method).

**Theorem 3.6** *Suppose that the solution* $\mathbf{x} = \mathbf{x}(t)$ *is the unique solution of Eq. (3.1). If* $\mathbf{x}(t) \in C^3[t_0, t_M]$ *and the points* $(t_k, \mathbf{x}_k)$ *for* $k = 0, 1, 2, \ldots, M - 1$ *is the sequence of the trapezoidal rule approximation of Eq. (3.1), then*

$$\|\boldsymbol{\varepsilon}_{k+1}\| = \|\mathbf{x}(t_{k+1}) - \mathbf{x}_k - h_{k+1}\boldsymbol{\Phi}(\mathbf{x}_k, t_k, \mathbf{p})\| = O(h_{k+1}^3) \qquad (3.56)$$

*where* $\boldsymbol{\Phi}(\mathbf{x}_k, t_k, \mathbf{p}) = \frac{1}{2}h_{k+1}[\mathbf{f}(\mathbf{x}_k, t_k, \mathbf{p}) + \mathbf{f}(\mathbf{x}_{k+1}, t_{k+1}, \mathbf{p})]$, *and the global error for* $h_k = h$ ($k = 1, 2, 3, \ldots$) *and* $t_k = kh$ *is*

$$\|\mathbf{e}_k\| \leq \frac{h^2 L}{12K}(e^{t_k K} - 1) \qquad (3.57)$$

*with*

$$L = \max_{k \in \{1,2,\ldots,M\}} \|D^2\mathbf{f}(\mathbf{x}(t_k^c), t_k^c, \mathbf{p})\|, \quad K = \max_{k \in \{1,2,\ldots,M\}} \|\mathbf{A}_k\|, \quad \text{and}$$
$$\mathbf{A}_k = D_{\mathbf{x}}\mathbf{f}(\boldsymbol{\xi}_k, t_k, \mathbf{p}) \quad \text{and} \quad \|\boldsymbol{\xi}_k\| \in (\|\mathbf{x}(t_k)\|, \|\mathbf{x}_k\|).$$
$$(3.58)$$

*For a fixed* $t_k = kh$, *as h goes to zero,*

$$\|\mathbf{e}_k\| = \|\mathbf{x}(t_k) - \mathbf{x}_k\| = O(h^2). \qquad (3.59)$$

*Proof* Consider the Taylor series of $\mathbf{x}(t_k)$ at point $(t_{k+1}, \mathbf{x}(t_{k+1}))$ as

$$\mathbf{x}(t_k) = \mathbf{x}(t_{k+1} - h_{k+1}) = \mathbf{x}(t_{k+1}) - h_{k+1}\dot{\mathbf{x}}(t_{k+1}) + \frac{1}{2}h_{k+1}^2\ddot{\mathbf{x}}(t_{k+1}) - \frac{1}{6}h_k^3\dddot{\mathbf{x}}(t_k^c)$$

$$= \mathbf{x}(t_{k+1}) - h_{k+1}\mathbf{f}(\mathbf{x}(t_{k+1}), t_k, \mathbf{p})$$

$$+ \frac{1}{2}h_{k+1}^2 D\mathbf{f}(\mathbf{x}(t_{k+1}), t_{k+1}, \mathbf{p}) - \frac{1}{6}h_{k+1}^3 D^2\mathbf{f}(\mathbf{x}(t_k^c), t_k^c, \mathbf{p})$$

and the Taylor series of $\mathbf{x}(t_{k+1})$ at point $(t_k, \mathbf{x}(t_k))$ is

$$\mathbf{x}(t_{k+1}) = \mathbf{x}(t_k + h_{k+1}) = \mathbf{x}(t_k) + h_{k+1}\dot{\mathbf{x}}(t_k) + \frac{1}{2}h_{k+1}^2\ddot{\mathbf{x}}(t_{k+1}) + \frac{1}{6}h_k^3\dddot{\mathbf{x}}(t_k^c)$$

$$= \mathbf{x}(t_k) + h_{k+1}\mathbf{f}(\mathbf{x}(t_k), t_k, \mathbf{p})$$

$$+ \frac{1}{2}h_{k+1}^2 D\mathbf{f}(\mathbf{x}(t_k), t_k, \mathbf{p}) + \frac{1}{6}h_{k+1}^3 D^2\mathbf{f}(\mathbf{x}(t_k^c), t_k^c, \mathbf{p}).$$

The summation of the Taylor series based on $\mathbf{x}(t_{k+1})$ at point $(t_k, \mathbf{x}(t_k))$ and $\mathbf{x}(t_k)$ at point $(t_{k+1}, \mathbf{x}(t_{k+1}))$ gives

$$\mathbf{x}(t_{k+1}) = \mathbf{x}(t_k) + \frac{1}{2}h_{k+1}[\mathbf{f}(\mathbf{x}(t_k), t_k, \mathbf{p}) + \mathbf{f}(\mathbf{x}(t_{k+1}), t_{k+1}, \mathbf{p})]$$

$$- \frac{1}{4}h_{k+1}^2[D\mathbf{f}(\mathbf{x}(t_{k+1}), t_{k+1}, \mathbf{p}) - D\mathbf{f}(\mathbf{x}(t_k), t_k, \mathbf{p})]$$

$$+ \frac{1}{6}h_{k+1}^3 D^2\mathbf{f}(\mathbf{x}(t_k^c), t_k^c, \mathbf{p}).$$

Because

$$D\mathbf{f}(\mathbf{x}(t_{k+1}), t_{k+1}, \mathbf{p}) - D\mathbf{f}(\mathbf{x}(t_k), t_k, \mathbf{p}) = h_{k+1}D^2\mathbf{f}(\mathbf{x}(t_k^c), t_k^c, \mathbf{p}),$$

we have

$$\mathbf{x}(t_{k+1}) = \mathbf{x}(t_k) + \frac{1}{2}h_{k+1}[\mathbf{f}(\mathbf{x}(t_k), t_k, \mathbf{p}) + \mathbf{f}(\mathbf{x}(t_{k+1}), t_{k+1}, \mathbf{p})]$$

$$- \frac{1}{12}h_{k+1}^3 D^2\mathbf{f}(\mathbf{x}(t_k^c), t_k^c, \mathbf{p}).$$

Using the trapezoidal rule approximation as

$$\mathbf{x}_{k+1} = \mathbf{x}_k + h_{k+1}\mathbf{\Phi}(\mathbf{x}_k, t_k, \mathbf{p})$$

$$= \mathbf{x}_k + \frac{1}{2}h_{k+1}[\mathbf{f}(\mathbf{x}_k, t_k, \mathbf{p}) + \mathbf{f}(\mathbf{x}_{k+1}, t_{k+1}, \mathbf{p})],$$

we can have a relation as

$$
\begin{aligned}
\mathbf{x}(t_{k+1}) - \mathbf{x}_{k+1} = {}& \mathbf{x}(t_k) - \mathbf{x}_k + \frac{1}{2} h_{k+1} [\mathbf{f}(\mathbf{x}(t_k), t_k, \mathbf{p}) - \mathbf{f}(\mathbf{x}_k, t_k, \mathbf{p})] \\
& + \frac{1}{2} h_{k+1} [\mathbf{f}(\mathbf{x}(t_{k+1}), t_{k+1}, \mathbf{p}) - \mathbf{f}(\mathbf{x}_{k+1}, t_{k+1}, \mathbf{p})] \\
& - \frac{1}{12} h_{k+1}^3 D^2 \mathbf{f}(\mathbf{x}(t_k^c), t_k^c, \mathbf{p})
\end{aligned}
$$

and the Lipschitz condition gives

$$
\begin{aligned}
\mathbf{x}(t_{k+1}) - \mathbf{x}_{k+1} = {}& \mathbf{x}(t_k) - \mathbf{x}_k + \frac{1}{2} h_{k+1} \mathbf{A}_k (\mathbf{x}(t_k) - \mathbf{x}_k) \\
& + \frac{1}{2} h_{k+1} \mathbf{A}_{k+1} (\mathbf{x}(t_{k+1}) - \mathbf{x}_{k+1}) - \frac{1}{12} h_{k+1}^3 D^2 \mathbf{f}(\mathbf{x}(t_k^c), t_k^c, \mathbf{p})
\end{aligned}
$$

where

$$
\begin{aligned}
\mathbf{A}_k &= D_{\mathbf{x}} \mathbf{f}(\boldsymbol{\xi}_k, t_k, \mathbf{p}) \quad \text{and} \quad \|\boldsymbol{\xi}_k\| \in (\|\mathbf{x}(t_k)\|, \|\mathbf{x}_k\|), \\
\mathbf{A}_{k+1} &= D_{\mathbf{x}} \mathbf{f}(\boldsymbol{\xi}_{k+1}, t_{k+1}, \mathbf{p}) \quad \text{and} \quad \|\boldsymbol{\xi}_{k+1}\| \in (\|\mathbf{x}(t_{k+1})\|, \|\mathbf{x}_{k+1}\|).
\end{aligned}
$$

(i)  If $\mathbf{x}(t_k) = \mathbf{x}_k$, we have the local error as

$$
\boldsymbol{\varepsilon}_{k+1} = \frac{1}{2} h_{k+1} \mathbf{A}_{k+1} \boldsymbol{\varepsilon}_{k+1} - \frac{1}{12} h_{k+1}^3 D^2 \mathbf{f}(\mathbf{x}(t_k^c), t_k^c, \mathbf{p}).
$$

That is

$$
\boldsymbol{\varepsilon}_{k+1} = -\frac{1}{12} h_{k+1}^3 \left( \mathbf{I} - \frac{1}{2} h_{k+1} \mathbf{A}_{k+1} \right)^{-1} D^2 \mathbf{f}(\mathbf{x}(t_k^c), t_k^c, \mathbf{p}).
$$

Therefore,

$$
\|\boldsymbol{\varepsilon}_{k+1}\| = \|\mathbf{x}(t_{k+1}) - \mathbf{x}_k - h_{k+1} \boldsymbol{\Phi}(\mathbf{x}_k, t_k, \mathbf{p})\| = O(h_{k+1}^3).
$$

(ii)  If $\mathbf{x}(t_k) \neq \mathbf{x}_k$, we have

$$
\mathbf{e}_{k+1} = \mathbf{e}_k + \frac{1}{2} h_{k+1} \mathbf{A}_k \mathbf{e}_k + \frac{1}{2} h_{k+1} \mathbf{A}_{k+1} \mathbf{e}_{k+1} - \frac{1}{12} h_{k+1}^3 D^2 \mathbf{f}(\mathbf{x}(t_k^c), t_k^c, \mathbf{p})
$$

giving

$$
\begin{aligned}
\|\mathbf{e}_{k+1}\| \leq {}& \|\mathbf{e}_k\| + \frac{1}{2} h_{k+1} \|\mathbf{A}_k\| \times \|\mathbf{e}_k\| \\
& + \frac{1}{2} h_{k+1} \|\mathbf{A}_{k+1}\| \times \|\mathbf{e}_{k+1}\| \\
& - \frac{1}{12} h_{k+1}^3 \|D^2 \mathbf{f}(\mathbf{x}(t_k^c), t_k^c, \mathbf{p})\|.
\end{aligned}
$$

For $h_k = h$ ($k = 1, 2, 3, \ldots$) and $t_k = kh$, letting

$$L = \max_{k \in \{1,2,\ldots,M\}} \|D^2\mathbf{f}(\mathbf{x}(t_k^c), t_k^c, \mathbf{p})\| \quad \text{and} \quad K = \max_{k \in \{1,2,\ldots,M\}} \|\mathbf{A}_k\|,$$

consider a simple discrete equation as

$$z_{k+1} = (1 + \frac{1}{2}hK)(1 - \frac{1}{2}hK)^{-1}z_k + \frac{1}{12}(1 - \frac{1}{2}hK)^{-1}h^3L \quad \text{with } z_0 = 0$$

and

$$\begin{aligned}
z_{k+1} &= \frac{1}{12}h^3L\sum_{j=0}^{k-1}(1 + \frac{1}{2}hK)^j(1 - \frac{1}{2}hK)^{-j} \\
&= \frac{h^2L}{12K}[(1 + \frac{1}{2}hK)^k(1 - \frac{1}{2}hK)^{-k} - 1] \quad \text{for } k = 1, 2, \ldots, M.
\end{aligned}$$

For $k > 0$, we have $1 + hK > 0$. If $z_k \geq \|\mathbf{e}_k\|$, then $z_{k+1} \geq \|\mathbf{e}_{k+1}\|$. Since $t_k = kh$, for $k \to \infty$, we have

$$\lim_{k\to\infty}[1 + \frac{1}{k}(\frac{1}{2}t_kK)]^k[1 - \frac{1}{k}(\frac{1}{2}t_kK)]^{-k} = e^{\frac{1}{2}t_kK}e^{\frac{1}{2}t_kK} = e^{t_kK},$$

so

$$\|\mathbf{e}_k\| \leq \frac{h^2L}{12K}(e^{t_kK} - 1).$$

For a fixed $t_k = kh$, as $h$ goes to zero, we have

$$\|\mathbf{e}_k\| = \|\mathbf{x}(t_k) - \mathbf{x}_k\| = O(h^2) \quad \text{for } k = 1, 2, \ldots, M.$$

This theorem is proved.                                                                      $\square$

Consider the global error through the following relation

$$\mathbf{e}_{k+1} = (\mathbf{I} - \frac{1}{2}h_{k+1}\mathbf{A}_{k+1})^{-1}(\mathbf{I} + \frac{1}{2}h_{k+1}\mathbf{A}_k)\mathbf{e}_k. \tag{3.60}$$

If $\mathbf{e}_{k+1} = \lambda\mathbf{e}_k$, then the foregoing equation becomes

$$[(\mathbf{I} - \frac{1}{2}h_{k+1}\mathbf{A}_{k+1})^{-1}(\mathbf{I} + \frac{1}{2}h_{k+1}\mathbf{A}_k) - \lambda\mathbf{I}]\mathbf{e}_k = \mathbf{0}. \tag{3.61}$$

The corresponding eigenvalues are generated by

$$\left| \left( \mathbf{I} - \frac{1}{2} h_{k+1} \mathbf{A}_{k+1} \right)^{-1} \left( \mathbf{I} + \frac{1}{2} h_{k+1} \mathbf{A}_k \right) - \lambda \mathbf{I} \right| = 0. \tag{3.62}$$

If all eigenvalues $|\lambda_j| < 1$ $(j = 1, 2, \ldots, n)$, then

$$\|\mathbf{e}_{k+1}\| < \|\mathbf{e}_k\|. \tag{3.63}$$

In other words, the global error will not be enlarged. The trapezoidal rule method gives a stable approximation. For one-dimensional systems, we have $\mathbf{A}_k = L_{k1}$ and $\mathbf{A}_{k+1} = L_{k2}$, and $\mathbf{I} = 1$. Equation (3.62) gives the stability interval for the trapezoidal discrete method as

$$\left| \frac{1 + \frac{1}{2} h_{k+1} L_{k1}}{1 - \frac{1}{2} h_{k+1} L_{k2}} \right| < 1. \tag{3.64}$$

Since $h_{k+1} > 0$, if $L_{k1} < 0$ and $L_{k2} < 0$, then the foregoing equation always exists.

The trapezoidal method is a simple, stable discrete method that is often used in practice. This method is also an implicit discrete method because the trapezoidal method uses the values at both ends of the vector field (right-hand side in a differential equation) in a discrete interval. For the trapezoidal method, the iterative method or Newton–Raphson method will be also used to determine values for each step. The trapezoidal method also possesses a large stability range like the backward Euler's method. For any positive step size, the amplification factor is less than one, and the errors will not be magnified. The computer round-off error for the trapezoidal rule method will get much improved.

For comparison of the forward Euler's method, backward Euler's method, and trapezoidal rule approximation, a geometrical illustration for three approximations during time interval $[t_k, t_{k+1}]$ is presented in Fig. 3.2. The final points obtained from the three methods are presented through $\mathbf{x}_{k+1}^{FE}, \mathbf{x}_{k+1}^{BE}$, and $\mathbf{x}_{k+1}^{TR}$. The forward Euler method is based on the point $\mathbf{x}_k$ with its tangential vector $\dot{\mathbf{x}}_k = \mathbf{f}(\mathbf{x}_k, t_k, \mathbf{p})$, and the final point for time $t_{k+1}$ is $\mathbf{x}_{k+1}^{FE}$ connected through a line with a slope $\mathbf{f}(\mathbf{x}_k, t_k, \mathbf{p})$. The backward Euler method is based on the point $\mathbf{x}_k$ with its tangential vector $\dot{\mathbf{x}}_{k+1} = \mathbf{f}(\mathbf{x}_{k+1}, t_{k+1}, \mathbf{p})$, and the final point for time $t_{k+1}$ is $\mathbf{x}_{k+1}^{BE}$ connected through a line with a slope $\mathbf{f}(\mathbf{x}_{k+1}, t_{k+1}, \mathbf{p})$. The trapezoidal rule approximation is based on the point $\mathbf{x}_k$ with the average of the tangential vector $\dot{\mathbf{x}}_{k+1} = \mathbf{f}(\mathbf{x}_{k+1}, t_{k+1}, \mathbf{p})$ and $\dot{\mathbf{x}}_k = \mathbf{f}(\mathbf{x}_k, t_k, \mathbf{p})$, and the final point for time $t_{k+1}$ is $\mathbf{x}_{k+1}^{TR}$ connected through a line with a average slope $\frac{1}{2} (\mathbf{f}(\mathbf{x}_k, t_k, \mathbf{p}) + \mathbf{f}(\mathbf{x}_{k+1}, t_{k+1}, \mathbf{p}))$. So, the local error of the forward and backward Euler methods is $O(h^2)$, where $h = t_{k+1} - t_k$. However, the local error of the trapezoidal rule approximation is $O(h^3)$, which can be observed in Fig. 3.2. As $h \to 0$, the global errors for three methods are $O(h)$ (FE and BE) and $O(h^2)$ (TR).

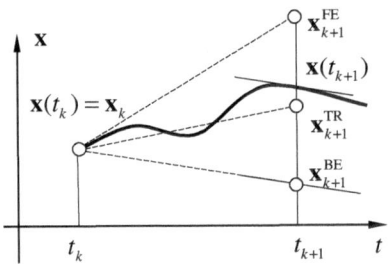

**Fig. 3.2** A geometrical illustration of forward Euler method (FE), backward Euler method (BE), and trapezoidal rule approximation (TR) during the time subinterval $t \in [t_k, t_{k+1}]$

### 3.2.4 Midpoint Method

If $h_k \to 0$, then $\mathbf{f}(\mathbf{x}_k^c, t_k^c, \mathbf{p}) \approx \mathbf{f}(\frac{1}{2}(\mathbf{x}_k + \mathbf{x}_{k+1}), t_{k+1/2}, \mathbf{p})$ and Eq. (3.21) becomes

$$\mathbf{x}_{k+1} = \mathbf{x}_k + h_{k+1}\mathbf{f}(\frac{1}{2}(\mathbf{x}_k + \mathbf{x}_{k+1}), t_{k+1/2}, \mathbf{p}) \quad \text{for } k = 0, 1, 2, \ldots, M - 1. \quad (3.65)$$

**Definition 3.12** The discretization of Eq. (3.1) during the time interval $[t_k, t_{k+1}]$ as

$$t_{k+1} = t_k + h_k, \quad \mathbf{X}_{1/2} = \frac{1}{2}(\mathbf{x}_k + \mathbf{x}_{k+1})$$
$$\mathbf{x}_{k+1} = \mathbf{x}_k + h_{k+1}\mathbf{f}(\mathbf{X}_{1/2}, t_{k+1/2}, \mathbf{p}) \quad \text{for } k = 0, 1, 2, \ldots, M - 1 \quad (3.66)$$

is called the midpoint discrete approximation.

**Theorem 3.7** *Suppose that the solution* $\mathbf{x} = \mathbf{x}(t)$ *is the unique solution of* Eq. (3.1). *If* $\mathbf{x}(t) \in C^3[t_0, t_M]$ *and the points* $(t_k, \mathbf{x}_k)$ *for* $k = 0, 1, 2, \ldots, M - 1$ *is the sequence of the midpoint approximation of* Eq. (3.1), *then*

$$\|\boldsymbol{\varepsilon}_{k+1}\| = \|\mathbf{x}(t_{k+1}) - \mathbf{x}_k - h_{k+1}\mathbf{f}(\frac{1}{2}(\mathbf{x}_k + \mathbf{x}_{k+1}), t_{k+1/2}, \mathbf{p})\| = O(h_{k+1}^2). \quad (3.67)$$

*and the global error for* $h_k = h$ $(k = 1, 2, 3, \ldots)$ *and* $t_k = kh$ *is*

$$\|\mathbf{e}_k\| \leq \frac{h^2 L}{24K}(e^{t_k K} - 1) \quad (3.68)$$

*with*

$$L = \max_{k \in \{1,2,\ldots,M\}} \|D^2\mathbf{f}(\mathbf{x}(t_{k+1}^c), t_{k+1}^c, \mathbf{p}) + D^2\mathbf{f}(\mathbf{x}(t_k^c), t_k^c, \mathbf{p})\| \quad \text{and}$$
$$K = \max_{k \in \{1,2,\ldots,M\}} \|\mathbf{A}_k\|. \quad (3.69)$$

*For a fixed $t_k = kh$, as $h$ goes to zero,*

$$\|\mathbf{e}_k\| = \|\mathbf{x}(t_k) - \mathbf{x}_k\| = O(h^2). \tag{3.70}$$

*Proof* Consider the Taylor series of $\mathbf{x}(t_k)$ and $\mathbf{x}(t_{k+1})$ at point $(t_{k+1/2}, \mathbf{x}(t_{k+1/2}))$ as

$$
\begin{aligned}
\mathbf{x}(t_k) &= \mathbf{x}(t_{k+1/2} - \tfrac{1}{2}h_{k+1}) = \mathbf{x}(t_{k+1/2}) - \tfrac{1}{2}h_{k+1}\dot{\mathbf{x}}(t_{k+1/2}) \\
&\quad + \tfrac{1}{8}h_{k+1}^2\ddot{\mathbf{x}}(t_{k+1/2}) - \tfrac{1}{24}h_{k+1}^3\dddot{\mathbf{x}}(t_{k+1/2}^c) \\
&= \mathbf{x}(t_{k+1/2}) - \tfrac{1}{2}h_{k+1}\mathbf{f}(\mathbf{x}(t_{k+1/2}), t_{k+1/2}, \mathbf{p}) \\
&\quad + \tfrac{1}{8}h_{k+1}^2 D\mathbf{f}(\mathbf{x}(t_{k+1/2}), t_{k+1/2}, \mathbf{p}) - \tfrac{1}{24}h_{k+1}^3 D^2\mathbf{f}(\mathbf{x}(t_k^c), t_k^c, \mathbf{p}), \\
\mathbf{x}(t_{k+1}) &= \mathbf{x}(t_{k+1} + \tfrac{1}{2}h_{k+1}) = \mathbf{x}(t_{k+1/2}) + \tfrac{1}{2}h_{k+1}\dot{\mathbf{x}}(t_{k+1}) \\
&\quad + \tfrac{1}{8}h_{k+1}^2\ddot{\mathbf{x}}(t_{k+1/2}) + \tfrac{1}{24}h_{k+1}^3\dddot{\mathbf{x}}(t_{k+1/2}^c) \\
&= \mathbf{x}(t_{k+1/2}) + \tfrac{1}{2}h_{k+1}\mathbf{f}(\mathbf{x}(t_{k+1/2}), t_{k+1/2}, \mathbf{p}) \\
&\quad + \tfrac{1}{8}h_{k+1}^2 D\mathbf{f}(\mathbf{x}(t_{k+1/2}), t_{k+1/2}, \mathbf{p}) + \tfrac{1}{24}h_{k+1}^3 D^2\mathbf{f}(\mathbf{x}(t_{k+1}^c), t_{k+1}^c, \mathbf{p}).
\end{aligned}
$$

Thus, from the foregoing equations, we have

$$
\begin{aligned}
\mathbf{x}(t_{k+1}) &= \mathbf{x}(t_k) + h_{k+1}\mathbf{f}(\mathbf{x}(t_{k+1/2}), t_{k+1/2}, \mathbf{p}) \\
&\quad + \tfrac{1}{24}h_{k+1}^3[D^2\mathbf{f}(\mathbf{x}(t_{k+1}^c), t_{k+1}^c, \mathbf{p}) + D^2\mathbf{f}(\mathbf{x}(t_k^c), t_k^c, \mathbf{p})], \\
\mathbf{x}(t_{k+1/2}) &= \tfrac{1}{2}[\mathbf{x}(t_k) + \mathbf{x}(t_{k+1})] - \tfrac{1}{4}h_{k+1}^2 D\mathbf{f}(\mathbf{x}(t_{k+1/2}), t_{k+1/2}, \mathbf{p}) + O(h_{k+1}^3).
\end{aligned}
$$

The midpoint approximation gives

$$\mathbf{x}_{k+1} = \mathbf{x}_k + h_{k+1}\mathbf{f}(\mathbf{X}_{1/2}, t_{k+1/2}, \mathbf{p}), \quad \mathbf{X}_{1/2} = \tfrac{1}{2}(\mathbf{x}_k + \mathbf{x}_{k+1}).$$

(i) If $\mathbf{x}(t_k) = \mathbf{x}_k$, then the local error is

$$
\begin{aligned}
\boldsymbol{\varepsilon}_{k+1} &= \mathbf{x}(t_{k+1}) - \mathbf{x}_k - h_{k+1}\mathbf{f}(\mathbf{X}_{1/2}, t_{k+1/2}, \mathbf{p}) \\
&= h_{k+1}[\mathbf{f}(\mathbf{x}(t_{k+1/2}), t_{k+1/2}, \mathbf{p}) - \mathbf{f}(\mathbf{x}_{k+1/2}, t_{k+1/2}, \mathbf{p})] \\
&\quad + \tfrac{1}{24}h_{k+1}^3[D^2\mathbf{f}(\mathbf{x}(t_{k+1}^c), t_{k+1}^c, \mathbf{p}) + D^2\mathbf{f}(\mathbf{x}(t_k^c), t_k^c, \mathbf{p})]
\end{aligned}
$$

with

$$\mathbf{f}(\mathbf{x}(t_{k+1/2}), t_{k+1/2}, \mathbf{p}) - \mathbf{f}(\mathbf{x}_{k+1/2}, t_{k+1/2}, \mathbf{p})$$
$$= D_{\mathbf{x}}\mathbf{f}(\mathbf{x}_{k+1/2}^c, t_{k+1/2}, \mathbf{p})(\mathbf{x}(t_{k+1/2}) - \mathbf{x}_{k+1/2})$$
$$= \frac{1}{2}\mathbf{A}_{k+1/2}[\mathbf{x}(t_{k+1}) - \mathbf{x}_{k+1} + \mathbf{x}(t_k) - \mathbf{x}_k]$$

where $\mathbf{A}_{k+1/2} = D_{\mathbf{x}}\mathbf{f}(\xi_{k+1/2}, t_{k+1/2}, \mathbf{p})$ and $\|\xi_{k+1/2}\| \in (\|\mathbf{x}(t_{k+1/2})\|, \|\mathbf{x}_{k+1/2}\|)$.
So

$$\varepsilon_{k+1} = \frac{1}{24}h_{k+1}^3\left(\mathbf{I} - \frac{1}{2}h_{k+1}\mathbf{A}_{k+1/2}\right)^{-1}$$
$$\times \left[D^2\mathbf{f}(\mathbf{x}(t_{k+1}^c), t_{k+1}^c, \mathbf{p}) + D^2\mathbf{f}(\mathbf{x}(t_k^c), t_k^c, \mathbf{p})\right] + O(h_{k+1}^4).$$

Thus, the local error is

$$\|\varepsilon_{k+1}\| = \|\mathbf{x}(t_{k+1}) - \mathbf{x}_k - h_{k+1}\mathbf{f}(\mathbf{x}_{k+1}, t_{k+1}, \mathbf{p})\| = O(h_{k+1}^3).$$

(ii)  If $\mathbf{x}(t_{k+1}) \neq \mathbf{x}_{k+1}$, then $\mathbf{f}(\mathbf{x}(t_{k+1}), t_{k+1}, \mathbf{p}) \neq \mathbf{f}(\mathbf{x}_{k+1}, t_{k+1}, \mathbf{p})$

$$\mathbf{x}(t_{k+1}) - \mathbf{x}_{k+1} = \mathbf{x}(t_k) - \mathbf{x}_k + h_{k+1}[\mathbf{f}(\mathbf{x}(t_{k+1/2}), t_{k+1/2}, \mathbf{p})$$
$$- \mathbf{f}(\mathbf{x}_{k+1/2}, t_{k+1/2}, \mathbf{p})]$$
$$+ \frac{1}{24}h_{k+1}^3[D^2\mathbf{f}(\mathbf{x}(t_{k+1}^c), t_{k+1}^c, \mathbf{p}) + D^2\mathbf{f}(\mathbf{x}(t_k^c), t_k^c, \mathbf{p})]$$

and the Lipschitz condition yields

$$\mathbf{x}(t_{k+1}) - \mathbf{x}_{k+1} = \mathbf{x}(t_k) - \mathbf{x}_k + \frac{1}{2}h_{k+1}\mathbf{A}_{k+1/2}[\mathbf{x}(t_{k+1}) - \mathbf{x}_{k+1} + \mathbf{x}(t_k) - \mathbf{x}_k]$$
$$+ \delta_{k+1}$$

where $\delta_{k+1} = \frac{1}{24}h_{k+1}^3[D^2\mathbf{f}(\mathbf{x}(t_{k+1}^c), t_{k+1}^c, \mathbf{p}) + D^2\mathbf{f}(\mathbf{x}(t_k^c), t_k^c, \mathbf{p})]$. So

$$\mathbf{e}_{k+1} = \mathbf{e}_k + \frac{1}{2}h_{k+1}\mathbf{A}_{k+1/2}(\mathbf{e}_{k+1} + \mathbf{e}_k) + \delta_{k+1}.$$

gives

$$\|\mathbf{e}_{k+1}\| \leq \|\mathbf{e}_k\| + \frac{1}{2}h_{k+1}\|\mathbf{A}_{k+1/2}\|(\|\mathbf{e}_{k+1}\| + \|\mathbf{e}_k\|) + \|\delta_{k+1}\|.$$

For $h_k = h$ $(k = 1, 2, 3, \ldots)$ and $t_k = kh$, letting

$$L = \max_{k \in \{1,2,\ldots,M\}} \|D^2\mathbf{f}(\mathbf{x}(t_{k+1}^c), t_{k+1}^c, \mathbf{p}) + D^2\mathbf{f}(\mathbf{x}(t_k^c), t_k^c, \mathbf{p})\| \quad \text{and}$$

$$K = \max_{k \in \{1,2,\ldots,M\}} \|\mathbf{A}_k\|,$$

consider a simple discrete equation as

$$z_{k+1} = z_k + \frac{1}{2}(hK)z_{k+1} + \frac{1}{2}(hK)z_k + \frac{1}{24}h^3 L \quad \text{with} \ z_0 = 0$$

and

$$z_k = \frac{1}{24}h^3 L \sum_{j=0}^{k-1} (1 - \frac{1}{2}hK)^{-j}(1 + \frac{1}{2}hK)^j$$

$$= \frac{h^2 L}{24K}[(1 - \frac{1}{2}hK)^{-k}(1 + \frac{1}{2}hK)^k - 1] \quad \text{for} \ k = 1, 2, \ldots, M.$$

For $k > 0$, we have $1 + hK > 0$. If $z_k \geq \|\mathbf{e}_k\|$, then $z_{k+1} \geq \|\mathbf{e}_{k+1}\|$. Since $t_k = kh$, for $k \to \infty$, we have

$$\lim_{k \to \infty} [1 - \frac{1}{k}(\frac{1}{2}t_k K)]^{-k}[1 + \frac{1}{k}(\frac{1}{2}t_k K)]^k = e^{\frac{1}{2}t_k K}e^{\frac{1}{2}t_k K} = e^{t_k K},$$

so

$$\|\mathbf{e}_k\| \leq \frac{h^2 L}{24K}(e^{t_k K} - 1).$$

For a fixed $t_k = kh$, as $h$ goes to zero, we have

$$\|\mathbf{e}_k\| = \|\mathbf{x}(t_k) - \mathbf{x}_k\| = O(h^2) \quad \text{for} \ k = 1, 2, \ldots, M.$$

This theorem is proved.                                                                       □

Consider the global error

$$\mathbf{e}_{k+1} = (\mathbf{I} - \frac{1}{2}h_{k+1}\mathbf{A}_{k+1/2})^{-1}(\mathbf{I} + \frac{1}{2}h_{k+1}\mathbf{A}_{k+1/2})\mathbf{e}_k. \tag{3.71}$$

If $\mathbf{e}_{k+1} = \lambda\mathbf{e}_k$, then the foregoing equation becomes

$$[(\mathbf{I} - \frac{1}{2}h_{k+1}\mathbf{A}_{k+1/2})^{-1}(\mathbf{I} + \frac{1}{2}h_{k+1}\mathbf{A}_{k+1/2}) - \lambda\mathbf{I}]\mathbf{e}_k = \mathbf{0}. \tag{3.72}$$

The corresponding eigenvalues are generated by

$$\left| (\mathbf{I} - \frac{1}{2}h_{k+1}\mathbf{A}_{k+1/2})^{-1}(\mathbf{I} + \frac{1}{2}h_{k+1}\mathbf{A}_{k+1/2}) - \lambda\mathbf{I} \right| = 0. \tag{3.73}$$

If all eigenvalues $|\lambda_j| < 1$ ($j = 1, 2, \ldots, n$), then

$$\|\mathbf{e}_{k+1}\| < \|\mathbf{e}_k\|. \tag{3.74}$$

In other words, the global error will not be enlarged. The midpoint method gives a stable approximation. For one-dimensional systems, we have $\mathbf{A}_{k+1/2} = L_{k+1/2}$ and $\mathbf{I} = 1$. Equation (3.73) gives the stability interval for the midpoint method as

$$\left| \frac{1 + \frac{1}{2}h_{k+1}L_{k+1/2}}{1 - \frac{1}{2}h_{k+1}L_{k+1/2}} \right| < 1. \tag{3.75}$$

Since $h_{k+1} > 0$, if $L_{k+1/2} < 0$, then the foregoing equation always exists.

The midpoint method is a simple, stable discrete method that is often used in practice. This method is also an implicit discrete method because the exact midpoint $\mathbf{x}(t_{k+1/2})$ is approximated by with $\mathbf{x}(t_{k+1/2}) \approx \frac{1}{2}[\mathbf{x}(t_k) + \mathbf{x}(t_{k+1})]$ based on two exact points $\mathbf{x}(t_{k+1})$ and $\mathbf{x}(t_k)$. In the discrete map, the midpoint is further approximated by $\mathbf{x}(t_{k+1/2}) \approx \mathbf{x}_{k+1/2} = \frac{1}{2}(\mathbf{x}_{k+1} + \mathbf{x}_k)$. The corresponding vector field at the midpoint $\mathbf{x}(t_{k+1/2})$ is approximated by

$$\begin{aligned} \mathbf{f}(\mathbf{x}(t_{k+1/2}), t_{k+1/2}, \mathbf{p}) &\approx \mathbf{f}(\mathbf{x}_{k+1/2}, t_{k+1/2}, \mathbf{p}) \\ &= \mathbf{f}(\frac{1}{2}(\mathbf{x}_k + \mathbf{x}_{k+1}), t_k + \frac{1}{2}h_{k+1}, \mathbf{p}) \end{aligned} \tag{3.76}$$

in a discrete interval. For this method, the iterative method or Newton–Raphson method will be also used to determine values for each step. The midpoint method, similar to the trapezoidal rule method, possesses a large stability range. For any positive step size, the amplification factor is less than one, and the errors will not be magnified.

## 3.3 Introduction to Runge–Kutta Methods

From the forward Euler method, backward Euler method, and the trapezoidal rule approximation, the different slopes are used for approximation. Runge–Kutta method is based on the initial estimate from the forward or back Euler method. From the initial estimate, the slope will be adjusted at the location $t_k$, which will be used for the next tentative step. At this point, compute the new slope that is used to further adjust the slope at the original location $(\mathbf{x}_k, t_k)$. Repeat the same procedure to adjust the slope to get desired. Finally, combine all the estimates to make the actual step to $(\mathbf{x}_{k+1}, t_k + h_{k+1})$. To determine the adjustments, the Runge–Kutta method uses a polynomial model for $\mathbf{x}(t)$ with the criterion that the Taylor series of the actual estimate $\mathbf{x}_{k+1}$ must agree, as much as possible, with the Taylor series expansion of $\mathbf{x}(t_k + h)$ based on the point $(\mathbf{x}_k, t_k)$. This Taylor series method will be discussed first.

### 3.3.1 Taylor Series Method

The Taylor series method is extensively used to compare the accuracy of various order numerical methods for solving the differential equations. The Taylor series method is based on the Taylor series expansion with specified degree of accuracy. The Taylor series theorem is presented in a form suitable for solve differential equation.

**Theorem 3.8** *Suppose the solution* $\mathbf{x} = \mathbf{x}(t)$ *is the unique solution of* Eq. (3.1). *If* $\mathbf{x}(t) \in C^{N+1}[t_0, t_M]$ *and at the points* $(t_k, \mathbf{x}_k)$ *for* $k = 0, 1, 2, \ldots, M - 1$, *an Nth-order Taylor series expansion of* Eq. (3.1) *is*

$$\mathbf{x}(t_k + h_{k+1}) = \mathbf{x}(t_k) + \sum_{j=1}^{N} \frac{1}{j!} D^{(j)} \mathbf{x}(t_k) h_{k+1}^j + \frac{1}{(N+1)!} D^{(N+1)} \mathbf{x}(t_k^c) h_{k+1}^{(N)}, \quad (3.77)$$

*where*

$$\begin{aligned} D &= \frac{\partial()}{\partial \mathbf{x}} \dot{\mathbf{x}} + \frac{\partial()}{\partial t} = \frac{\partial()}{\partial \mathbf{x}} \mathbf{f}(\mathbf{x}(t), t, \mathbf{p}) + \frac{\partial()}{\partial t}, \\ \mathbf{x}^{(j)} &= D^{(j-1)} \circ D\mathbf{x} = D^{(j-1)} \mathbf{f}(\mathbf{x}(t), t, \mathbf{p}) \end{aligned} \quad (3.78)$$

*with*

$$\begin{aligned} \mathbf{x}^{(1)} &= \dot{\mathbf{x}} = \mathbf{f}, \\ \mathbf{x}^{(2)} &= \ddot{\mathbf{x}} = \mathbf{f}_t + \mathbf{f}_\mathbf{x} \dot{\mathbf{x}} = \mathbf{f}_t + \mathbf{f}_\mathbf{x} \mathbf{f}, \\ \mathbf{x}^{(3)} &= \mathbf{f}_{tt} + 2\mathbf{f}_{t\mathbf{x}} \dot{\mathbf{x}} + \mathbf{f}_\mathbf{x} \ddot{\mathbf{x}} + \mathbf{f}_{\mathbf{x}\mathbf{x}} \dot{\mathbf{x}}^2 \\ &= \mathbf{f}_{tt} + 2\mathbf{f}_{t\mathbf{x}} \mathbf{f} + \mathbf{f}_\mathbf{x}(\mathbf{f}_t + \mathbf{f}_\mathbf{x} \mathbf{f}) + \mathbf{f}_{\mathbf{x}\mathbf{x}} \mathbf{f}^2, \\ \mathbf{x}^{(4)} &= (\mathbf{f}_{ttt} + 3\mathbf{f}_{tt\mathbf{x}} \dot{\mathbf{x}} + 3\mathbf{f}_{t\mathbf{x}\mathbf{x}} \mathbf{f}^2 + \mathbf{f}_{\mathbf{x}\mathbf{x}\mathbf{x}}) + \mathbf{f}_\mathbf{x}(\mathbf{f}_{tt} + 2\mathbf{f}_{t\mathbf{x}} \mathbf{f} + \mathbf{f}_{\mathbf{x}\mathbf{x}} \mathbf{f}^2) \\ &\quad + 3(\mathbf{f}_t + \mathbf{f}_\mathbf{x} \mathbf{f})(\mathbf{f}_{t\mathbf{x}} + \mathbf{f}_{\mathbf{x}\mathbf{x}} \mathbf{f}) + \mathbf{f}_\mathbf{x}^2(\mathbf{f}_t + \mathbf{f}_\mathbf{x} \mathbf{f}), \end{aligned} \quad (3.79)$$

$$\vdots$$

**Definition 3.13** The discretization of Eq. (3.1) during the time interval $[t_k, t_{k+1}]$ as

$$\begin{aligned} t_{k+1} &= t_k + h_k, \\ \mathbf{x}_{k+1} &= \mathbf{x}_k + h_{k+1} \mathbf{f}(\mathbf{x}_k, t_k, \mathbf{p}) + \frac{1}{2!} h_{k+1}^2 D\mathbf{f}(\mathbf{x}_k, t_k, \mathbf{p}) + \cdots \\ &\quad + \frac{1}{N!} h_{k+1}^N D^{(N-1)} \mathbf{f}(\mathbf{x}_k, t_k, \mathbf{p}) \quad \text{for} \quad k = 0, 1, 2, \cdots, M - 1 \end{aligned} \quad (3.80)$$

is called the Nth-order Taylor series approximation (or Taylor series method).

**Theorem 3.9** *Suppose the solution* $\mathbf{x} = \mathbf{x}(t)$ *is the unique solution of* Eq. (3.1). *If* $\mathbf{x}(t) \in C^{N+1}[t_0, t_M]$ *and the points* $(t_k, \mathbf{x}_k)$ *for* $k = 0, 1, 2, \ldots, M - 1$ *is the sequence of the Nth-order Taylor series approximation of* Eq. (3.1), *then*

$$\|\boldsymbol{\varepsilon}_{k+1}\| = \|\mathbf{x}(t_{k+1}) - \mathbf{x}_k - h_{k+1}\boldsymbol{\Phi}(\mathbf{x}_k, t_k, \mathbf{p})\| = O(h_{k+1}^{N+1}), \qquad (3.81)$$

*where*

$$\begin{aligned}
\boldsymbol{\Phi}(\mathbf{x}_k, t_k, \mathbf{p}) &= \mathbf{f}(\mathbf{x}_k, t_k, \mathbf{p}) + \frac{1}{2!}h_{k+1}D\mathbf{f}(\mathbf{x}_k, t_k, \mathbf{p}) \\
&\quad + \cdots + \frac{1}{N!}h_{k+1}^{N-1}D^{(N-1)}\mathbf{f}(\mathbf{x}_k, t_k, \mathbf{p})
\end{aligned} \qquad (3.82)$$

*and the global error for* $h_k = h$ $(k = 1, 2, 3, \ldots)$ *and* $t_k = kh$ *is*

$$\|\mathbf{e}_k\| \le \frac{h^N L}{(N+1)!K}(e^{t_k K} - 1) \qquad (3.83)$$

*with*

$$\begin{aligned}
L &= \max_{k \in \{1,2,\ldots,M\}} \|D^{N+1}\mathbf{f}(\mathbf{x}(t_k^c), t_k^c, \mathbf{p})\|, \\
K &= \max_{k \in \{1,2,\ldots,M\}} \|\sum_{j=1}^{N} h^{j-1}\mathbf{A}_k^{(j)}\|, \quad and \\
\mathbf{A}_k^{(j)} &= D_{\mathbf{x}} \circ D^{(j-1)}\mathbf{f}(\boldsymbol{\xi}_k^{(j)}, t_k, \mathbf{p}) \quad and \quad \|\boldsymbol{\xi}_k^{(j)}\| \in (\|\mathbf{x}(t_k)\|, \|\mathbf{x}_k\|).
\end{aligned} \qquad (3.84)$$

*For a fixed* $t_k = kh$, *as* $h$ *goes to zero,*

$$\|\mathbf{e}_k\| = \|\mathbf{x}(t_k) - \mathbf{x}_k\| = O(h^N). \qquad (3.85)$$

*Proof* Consider the Taylor series of $\mathbf{x}(t_{k+1})$ at point $(t_k, \mathbf{x}(t_k))$ as

$$\begin{aligned}
\mathbf{x}(t_{k+1}) &= \mathbf{x}(t_k + h_{k+1}) \\
&= \mathbf{x}(t_k) + \sum_{j=1}^{N} \frac{1}{j!}D^{(j)}\mathbf{x}(t_k)h_{k+1}^j + \frac{1}{(N+1)!}D^{(N+1)}\mathbf{x}(t_k^c)h_{k+1}^{N+1} \\
&= \mathbf{x}(t_k) + h_{k+1}\mathbf{f}(\mathbf{x}(t_k), t_k, \mathbf{p}) + \frac{1}{2!}h_{k+1}^2 D\mathbf{f}(\mathbf{x}(t_k), t_k, \mathbf{p}) \\
&\quad + \cdots + \frac{1}{N!}h_{k+1}^N D^{(N-1)}\mathbf{f}(\mathbf{x}(t_k), t_k, \mathbf{p}) + \frac{1}{(N+1)!}h_{k+1}^{N+1}D^{(N)}\mathbf{f}(\mathbf{x}(t_k^c), t_k^c, \mathbf{p})
\end{aligned}$$

and the $N$th-order Taylor series approximation gives

$$\mathbf{x}_{k+1} = \mathbf{x}_k + h_{k+1}\mathbf{f}(\mathbf{x}_k, t_k, \mathbf{p}) + \frac{1}{2!}h_{k+1}^2 D\mathbf{f}(\mathbf{x}_k, t_k, \mathbf{p})$$
$$+ \cdots + \frac{1}{N!}h_{k+1}^N D^{(N-1)}\mathbf{f}(\mathbf{x}_k, t_k, \mathbf{p}).$$

If $\mathbf{x}(t_k) = \mathbf{x}_k$, then $\mathbf{f}(\mathbf{x}(t_k), t_k, \mathbf{p}) = \mathbf{f}(\mathbf{x}_k, t_k, \mathbf{p})$. Therefore, the local error is

$$\boldsymbol{\varepsilon}_{k+1} = \mathbf{x}(t_{k+1}) - \mathbf{x}_k - h_{k+1}\boldsymbol{\Phi}(\mathbf{x}_k, t_k, \mathbf{p}) = \frac{1}{N!}h_{k+1}^{N+1}D^{(N)}\mathbf{f}(\mathbf{x}(t_k^c), t_k^c, \mathbf{p}).$$

Thus, the local error at the $(k + 1)$th step is

$$\|\boldsymbol{\varepsilon}_{k+1}\| = \|\mathbf{x}(t_{k+1}) - \mathbf{x}_k - h_{k+1}\boldsymbol{\Phi}(\mathbf{x}_k, t_k, \mathbf{p})\| = O(h_{k+1}^{N+1}).$$

If $\mathbf{x}(t_k) \neq \mathbf{x}_k$, then $\mathbf{f}(\mathbf{x}(t_k), t_k, \mathbf{p}) \neq \mathbf{f}(\mathbf{x}_k, t_k, \mathbf{p})$, thus

$$\mathbf{x}(t_{k+1}) - \mathbf{x}_{k+1} = \mathbf{x}(t_k) - \mathbf{x}_k + h_{k+1}(\mathbf{f}(\mathbf{x}(t_k), t_k, \mathbf{p}) - \mathbf{f}(\mathbf{x}_k, t_k, \mathbf{p}))$$
$$+ \frac{1}{2!}h_{k+1}^2(D\mathbf{f}(\mathbf{x}(t_k), t_k, \mathbf{p}) - D\mathbf{f}(\mathbf{x}_k, t_k, \mathbf{p}))$$
$$+ \cdots + \frac{1}{N!}h_{k+1}^N(D^{(N-1)}\mathbf{f}(\mathbf{x}(t_k), t_k, \mathbf{p}) - D^{(N-1)}\mathbf{f}(\mathbf{x}_k, t_k, \mathbf{p}))$$
$$+ \frac{1}{(N+1)!}h_{k+1}^{N+1}D^{(N)}\mathbf{f}(\mathbf{x}(t_k^c), t_k^c, \mathbf{p})$$

and the Lipschitz conditions give

$$\mathbf{x}(t_{k+1}) - \mathbf{x}_{k+1} = (\mathbf{I} + \sum_{j=1}^{N} h_{k+1}^j \mathbf{A}_k^{(j)})(\mathbf{x}(t_k) - \mathbf{x}_k) + \boldsymbol{\varepsilon}_{k+1}$$

where

$$\mathbf{A}_k^{(j)} = D_{\mathbf{x}} \circ D^{(j-1)}\mathbf{f}(\xi_k^{(j)}, t_k, \mathbf{p}) \quad \text{and} \quad \|\xi_k^{(j)}\| \in (\|\mathbf{x}(t_k)\|, \|\mathbf{x}_k\|).$$

So

$$\mathbf{e}_{k+1} = (\mathbf{I} + \sum_{j=1}^{N} h_{k+1}^j \mathbf{A}_k^{(j)})\mathbf{e}_k + \boldsymbol{\varepsilon}_{k+1}$$

gives

$$\|\mathbf{e}_{k+1}\| \leq \|(\mathbf{I} + h_{k+1}\sum_{j=1}^{N} h_{k+1}^{j-1}\mathbf{A}_k^{(j)})\| \times \|\mathbf{e}_k\| + \|\boldsymbol{\varepsilon}_{k+1}\|.$$

For $h_k = h$ $(k = 1, 2, 3, \ldots)$ and $t_k = kh$, letting

$$L = \max_{k \in \{1,2,\ldots,M\}} \left\| D^{N+1} \mathbf{f}(\mathbf{x}(t_k^c), t_k^c, \mathbf{p}) \right\| \quad \text{and}$$

$$K = \max_{k \in \{1,2,\ldots,M\}} \left\| \sum_{j=1}^{N} h^{j-1} \mathbf{A}_k^{(j)} \right\|,$$

consider a simple discrete equation as

$$z_{k+1} = (1 + hK)z_k + \frac{1}{(N+1)!} h^{N+1} L \quad \text{with } z_0 = 0$$

and

$$z_k = \frac{1}{(N+1)!} h^{N+1} Lt \Big( \sum_{l=0}^{k-1} (1 + hK)^l \Big)$$

$$= \frac{h^{N+1} L}{(N+1)! K} [(1 + hK)^k - 1] \quad \text{for } k = 1, 2, \ldots, M.$$

For $k > 0$, we have $1 + hK > 0$. If $z_k \geq \|\mathbf{e}_k\|$, then $z_{k+1} \geq \|\mathbf{e}_{k+1}\|$. Since $t_k = kh$, for $k \to \infty$ and $h \to 0$, we have

$$\lim_{k \to \infty} \Big( 1 + \frac{t_k}{k} K \Big)^k = e^{t_k K},$$

so

$$\|\mathbf{e}_k\| \leq \frac{h^N L}{(N+1)! K} (e^{t_k K} - 1).$$

For a fixed $t_k = kh$, as $h$ goes to zero, we have

$$\|\mathbf{e}_k\| = \|\mathbf{x}(t_k) - \mathbf{x}_k\| = O(h^N) \quad \text{for } k = 1, 2, \ldots, M.$$

This theorem is proved.                                                 □

Consider the global error at

$$\mathbf{e}_{k+1} = (\mathbf{I} + h_{k+1} \mathbf{B}_k) \mathbf{e}_k, \tag{3.86}$$

where $\mathbf{B}_k = \sum_{j=1}^{N} h_{k+1}^j \mathbf{A}_k^{(j)}$. If $\mathbf{e}_{k+1} = \lambda \mathbf{e}_k$, then the foregoing equation becomes

$$[(\mathbf{I} + h_{k+1} \mathbf{B}_k) - \lambda \mathbf{I}] \mathbf{e}_k = \mathbf{0}. \tag{3.87}$$

The corresponding eigenvalues are generated by

$$|\mathbf{I} + h_{k+1}\mathbf{B}_k - \lambda\mathbf{I}| = \mathbf{0}. \tag{3.88}$$

If all eigenvalues $|\lambda_j| < 1$ $(j = 1, 2, \ldots, n)$, then

$$\|\mathbf{e}_{k+1}\| < \|\mathbf{e}_k\|. \tag{3.89}$$

So the global error will not be enlarged. Thus, under such a condition, the Taylor series method gives a stable approximation. For one-dimensional systems, we have $\mathbf{B}_k = L_k$ and $\mathbf{I} = 1$. Equation (3.89) gives the stability interval for the Taylor series method as

$$|1 + h_{k+1}L_k| < 1. \tag{3.90}$$

Since $h_{k+1} > 0$, if $L_k < 0$, then the foregoing equation always exists.

The Taylor series method can be implemented as follows. From the Taylor series, we can construct the following relations

$$
\begin{aligned}
\mathbf{x}_{k+1} &= \mathbf{x}_k + C_1^0 h_{k+1}\mathbf{x}_k^{(1)} + C_2^0 \frac{1}{2!} h_{k+1}^2 \mathbf{x}_k^{(2)} + \cdots \\
&\quad + C_N^0 \frac{1}{N!} h_{k+1}^N \mathbf{x}_k^{(N)} + C_{N+1}^0 \frac{1}{(N+1)!} h_{k+1}^{N+1} \mathbf{x}_k^{(N+1)} + \cdots, \\
h_{k+1}\mathbf{x}_{k+1}^{(1)} &= C_1^1 h_{k+1}^2 \mathbf{x}_k^{(1)} + C_2^1 \frac{1}{2!} h_{k+1}^3 \mathbf{x}_k^{(2)} + \cdots \\
&\quad + C_N^1 \frac{1}{N!} h_{k+1}^N \mathbf{x}_k^{(N)} + C_{N+1}^1 \frac{1}{(N+1)!} h_{k+1}^{N+1} \mathbf{x}_k^{(N)} + \cdots, \\
\frac{1}{2!} h_{k+1}^2 \mathbf{x}_{k+1}^{(2)} &= C_2^2 \frac{1}{2!} h_{k+1}^3 \mathbf{x}_k^{(2)} + \cdots \\
&\quad + C_N^2 \frac{1}{N!} h_{k+1}^N \mathbf{x}_k^{(N)} + C_{N+1}^2 \frac{1}{(N+1)!} h_{k+1}^{N+1} \mathbf{x}_k^{(N)} + \cdots, \\
&\quad \vdots \\
\frac{1}{N!} h_{k+1}^N \mathbf{x}_{k+1}^{(N)} &= C_N^N \frac{1}{N!} h_{k+1}^N \mathbf{x}_k^{(N)} + C_{N+1}^N \frac{1}{(N+1)!} h_{k+1}^{N+1} \mathbf{x}_k^{(N)} + \cdots.
\end{aligned} \tag{3.91}
$$

From the foregoing equation, we have

$$\mathbf{y}_{k+1} = \mathbf{C}\mathbf{y}_k + \mathbf{B}\left(\frac{1}{(N+1)!} h_{k+1}^{N+1} \mathbf{x}_k^{(N+1)}\right), \tag{3.92}$$

where

$$\mathbf{y}_{k+1} = \left(\mathbf{x}_{k+1}, h_{k+1}\mathbf{x}_{k+1}^{(1)}, \frac{1}{2!} h_{k+1}^2 \mathbf{x}_{k+1}^{(2)}, \ldots, \frac{1}{N!} h_{k+1}^N \mathbf{x}_{k+1}^{(N)}\right)^{\mathrm{T}},$$

$$\mathbf{y}_k = \left(\mathbf{x}_k, h_{k+1}\mathbf{x}_k^{(1)}, \frac{1}{2!}h_{k+1}^2\mathbf{x}_k^{(2)}, \dots, \frac{1}{N!}h_{k+1}^N\mathbf{x}_k^{(N)}\right)^\mathrm{T},$$

$$\mathbf{C} = \begin{pmatrix} 1 & C_1^0 & C_2^0 & \cdots & C_N^0 \\ 0 & C_1^1 & C_2^1 & \cdots & C_N^1 \\ 0 & 0 & C_2^2 & \cdots & C_N^2 \\ \vdots & \vdots & \vdots & \cdots & \vdots \\ 0 & 0 & 0 & 0 & C_N^N \end{pmatrix}, \tag{3.93}$$

$$\mathbf{B} = \left(C_{N+1}^0 \mathbf{I}_{n\times n}, C_{N+1}^1 \mathbf{I}_{n\times n}, C_{N+1}^2 \mathbf{I}_{n\times n}, \dots, C_{N+1}^N \mathbf{I}_{n\times n}\right)^\mathrm{T},$$

and

$$C_N^k = \frac{N(N-1)\cdots(N-k+1)}{k!} \quad \text{and} \quad k! = k(k-1)\times\cdots\times 2 \times 1. \tag{3.94}$$

Thus

$$\mathbf{y}_{k+1} = \mathbf{C}\mathbf{y}_k. \tag{3.95}$$

Further we can obtain all points $\mathbf{x}_k$ ($k = 1, 2, 3, \dots$) with the control error

$$\frac{1}{(N+1)!}h_{k+1}^{N+1}\|\mathbf{B}D^{(N)}\mathbf{f}(\mathbf{x}_k, t_k, p)\| < \varepsilon \tag{3.96}$$

with $\mathbf{x}_k^{(N+1)} = D^{(N)}\mathbf{f}(\mathbf{x}_k, t_k, p)$.

### 3.3.2 Runge–Kutta Method of Order 2

From the midpoint discrete method, using the Euler's method

$$\mathbf{x}_{k+1} = \mathbf{x}_k + h_{k+1}\mathbf{f}(\mathbf{x}_k, t_k, \mathbf{p}) \tag{3.97}$$

we have

$$\mathbf{x}_{1/2} = \frac{1}{2}(\mathbf{x}_k + \mathbf{x}_{k+1}) \approx \mathbf{x}_k + \frac{1}{2}h_{k+1}\mathbf{f}(\mathbf{x}_k, t_k, \mathbf{p}). \tag{3.98}$$

Thus, letting $\mathbf{X}_2 \approx \mathbf{x}_{1/2}$, Eq. (3.66) becomes

$$\begin{aligned} t_{k+1} &= t_k + h_k, \quad \mathbf{X}_2 = \mathbf{x}_k + \frac{1}{2}h_{k+1}\mathbf{f}(\mathbf{x}_k, t_k, \mathbf{p}), \\ \mathbf{x}_{k+1} &= \mathbf{x}_k + h_{k+1}\mathbf{f}(\mathbf{X}_2, t_k + \frac{1}{2}h_{k+1}, \mathbf{p}) \quad \text{for } k = 0, 1, 2, \dots, M-1. \end{aligned} \tag{3.99}$$

The foregoing equations can be generalized as

$$
\begin{aligned}
\mathbf{X}_1 &= \mathbf{x}_k, \\
\mathbf{X}_2 &= \mathbf{x}_k + h_{k+1} a_{21} \mathbf{f}(\mathbf{X}_1, t_k, \mathbf{p}), \\
\mathbf{x}_{k+1} &= \mathbf{x}_k + h_{k+1} \sum_{i=1}^{2} b_i \mathbf{f}(\mathbf{X}_i, t_k + c_i h_{k+1}, \mathbf{p}) \\
&\quad \text{for } k = 0, 1, 2, \ldots, M - 1.
\end{aligned}
\tag{3.100}
$$

Thus, the Runge–Kutta method of order 2 is presented as follows.

**Definition 3.13** The discretization of Eq. (3.1) during the time interval $[t_k, t_{k+1}]$ as

$$
\begin{aligned}
t_{k+1} &= t_k + h_k, \\
\mathbf{x}_{k+1} &= \mathbf{x}_k + h_{k+1} \mathbf{\Phi}(\mathbf{x}_k, t_k, \mathbf{p}) \qquad \text{for } k = 0, 1, 2, \ldots, M - 1
\end{aligned}
\tag{3.101}
$$

where

$$
\begin{aligned}
\mathbf{\Phi}(\mathbf{x}_k, t_k, \mathbf{p}) &= b_1 \mathbf{f}(\mathbf{X}_1, t_k, \mathbf{p}) + b_2 \mathbf{f}(\mathbf{X}_2, t_k + c_2 h_{k+1}, \mathbf{p}), \\
\mathbf{X}_1 &= \mathbf{x}_k, \quad \mathbf{X}_2 = \mathbf{x}_k + h_{k+1} a_{21} \mathbf{f}(\mathbf{X}_1, t_k, \mathbf{p}),
\end{aligned}
\tag{3.102}
$$

is called the two-stage Runge–Kutta method (or the Runge–Kutta method of order 2, or the second-order Runge–Kutta method).

Consider the Taylor series of $\mathbf{f}(\mathbf{X}_2, t_k + c_2 h_{k+1}, \mathbf{p})$ as

$$
\begin{aligned}
\mathbf{f}(\mathbf{X}_2, t_k + c_2 h_{k+1}, \mathbf{p}) &= \mathbf{f}(\mathbf{x}_k + a_{12} h_{k+1} \mathbf{f}, t_k + c_2 h_{k+1}, \mathbf{p}) \\
&= \mathbf{f} + h_{k+1}(a_{21} \mathbf{f}_{\mathbf{x}} \mathbf{f} + c_2 \mathbf{f}_t) + O(h_{k+1}^2).
\end{aligned}
\tag{3.103}
$$

Using

$$
\mathbf{x}_{k+1} = \mathbf{x}_k + h_{k+1}[b_1 \mathbf{f}(\mathbf{X}_1, t_k, \mathbf{p}) + b_2 \mathbf{f}(\mathbf{X}_2, t_k + c_2 h_{k+1}, \mathbf{p})],
\tag{3.104}
$$

we have

$$
\begin{aligned}
\mathbf{x}_{k+1} &= \mathbf{x}_k + h_{k+1}(b_1 + b_2) \mathbf{f} \\
&\quad + b_2(a_{21} h_{k+1} \mathbf{f}_{\mathbf{x}} \mathbf{f} + c_2 h_{k+1} \mathbf{f}_t) h_{k+1}^2 + O(h_{k+1}^3).
\end{aligned}
\tag{3.105}
$$

From the Taylor series, we have

$$
\begin{aligned}
\mathbf{x}(t_k + h_{k+1}) &= \mathbf{x}(t_k) + \dot{\mathbf{x}}(t_k) h_{k+1} + \frac{h_{k+1}^2}{2!} \ddot{\mathbf{x}}(t_k) + \frac{h_{k+1}^3}{3!} \mathbf{x}^{(3)}(t_k^c) \\
&= \mathbf{x}(t_k) + h_{k+1} \mathbf{f} + \frac{h_{k+1}^2}{2!}(\mathbf{f}_{\mathbf{x}} \mathbf{f} + \mathbf{f}_t) + \frac{h_{k+1}^3}{3!} D^{(2)} \mathbf{f}(\mathbf{x}(t_k^c), t_k^c, \mathbf{p})
\end{aligned}
\tag{3.106}
$$

where

$$Df(\mathbf{x}_k, t_k, \mathbf{p}) = \mathbf{f}_x \mathbf{f} + \mathbf{f}_t, \quad D^{(2)}\mathbf{f}(\mathbf{x}_k, t_k, \mathbf{p}) = D \circ Df(\mathbf{x}_k, t_k, \mathbf{p}). \tag{3.107}$$

Comparison of Eqs. (3.105) and (3.106) gives the following:

$$\begin{aligned} h_{k+1} \Rightarrow \mathbf{f} : & \quad b_1 + b_2 = 1, \\ h_{k+1}^2 \Rightarrow \mathbf{f}_x \mathbf{f} : & \quad b_2 a_{21} = \frac{1}{2}, \\ h_{k+1}^2 \Rightarrow \mathbf{f}_t : & \quad b_2 c_2 = \frac{1}{2}. \end{aligned} \tag{3.108}$$

If $b_2 = 0$, we have $b_1 = 1$ and the last two equations of Eq. (3.108) cannot be satisfied. Thus, Eq. (3.100) becomes

$$\begin{aligned} t_{k+1} &= t_k + h_k, \\ \mathbf{x}_{k+1} &= \mathbf{x}_k + h_{k+1}\mathbf{f}(\mathbf{x}_k, t_k, \mathbf{p}) \quad \text{for } k = 0, 1, 2, \dots, M-1 \end{aligned} \tag{3.109}$$

which is the forward Euler method. This case gives the Runge–Kutta method of first order.

If $b_2 \neq 0$, choosing $c_2$, Eq. (3.108) becomes

$$b_2 = \frac{1}{2c_2}, \quad b_1 = 1 - b_2, \quad a_{21} = \frac{1}{2b_2}. \tag{3.110}$$

This case gives the Runge–Kutta method of the second order. There are three popular choices as follows:

(i) If $c_2 = \frac{1}{2}$, we have $b_2 = 1$, $b_1 = 0$ and $a_{21} = \frac{1}{2}$, thus, Eq. (3.100) becomes

$$\begin{aligned} t_{k+1} &= t_k + h_k, \\ \mathbf{x}_{k+1} &= \mathbf{x}_k + h_{k+1}\mathbf{f}\left(\mathbf{X}_2, t_k + \frac{1}{2}h_{k+1}, \mathbf{p}\right) \\ \mathbf{X}_2 &= \mathbf{x}_k + \frac{1}{2}h_{k+1}\mathbf{f}(\mathbf{x}_k, t_k, \mathbf{p}) \quad \text{for } k = 0, 1, 2, \dots, M-1. \end{aligned} \tag{3.111}$$

This scheme is called the modified Euler–Cauchy method. Using the Euler approximation, we have $\mathbf{x}_{1/2} \approx \mathbf{X}_2$. Thus, this method is the approximation of the midpoint discrete scheme. A table of the coefficients in the Runge–Kutta method is called the Butcher tableau. From the Butcher's condensed nomenclature, we have the following array form to arrange the coefficients.

$$\begin{array}{c|cc} c_1 & a_{11} & a_{12} \\ c_2 & a_{21} & a_{22} \\ \hline & b_1 & b_2 \end{array} \xrightarrow{c_2 = \frac{1}{2}} \begin{array}{c|cc} 0 & 0 & 0 \\ \frac{1}{2} & \frac{1}{2} & 0 \\ \hline & 0 & 1 \end{array} \tag{3.112}$$

Note that $a_{ij} = 0$ for $(i \leq j)$ with $c_i = \Sigma_{j=1}^2 a_{ij}$ for explicit Runge–Kutta method.

(ii) If $b_2 = \frac{1}{2}$, we have $b_1 = \frac{1}{2}$ and $a_{21} = c_2 = 1$, thus, Eq. (3.100) becomes

$$t_{k+1} = t_k + h_k,$$

$$\mathbf{x}_{k+1} = \mathbf{x}_k + \frac{1}{2} h_{k+1} [\mathbf{f}(\mathbf{x}_k, t_k, \mathbf{p}) + \mathbf{f}(\mathbf{x}_{k+1}, t_{k+1}, \mathbf{p})] \qquad (3.113)$$

$$\text{for } k = 0, 1, 2, \ldots, M - 1.$$

Thus, the trapezoidal discrete approximation is recovered. The Butcher table is given as

$$
\begin{array}{c|cc}
0 & 0 & 0 \\
c_2 & a_{21} & 0 \\
\hline
& b_1 & b_2
\end{array}
\quad \xrightarrow{c_2 = 1} \quad
\begin{array}{c|cc}
0 & 0 & 0 \\
1 & 1 & 0 \\
\hline
& \frac{1}{2} & \frac{1}{2}
\end{array}
\qquad (3.114)
$$

(iii) If $b_2 = \frac{3}{4}$, we have $b_1 = \frac{1}{4}$ and $a_{21} = c_2 = \frac{2}{3}$, thus, Eq. (3.100) becomes

$$t_{k+1} = t_k + h_k,$$

$$\mathbf{x}_{k+1} = \mathbf{x}_k + \frac{1}{4} h_{k+1} [\mathbf{f}(\mathbf{x}_k, t_k, \mathbf{p}) + 3\mathbf{f}(\mathbf{X}_2, t_k + \frac{2}{3} h_{k+1}, \mathbf{p})], \qquad (3.115)$$

$$\mathbf{X}_2 = \mathbf{x}_k + \frac{2}{3} h_{k+1} \mathbf{f}(\mathbf{x}_k, t_k, \mathbf{p}) \quad \text{for } k = 0, 1, 2, \ldots, M - 1.$$

The Butcher table is given as

$$
\begin{array}{c|cc}
0 & 0 & 0 \\
c_2 & a_{21} & 0 \\
\hline
& b_1 & b_2
\end{array}
\quad \xrightarrow{c_2 = \frac{2}{3}} \quad
\begin{array}{c|cc}
0 & 0 & 0 \\
\frac{2}{3} & \frac{2}{3} & 0 \\
\hline
& \frac{3}{4} & \frac{1}{4}
\end{array}
\qquad (3.116)
$$

For this case, the local and global error can be similarly discussed as in the previous section, or it can follow the results presented in the general Runge–Kutta method in next section. This scheme is explicit like the forward Euler method, and its stability is also similar to the stability of the forward Euler method.

## 3.4 Explicit Runge–Kutta Methods

From the previous discussion, the general theoretical frame for explicit Runge–Kutta method will be discussed.

**Definition 3.14** The discretization of Eq. (3.1) during the time interval $[t_k, t_{k+1}]$ as

$$t_{k+1} = t_k + h_k,$$
$$\mathbf{x}_{k+1} = \mathbf{x}_k + h_{k+1}\boldsymbol{\Phi}(\mathbf{x}_k, t_k, \mathbf{p}) \quad \text{for } k = 0, 1, 2, \ldots, M-1 \tag{3.117}$$

where

$$\boldsymbol{\Phi}(\mathbf{x}_k, t_k, \mathbf{p}) = \sum_{i=1}^{s} b_i \mathbf{f}(\mathbf{X}_i, t_k + c_i h_{k+1}, \mathbf{p}),$$

$$\mathbf{X}_i = \mathbf{x}_k + h_{k+1} \sum_{j=1}^{s} a_{ij}\mathbf{f}(\mathbf{X}_j, t_k + c_j h_{k+1}, \mathbf{p}), \tag{3.118}$$

$$a_{ij} = 0 \quad \text{for } (i \le j)$$

is called the $s$-stage Runge–Kutta method (or Runge–Kutta method of order $s$, or the $s$th-order Runge–Kutta).

From the Taylor series and the Runge–Kutta expansion, the expression of the true $\mathbf{x}(t_{k+1})$ at time $t_{k+1}$ based on the point $\mathbf{x}(t_k)$ at time $t_k$ is

$$\mathbf{x}(t_{k+1}) = \mathbf{x}(t_k) + h_{k+1}\boldsymbol{\Phi}(\mathbf{x}(t_k), t_k, \mathbf{p}) + \frac{1}{(s+1)!} h_{k+1}^{s+1} \boldsymbol{\delta}^{(s+1)}(\mathbf{x}_k^c, t_k^c, \mathbf{p}), \tag{3.119}$$

where $\boldsymbol{\delta}^{(s+1)}(\mathbf{x}_k^c, t_k^c, \mathbf{p}) \ne D^{(s)}\mathbf{f}(\mathbf{x}_k^c, t_k^c, \mathbf{p})$ derived from the Runge–Kutta scheme. The local error for the Runge–Kutta method of the $s$th-order is based on the $\mathbf{x}(t_{k+1}) = \mathbf{x}_k$ and $\boldsymbol{\Phi}(\mathbf{x}(t_k), t_k, \mathbf{p}) = \boldsymbol{\Phi}(\mathbf{x}_k, t_k, \mathbf{p})$, thus we have the local error is

$$\boldsymbol{\varepsilon}_{k+1} = \mathbf{x}(t_{k+1}) - \mathbf{x}_k - h_{k+1}\boldsymbol{\Phi}(\mathbf{x}_k, t_k, \mathbf{p}) = \frac{1}{(s+1)!} h_{k+1}^{s+1} \boldsymbol{\delta}^{(s+1)}(\mathbf{x}_k^c, t_k^c, \mathbf{p}). \tag{3.120}$$

**Theorem 3.10** *Suppose the solution* $\mathbf{x} = \mathbf{x}(t)$ *is the unique solution of* Eq. (3.1). *If* $\mathbf{x}(t) \in C^{s+1}[t_0, t_M]$ *and the points* $(t_k, \mathbf{x}_k)$ *for* $k = 0, 1, 2, \ldots, M-1$ *is the sequence of the* $s$th-*stage Runge–Kutta discrete approximation of* Eq. (3.1), *then the corresponding local error satisfies*

$$\|\boldsymbol{\varepsilon}_{k+1}\| = \|\mathbf{x}(t_{k+1}) - \mathbf{x}_k - h_{k+1}\boldsymbol{\Phi}(\mathbf{x}_k, t_k, \mathbf{p})\| = O(h_{k+1}^{s+1}) \tag{3.121}$$

*where*

$$\boldsymbol{\Phi}(\mathbf{x}_k, t_k, \mathbf{p}) = \sum_{i=1}^{s} b_i \mathbf{f}(\mathbf{X}_i, t_k + c_i h_{k+1}, \mathbf{p}),$$

$$\mathbf{X}_i = \mathbf{x}_k + h_{k+1} \sum_{j=1}^{s} a_{ij}\mathbf{f}(\mathbf{X}_j, t_k + c_j h_{k+1}, \mathbf{p}), \tag{3.122}$$

$$a_{ij} = 0 \quad \text{for } (i \le j)$$

*and the global error for $h_k = h$ $(k = 1, 2, 3, \ldots)$ and $t_k = kh$ is*

$$\|\mathbf{e}_k\| \leq \frac{h^s L}{(s+1)!K}(e^{t_k K} - 1) \tag{3.123}$$

*with*

$$L = \max_{k \in \{1,2,\ldots,M\}} \|\boldsymbol{\delta}^{(s)}(\mathbf{x}_k^c, t_k^c, \mathbf{p})\|,$$

$$K = \max_{k \in \{1,2,\ldots,M\}} \|\mathbf{A}_k\|, \quad and \tag{3.124}$$

$$\mathbf{A}_k = D_{\mathbf{x}}\boldsymbol{\Phi}(\boldsymbol{\xi}_k, t_k, \mathbf{p}) \quad and \quad \|\boldsymbol{\xi}_k\| \in (\|\mathbf{x}(t_k)\|, \|\mathbf{x}_k\|).$$

*For a fixed $t_k = kh$, as h goes to zero,*

$$\|\mathbf{e}_k\| = \|\mathbf{x}(t_k) - \mathbf{x}_k\| = O(h^s). \tag{3.125}$$

*Proof* Based on the Taylor series, the Runge–Kutta expansion of $\mathbf{x}(t_{k+1})$ at point $(t_k, \mathbf{x}(t_k))$ as

$$\mathbf{x}(t_{k+1}) = \mathbf{x}(t_k + h_{k+1})$$

$$= \mathbf{x}(t_k) + h_{k+1}\boldsymbol{\Phi}(\mathbf{x}(t_k), t_k, \mathbf{p}) + \frac{1}{(s+1)!}h_{k+1}^{s+1}\boldsymbol{\delta}^{(s+1)}(\mathbf{x}_k^c, t_k^c, \mathbf{p}).$$

The Runge–Kutta approximation of the $s^{\text{th}}$-order gives

$$\mathbf{x}_{k+1} = \mathbf{x}_k + h_{k+1}\boldsymbol{\Phi}(\mathbf{x}_k, t_k, \mathbf{p}).$$

Consider the difference between the true and approximate solutions as

$$\mathbf{x}(t_{k+1}) - \mathbf{x}_{k+1} = \mathbf{x}(t_k) - \mathbf{x}_k + h_{k+1}[\boldsymbol{\Phi}(\mathbf{x}(t_k), t_k, \mathbf{p}) - \boldsymbol{\Phi}(\mathbf{x}_k, t_k, \mathbf{p})]$$

$$+ \frac{1}{(s+1)!}h_{k+1}^{(s+1)}\boldsymbol{\delta}^{(s+1)}(\mathbf{x}_k^c, t_k^c, \mathbf{p}).$$

If $\mathbf{x}(t_k) = \mathbf{x}_k$, then $\boldsymbol{\Phi}(\mathbf{x}(t_k), t_k, \mathbf{p}) = \boldsymbol{\Phi}(\mathbf{x}_k, t_k, \mathbf{p})$. Therefore, the local error is

$$\boldsymbol{\varepsilon}_{k+1} = \mathbf{x}(t_{k+1}) - \mathbf{x}_k - h_{k+1}\boldsymbol{\Phi}(\mathbf{x}_k, t_k, \mathbf{p}) = \frac{1}{(s+1)!}h_{k+1}^{s+1}\boldsymbol{\delta}^{(s+1)}(\mathbf{x}_k^c, t_k^c, \mathbf{p}).$$

Thus, the local error of the Runge–Kutta method of the $s$th order at the $(k+1)$th step is

$$\|\boldsymbol{\varepsilon}_{k+1}\| = \|\mathbf{x}(t_{k+1}) - \mathbf{x}_k - h_{k+1}\boldsymbol{\Phi}(\mathbf{x}_k, t_k, \mathbf{p})\| = O(h_{k+1}^s).$$

If $\mathbf{x}(t_k) \neq \mathbf{x}_k$, then $\boldsymbol{\Phi}(\mathbf{x}(t_k), t_k, \mathbf{p}) \neq \boldsymbol{\Phi}(\mathbf{x}_k, t_k, \mathbf{p})$, thus,

$$
\begin{aligned}
\mathbf{x}(t_{k+1}) - \mathbf{x}_{k+1} &= \mathbf{x}(t_k) - \mathbf{x}_k + h_{k+1}[\boldsymbol{\Phi}(\mathbf{x}(t_k), t_k, \mathbf{p}) - \boldsymbol{\Phi}(\mathbf{x}_k, t_k, \mathbf{p})] \\
&\quad + \frac{1}{(s+1)!} h_{k+1}^{s+1} \boldsymbol{\delta}^{(s+1)}(\mathbf{x}_k^c, t_k^c, \mathbf{p}),
\end{aligned}
$$

and the Lipschitz conditions produce

$$
\mathbf{x}(t_{k+1}) - \mathbf{x}_{k+1} = (\mathbf{I} + h_{k+1}\mathbf{A}_k)(\mathbf{x}(t_k) - \mathbf{x}_k) + \boldsymbol{\varepsilon}_{k+1}
$$

where

$$
\mathbf{A}_k = D_{\mathbf{x}}\boldsymbol{\Phi}(\boldsymbol{\xi}_k, t_k, \mathbf{p}) \quad \text{and} \quad \|\boldsymbol{\xi}_k\| \in (\|\mathbf{x}(t_k)\|, \|\mathbf{x}_k\|).
$$

So

$$
\mathbf{e}_{k+1} = (\mathbf{I} + h_{k+1}\mathbf{A}_k)\mathbf{e}_k + \boldsymbol{\varepsilon}_{k+1}
$$

gives

$$
\|\mathbf{e}_{k+1}\| \leq \|(\mathbf{I} + h_{k+1}\mathbf{A}_k)\| \times \|\mathbf{e}_k\| + \|\boldsymbol{\varepsilon}_{k+1}\|.
$$

For $h_k = h$ ($k = 1, 2, 3, \ldots$) and $t_k = kh$, letting

$$
L = \max_{k \in \{1,2,\ldots,M\}} \|\boldsymbol{\delta}^{(s)}(\mathbf{x}_k^c, t_k^c, \mathbf{p})\| \quad \text{and} \quad K = \max_{k \in \{1,2,\ldots,M\}} \|\mathbf{A}_k\|,
$$

consider a simple discrete equation as

$$
z_{k+1} = (1 + hK)z_k + \frac{1}{(s+1)!} h^{s+1} L \quad \text{with } z_0 = 0
$$

and

$$
\begin{aligned}
z_k &= \frac{1}{(s+1)!} h^{s+1} L \left( \sum_{l=0}^{k-1} (1 + hK)^l \right) \\
&= \frac{h^{s+1} L}{(s+1)! K} [(1 + hK)^k - 1] \text{ for } k = 1, 2, \ldots, M.
\end{aligned}
$$

For $k > 0$, we have $1 + hK > 0$. If $z_k \geq \|\mathbf{e}_k\|$, then $z_{k+1} \geq \|\mathbf{e}_{k+1}\|$. Since $t_k = kh$, for $k \rightarrow \infty$ and $h \rightarrow 0$, we have

$$
\lim_{k \rightarrow \infty} \left(1 + \frac{t_k}{k} K\right)^k = e^{t_k K}.
$$

so

$$\|\mathbf{e}_k\| \le \frac{h^s L}{(s+1)!K}(e^{t_k K} - 1).$$

For a fixed $t_k = kh$, as $h$ goes to zero, we have

$$\|\mathbf{e}_k\| = \|\mathbf{x}(t_k) - \mathbf{x}_k\| = O(h^s) \quad \text{for } k = 1, 2, \dots, M.$$

This theorem is proved. □

Consider the global error at the

$$\mathbf{e}_{k+1} = (\mathbf{I} + h_{k+1}\mathbf{A}_k)\mathbf{e}_k. \tag{3.126}$$

If $\mathbf{e}_{k+1} = \lambda\mathbf{e}_k$, then the foregoing equation becomes

$$[(\mathbf{I} + h_{k+1}\mathbf{A}_k) - \lambda\mathbf{I}]\mathbf{e}_k = \mathbf{0}. \tag{3.127}$$

The corresponding eigenvalues are generated by

$$|\mathbf{I} + h_{k+1}\mathbf{A}_k - \lambda\mathbf{I}| = 0. \tag{3.128}$$

If all eigenvalues $|\lambda_j| < 1$ ($j = 1, 2, \dots, n$), then

$$\|\mathbf{e}_{k+1}\| < \|\mathbf{e}_k\|. \tag{3.129}$$

So the global error will not be enlarged. Thus, under such a condition, the Runge–Kutta method of order $s$ gives a stable approximation. For one-dimensional systems, we have $\mathbf{A}_k = K_k$ and $\mathbf{I} = 1$. Equation (3.128) gives the stability interval as

$$|1 + h_{k+1}K_k| < 1. \tag{3.130}$$

Since $h_{k+1} > 0$, if $K_k < 0$, then the foregoing equation always exists.

The Butcher tableau for the coefficients in the Runge–Kutta method is presented as follows:

$$
\begin{array}{c|cccccc}
c_1 & 0 & 0 & \cdots & 0 & 0 \\
c_2 & a_{21} & 0 & \cdots & 0 & 0 \\
c_3 & a_{31} & a_{32} & \cdots & 0 & 0 \\
\vdots & \vdots & \vdots & \vdots & \vdots & \vdots \\
c_s & a_{s1} & a_{s2} & \cdots & a_{s(s-1)} & 0 \\
\hline
 & b_1 & b_2 & \cdots & b_{s-1} & b_s
\end{array}
\tag{3.131}
$$

The condition for coefficients $a_{ij}$ and $c_i$ is often assumed as

$$c_1 = 0 \quad \text{and} \quad c_i = \sum_{j=1}^{i-1} a_{ij} \quad \text{for } i = 2, \ldots, s. \tag{3.132}$$

Consider the $s$th-order Taylor series of $\mathbf{f}(\mathbf{X}_i, t_k + c_i h_{k+1}, \mathbf{p})$ as

$$\begin{aligned}
\mathbf{f}_i^{(s)} &\equiv \mathbf{f}(\mathbf{X}_i, t_k + c_i h_{k+1}, \mathbf{p}) \\
&= \mathbf{f}_1 + \sum_{m=1}^{s} h_{k+1}^m \sum_{l=0}^{m} C_m^{m-l} D_{\mathbf{x}}^{(l)} \mathbf{f}_1 \cdot D_t^{(m-l)} \mathbf{f}_1 \cdot \left( \sum_{j=1}^{i-1} a_{ij} \mathbf{f}_j^{(l)} \right)^l c_i^{m-l} \\
&\quad + h_{k+1}^{s+1} \sum_{l=0}^{s+1} C_{s+1}^{s+1-l} D_{\mathbf{x}}^{(l)} \mathbf{f}_1 \cdot D_t^{(s+1-l)} \mathbf{f}_1 \cdot \left( \sum_{j=1}^{i-1} a_{ij} \mathbf{f}_j^{(l)} \right)^l c_i^{s+1-l}
\end{aligned} \tag{3.133}$$

where

$$\begin{aligned}
\mathbf{f}_1 &\equiv \mathbf{f}(\mathbf{x}_k, t_k, \mathbf{p}), \quad D_{\mathbf{x}}^{(p)} \mathbf{f}_1 \equiv D_{\mathbf{x}}^{(p)} \mathbf{f}(\mathbf{x}_k, t_k, \mathbf{p}), \\
D_t^{(p)} \mathbf{f}_1 &\equiv D_t^{(p)} \mathbf{f}(\mathbf{x}_k, t_k, \mathbf{p}).
\end{aligned} \tag{3.134}$$

We have

$$\begin{aligned}
\mathbf{X}_i - \mathbf{x}_k &= h_{k+1} \sum_{j=1}^{s} a_{ij} \mathbf{f}(\mathbf{X}_j, t_k + c_j h_{k+1}, \mathbf{p}) \\
&= h_{k+1} \sum_{j=1}^{s} a_{ij} (\mathbf{f} + h_{k+1} \mathbf{f}(\mathbf{X}_j, t_k + c_j h_{k+1}, \mathbf{p})), \\
\mathbf{f}(\mathbf{X}_i, t_k + c_i h_{k+1}, \mathbf{p}) &= \mathbf{f} + \sum_{m=1}^{s} h_{k+1}^m \sum_{l=0}^{m} C_m^{m-l} D_{\mathbf{x}}^{(l)} \mathbf{f} \cdot D_t^{(m-l)} \mathbf{f} \cdot \left( \sum_{j=1}^{i-1} a_{ij} \mathbf{f}_j^{(l)} \right)^l c_i^{m-l} \\
&\quad + h_{k+1}^{s+1} \sum_{l=0}^{s+1} C_{s+1}^{s+1-l} D_{\mathbf{x}}^{(l)} \mathbf{f} \cdot D_t^{(s+1-l)} \mathbf{f} \cdot \left( \sum_{j=1}^{i-1} a_{ij} \mathbf{f}_j^{(l)} \right)^l c_i^{s+1-l}.
\end{aligned} \tag{3.135}$$

Therefore, the $s$-stage Runge–Kutta discrete scheme is given by

$$\mathbf{x}_{k+1} = \mathbf{x}_k + h_{k+1} \boldsymbol{\Phi}(\mathbf{x}_k, t_k, \mathbf{p}) \tag{3.136}$$

where

$$\boldsymbol{\Phi}(\mathbf{x}_k, t_k, \mathbf{p}) = \sum_{i=1}^{s} b_i \mathbf{f}(\mathbf{X}_i, t_k + c_i h_{k+1}, \mathbf{p}),$$

$$\mathbf{X}_i = \mathbf{x}_k + h_{k+1} \sum_{j=1}^{s} a_{ij}\mathbf{f}(\mathbf{X}_j, t_k + c_j h_{k+1}, \mathbf{p}),$$

$$a_{ij} = 0 \quad \text{for } i \leq j. \tag{3.137}$$

Consider the Taylor series of $\mathbf{x}(t_{k+1})$ at the point $\mathbf{x}(t_k)$ as

$$\mathbf{x}(t_{k+1}) = \mathbf{x}(t_k + h_{k+1})$$

$$= \mathbf{x}(t_k) + \sum_{m=1}^{s} \frac{1}{m!} h_{k+1}^m D^{(m-1)}\mathbf{f}(\mathbf{x}(t_k), t_k, \mathbf{p}) \tag{3.138}$$

$$+ \frac{1}{(s+1)!} h_{k+1}^{s+1} D^{(s)}\mathbf{f}(\mathbf{x}(t_k), t_k, \mathbf{p})$$

where

$$D\mathbf{f} \equiv [D_{\mathbf{x}}(\cdot) \cdot \mathbf{f} + D_t(\cdot)]\mathbf{f} = D_{\mathbf{x}}\mathbf{f} \cdot \mathbf{f} + D_t\mathbf{f},$$

$$D^{(2)}\mathbf{f} = D_{\mathbf{x}}^{(2)}\mathbf{f} \cdot \mathbf{f}^2 + 2D_{\mathbf{x}t}\mathbf{f} \cdot \mathbf{f} + (D_{\mathbf{x}}\mathbf{f})^2 \cdot \mathbf{f} + D_{\mathbf{x}}\mathbf{f}D_t\mathbf{f} + D_t^{(2)}\mathbf{f}$$

$$\vdots \tag{3.139}$$

$$D^{(m)}\mathbf{f} = D(D^{(m-1)}\mathbf{f});$$

with

$$D(\cdot) \equiv [D_{\mathbf{x}}(\cdot) \cdot \mathbf{f} + D_t(\cdot)],$$

$$D^m(\cdot) = [D_{\mathbf{x}}(\cdot) \cdot \mathbf{f} + D_t(\cdot)]^m = \sum_{k=0}^{m} C_m^k D_{\mathbf{x}^k t^{m-k}}(\cdot) \cdot \mathbf{f}^k,$$

$$D(D^m(\cdot)) = \sum_{k=0}^{m} C_m^k [D_{\mathbf{x}^{k+1} t^{m-k}}(\cdot) \cdot \mathbf{f} + D_{\mathbf{x}^k t^{m-k+1}}(\cdot)] \cdot \mathbf{f}^k \tag{3.140}$$

$$+ \sum_{k=0}^{m} k C_m^k D_{\mathbf{x}^k t^{m-k}}(\cdot) \cdot [D_{\mathbf{x}}(\cdot) \cdot \mathbf{f} + D_t(\cdot)] \cdot \mathbf{f}^{k-1}.$$

The error function is given by

$$\varepsilon_{k+1} = \frac{h_{k+1}^{s+1}}{(s+1)!} \boldsymbol{\delta}^{(s+1)}(\mathbf{x}_k^c, t_k^c, \mathbf{p}), \tag{3.141}$$

and

$$\boldsymbol{\delta}^{(s+1)}(\mathbf{x}_k^c, t_k^c, \mathbf{p}) = D^{(s)}\mathbf{f} - \frac{(s+1)!}{h_{k+1}^{s+1}} \frac{\partial^s \boldsymbol{\Phi}(\mathbf{x}_k, t_k, \mathbf{p})}{\partial h_{k+1}^s}$$

$$= D^{(s)}\mathbf{f} - \frac{(s+1)!}{h_{k+1}^{s+1}} \sum_{i=1}^{s} b_i \frac{\partial^s}{\partial h_{k+1}^s} \mathbf{f}(\mathbf{X}_i, t_k + c_i h_{k+1}, \mathbf{p}), \tag{3.142}$$

or

$$\delta^{(s+1)}(\mathbf{x}_k^c, t_k^c, \mathbf{p}) = \frac{(s+1)!}{h_{k+1}^{s+1}} (\sum_{m=0}^{s} \frac{1}{(m+1)!} h_{k+1}^{m+1} D^{(i)}\mathbf{f}$$
$$- \{\sum_{i=1}^{s} b_i [\sum_{m=1}^{s} h_{k+1}^{m+1} \sum_{l=0}^{m} C_m^{m-l} D_{\mathbf{x}^l t^{m-l}}^m \mathbf{f} \cdot (\sum_{j=1}^{i-1} a_{ij}\mathbf{f}_j^{(l)})^l c_i^{m-l}]\}). \tag{3.143}$$

To reduce the local errors of the $s$th-order Runge–Kutta method, one tried to minimization of $|\delta^{(s+1)}(\mathbf{x}_k^c, t_k^c, \mathbf{p})|$ through the rough estimate as

$$\|\delta^{(s+1)}(\mathbf{x}_k^c, t_k^c, \mathbf{p})\| < CMK^s,$$
$$\|\mathbf{f}\| < M \quad \text{and} \quad \|D_{\mathbf{x}^l t^{m-l}}^m \mathbf{f}\| < \frac{K^m M}{M^l} = \frac{K^m}{M^{l-1}}. \tag{3.144}$$

Note the constants definitions do not make any physical reason, which just make them simple for rough estimates of the error function.

Consider a second-order Runge–Kutta method as an example to compute the error function. The third-order Taylor series of $\mathbf{f}(\mathbf{X}_2, t_k + c_2 h_{k+1}, \mathbf{p})$ should be computed, i.e.,

$$\mathbf{f}(\mathbf{X}_2, t_k + c_2 h_{k+1}, \mathbf{p}) = \mathbf{f}(\mathbf{x}_k + a_{12} h_{k+1}\mathbf{f}, t_k + c_2 h_{k+1}, \mathbf{p})$$
$$= \mathbf{f} + h_{k+1}(a_{21}\mathbf{f}_{\mathbf{x}}\mathbf{f} + c_2\mathbf{f}_t)$$
$$+ \frac{1}{2!} h_{k+1}^2 (a_{21}^2 \mathbf{f}_{\mathbf{xx}} \cdot \mathbf{f}^2 + 2a_{21} c_2 \mathbf{f}_{\mathbf{x}t} \cdot \mathbf{f} + c_2^2 \mathbf{f}_{tt}) + O(h_{k+1}^3). \tag{3.145}$$

Thus, the function $\mathbf{\Phi}(\mathbf{x}_k, t_k, \mathbf{p})$ is re-expressed with one more higher order as

$$\mathbf{\Phi}(\mathbf{x}_k, t_k, \mathbf{p}) = b_1\mathbf{f}(\mathbf{X}_1, t_k, \mathbf{p}) + b_2\mathbf{f}(\mathbf{X}_2, t_k + c_2 h_{k+1}, \mathbf{p})$$
$$= b_1\mathbf{f} + b_2[h_{k+1}(a_{21}\mathbf{f}_{\mathbf{x}}\mathbf{f} + c_2\mathbf{f}_t)$$
$$+ \frac{1}{2!} h_{k+1}^2 (a_{21}^2 \mathbf{f}_{\mathbf{xx}} \cdot \mathbf{f}^2 + 2a_{21} c_2 \mathbf{f}_{\mathbf{x}t} \cdot \mathbf{f} + c_2^2 \mathbf{f}_{tt})] + O(h_{k+1}^3) \tag{3.146}$$

and the third-order derivative of $\mathbf{x}(t_{k+1})$ at point $\mathbf{x}(t_{k+1})$ in the Taylor series is

$$\mathbf{x}^{(3)}(t_k) = D^{(2)}\mathbf{f}(\mathbf{x}_k, t_k, \mathbf{p})$$
$$= \mathbf{f}_{\mathbf{xx}} \cdot \mathbf{f}^2 + 2\mathbf{f}_{\mathbf{x}t} \cdot \mathbf{f} + \mathbf{f}_{\mathbf{x}}^2 \cdot \mathbf{f} + \mathbf{f}_{\mathbf{x}} \cdot \mathbf{f}_t + \mathbf{f}_{tt}. \tag{3.147}$$

The error function is computed as

$$\delta^{(3)}(\mathbf{x}_k^c, t_k^c, \mathbf{p}) = (1 - \frac{3!}{2!} b_2 c_2^2)(\mathbf{f}_{\mathbf{xx}} \cdot \mathbf{f}^2 + 2\mathbf{f}_{\mathbf{x}t} \cdot \mathbf{f} + \mathbf{f}_{tt}) + (\mathbf{f}_{\mathbf{x}}^2 \cdot \mathbf{f} + \mathbf{f}_{\mathbf{x}} \cdot \mathbf{f}_t), \tag{3.148}$$

and the local error function is

$$\varepsilon_{k+1} = \frac{h_{k+1}^3}{3!}\left[\left(1 - 3b_2c_2^2\right)\left(\mathbf{f_{xx}} \cdot \mathbf{f}^2 + 2\mathbf{f_{xt}} \cdot \mathbf{f} + \mathbf{f_{tt}}\right) + \left(\mathbf{f_x^2} \cdot \mathbf{f} + \mathbf{f_x} \cdot \mathbf{f_t}\right)\right]. \quad (3.149)$$

With $b_2c_2 = \frac{1}{2}$ in Eq. (3.108), we have

$$\varepsilon_{k+1} = \frac{h_{k+1}^3}{3!}\left[(1 - \frac{3}{2}c_2)(\mathbf{f_{xx}} \cdot \mathbf{f}^2 + 2\mathbf{f_{xt}} \cdot \mathbf{f} + \mathbf{f_{tt}}) + (\mathbf{f_x^2} \cdot \mathbf{f} + \mathbf{f_x} \cdot \mathbf{f_t})\right]. \quad (3.150)$$

Using Eq. (3.145), the foregoing equation gives

$$\|\varepsilon_{k+1}\| = \frac{h_{k+1}^3}{3!}\left[4\left|1 - \frac{3}{2}c_2\right| + 2\right]MK^2. \quad (3.151)$$

So we have $c_2 = 2/3$ to make $\|\varepsilon_{k+1}\|$ minimized. Therefore, the local errors for the second-order Runge–Kutta method can be computed as follows:

$$\begin{aligned}
\|\varepsilon_{k+1}\| &= \frac{1}{2}h_{k+1}^3 MK^2 \quad \text{for} \quad c_2 = \frac{1}{2}, \\
\|\varepsilon_{k+1}\| &= \frac{1}{3}h_{k+1}^3 MK^2 \quad \text{for} \quad c_2 = \frac{2}{3}, \\
\|\varepsilon_{k+1}\| &= \frac{2}{3}h_{k+1}^3 MK^2 \quad \text{for} \quad c_2 = 1.
\end{aligned} \quad (3.152)$$

### 3.4.1 Runge–Kutta Method of Order 3

To demonstrate the general procedure to develop the Runge–Kutta scheme, herein consider the third-order Runge–Kutta method as

$$\begin{aligned}
\mathbf{x}_{k+1} &= \mathbf{x}_k + h_{k+1}(b_1\mathbf{f}_1 + b_2\mathbf{f}_2 + b_3\mathbf{f}_3); \\
\mathbf{f}_1 &= \mathbf{f}(\mathbf{x}_k, t_k, \mathbf{p}), \\
\mathbf{f}_2 &= \mathbf{f}(\mathbf{x}_k + a_{21}\mathbf{f}_1, t_k + c_2h_{k+1}, \mathbf{p}), \\
\mathbf{f}_3 &= \mathbf{f}(\mathbf{x}_k + a_{31}\mathbf{f}_1 + a_{32}\mathbf{f}_2, t_k + c_3h_{k+1}, \mathbf{p}).
\end{aligned} \quad (3.153)$$

The second-order Taylor series of $\mathbf{f}(\mathbf{X}_2, t_k + c_2h_{k+1}, \mathbf{p})$ is

$$\begin{aligned}
\mathbf{f}_2 &= \mathbf{f}(\mathbf{X}_2, t_k + c_2h_{k+1}, \mathbf{p}) \\
&= \mathbf{f} + h_{k+1}(a_{21}\mathbf{f_x} \cdot \mathbf{f} + c_2\mathbf{f_t}) \\
&\quad + \frac{1}{2!}h_{k+1}^2(a_{21}^2\mathbf{f_{xx}} \cdot \mathbf{f}^2 + 2a_{21}c_2\mathbf{f_{xt}} \cdot \mathbf{f} + c_2^2\mathbf{f_{tt}}) + O(h_{k+1}^3)
\end{aligned} \quad (3.154)$$

and the second-order Taylor series of $\mathbf{f}(\mathbf{X}_3, t_k + c_3 h_{k+1}, \mathbf{p})$

$$
\begin{aligned}
\mathbf{f}_3 &= \mathbf{f}(\mathbf{x}_k + a_{31}\mathbf{f} + a_{32}\mathbf{f}_2, t_k + c_3 h_{k+1}, \mathbf{p}) \\
&= \mathbf{f} + h_{k+1}(\mathbf{f}_\mathbf{x} \cdot (a_{31}\mathbf{f} + a_{32}\mathbf{f}_2) + c_3 \mathbf{f}_t) \\
&\quad + \frac{1}{2!} h_{k+1}^2 [\mathbf{f}_{\mathbf{xx}} \cdot (a_{31}\mathbf{f} + a_{32}\mathbf{f}_2)^2 + 2\mathbf{f}_{\mathbf{x}t}(a_{31}\mathbf{f} + a_{32}\mathbf{f}_2)c_3 + c_3^2 \mathbf{f}_{tt}] + O(h_{k+1}^3) \\
&= \mathbf{f} + h_{k+1}[(a_{31} + a_{32})\mathbf{f}_\mathbf{x} \cdot \mathbf{f} + c_3 \mathbf{f}_t] \\
&\quad + h_{k+1}^2 [a_{32}(a_{21}\mathbf{f}_\mathbf{x}^2 \cdot \mathbf{f} + c_2 \mathbf{f}_\mathbf{x} \cdot \mathbf{f}_t) + \frac{1}{2!}(a_{31} + a_{32})^2 \mathbf{f}_{\mathbf{xx}} \cdot \mathbf{f}^2 \\
&\quad + (a_{31} + a_{32})c_3 \mathbf{f}_{\mathbf{x}t} \cdot \mathbf{f} + \frac{1}{2!} c_3^2 \mathbf{f}_{tt}] + O(h_{k+1}^3).
\end{aligned}
\tag{3.155}
$$

Thus, we have

$$
\begin{aligned}
\mathbf{x}_{k+1} &= \mathbf{x}_k + h_{k+1}(b_1 \mathbf{f} + b_2 \mathbf{f}_2 + b_3 \mathbf{f}_3) \\
&= \mathbf{x}_k + h_{k+1}(b_1 + b_2 + b_3)\mathbf{f} \\
&\quad + h_{k+1}^2 \{[b_2 a_{21} + b_3(a_{31} + a_{32})]\mathbf{f}_\mathbf{x} \cdot \mathbf{f} + (b_2 c_2 + b_3 c_3)\mathbf{f}_t\} \\
&\quad + h_{k+1}^3 \{\frac{1}{2!}(b_2 a_{21}^2 + b_3(a_{31} + a_{32})^2)\mathbf{f}_{\mathbf{xx}} \cdot \mathbf{f}^2 \\
&\quad + [b_2 a_{21} c_2 + b_3(a_{31} + a_{32})c_3]\mathbf{f}_{\mathbf{x}t} \cdot \mathbf{f} + \frac{1}{2!}(b_2 c_2^2 + b_3 c_3^2)\mathbf{f}_{tt} \\
&\quad + b_3 a_{32} a_{21}\mathbf{f}_\mathbf{x}^2 \cdot \mathbf{f} + b_3 a_{32} c_2 \mathbf{f}_\mathbf{x} \cdot \mathbf{f}_t\} + O(h_{k+1}^4).
\end{aligned}
\tag{3.156}
$$

From the third-order Taylor series of $\mathbf{x}(t_k + h_{k+1})$, we have

$$
\begin{aligned}
\mathbf{x}(t_k + h_{k+1}) &= \mathbf{x}(t_k) + \dot{\mathbf{x}}(t_k)h_{k+1} + \frac{1}{2!}h_{k+1}^2 \ddot{\mathbf{x}}(t_k) + \frac{1}{3!}h_{k+1}^3 \mathbf{x}^{(3)}(t_k) + O(h_{k+1}^4) \\
&= \mathbf{x}(t_k) + h_{k+1}\mathbf{f} + \frac{1}{2!}h_{k+1}^2(\mathbf{f}_\mathbf{x} \cdot \mathbf{f} + \mathbf{f}_t) \\
&\quad + \frac{1}{3!}h_{k+1}^3(\mathbf{f}_{\mathbf{xx}} \cdot \mathbf{f}^2 + 2\mathbf{f}_{\mathbf{x}t} \cdot \mathbf{f} + \mathbf{f}_\mathbf{x}^2 \cdot \mathbf{f} + \mathbf{f}_\mathbf{x} \cdot \mathbf{f}_t + \mathbf{f}_{tt}) + O(h_{k+1}^4).
\end{aligned}
\tag{3.157}
$$

Using $\mathbf{x}(t_k + h_{k+1}) = \mathbf{x}_{k+1}$, comparison of Eqs. (3.156) and (3.157) at $\mathbf{x}(t_k) = \mathbf{x}_k$ gives

$$
\begin{aligned}
h_{k+1} &\Rightarrow \mathbf{f}: & b_1 + b_2 + b_3 &= 1, \\
h_{k+1}^2 &\Rightarrow \mathbf{f}_\mathbf{x}\mathbf{f}: & b_2 a_{21} + b_3(a_{31} + a_{32}) &= \frac{1}{2}, \\
h_{k+1}^2 &\Rightarrow \mathbf{f}_t: & b_2 c_2 + b_3 c_3 &= \frac{1}{2},
\end{aligned}
$$

$$h^3_{k+1} \Rightarrow \mathbf{f_{xx}f}^2 : \quad \frac{1}{2}b_2a^2_{21} + \frac{1}{2}b_3(a_{31} + a_{32})^2 = \frac{1}{6},$$

$$h^3_{k+1} \Rightarrow \mathbf{f_{xt}f} : \quad b_2a_{21}c_2 + b_3(a_{31} + a_{32})c_3 = \frac{1}{3},$$

$$h^3_{k+1} \Rightarrow \mathbf{f_{tt}} : \quad b_2c^2_2 + b_3c^2_3 = \frac{1}{3}, \tag{3.158}$$

$$h^3_{k+1} \Rightarrow \mathbf{f^2_x f} : \quad b_3a_{32}a_{21} = \frac{1}{6},$$

$$h^3_{k+1} \Rightarrow \mathbf{f_x f_t} : \quad b_3a_{32}c_2 = \frac{1}{6}.$$

Hence, the 7th and 8th equations with 3rd and 4th equations of Eq. (3.158) become as follows:

$$a_{21} = c_2, \quad c_3 = a_{31} + a_{32}. \tag{3.159}$$

From the foregoing equation, Eq. (3.158) becomes

$$a_{21} = c_2,$$
$$a_{31} + a_{32} = c_3, \tag{3.160}$$
$$b_1 + b_2 + b_3 = 1,$$

and

$$b_2c_2 + b_3c_3 = \frac{1}{2},$$
$$b_2c^2_2 + b_3c^2_3 = \frac{1}{3}, \tag{3.161}$$
$$b_3a_{32}c_2 = \frac{1}{6}.$$

Equation (3.161) can be deformed as

$$\begin{pmatrix} c_2 & c_3 & -\frac{1}{2} \\ c^2_2 & c^2_3 & -\frac{1}{3} \\ 0 & a_{32}c_2 & -\frac{1}{6} \end{pmatrix} \begin{pmatrix} b_2 \\ b_3 \\ 1 \end{pmatrix} = \begin{pmatrix} 0 \\ 0 \\ 0 \end{pmatrix}. \tag{3.162}$$

Thus, the condition for existence of Eq. (3.162) requires

$$\begin{vmatrix} c_2 & c_3 & -\frac{1}{2} \\ c^2_2 & c^2_3 & -\frac{1}{3} \\ 0 & a_{32}c_2 & -\frac{1}{6} \end{vmatrix} = 0 \tag{3.163}$$

from which we have

$$c_3(c_3 - c_2) = a_{32}c_2(3c_2 - 1). \tag{3.164}$$

In summary, for chosen $c_2$ and $c_3$, we have $a_{21}$, $a_{31}$, and $a_{32}$ given by

$$\begin{aligned}
a_{21} &= c_2, \\
a_{31} + a_{32} &= c_3, \\
c_3(c_3 - c_2) &= a_{32}c_2(3c_2 - 1).
\end{aligned} \tag{3.165}$$

and the coefficients $b_1$, $b_2$, and $b_3$ are determined from the chosen $c_2$ and $c_3$ via the following equations:

$$\begin{aligned}
b_2c_2 + b_3c_3 &= \frac{1}{2}, \\
b_2c_2^2 + b_3c_3^2 &= \frac{1}{3}, \\
b_1 + b_2 + b_3 &= 1.
\end{aligned} \tag{3.166}$$

Since the two coefficients should be arbitrarily selected, there are infinite solutions for coefficients. Herein, consider a few special cases.

(i)   For $c_2 = 1/2$ and $c_3 = 1$, we have

$$
\begin{array}{c|ccc}
0 & 0 & 0 & 0 \\
c_2 & a_{21} & 0 & 0 \\
c_3 & a_{31} & a_{32} & 0 \\
\hline
 & b_1 & b_2 & b_3
\end{array}
\quad \xrightarrow{c_2 = \frac{1}{2},\, c_3 = 1} \quad
\begin{array}{c|ccc}
0 & 0 & 0 & 0 \\
\frac{1}{2} & \frac{1}{2} & 0 & 0 \\
1 & -1 & 2 & 0 \\
\hline
 & \frac{1}{6} & \frac{2}{3} & \frac{1}{6}
\end{array}
\tag{3.167}
$$

which is called the classic third-order Runge–Kutta method. In other words, we have

$$\begin{aligned}
\mathbf{x}_{k+1} &= \mathbf{x}_k + \frac{1}{6}h_{k+1}(\mathbf{f}_1 + 4\mathbf{f}_2 + \mathbf{f}_3); \\
\mathbf{f}_1 &= \mathbf{f}(\mathbf{x}_k, t_k, \mathbf{p}), \\
\mathbf{f}_2 &= \mathbf{f}(\mathbf{x}_k + \frac{1}{2}\mathbf{f}_1, t_k + \frac{1}{2}h_{k+1}, \mathbf{p}), \\
\mathbf{f}_3 &= \mathbf{f}(\mathbf{x}_k - \mathbf{f}_1 + 2\mathbf{f}_2, t_k + h_{k+1}, \mathbf{p}).
\end{aligned} \tag{3.168}$$

(ii)  For $c_2 = c_3 = 2/3$, we have

$$
\begin{array}{c|ccc}
0 & 0 & 0 & 0 \\
c_2 & a_{21} & 0 & 0 \\
c_3 & a_{31} & a_{32} & 0 \\
\hline
 & b_1 & b_2 & b_3
\end{array}
\quad \xrightarrow{c_2 = c_3 = \frac{2}{3}} \quad
\begin{array}{c|ccc}
0 & 0 & 0 & 0 \\
\frac{2}{3} & \frac{2}{3} & 0 & 0 \\
\frac{2}{3} & 0 & \frac{2}{3} & 0 \\
\hline
 & \frac{1}{4} & \frac{3}{8} & \frac{3}{8}
\end{array}
\tag{3.169}
$$

which is called the Nystrom form related to third-order Runge–Kutta method. That is,

$$
\begin{aligned}
\mathbf{x}_{k+1} &= \mathbf{x}_k + \frac{1}{8}h_{k+1}(2\mathbf{f}_1 + 3\mathbf{f}_2 + 3\mathbf{f}_3); \\
\mathbf{f}_1 &= \mathbf{f}(\mathbf{x}_k, t_k, \mathbf{p}), \\
\mathbf{f}_2 &= \mathbf{f}(\mathbf{x}_k + \frac{1}{3}\mathbf{f}_1, t_k + \frac{2}{3}h_{k+1}, \mathbf{p}), \\
\mathbf{f}_3 &= \mathbf{f}(\mathbf{x}_k + \frac{1}{3}\mathbf{f}_2, t_k + \frac{2}{3}h_{k+1}, \mathbf{p}).
\end{aligned}
\tag{3.170}
$$

(iii)  For $c_2 = 1/3$ and $c_3 = 2/3$, we have

$$
\begin{array}{c|ccc}
0 & 0 & 0 & 0 \\
c_2 & a_{21} & 0 & 0 \\
c_3 & a_{31} & a_{32} & 0 \\
\hline
 & b_1 & b_2 & b_3
\end{array}
\quad \xrightarrow{c_2 = \frac{1}{3},\, c_3 = \frac{2}{3}} \quad
\begin{array}{c|ccc}
0 & 0 & 0 & 0 \\
\frac{1}{3} & \frac{1}{3} & 0 & 0 \\
\frac{2}{3} & 0 & \frac{2}{3} & 0 \\
\hline
 & \frac{1}{4} & 0 & \frac{3}{4}
\end{array}
\tag{3.171}
$$

which is called the Heun form related to the third-order Runge–Kutta method. That is,

$$
\begin{aligned}
\mathbf{x}_{k+1} &= \mathbf{x}_k + \frac{1}{4}h_{k+1}(\mathbf{f}_1 + \mathbf{f}_3); \\
\mathbf{f}_1 &= \mathbf{f}(\mathbf{x}_k, t_k, \mathbf{p}), \\
\mathbf{f}_2 &= \mathbf{f}(\mathbf{x}_k + \frac{1}{3}\mathbf{f}_1, t_k + \frac{1}{3}h_{k+1}, \mathbf{p}), \\
\mathbf{f}_3 &= \mathbf{f}(\mathbf{x}_k + \frac{2}{3}\mathbf{f}_2, t_k + \frac{2}{3}h_{k+1}, \mathbf{p}).
\end{aligned}
\tag{3.172}
$$

For the error analysis, the third-order Taylor series of $\mathbf{f}(\mathbf{X}_2, t_k + c_2 h_{k+1}, \mathbf{p})$ is given by

$$
\begin{aligned}
\mathbf{f}_2 &= \mathbf{f}(\mathbf{x}_k + a_{21}\mathbf{f}, t_k + c_2 h_{k+1}, \mathbf{p}) \\
&= \mathbf{f} + h_{k+1}(a_{21}\mathbf{f}_\mathbf{x} \cdot \mathbf{f} + c_2 \mathbf{f}_t) + \frac{1}{2!}h_{k+1}^2(a_{21}^2 \mathbf{f}_{\mathbf{xx}} \cdot \mathbf{f}^2 + 2a_{21}c_2 \mathbf{f}_{\mathbf{x}t} \cdot \mathbf{f} + c_2^2 \mathbf{f}_{tt}) \\
&\quad + \frac{1}{3!}h_{k+1}^3(a_{21}^3 \mathbf{f}_{\mathbf{xxx}} \cdot \mathbf{f}^3 + 3a_{21}^2 c_2 \mathbf{f}_{\mathbf{xx}t} \cdot \mathbf{f}^2 + 3a_{21}c_2^2 \mathbf{f}_{\mathbf{x}tt} \cdot \mathbf{f} + c_2^3 \mathbf{f}_{ttt}) + O(h_{k+1}^4)
\end{aligned}
\tag{3.173}
$$

and the third-order Taylor series of $\mathbf{f}(\mathbf{X}_3, t_k + c_3 h_{k+1}, \mathbf{p})$ is given by

$$
\begin{aligned}
\mathbf{f}_3 &= \mathbf{f}(\mathbf{x}_k + a_{31}\mathbf{f} + a_{32}\mathbf{f}_2, t_k + c_3 h_{k+1}, \mathbf{p}) \\
&= \mathbf{f} + h_{k+1}[\mathbf{f}_\mathbf{x} \cdot (a_{31}\mathbf{f} + a_{32}\mathbf{f}_2) + c_3 \mathbf{f}_t] \\
&\quad + \frac{1}{2!}h_{k+1}^2[\mathbf{f}_{\mathbf{xx}} \cdot (a_{31}\mathbf{f} + a_{32}\mathbf{f}_2)^2 + 2\mathbf{f}_{\mathbf{x}t} \cdot (a_{31}\mathbf{f} + a_{32}\mathbf{f}_2)c_3 + c_3^2 \mathbf{f}_{tt}]
\end{aligned}
$$

$$+ \frac{1}{3!} h_{k+1}^3 [\mathbf{f_{xxx}} \cdot (a_{31}\mathbf{f} + a_{32}\mathbf{f_2})^3 + 3c_3 \mathbf{f_{xxt}} \cdot (a_{31}\mathbf{f} + a_{32}\mathbf{f_2})^2$$
$$+ 3c_3^2 \mathbf{f_{xxt}} \cdot (a_{31}\mathbf{f} + a_{32}\mathbf{f_2}) + c_3^3 \mathbf{f_{ttt}}] + O(h_{k+1}^4). \tag{3.174}$$

We have

$$\mathbf{f_3} = \mathbf{f} + h_{k+1}[(a_{31} + a_{32})\mathbf{f_x} \cdot \mathbf{f} + c_3 \mathbf{f_t}] + h_{k+1}^2[a_{32}\mathbf{f_x} \cdot (a_{21}\mathbf{f_x} \cdot \mathbf{f} + c_2\mathbf{f_t})$$
$$+ \frac{1}{2!}(a_{31} + a_{32})^2 \mathbf{f_{xx}} \cdot \mathbf{f}^2 + (a_{31} + a_{32})c_3 \mathbf{f_{xt}} \cdot \mathbf{f} + \frac{1}{2!}c_3^2 \mathbf{f_{tt}}]$$
$$+ h_{k+1}^3 [\frac{1}{2!} a_{32}\mathbf{f_x} \cdot (a_{21}^2 \mathbf{f_{xx}} \cdot \mathbf{f}^2 + 2a_{21}c_2\mathbf{f_{xt}} \cdot \mathbf{f} + c_2^2 \mathbf{f_{tt}})$$
$$+ a_{32}c_3 D\mathbf{f_x} \cdot (a_{21}\mathbf{f} + c_2\mathbf{f_t})] + \frac{1}{3!} h_{k+1}^3 [\mathbf{f_{xxx}}(a_{31} + a_{32})^3 \cdot \mathbf{f}^3$$
$$+ 3c_3(a_{31} + a_{32})^2 \mathbf{f_{xxt}} \cdot \mathbf{f}^2 + 3c_3^2(a_{31} + a_{32})\mathbf{f_{xtt}} \cdot \mathbf{f} + c_3^3 \mathbf{f_{ttt}}] + O(h_{k+1}^4). \tag{3.175}$$

Thus, the $h_{k+1}^3$ terms of $b_1\mathbf{f} + b_2\mathbf{f_2} + b_3\mathbf{f_3}$ is given by $\partial^3(b_1\mathbf{f} + b_2\mathbf{f_2} + b_3\mathbf{f_3})/\partial h_{k+1}^3$, i.e.,

$$\frac{\partial^3}{\partial h_{k+1}^3}(b_1\mathbf{f} + b_2\mathbf{f_2} + b_3\mathbf{f_3})$$
$$= b_2[\frac{1}{3!}(a_{21}^3 \mathbf{f_{xxx}}\mathbf{f}^3 + 3a_{21}^2 c_2 \mathbf{f_{xxt}}\mathbf{f}^2 + 3a_{21}c_2^2 \mathbf{f_{xtt}}\mathbf{f} + c_2^3 \mathbf{f_{ttt}})]$$
$$+ b_3\{\frac{1}{2!} a_{32}\mathbf{f_x}(a_{21}^2 \mathbf{f_{xx}}\mathbf{f}^2 + 2a_{21}c_2\mathbf{f_{xt}}\mathbf{f} + c_2^2 \mathbf{f_{tt}}) + a_{32}c_3 D\mathbf{f_x} \cdot (a_{21}\mathbf{f} + c_2\mathbf{f_t})$$
$$+ \frac{1}{3!}[(a_{31} + a_{32})^3 \mathbf{f_{xxx}}\mathbf{f}^3 + 3c_3(a_{31} + a_{32})^2 \mathbf{f_{xxt}}\mathbf{f}^2 + 3c_3^2(a_{31} + a_{32})\mathbf{f_{xtt}}\mathbf{f} + c_3^3 \mathbf{f_{ttt}}]\}$$
$$= \frac{1}{3!}([b_3 c_3^3 + b_2 a_{21}^3]\mathbf{f_{xxx}}\mathbf{f}^3 + 3[b_2 c_2^3 + b_3 c_3^3]\mathbf{f_{xxt}}\mathbf{f}^2 + 3[b_2 c_2^3 + b_3 c_3^3]\mathbf{f_{xtt}}\mathbf{f}$$
$$+ [b_3 c_3^3 + b_2 c_2^3]\mathbf{f_{ttt}}) + \frac{1}{2!} a_{32}b_3\mathbf{f_x}(a_{21}^2 \mathbf{f_{xx}}\mathbf{f}^2 + 2a_{21}c_2\mathbf{f_{xt}}\mathbf{f} + c_2^2 \mathbf{f_{tt}})$$
$$+ b_3 a_{32}c_3 D\mathbf{f_x} \cdot (a_{21}\mathbf{f} + c_2\mathbf{f_t})$$
$$= \frac{1}{3!}[b_3 c_3^3 + b_2 c_2^3]D^3\mathbf{f} + \frac{1}{2!} a_{32}b_3 c_3^2 \mathbf{f_x}D^2\mathbf{f} + b_3 a_{32}c_3 c_2 D\mathbf{f_x} \cdot D\mathbf{f}. \tag{3.176}$$

The fourth-order Taylor series of $\mathbf{x}(t_k + h_{k+1})$ is given by

$$\mathbf{x}(t_k + h_{k+1})$$
$$= \mathbf{x}(t_k) + \dot{\mathbf{x}}(t_k)h_{k+1} + \frac{h_{k+1}^2}{2!}\ddot{\mathbf{x}}(t_k) + \frac{h_{k+1}^3}{3!}\mathbf{x}^{(3)}(t_k) + \frac{h_{k+1}^4}{4!}\mathbf{x}^{(4)}(t_k^c)$$
$$= \mathbf{x}(t_k) + h_{k+1}\mathbf{f} + \frac{h_{k+1}^2}{2!}(\mathbf{f_x}\mathbf{f} + \mathbf{f_t}) + \frac{h_{k+1}^3}{3!}(\mathbf{f_{xx}}\mathbf{f}^2 + 2\mathbf{f_{xt}}\mathbf{f} + \mathbf{f_x^2}\mathbf{f} + \mathbf{f_x}\mathbf{f_t} + \mathbf{f_{tt}})$$

$$+ \frac{h_{k+1}^4}{4!} \big[ \underbrace{(\mathbf{f_{xxx}} \cdot \mathbf{f} + \mathbf{f_{xxt}}) \cdot \mathbf{f}^2 + 2\mathbf{f_{xx}} \cdot D\mathbf{f} \cdot \mathbf{f}^2}_{D(\mathbf{f_{xx}f^2})} + \underbrace{2(\mathbf{f_{xxt}} \cdot \mathbf{f} + \mathbf{f_{xtt}}) \cdot \mathbf{f} + 2\mathbf{f_{xt}} \cdot D\mathbf{f}}_{D(2\mathbf{f_{xt}f})}$$

$$+ \underbrace{2\mathbf{f_x} \cdot (\mathbf{f_{xx}} \cdot \mathbf{f} + \mathbf{f_{xt}}) \cdot \mathbf{f} + \mathbf{f_x^2} \cdot D\mathbf{f}}_{D(\mathbf{f_x^2 f})} + \underbrace{D\mathbf{f_x} \cdot \mathbf{f}_t + \mathbf{f_x} \cdot (\mathbf{f_{xt}} \cdot \mathbf{f} + \mathbf{f_{tt}})}_{D(\mathbf{f_x f})} + \underbrace{\mathbf{f_{xtt}} \cdot \mathbf{f} + \mathbf{f_{ttt}}}_{D\mathbf{f}_{tt}} \big]$$

$$(3.177)$$

Simplification of the foregoing equation gives

$$\mathbf{x}(t_k + h_{k+1}) = \mathbf{x}(t_k) + h_{k+1}\mathbf{f} + \frac{h_{k+1}^2}{2!} D\mathbf{f} + \frac{h_{k+1}^3}{3!}(D^2\mathbf{f} + D\mathbf{f} \cdot \mathbf{f_x})$$
$$+ \frac{h_{k+1}^4}{4!}(D^3\mathbf{f} + D^2\mathbf{f} \cdot \mathbf{f_x} + 3D\mathbf{f} \cdot D\mathbf{f_x} + \mathbf{f_x^2} \cdot D\mathbf{f}). \tag{3.178}$$

Thus the fourth-order error function compared to the Taylor series is given by

$$\boldsymbol{\delta}^{(4)}(\mathbf{x}_k^c, t_k^c, \mathbf{p}) = [1 - 4(c_2^3 b_2 + c_3^3 b_3)D^3\mathbf{f}] + (1 - 12c_2^2 a_{32} b_3)\mathbf{f_x}D^2\mathbf{f}$$
$$+ (3 - 24c_2 c_3 a_{32} b_3)D\mathbf{f_x}D\mathbf{f} + \mathbf{f_x^2}D\mathbf{f}, \tag{3.179}$$
$$\text{with } D^m(\cdot) = [D_{\mathbf{x}}(\cdot) \cdot \mathbf{f} + D_t(\cdot)]^m = \sum_{k=0}^{m} C_m^k D_{\mathbf{x}^k t^{m-k}}(\cdot) \cdot \mathbf{f}^k.$$

If $\mathbf{x}(t_k) = \mathbf{x}_k$, then $\boldsymbol{\Phi}(\mathbf{x}(t_k), t_k, \mathbf{p}) = \boldsymbol{\Phi}(\mathbf{x}_k, t_k, \mathbf{p})$. Therefore, the local error is

$$\boldsymbol{\varepsilon}_{k+1} = \frac{1}{4!} h_{k+1}^4 \boldsymbol{\delta}^{(4)}(\mathbf{x}_k^c, t_k^c, \mathbf{p}) = O(h_{k+1}^4). \tag{3.180}$$

Using the notations of Eq. (3.145), the local error becomes

$$\|\boldsymbol{\varepsilon}_{k+1}\| = \frac{1}{4!} h_{k+1}^4 \big[ 8|1 - (c_2^3 b_2 + c_3^3 b_3)| $$
$$+ 4|1 - 12c_2^2 a_{32} b_3| + 4|3 - 24c_2 c_3 a_{32} b_3| + 2 \big] MK^3. \tag{3.181}$$

Similarly, one can obtain $c_2 = 1/2$ and $c_3 = 3/4$ for the optimized third-order Runge–Kutta method, which is given by

$$\begin{array}{c|ccc}
0 & 0 & 0 & 0 \\
c_2 & a_{21} & 0 & 0 \\
c_3 & a_{31} & a_{32} & 0 \\
\hline
 & b_1 & b_2 & b_3
\end{array}
\quad \xrightarrow{c_2 = \frac{1}{2},\, c_3 = \frac{3}{4}} \quad
\begin{array}{c|ccc}
0 & 0 & 0 & 0 \\
\frac{1}{2} & \frac{1}{2} & 0 & 0 \\
\frac{3}{4} & 0 & \frac{3}{4} & 0 \\
\hline
 & \frac{2}{9} & \frac{1}{3} & \frac{4}{9}
\end{array}
\tag{3.182}$$

The optimized third-order Runge–Kutta scheme is given by

$$
\begin{aligned}
\mathbf{x}_{k+1} &= \mathbf{x}_k + \frac{1}{9}h_{k+1}(2\mathbf{f}_1 + 3\mathbf{f}_2 + 4\mathbf{f}_3); \\
\mathbf{f}_1 &= \mathbf{f}(\mathbf{x}_k, t_k, \mathbf{p}), \\
\mathbf{f}_2 &= \mathbf{f}(\mathbf{x}_k + \frac{1}{2}\mathbf{f}_1, t_k + \frac{1}{2}h_{k+1}, \mathbf{p}), \\
\mathbf{f}_3 &= \mathbf{f}(\mathbf{x}_k + \frac{3}{4}\mathbf{f}_2, t_k + \frac{3}{4}h_{k+1}, \mathbf{p}).
\end{aligned}
\tag{3.183}
$$

Therefore, the local errors for third-order Runge–Kutta method can be computed as

$$
\begin{aligned}
\|\boldsymbol{\varepsilon}_{k+1}\| &= \frac{1}{2}h_{k+1}^4 M K^3 \quad \text{for} \quad c_2 = \frac{1}{2}, c_3 = 1, \\
\|\boldsymbol{\varepsilon}_{k+1}\| &= \frac{1}{4}h_{k+1}^4 M K^3 \quad \text{for} \quad c_2 = c_3 = \frac{1}{4}, \\
\|\boldsymbol{\varepsilon}_{k+1}\| &= \frac{1}{9}h_{k+1}^4 M K^3 \quad \text{for} \quad c_2 = \frac{1}{2}, c_3 = \frac{3}{4}, \\
\|\boldsymbol{\varepsilon}_{k+1}\| &= \frac{1}{9}h_{k+1}^4 M K^3 \quad \text{for} \quad c_2 = \frac{1}{3}, c_3 = \frac{2}{3}.
\end{aligned}
\tag{3.184}
$$

This optimization based on Eq. (3.145) is a very rough estimate, and sometimes, such estimates are very conservative. Thus, one does not use such a way to get the optimized Runge–Kutta scheme.

### 3.4.2 Runge–Kutta Method of Order 4

Consider the fourth-order Runge–Kutta method as

$$
\begin{aligned}
\mathbf{x}_{k+1} &= \mathbf{x}_k + h_{k+1}(b_1\mathbf{f}_1 + b_2\mathbf{f}_2 + b_3\mathbf{f}_3 + b_4\mathbf{f}_3); \\
\mathbf{f}_1 &= \mathbf{f}(\mathbf{x}_k, t_k, \mathbf{p}), \\
\mathbf{f}_2 &= \mathbf{f}(\mathbf{x}_k + a_{21}\mathbf{f}_1, t_k + c_2 h_{k+1}, \mathbf{p}), \\
\mathbf{f}_3 &= \mathbf{f}(\mathbf{x}_k + a_{31}\mathbf{f}_1 + a_{32}\mathbf{f}_2, t_k + c_3 h_{k+1}, \mathbf{p}), \\
\mathbf{f}_4 &= \mathbf{f}(\mathbf{x}_k + a_{41}\mathbf{f}_1 + a_{42}\mathbf{f}_2 + a_{43}\mathbf{f}_3, t_k + c_4 h_{k+1}, \mathbf{p}).
\end{aligned}
\tag{3.185}
$$

Using

$$
c_2 = a_{21}, \quad c_3 = a_{31} + a_{32}, \quad c_4 = a_{41} + a_{42} + a_{43}, \tag{3.186}
$$

the third-order Taylor series of $\mathbf{f}(\mathbf{X}_2, t_k + c_2 h_{k+1}, \mathbf{p})$ is given in Eq. (3.173)

$$\mathbf{f}_2 = \mathbf{f}(\mathbf{x}_k + a_{21}\mathbf{f}, t_k + c_2 h_{k+1}, \mathbf{p})$$
$$= \mathbf{f} + h_{k+1}c_2 D\mathbf{f} + \frac{1}{2!}h_{k+1}^2 c_2^2 D^2\mathbf{f} + \frac{1}{3!}h_{k+1}^3 c_2^3 D^3\mathbf{f} + O(h_{k+1}^4) \tag{3.187}$$

and the third-order Taylor series of $\mathbf{f}(\mathbf{X}_3, t_k + c_3 h_{k+1}, \mathbf{p})$ is given in Eq. (3.175)

$$\mathbf{f}_3 = \mathbf{f} + h_{k+1}c_3 D\mathbf{f} + h_{k+1}^2 [a_{32}c_2\mathbf{f_x} \cdot D\mathbf{f} + \frac{1}{2!}c_3^2 D^2\mathbf{f}]$$
$$+ h_{k+1}^3 \{\frac{1}{2!}a_{32}c_2^2\mathbf{f_x} \cdot D\mathbf{f} + a_{32}c_3 c_2 D\mathbf{f_x} \cdot D\mathbf{f}\} + \frac{1}{3!}h_{k+1}^3 c_3^3 D^3\mathbf{f} + O(h_{k+1}^4) \tag{3.188}$$

and

$$\mathbf{f}_4 = \mathbf{f} + h_{k+1}[\mathbf{f_x} \cdot (a_{41}\mathbf{f} + a_{42}\mathbf{f}_2 + a_{43}\mathbf{f}_3) + c_4\mathbf{f}_t]$$
$$+ \frac{1}{2!}h_{k+1}^2 [\mathbf{f_{xx}} \cdot (a_{41}\mathbf{f} + a_{42}\mathbf{f}_2 + a_{43}\mathbf{f}_3)^2 + 2c_4\mathbf{f}_{xt} \cdot (a_{41}\mathbf{f} + a_{42}\mathbf{f}_2 + a_{43}\mathbf{f}_3) + c_4^2\mathbf{f}_{tt}]$$
$$+ \frac{1}{3!}h_{k+1}^3 [\mathbf{f_{xxx}} \cdot (a_{41}\mathbf{f} + a_{42}\mathbf{f}_2 + a_{43}\mathbf{f}_3)^3 + 3c_4\mathbf{f}_{xxt} \cdot (a_{41}\mathbf{f} + a_{42}\mathbf{f}_2 + a_{43}\mathbf{f}_3)^2$$
$$+ 3c_4^2\mathbf{f}_{xtt} \cdot (a_{41}\mathbf{f} + a_{42}\mathbf{f}_2 + a_{43}\mathbf{f}_3)^2 + c_4^3\mathbf{f}_{tt}] + O(h_{k+1}^4)$$
$$= \mathbf{f} + h_{k+1}c_4 D\mathbf{f} + h_{k+1}^2 [(a_{42}c_2 + a_{43}c_3)\mathbf{f_x} \cdot D\mathbf{f} + \frac{1}{2!}c_4^2 D^2\mathbf{f}]$$
$$+ h_{k+1}^3 [\frac{1}{2!}(a_{42}c_2^2 + a_{43}c_3^2)\mathbf{f_x} \cdot D^2\mathbf{f} + c_4(c_2 a_{42} + c_3 a_{43})D\mathbf{f_x} \cdot D\mathbf{f}]$$
$$+ \frac{1}{3!}h_{k+1}^3 [c_4^3 D^3\mathbf{f} + c_2 a_{32} a_{43}\mathbf{f_x}^2 \cdot D\mathbf{f}] + O(h_{k+1}^4). \tag{3.189}$$

Thus, we have

$$\mathbf{x}_{k+1} = \mathbf{x}_k + h_{k+1}(b_1\mathbf{f} + b_2\mathbf{f}_2 + b_3\mathbf{f}_3 + b_4\mathbf{f}_4)$$
$$= \mathbf{x}_k + h_{k+1}(b_1 + b_2 + b_3 + b_4)\mathbf{f} + h_{k+1}^2(b_2 c_2 + b_3 c_3 + b_4 c_4)D\mathbf{f}$$
$$+ h_{k+1}^3 \{\frac{1}{2!}(b_2 c_2^2 + b_3 c_3^2 + b_4 c_4^2)D^2\mathbf{f} + [b_3 a_{32}c_2 + b_4(c_2 a_{42} + c_3 a_{43})]\mathbf{f_x} \cdot D\mathbf{f}\}$$
$$+ h_{k+1}^4 \{\frac{1}{3!}(b_3 c_3^3 + b_2 c_2^3 + b_4 c_4^3)D^3\mathbf{f}$$
$$+ \frac{1}{2!}[b_3 a_{32}c_2^2 + b_4(c_2^2 a_{42} + c_3^2 a_{43})a_{32}a_{21}]\mathbf{f_x}D^2\mathbf{f}$$
$$+ [b_3 a_{32}c_3 c_2 + b_4 c_4(c_2 a_{42} + c_3 a_{43})]D\mathbf{f_x} \cdot D\mathbf{f}$$
$$+ b_4 c_1 a_{32} a_{43}\mathbf{f_x}^2 D\mathbf{f}\} + O(h_{k+1}^5). \tag{3.190}$$

The fourth-order Taylor series of $\mathbf{x}(t_k + h_{k+1})$ is given in Eq. (3.178), i.e.

$$\mathbf{x}(t_k + h_{k+1}) = \mathbf{x}(t_k) + h_{k+1}\mathbf{f} + \frac{h_{k+1}^2}{2!}D\mathbf{f} + \frac{h_{k+1}^3}{3!}(D^2\mathbf{f} + D\mathbf{f}\cdot\mathbf{f_x})$$

$$+ \frac{h_{k+1}^4}{4!}(D^3\mathbf{f} + D^2\mathbf{f}\cdot\mathbf{f_x} + 3D\mathbf{f}\cdot D\mathbf{f_x} + \mathbf{f_x}^2\cdot D\mathbf{f}). \tag{3.191}$$

Using $\mathbf{x}(t_k + h_{k+1}) = \mathbf{x}_{k+1}$, comparison of Eqs. (3.190) and (3.191) at $\mathbf{x}(t_k) = \mathbf{x}_k$ gives

$$
\begin{aligned}
h_{k+1} &\Rightarrow \mathbf{f}: & b_1 + b_2 + b_3 + b_4 &= 1, \\[4pt]
h_{k+1}^2 &\Rightarrow D\mathbf{f}: & b_2c_2 + b_3c_3 + b_4c_4 &= \frac{1}{2}, \\[4pt]
h_{k+1}^3 &\Rightarrow D^2\mathbf{f}: & b_2c_2^2 + b_3c_3^2 + b_4c_4^2 &= \frac{1}{3}, \\[4pt]
h_{k+1}^3 &\Rightarrow \mathbf{f_x}D\mathbf{f}: & b_3a_{32}c_2 + b_4(c_2a_{42} + c_3a_{43}) &= \frac{1}{6}, \\[4pt]
h_{k+1}^4 &\Rightarrow D^3\mathbf{f}: & b_3c_3^3 + b_2c_2^3 + b_4c_4^3 &= \frac{1}{4}, \\[4pt]
h_{k+1}^4 &\Rightarrow \mathbf{f_x}D^2\mathbf{f}: & b_3a_{32}c_2^2 + b_4(c_2^2a_{42} + c_3^2a_{43})a_{32}a_{21} &= \frac{1}{12}, \\[4pt]
h_{k+1}^4 &\Rightarrow D\mathbf{f_x}\cdot D\mathbf{f}: & b_3a_{32}c_3c_2 + b_4c_4[c_2a_{42} + c_3a_{43}] &= \frac{1}{8}, \\[4pt]
h_{k+1}^4 &\Rightarrow \mathbf{f_x}^2 D\mathbf{f}: & b_4c_1a_{32}a_{43} &= \frac{1}{24}.
\end{aligned}
\tag{3.192}
$$

From Eqs. (3.186) and (3.192), there are 11 equations with 13 unknowns. Thus, two unknowns should be selected arbitrarily. Herein, consider a few special cases.

(i)  For $c_2 = 1/2$ and $c_3 = 1/2$, we have

$$
\begin{array}{c|cccc}
0 & 0 & 0 & 0 & 0 \\
c_2 & a_{21} & 0 & 0 & 0 \\
c_3 & a_{31} & a_{32} & 0 & 0 \\
c_4 & a_{41} & a_{42} & a_{43} & 0 \\
\hline
 & b_1 & b_2 & b_3 & b_4
\end{array}
\quad
\xrightarrow[c_4=1]{c_2=\frac{1}{2},\,c_3=\frac{1}{2}}
\quad
\begin{array}{c|cccc}
0 & 0 & 0 & 0 & 0 \\
\frac{1}{2} & \frac{1}{2} & 0 & 0 & 0 \\
\frac{1}{2} & 0 & \frac{1}{2} & 0 & 0 \\
1 & 0 & 0 & 1 & 0 \\
\hline
 & \frac{1}{6} & \frac{1}{3} & \frac{1}{3} & \frac{1}{6}
\end{array}
\tag{3.193}
$$

which is called the classic fourth-order Runge–Kutta method. In other words, we have

$$\mathbf{x}_{k+1} = \mathbf{x}_k + \frac{1}{6}h_{k+1}(\mathbf{f}_1 + 2\mathbf{f}_2 + 2\mathbf{f}_3 + \mathbf{f}_4);$$

$$\mathbf{f}_1 = \mathbf{f}(\mathbf{x}_k, t_k, \mathbf{p}),$$

$$\mathbf{f}_2 = \mathbf{f}(\mathbf{x}_k + \frac{1}{2}\mathbf{f}_1, t_k + \frac{1}{2}h_{k+1}, \mathbf{p}),$$

$$\mathbf{f}_3 = \mathbf{f}(\mathbf{x}_k + \frac{1}{2}\mathbf{f}_2, t_k + \frac{1}{2}h_{k+1}, \mathbf{p}),$$  (3.194)

$$\mathbf{f}_4 = \mathbf{f}(\mathbf{x}_k + \mathbf{f}_3, t_k + h_{k+1}, \mathbf{p}).$$

(ii)   For $c_2 = 1/3$, $c_3 = 2/3$, we have

$$
\begin{array}{c|cccc}
0 & 0 & 0 & 0 & 0 \\
c_2 & a_{21} & 0 & 0 & 0 \\
c_3 & a_{31} & a_{32} & 0 & 0 \\
c_4 & a_{41} & a_{42} & a_{43} & 0 \\
\hline
& b_1 & b_2 & b_3 & b_4
\end{array}
\quad \xrightarrow[c_4=1]{c_2=\frac{1}{3},\, c_3=\frac{2}{3}} \quad
\begin{array}{c|cccc}
0 & 0 & 0 & 0 & 0 \\
\frac{1}{3} & \frac{1}{3} & 0 & 0 & 0 \\
\frac{2}{3} & -\frac{1}{3} & 1 & 0 & 0 \\
1 & 1 & -1 & 1 & 0 \\
\hline
& \frac{1}{8} & \frac{3}{8} & \frac{3}{8} & \frac{1}{8}
\end{array}
$$  (3.195)

which is called the Kutta form relative to the fourth-order Runge–Kutta method.

$$\mathbf{x}_{k+1} = \mathbf{x}_k + \frac{1}{8}h_{k+1}(\mathbf{f}_1 + 3\mathbf{f}_2 + 3\mathbf{f}_3 + \mathbf{f}_4);$$

$$\mathbf{f}_1 = \mathbf{f}(\mathbf{x}_k, t_k, \mathbf{p}),$$

$$\mathbf{f}_2 = \mathbf{f}(\mathbf{x}_k + \frac{1}{3}\mathbf{f}_1, t_k + \frac{1}{3}h_{k+1}, \mathbf{p}),$$  (3.196)

$$\mathbf{f}_3 = \mathbf{f}(\mathbf{x}_k - \frac{1}{3}\mathbf{f}_1 + \mathbf{f}_2, t_k + \frac{2}{3}h_{k+1}, \mathbf{p}),$$

$$\mathbf{f}_4 = \mathbf{f}(\mathbf{x}_k + \mathbf{f}_1 - \mathbf{f}_2 + \mathbf{f}_3, t_k + h_{k+1}, \mathbf{p}).$$

(iii)   For $c_2 = 1/2$ and $c_3 = 1/2$, we have

$$
\begin{array}{c|cccc}
0 & 0 & 0 & 0 & 0 \\
c_2 & a_{21} & 0 & 0 & 0 \\
c_3 & a_{31} & a_{32} & 0 & 0 \\
c_4 & a_{41} & a_{42} & a_{43} & 0 \\
\hline
& b_1 & b_2 & b_3 & b_4
\end{array}
\quad \xrightarrow[c_4=1]{c_2=\frac{1}{2},\, c_3=\frac{1}{2}} \quad
\begin{array}{c|cccc}
0 & 0 & 0 & 0 & 0 \\
\frac{1}{2} & \frac{1}{2} & 0 & 0 & 0 \\
\frac{1}{2} & \frac{\sqrt{2}-1}{2} & \frac{2-\sqrt{2}}{2} & 0 & 0 \\
1 & 0 & -\frac{\sqrt{2}}{2} & 1+\frac{\sqrt{2}}{2} & 0 \\
\hline
& \frac{1}{6} & \frac{2-\sqrt{2}}{6} & \frac{2+\sqrt{2}}{6} & \frac{1}{6}
\end{array}
$$  (3.197)

which is called the Gill form related to the fourth-order Runge–Kutta method, which is based on minimizing the round-off error. That is,

$$\mathbf{x}_{k+1} = \mathbf{x}_k + \frac{1}{6}h_{k+1}[\mathbf{f}_1 + (2 - \sqrt{2})\mathbf{f}_2 + (2 + \sqrt{2})\mathbf{f}_3 + \mathbf{f}_4];$$

$$\mathbf{f}_1 = \mathbf{f}(\mathbf{x}_k, t_k, \mathbf{p}),$$

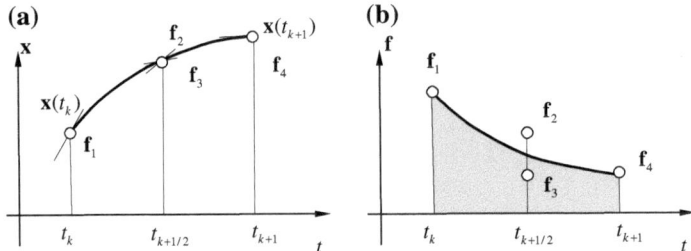

**Fig. 3.3** **a** Solution curve and slopes at three points, **b** integrations of vector fields during the time subinterval $t \in [t_k, t_{k+1}]$

$$\mathbf{f}_2 = \mathbf{f}(\mathbf{x}_k + \frac{1}{2}\mathbf{f}_1, t_k + \frac{1}{2}h_{k+1}, \mathbf{p}),$$

$$\mathbf{f}_3 = \mathbf{f}(\mathbf{x}_k + \frac{\sqrt{2}-1}{2}\mathbf{f}_1 + \frac{2-\sqrt{2}}{2}\mathbf{f}_2, t_k + \frac{1}{2}h_{k+1}, \mathbf{p}),$$   (3.198)

$$\mathbf{f}_4 = \mathbf{f}(\mathbf{x}_k - \frac{\sqrt{2}}{2}\mathbf{f}_2 + (1 + \frac{\sqrt{2}}{2})\mathbf{f}_3, t_k + h_{k+1}, \mathbf{p}).$$

Suppose there is a curve given by dynamical system during time $t \in [t_k, t_{k+1}]$. Consider three points at $t = t_k, t_{k+1/2}, t_{k+1}$ with four slopes, and the integration of vector field during time interval $t \in [t_k, t_{k+1}]$, as shown in Fig. 3.3. The integration of Eq. (3.1) gives

$$\mathbf{x}_{k+1} - \mathbf{x}_k = \int_{t_k}^{t_{k+1}} \mathbf{f}(\mathbf{x}, t, \mathbf{p}) dt.$$   (3.199)

If the Simpson's rule is used with three points, the approximation of (3.199) is

$$\mathbf{x}_{k+1} - \mathbf{x}_k \approx \frac{1}{6}h_{k+1}[\mathbf{f}(\mathbf{x}(t_k), t_k, \mathbf{p}) + 4\mathbf{f}(\mathbf{x}(t_{k+1/2}), t_{k+1/2}, \mathbf{p}) + \mathbf{f}(\mathbf{x}(t_{k+1}), t_{k+1}, \mathbf{p})].$$   (3.200)

Setting

$$\mathbf{f}(\mathbf{x}(t_k), t_k, \mathbf{p}) = \mathbf{f}_1,$$
$$\mathbf{f}(\mathbf{x}(t_{k+1}), t_{k+1}, \mathbf{p}) \approx \mathbf{f}_4,$$   (3.201)
$$\mathbf{f}(\mathbf{x}(t_{k+1/2}), t_{k+1/2}, \mathbf{p}) \approx \frac{1}{2}(\mathbf{f}_2 + \mathbf{f}_3).$$

Thus, we have

$$\mathbf{x}_{k+1} - \mathbf{x}_k \approx \frac{1}{6}h_{k+1}[\mathbf{f}_1 + 2(\mathbf{f}_2 + \mathbf{f}_3) + \mathbf{f}_4] \tag{3.202}$$

which gives the classic Runge–Kutta method. From the Simpson approximation, the local error function is

$$\|\boldsymbol{\varepsilon}_{k+1}\| = \frac{1}{2880}h_{k+1}^5\mathbf{f}^{(4)}(\mathbf{x}_k^c, t_k^c, \mathbf{p}) = O(h_{k+1}^5). \tag{3.203}$$

Similar to the third-order Runge–Kutta method, the fourth-order local error function is

$$\begin{aligned}
\boldsymbol{\delta}^{(5)}(\mathbf{x}_k^c, t_k^c, \mathbf{p}) = {}& [1 - 5(c_2^4 b_2 + c_3^4 b_3 + c_4^4 b_4)]D^4\mathbf{f} \\
& + \{6 - 60[c_2 c_3^2 a_{32} b_3 + b_4 c_4^2(a_{42} c_2 + a_{43} c_3)]\}D^2\mathbf{f_x}D\mathbf{f} \\
& + \{4 - 60[c_3 c_2^2 a_{32} b_3 + b_4 c_4(a_{42} c_2^2 + a_{43} c_3^2)]\}D\mathbf{f_x}D^2\mathbf{f} \\
& + (1 - 60 b_4 a_{43} a_{32} c_2^2)\mathbf{f_x^2}D^2\mathbf{f} \\
& + \{3 - 60[b_3 a_{32}^2 c_2^2 + b_4(a_{43} c_3 + a_{42} c_2)^2]\}\mathbf{f_{xx}}D^2\mathbf{f} \\
& + \{1 - 20[b_3 a_{32} c_2^2 + b_4(a_{42} c_2^2 + a_{43} c_3^2)]\}\mathbf{f_x}D^3\mathbf{f} \\
& + [7 - b_4 a_{43} a_{32} c_2(c_3 + c_4)]\mathbf{f_x}D\mathbf{f_x}D\mathbf{f} \\
& + \mathbf{f_x^3}D\mathbf{f}.
\end{aligned} \tag{3.204}$$

If $\mathbf{x}(t_k) = \mathbf{x}_k$, then $\boldsymbol{\Phi}(\mathbf{x}(t_k), t_k, \mathbf{p}) = \boldsymbol{\Phi}(\mathbf{x}_k, t_k, \mathbf{p})$. Therefore, the local error is

$$\boldsymbol{\varepsilon}_{k+1} = \frac{1}{5!}h_{k+1}^5\boldsymbol{\delta}^{(5)}(\mathbf{x}_k^c, t_k^c, \mathbf{p}) = O(h_{k+1}^5). \tag{3.205}$$

Other explicit Runge–Kutta methods can be referred to other reference books (e.g., Lapidus and Seinfeld 1971; Haier 1987).

## 3.5 Implicit Runge–Kutta Methods

After the explicit Runge–Kutta methods, the implicit Runge–Kutta method will be discussed herein. The Definition 3.14 is redefined for implicit Runge–Kutta method.

**Definition 3.15** The discretization of Eq. (3.1) during the time interval $[t_k, t_{k+1}]$ is given by

$$\begin{aligned}
& t_{k+1} = t_k + h_{k+1}, \\
& \mathbf{x}_{k+1} = \mathbf{x}_k + h_{k+1}\boldsymbol{\Phi}(\mathbf{x}_k, t_k, \mathbf{p}) \quad \text{for } k = 0, 1, 2, \ldots, M-1
\end{aligned} \tag{3.206}$$

where

$$\boldsymbol{\Phi}(\mathbf{x}_k, t_k, \mathbf{p}) = \sum_{i=1}^{s} b_i \mathbf{f}(\mathbf{X}_i, t_k + c_i h_{k+1}, \mathbf{p}),$$

$$\mathbf{X}_i = \mathbf{x}_k + h_{k+1} \sum_{j=1}^{s} a_{ij} \mathbf{f}(\mathbf{X}_j, t_k + c_j h_{k+1}, \mathbf{p}).$$

(3.207)

For one of $a_{ij} \neq 0$ $(j \geq i = 1, 2, \ldots, s)$, the aforesaid discretization is called the $s$-stage implicit Runge–Kutta method (or implicit Runge–Kutta method of order $s$, or the $s$th-order implicit Runge–Kutta method).

The Butcher tableau for the coefficients in the implicit Runge–Kutta method is presented as follows:

$$
\begin{array}{c|ccccc}
c_1 & a_{11} & a_{12} & \cdots & a_{1(s-1)} & a_{1s} \\
c_2 & a_{21} & a_{22} & \cdots & a_{2(s-1)} & a_{2s} \\
c_3 & a_{31} & a_{32} & \cdots & a_{3(s-1)} & a_{3s} \\
\vdots & \vdots & \vdots & & \vdots & \vdots \\
c_s & a_{s1} & a_{s2} & \cdots & a_{s(s-1)} & a_{ss} \\
\hline
 & b_1 & b_2 & \cdots & b_{s-1} & b_s
\end{array}
$$

(3.208)

For $a_{ij} = 0$ $(i < j)$, the aforesaid Runge–Kutta method is called the $s$-stage semi-implicit Runge–Kutta method (or semi-implicit Runge–Kutta method of order $s$, or the $s$th-order semi-implicit Runge–Kutta method).

$$
\begin{array}{c|ccccc}
c_1 & a_{11} & 0 & \cdots & 0 & 0 \\
c_2 & a_{21} & a_{22} & \cdots & 0 & 0 \\
c_3 & a_{31} & a_{32} & \cdots & 0 & 0 \\
\vdots & \vdots & \vdots & & \vdots & \vdots \\
c_s & a_{s1} & a_{s2} & \cdots & a_{s(s-1)} & a_{ss} \\
\hline
 & b_1 & b_2 & \cdots & b_{s-1} & b_s
\end{array}
$$

(3.209)

### 3.5.1 Polynomial Interpolation

Dynamical system in Eq. (3.1) can be converted into an integral equation. Integration of Eq. (3.1) over the interval $[t_k, t]$ gives

$$\mathbf{x}(t) = \mathbf{x}(t_k) + \int_{t_k}^{t} \mathbf{f}(\mathbf{x}, t, \mathbf{p}) \mathrm{d}t.$$

(3.210)

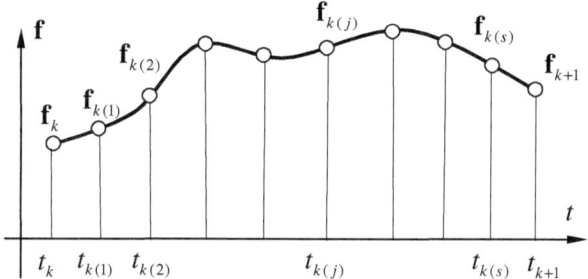

**Fig. 3.4** The node points at $t_{k(j)} = t_k + c_j h_{k+1}$ $(j = 1, 2, \ldots, s)$ with $t_{k(j)} \in [t_k, t_{k+1}]$ for Lagrange interpolation polynomial

Let $\mathbf{x}(t_k) \approx \mathbf{x}_k$ and the time interval $[t_k, t_{k+1}]$ are divided into $s$-segments as

$$t_{k(j)} = t_k + c_j h_{k+1} \qquad \text{for } j = 1, 2, \ldots, s$$
$$\text{with} \quad 0 \leq c_1 < c_2 < \cdots < c_s \leq 1 \tag{3.211}$$

where $h_{k+1} = t_{k+1} - t_k$. The integrand on the right hand of Eq. (3.210) is approximated with an interpolation polynomial $\mathbf{P}(t)$ of degree less than $s$, which interpolates $\mathbf{f}(\mathbf{x}(t), t, \mathbf{p})$ at node points $t_{k(j)}$ $(j = 1, 2, \ldots, s)$ on the time interval $[t_k, t_{k+1}]$. In Fig. 3.4, node points for $\mathbf{f}(\mathbf{x}(t), t, \mathbf{p})$ is presented with

$$\mathbf{P}(t_{k(j)}) = \mathbf{f}(\mathbf{x}(t_{k(j)}), t_{k(j)}, \mathbf{p}) = \mathbf{f}_{k(j)}. \tag{3.212}$$

Consider $\mathbf{P}(t)$ as a general polynomial of degree $(s - 1)$ with $s$-independent parameters $\mathbf{a}_0, \mathbf{a}_1, \ldots, \mathbf{a}_s$ as

$$\mathbf{P}(t) = \mathbf{a}_0 + \mathbf{a}_1 t + \cdots + \mathbf{a}_s t^{s-1}. \tag{3.213}$$

Using Eq. (3.213), we have

$$\mathbf{a}_0 + \mathbf{a}_1 t_{k(1)} + \cdots + \mathbf{a}_{s-1} t_{k(1)}^{s-1} = \mathbf{f}_{k(1)},$$
$$\mathbf{a}_0 + \mathbf{a}_1 t_{k(2)} + \cdots + \mathbf{a}_{s-1} t_{k(2)}^{s-1} = \mathbf{f}_{k(2)},$$
$$\vdots \tag{3.214}$$
$$\mathbf{a}_0 + \mathbf{a}_1 t_{k(s)} + \cdots + \mathbf{a}_{s-1} t_{k(s)}^{s-1} = \mathbf{f}_{k(s)}.$$

The foregoing equation can be rewritten as

$$\mathbf{M z} = \mathbf{F} \tag{3.215}$$

where

$$\mathbf{z} = (\mathbf{a}_0, \mathbf{a}_1, \ldots, \mathbf{a}_{s-1})^{\mathrm{T}},$$

$$\mathbf{F} = (\mathbf{f}_{k(1)}, \mathbf{f}_{k(2)}, \ldots, \mathbf{f}_{k(s)})^{\mathrm{T}},$$

$$\mathbf{M} = \begin{bmatrix} 1 & t_{k(1)} & \cdots & t_{k(1)}^{s-1} \\ 1 & t_{k(2)} & \cdots & t_{k(2)}^{s-1} \\ \vdots & \vdots & & \vdots \\ 1 & t_{k(s)} & \cdots & t_{k(s)}^{s-1} \end{bmatrix}. \tag{3.216}$$

The determinant of $\mathbf{M}$ is given by

$$\det(\mathbf{M}) = \prod_{0 \le i < j \le s-1} (t_{k(j)} - t_{k(i)}). \tag{3.217}$$

Without loss of generality, it is assumed that

$$\mathbf{f}_{k(i)} = \mathbf{0}, \quad \mathbf{f}_{k(j)} = \mathbf{f}_{k(j)} \quad \text{for } j \neq i, \tag{3.218}$$

where $\mathbf{0} = (0, 0, \ldots, 0)^{\mathrm{T}}$. Consider a special interpolation problem

$$\mathbf{P}_{(j)}(t) = \mathbf{c}_j \mathbf{f}_j \prod_{j=1,(j \neq i)}^{s} (t - t_{k(i)}), \tag{3.219}$$

where $\mathbf{c}_j = \mathrm{diag}(c_j^{(1)}, c_j^{(2)}, \ldots, c_j^{(n)})$.

For the polynomial in Eq. (3.213), there are $(s-1)$ zero points $t_{k(i)}(i \neq j)$. Using Eq. (3.219), we have

$$c_j^{(l)} = \prod_{j \neq i, j=1}^{s} \frac{1}{t_{k(j)} - t_{k(i)}}, \quad l = 1, 2, \ldots, n. \tag{3.220}$$

Thus

$$\mathbf{P}_{(j)}(t) = \mathbf{f}(\mathbf{x}_j, t_j, \mathbf{p}) l_j(t) \tag{3.221}$$

where

$$l_j(t) = \prod_{j \neq i, j=1}^{s} \frac{t - t_{k(i)}}{t_{k(j)} - t_{k(i)}}, \quad j = 1, 2, \ldots, s. \tag{3.222}$$

To solve the general interpolation, we have

$$\mathbf{P}_s(t) = \sum_{j=1}^{s} \mathbf{P}_{(j)}(t) = \sum_{j=1}^{s} \mathbf{f}(\mathbf{x}_j, t_j, \mathbf{p}) l_j(t). \tag{3.223}$$

For $t \in [t_k, t_{k+1}]$, the error of the interpolation is computed by

$$\mathbf{e} = \mathbf{f}(\mathbf{x}, t, \mathbf{p}) - \sum_{j=1}^{s} \mathbf{f}(\mathbf{x}_j, t_j, \mathbf{p}) l_j(t) = \frac{1}{s!} \prod_{j=1}^{s} (t - t_{k(j)}) \mathbf{f}(\mathbf{x}_k^c, t_k^c, \mathbf{p}). \tag{3.224}$$

where $t_k^c \in [t_k, t_{k+1}]$ and the corresponding points $\mathbf{x}_k^c \in (\mathbf{x}_k \mapsto \mathbf{x}_{k+1})$ which means $\mathbf{x}_k^c$ is on the flow from $\mathbf{x}_k$ to $\mathbf{x}_{k+1}$.

### 3.5.2 Implicit Runge–Kutta Methods

As in Sect. 3.2, from the implicit Euler (backforward Euler) method, the Butcher tableau of the implicit Runge–Kutta method is

$$\begin{array}{c|c} c_1 & a_{11} \\ \hline & b_1 \end{array} \quad \xrightarrow{c_1=1} \quad \begin{array}{c|c} 1 & 1 \\ \hline & 1 \end{array} \tag{3.225}$$

From the implicit midpoint rule, the Butcher tableau of the implicit Runge–Kutta method is

$$\begin{array}{c|c} c_1 & a_{11} \\ \hline & b_1 \end{array} \quad \xrightarrow{c_1=\frac{1}{2}} \quad \begin{array}{c|c} \frac{1}{2} & \frac{1}{2} \\ \hline & 1 \end{array} \tag{3.226}$$

From the Hammer method, the Butcher tableau of the implicit Runge–Kutta method is

$$\begin{array}{c|cc} 0 & 0 & 0 \\ c_2 & a_{21} & a_{22} \\ \hline & b_1 & b_2 \end{array} \quad \xrightarrow{c_1=0,\, c_2=\frac{2}{3}} \quad \begin{array}{c|cc} 0 & 0 & 0 \\ \frac{2}{3} & \frac{1}{3} & \frac{1}{3} \\ \hline & \frac{1}{4} & \frac{3}{4} \end{array} \tag{3.227}$$

Discussion of the collocation method is presented as follows. If the vector function can be approximated by the $\mathbf{f}(\mathbf{x}(r), r, \mathbf{p}) \approx \mathbf{P}_s(r)$, thus the equation in Eq. (3.210) becomes

$$\mathbf{x}(t) \approx \mathbf{x}_k + \int_{t_k}^{t} \mathbf{P}_s(r) dr. \tag{3.228}$$

Using the Lagrange interpolation polynomial as

$$\mathbf{P}_s(r) = \sum_{j=1}^{s} \mathbf{f}(\mathbf{x}(t_{k(j)}), t_{k(j)}, \mathbf{p}) l_j(r), \tag{3.229}$$

where

$$l_j(t) = \prod_{i \neq j} \left( \frac{t - t_{k(i)}}{t_{k(j)} - t_{k(i)}} \right), \quad j = 1, 2, \ldots, s. \tag{3.230}$$

Thus we have

$$\mathbf{x}(t) \approx \mathbf{x}_k + \sum_{j=1}^{s} \mathbf{f}(\mathbf{x}(t_{k(j)}), t_{k(j)}, \mathbf{p}) \int_{t_k}^{t} l_j(r) \mathrm{d}r. \tag{3.231}$$

At node points $t_{k(i)}(i = 1, 2, \ldots, s)$ on the time interval $[t_k, t_{k+1}]$, we have

$$\mathbf{x}_{k(i)} \approx \mathbf{x}_k + \sum_{j=1}^{s} \mathbf{f}(\mathbf{x}_{k(j)}, t_{k(j)}, \mathbf{p}) \int_{t_k}^{t_{k(i)}} l_j(r) \mathrm{d}r. \tag{3.232}$$

For $t_{k(s)} = t_{k+1}$, we have $\mathbf{x}_{k+1} = \mathbf{x}_{k(s)}$.

$$\mathbf{x}_{k+1} = \mathbf{x}_k + \sum_{j=1}^{s} \mathbf{f}(\mathbf{x}_{k(j)}, t_{k(j)}, \mathbf{p}) \int_{t_k}^{t_{k+1}} l_j(r) \mathrm{d}r. \tag{3.233}$$

It is assumed that

$$b_j = \int_{t_k}^{t_{k+1}} l_j(r) \mathrm{d}r \quad \text{and} \quad a_{ij} = \int_{t_k}^{t_{k(i)}} l_j(r) \mathrm{d}r \tag{3.234}$$

with

$$\begin{aligned} &\mathbf{X}_i = \mathbf{x}_{k(i)}, \quad \mathbf{X}_j = \mathbf{x}_{k(j)}; \quad t_{k(j)} = t_k + c_j h_{k+1} \\ &\text{for} \quad i, j = 1, 2, \ldots, s. \end{aligned} \tag{3.235}$$

Thus, with $t_{k+1} = t_k + h_k$ ($k = 0, 1, 2, \ldots, M - 1$), Eq. (3.233) becomes

$$\mathbf{x}_{k+1} = \mathbf{x}_k + h_{k+1} \sum_{i=1}^{s} b_i \mathbf{f}(\mathbf{X}_i, t_k + c_i h_{k+1}, \mathbf{p}) \tag{3.236}$$

and Eq. (3.232) becomes

$$\mathbf{X}_i = \mathbf{x}_k + h_{k+1} \sum_{j=1}^{s} a_{ij} \mathbf{f}(\mathbf{X}_j, t_k + c_j h_{k+1}, \mathbf{p}) \qquad (3.237)$$

for $i = 1, 2, \ldots, s$. Equations (3.236) and (3.237) are identical to Eqs. (3.206) and (3.207). The aforementioned method is called collocation as the approximate polynomial function satisfies the specific values at the selected node points. The points $t_{k(i)}$ at which the true vector field functions are used are called the collocation points. The Runge–Kutta method is not collocation method. The above discussion is summarized. As in Hairer et al. (1987), a collocation polynomial can be defined as follows.

**Definition 3.16** For a time interval $[t_k, t_{k+1}]$ for dynamical system in Eq. (3.1), there is a set of distinct $c_j \in [0, 1]$ ($j = 1, 2, \ldots, s$) with $c_j < c_{j+1}$ and $t_{k(j)} = t_k + c_j h_{k+1}$. The collocation polynomial $\mathbf{X}(t)$ of degree $s$ is defined by

$$\begin{aligned} \mathbf{X}(t_k) &= \mathbf{x}_k, \\ \dot{\mathbf{X}}(t_{k(j)}) &= \mathbf{f}(\mathbf{X}(t_{k(j)}), t_{k(j)}, \mathbf{p}) \qquad \text{for } j = 1, 2, \ldots, s. \end{aligned} \qquad (3.238)$$

The numerical solution $\mathbf{x}_{k+1}$ at $t_{k+1} = t_k + h_{k+1}$ is given by

$$\mathbf{x}_{k+1} = \mathbf{X}(t_k + h_{k+1}). \qquad (3.239)$$

**Theorem 3.11** *For a time interval $[t_k, t_{k+1}]$ for dynamical system in* Eq. (3.1)*, the node points is at $t_{k(j)} = t_k + c_j h_{k+1}$ with $c_j \in [0, 1]$ ($j = 1, 2, \ldots, s$). Based on the collocation method in* Eq. (3.232)*, the implicit Runge–Kutta method in* Eqs. (3.206) *and* (3.207) *requires* Eq. (3.234)*, i.e.,*

$$b_j = \int_{t_k}^{t_{k+1}} l_j(r) \, dr \quad \text{and} \quad a_{ij} = \int_{t_k}^{t_{k(i)}} l_j(r) \, dr \qquad (3.240)$$

*with the Lagrange polynomials $l_j(t)$, i.e.,*

$$l_j(t) = \prod_{i \neq j} \left( \frac{t - t_{k(i)}}{t_{k(j)} - t_{k(i)}} \right), \quad j = 1, 2, \ldots, s. \qquad (3.241)$$

*Proof* In a time interval $[t_k, t_{k+1}]$, the dynamical system in Eq. (3.1) can be approximated by the Lagrange polynomial

$$\dot{\mathbf{X}}(t) = \sum_{j=1}^{s} \mathbf{f}(\mathbf{X}(t_{k(j)}), t_{k(j)}, \mathbf{p}) l_j(t)$$

where

$$\mathbf{X}(t_k) = \mathbf{x}_k \quad \text{and}$$
$$\mathbf{f}_j = \mathbf{f}(\mathbf{X}(t_{k(j)}), t_{k(j)}, \mathbf{p}) \quad \text{for } j = 1, 2, \ldots, s.$$

The integration of $\dot{\mathbf{X}}(t)$ gives

$$\mathbf{X}(t) = \sum_{j=1}^{s} \mathbf{f}(\mathbf{X}(t_{k(j)}), t_{k(j)}, \mathbf{p}) \int_{t_k}^{t} l_j(r) \mathrm{d}r.$$

For $t = t_{k(i)} = t_k + c_i h_{k+1}$ and $\mathbf{X}(t_{k(j)}) = \mathbf{X}_j$, we have

$$\mathbf{X}(t_{k(i)}) = \sum_{j=1}^{s} \mathbf{f}(\mathbf{X}(t_{k(j)}), t_{k(j)}, \mathbf{p}) \int_{t_k}^{t_{k(i)}} l_j(r) \mathrm{d}r.$$

Compared to Eq. (3.206), we have

$$\mathbf{X}_i = \sum_{j=1}^{s} \mathbf{f}(\mathbf{X}_j, t_k + c_j h_{k+1}, \mathbf{p}) \int_{t_k}^{t_{k(i)}} l_j(t) \mathrm{d}t = \sum_{j=1}^{s} a_{ij} \mathbf{f}(\mathbf{X}_j, t_k + c_j h_{k+1}, \mathbf{p})$$

with

$$a_{ij} = \int_{t_k}^{t_{k(i)}} l_j(r) \mathrm{d}r.$$

For $t = t_{k+1}$ and $\mathbf{X}(t_{k+1}) = \mathbf{x}_{k+1}$, we have

$$\mathbf{x}_{k+1} = \mathbf{X}(t_{k+1}) = \sum_{j=1}^{s} \mathbf{f}(\mathbf{X}_j, t_{k(j)}, \mathbf{p}) \int_{t_k}^{t_{k+1}} l_j(r) \mathrm{d}r = \sum_{j=1}^{s} b_j \mathbf{f}(\mathbf{X}_j, t_k + c_j h_{k+1}, \mathbf{p})$$

where

$$b_j = \int_{t_k}^{t_{k+1}} l_j(r) \mathrm{d}r.$$

The proof is completed.                                                     ☐

Consider two nodes $t_k \leq t_{k(1)} < t_{k(2)} \leq t_{k+1}$ in the time interval $[t_k, t_{k+1}]$

$$t_{k(1)} = t_k + c_1 h_{k+1} \quad \text{and} \quad t_{k(2)} = t_k + c_2 h_{k+1} \tag{3.242}$$

and we have the polynomial as

$$\mathbf{P}_2(r) = l_1(r)\mathbf{f}(\mathbf{x}(t_{k(1)}), t_{k(1)}, \mathbf{p}) + l_2(r)\mathbf{f}(\mathbf{x}(t_{k(2)}), t_{k(2)}, \mathbf{p}) \tag{3.243}$$

where

$$l_1(r) = \frac{r - t_{k(2)}}{t_{k(1)} - t_{k(2)}}, \quad \text{and} \quad l_2(r) = \frac{r - t_{k(1)}}{t_{k(2)} - t_{k(1)}}. \tag{3.244}$$

The coefficients for implicit Runge–Kutta method are computed by

$$b_1 = \int_{t_k}^{t_{k+1}} l_1(r)dr = \frac{1}{2(c_2 - c_1)}[c_2^2 - (1 - c_2)^2]h_{k+1},$$

$$b_2 = \int_{t_k}^{t_{k+1}} l_2(r)dr = \frac{1}{2(c_2 - c_1)}[(1 - c_1)^2 - c_1^2]h_{k+1}; \tag{3.245}$$

and

$$a_{11} = \int_{t_k}^{t_{k(1)}} l_1(r)dr = \frac{1}{2(c_2 - c_1)}[c_2^2 - (c_2 - c_1)^2]h_{k+1},$$

$$a_{12} = \int_{t_k}^{t_{k(1)}} l_2(r)dr = -\frac{c_1^2}{2(c_2 - c_1)}h_{k+1};$$

$$a_{21} = \int_{t_k}^{t_{k(2)}} l_1(r)dr = \frac{c_2^2}{2(c_2 - c_1)}h_{k+1}, \tag{3.246}$$

$$a_{22} = \int_{t_k}^{t_{k(2)}} l_2(r)dr = \frac{1}{2(c_2 - c_1)}[(c_2 - c_1)^2 - c_1^2]h_{k+1}.$$

Consider a special case of $t_{k(1)} = t_k$ and $t_{k(2)} = t_{k+1}$. We have $c_1 = 0$ and $c_2 = 1$. The coefficients for the implicit Runge–Kutta method are computed by

$$b_1 = \frac{1}{2}, \quad b_2 = \frac{1}{2};$$

$$a_{11} = 0, \quad a_{12} = 0; \quad a_{21} = \frac{1}{2}, \quad a_{22} = \frac{1}{2} \tag{3.247}$$

and

$$X_1 = x_k,$$
$$X_2 = x_k + h_{k+1}\frac{1}{2}[f(X_1, t_k + c_1 h_{k+1}, p) + f(X_2, t_k + c_2 h_{k+1}, p)]. \tag{3.248}$$

Due to $X_2 = x_{k+1}$, the foregoing equation becomes

$$x_{k+1} = x_k + h_{k+1}\frac{1}{2}[f(x_k, t_k, p) + f(x_{k+1}, t_k + h_{k+1}, p)] \tag{3.249}$$

which is the trapezoidal method. The Butcher tableau for the trapezoidal method of order 2 is expressed by

$$\begin{array}{c|cc}
c_1 & a_{11} & a_{12} \\
c_2 & a_{21} & a_{22} \\
\hline
& b_1 & b_2
\end{array}
\quad \xrightarrow{c_1 = 0,\, c_2 = 1} \quad
\begin{array}{c|cc}
0 & 0 & 0 \\
1 & \frac{1}{2} & \frac{1}{2} \\
\hline
& \frac{1}{2} & \frac{1}{2}
\end{array} \tag{3.250}$$

For the construction of implicit Runge–Kutta methods, the simplification assumptions of Butcher are adopted, i.e.,

$$B(p): \quad \sum_{i=1}^{s} b_i c_i^{q-1} = \frac{1}{q}, \quad (q = 1, 2, \ldots, p);$$

$$C(\eta): \quad \sum_{j=1}^{s} a_{ij} c_j^{q-1} = \frac{c_i^q}{q}, \quad (i = 1, 2, \ldots, s;\ q = 1, 2, \ldots, \eta);$$

$$D(\xi): \quad \sum_{i=1}^{s} b_i c_i^{q-1} a_{ij} = \frac{b_j}{q}(1 - c_j^q), \quad (j = 1, 2, \ldots, s;\ q = 1, 2, \ldots, \xi). \tag{3.251}$$

Condition $B(p)$ gives the quadrature formula on $t \in [t_k, t_{k+1}]$

$$\int_{t_k}^{t_{k+1}} f(x, t, p)\mathrm{d}t \approx h_{k+1} \sum_{j=1}^{s} b_j f(X_j, t_k + c_j h_{k+1}, p) \tag{3.252}$$

based on polynomial of degree less than $p$, from which the Runge–Kutta method is of order $p$. Condition $C(\eta)$ gives the corresponding quadrature formulas on $t \in [t_k, t_k + c_i h_{k+1}]$, namely

$$\int_{t_k}^{t_k + c_j h_{k+1}} f(x, t, p)\mathrm{d}t \approx h_{k+1} \sum_{j=1}^{s} a_{ij} f(X_j, t_k + c_j h_{k+1}, p) \tag{3.253}$$

based on polynomial of degree less than $\eta$, from which the Runge–Kutta method is of order $p$. The importance of the simplification assumptions is given as follows.

**Theorem 3.12** (Butcher 1964). *If the coefficients $b_i$, $c_i$, $a_{ij}$ of the Runge–Kutta method are determined by $B(p)$, $C(\eta)$, and $D(\xi)$ with $p \leq \eta + \xi + 1$ and $p \leq 2\eta + 2$, the order of the method is $p$.*

*Proof* The proof can be referred to Butcher (1964). □

### 3.5.3 Gauss Method

From the foregoing analysis, $c_j (j = 1, 2, \ldots, s)$ for $t_{k(j)} = t_k + c_j h_{k+1}$ is arbitrarily selected. To find the better approximation of the integration, the Gauss–Legendre polynomial is used to determine $c_j \in [0, 1]$ $(j = 1, 2, \ldots, s)$, i.e.,

$$\frac{d^s}{dx^s}[x^s(1-x)^s] = 0 \Rightarrow x_j \quad (j = 1, 2, \ldots, s) \tag{3.254}$$

and the root of the Gauss–Legendre polynomial is assigned to $c_j \in [0, 1]$, i.e.,

$$c_j = x_j \quad (j = 1, 2, \ldots, s). \tag{3.255}$$

For the Gauss method, we have $p = 2s$ and $\eta = \xi = s$. For $s = 1$ (order 2), Eq. (3.254) gives

$$\frac{d}{dx}[x(1-x)] = 1 - 2x = 0 \Rightarrow x_1 = \frac{1}{2}. \tag{3.256}$$

Thus, $c_1 = \frac{1}{2}$ is selected with $t_{k(1)} = t_k + \frac{1}{2}h_{k+1}$. From $B(2s)$, $C(s)$, and $D(s)$,

$$\sum_{i=1}^{1} b_i = 1, \quad \sum_{i=1}^{1} b_i c_i = \frac{1}{2}, \quad \sum_{j=1}^{1} a_{1j} = c_1, \quad \sum_{i=1}^{1} b_i a_{i1} = b_1(1 - c_1) \tag{3.257}$$

from which we have $b_1 = 1$ and $a_{11} = \frac{1}{2}$.

$$X_1 = x_k + \frac{1}{2}h_{k+1}f(X_1, t_k + \frac{1}{2}h_{k+1}, p),$$
$$x_{k+1} = x_k + h_{k+1}f(X_1, t_k + \frac{1}{2}h_{k+1}, p). \tag{3.258}$$

Further, we have

$$X_1 = \frac{1}{2}(x_k + x_{k+1}), \tag{3.259}$$

and

$$\mathbf{x}_{k+1} = \mathbf{x}_k + h_{k+1}\mathbf{f}(\frac{1}{2}(\mathbf{x}_k + \mathbf{x}_{k+1}), t_k + \frac{1}{2}h_{k+1}, \mathbf{p}). \qquad (3.260)$$

The Butcher tableau for the Gauss method of order 2 ($s = 1$) is given by

$$\begin{array}{c|c} c_1 & a_{11} \\ \hline & b_1 \end{array} \quad \xrightarrow{c_1 = \frac{1}{2}} \quad \begin{array}{c|c} \frac{1}{2} & \frac{1}{2} \\ \hline & 1 \end{array} \qquad (3.261)$$

For $s = 2$ (order 4), Eq. (3.254) gives

$$\frac{\mathrm{d}^2}{\mathrm{d}x^2}[x^2(1-x)^2] = 2 - 12x + 12x^2 = 0 \Rightarrow x_1 = \frac{3-\sqrt{3}}{6}, \quad x_2 = \frac{3+\sqrt{3}}{6}. \quad (3.262)$$

Thus, we have $c_1 = \frac{1}{2} - \frac{1}{6}\sqrt{3}$ and $c_2 = \frac{1}{2} + \frac{1}{6}\sqrt{3}$ with $t_{k(1)} = t_k + (\frac{1}{2} - \frac{1}{6}\sqrt{3})h_{k+1}$ and $t_{k(2)} = t_k + (\frac{1}{2} + \frac{1}{6}\sqrt{3})h_{k+1}$. From $B(2s)$, $C(s)$, and $D(s)$, we have

$$\sum_{i=1}^{2} b_i = 1, \quad \sum_{i=1}^{2} b_i c_i = \frac{1}{2}, \quad \sum_{i=1}^{2} b_i c_i^2 = \frac{1}{3}, \quad \sum_{i=1}^{2} b_i c_i^3 = \frac{1}{4};$$

$$\sum_{j=1}^{2} a_{1j} = c_1, \quad \sum_{j=1}^{2} a_{1j} c_j = \frac{c_1^2}{2};$$

$$\sum_{j=1}^{2} a_{2j} = c_2, \quad \sum_{j=1}^{2} a_{2j} c_j = \frac{c_2^2}{2}; \qquad (3.263)$$

$$\sum_{i=1}^{2} b_i a_{i1} = b_1(1 - c_1), \quad \sum_{i=1}^{2} b_i c_i a_{i1} = \frac{b_1}{2}(1 - c_1^2);$$

$$\sum_{i=1}^{2} b_i a_{i2} = b_2(1 - c_2), \quad \sum_{i=1}^{2} b_i c_i a_{i2} = \frac{b_2}{2}(1 - c_2^2).$$

The coefficients for implicit Runge–Kutta method are given by

$$b_1 = \frac{1}{2}, \quad b_2 = \frac{1}{2};$$

$$a_{11} = \frac{1}{4}, \quad a_{12} = \frac{1}{3} - \frac{\sqrt{3}}{6}; \quad a_{21} = \frac{1}{3} + \frac{\sqrt{3}}{6}, \quad a_{22} = \frac{1}{4} \qquad (3.264)$$

and

$$\mathbf{X}_i = \mathbf{x}_k + \sum_{j=1}^{2} a_{ij}\mathbf{f}(\mathbf{X}_j, t_k + c_j h_{k+1}, \mathbf{p}), \quad i = 1, 2. \qquad (3.265)$$

The implicit Runge–Kutta method ($s = 2$ and order 4) is given by

$$\mathbf{x}_{k+1} = \mathbf{x}_k + h_{k+1}\frac{1}{2}[\mathbf{f}(\mathbf{X}_1, t_{k(1)}, \mathbf{p}) + \mathbf{f}(\mathbf{X}_2, t_{k(2)}, \mathbf{p})]. \tag{3.266}$$

The Butcher tableau for the Gauss method of order 4 ($s = 2$) is expressed by

$$
\begin{array}{c|cc}
c_1 & a_{11} & a_{12} \\
c_2 & a_{21} & a_{22} \\
\hline
 & b_1 & b_2
\end{array}
\quad
\xrightarrow{c_1 = \frac{1}{2} - \frac{1}{6}\sqrt{3},\, c_2 = \frac{1}{2} + \frac{1}{6}\sqrt{3}}
\quad
\begin{array}{c|cc}
\frac{1}{2} - \frac{\sqrt{3}}{6} & \frac{1}{4} & \frac{1}{3} - \frac{\sqrt{3}}{6} \\
\frac{1}{2} + \frac{\sqrt{3}}{6} & \frac{1}{3} + \frac{\sqrt{3}}{6} & \frac{1}{4} \\
\hline
 & \frac{1}{2} & \frac{1}{2}
\end{array}
\tag{3.267}
$$

For $s = 3$ (order 6), Eq. (3.254) gives

$$\frac{d^3}{dx^3}[x^3(1 - x)^3] = 6 - 48x + 180x^2 + 120x^3 = 0$$
$$\Rightarrow x_1 = \frac{5 - \sqrt{15}}{10}, \quad x_2 = \frac{1}{2}, \quad x_3 = \frac{5 + \sqrt{15}}{10}. \tag{3.268}$$

Choosing $c_i = x_i (i = 1, 2, 3)$, using $B(2s)$, $C(s)$, and $D(s)$ gives $b_i$ and $a_{ij}$, and the Butcher tableau for the Gauss method of order 6 ($s = 3$) is expressed by

$$
\begin{array}{c|ccc}
c_1 & a_{11} & a_{12} & a_{13} \\
c_2 & a_{21} & a_{22} & a_{23} \\
c_3 & a_{31} & a_{32} & a_{33} \\
\hline
 & b_1 & b_2 & b_3
\end{array}
\quad
\xrightarrow[c_3 = \frac{1}{2} + \frac{1}{10}\sqrt{15}]{c_1 = \frac{1}{2} - \frac{1}{10}\sqrt{15},\, c_2 = \frac{1}{2}}
\quad
\begin{array}{c|ccc}
\frac{5-\sqrt{15}}{10} & \frac{5}{36} & \frac{2}{9} - \frac{\sqrt{15}}{5} & \frac{5}{36} - \frac{\sqrt{15}}{30} \\
\frac{1}{2} & \frac{5}{36} + \frac{\sqrt{15}}{24} & \frac{2}{9} & \frac{5}{36} - \frac{\sqrt{15}}{24} \\
\frac{5+\sqrt{15}}{10} & \frac{5}{36} + \frac{\sqrt{15}}{30} & \frac{2}{9} + \frac{\sqrt{15}}{5} & \frac{5}{36} \\
\hline
 & \frac{5}{18} & \frac{4}{9} & \frac{5}{18}
\end{array}
\tag{3.269}
$$

For $s = 4$ (order 8), using Eq. (3.254) generates

$$c_1 = \frac{1}{2} - \left(\frac{15 + 2\sqrt{30}}{35}\right)^{1/2}, \quad c_2 = \frac{1}{2} - \left(\frac{15 - 2\sqrt{30}}{35}\right)^{1/2},$$
$$c_3 = \frac{1}{2} + \left(\frac{15 - 2\sqrt{30}}{35}\right)^{1/2}, \quad c_4 = \frac{1}{2} + \left(\frac{15 + 2\sqrt{30}}{35}\right)^{1/2}. \tag{3.270}$$

Using $B(2s)$, $C(s)$, and $D(s)$ gives $b_i$ and $a_{ij}$, and the Butcher tableau for the Gauss method of order 8 ($s = 4$) is expressed by

$$
\begin{array}{c|cccc}
c_1 & a_{11} & a_{12} & a_{13} & a_{14} \\
c_2 & a_{21} & a_{22} & a_{23} & a_{24} \\
c_3 & a_{31} & a_{32} & a_{33} & a_{34} \\
c_4 & a_{41} & a_{42} & a_{43} & a_{44} \\
\hline
 & b_1 & b_2 & b_3 & b_4
\end{array}
\tag{3.271}
$$

where

$$b_1 = 2\omega_1, b_2 = 2\omega_1', \ b_3 = 2\omega_1', b_1 = 2\omega_1;$$
$$a_{11} = \omega_1, a_{12} = \omega_1' - \omega_3 + \omega_4', a_{13} = \omega_1' - \omega_3 - \omega_4', a_{14} = \omega_1 - \omega_5;$$
$$a_{21} = \omega_1 - \omega_3' + \omega_4, a_{22} = \omega_1', a_{23} = \omega_1' - \omega_5', a_{24} = \omega_1 - \omega_3' - \omega_4; \quad (3.272)$$
$$a_{31} = \omega_1 + \omega_3' + \omega_4, a_{32} = \omega_1' + \omega_5', a_{33} = \omega_1', a_{34} = \omega_1 + \omega_3' - \omega_4;$$
$$a_{41} = \omega_1 + \omega_5, a_{42} = \omega_1' + \omega_3 + \omega_4', a_{43} = \omega_1' + \omega_3 - \omega_4', a_{44} = \omega_1$$

and

$$\omega_1 = \frac{1}{8} - \frac{\sqrt{30}}{144}, \quad \omega_1' = \frac{1}{8} + \frac{\sqrt{30}}{144};$$

$$\omega_2 = \frac{1}{2}\left(\frac{15 + 2\sqrt{30}}{35}\right)^{1/2}, \quad \omega_2' = \frac{1}{2}\left(\frac{15 - 2\sqrt{30}}{35}\right)^{1/2};$$

$$\omega_3 = \omega_2\left(\frac{1}{6} + \frac{\sqrt{30}}{24}\right), \quad \omega_3' = \omega_2\left(\frac{1}{6} - \frac{\sqrt{30}}{24}\right); \quad (3.273)$$

$$\omega_4 = \omega_2\left(\frac{1}{21} + \frac{5\sqrt{30}}{168}\right), \quad \omega_4' = \omega_2\left(\frac{1}{21} - \frac{5\sqrt{30}}{168}\right).$$

For $s = 5$ (order 10), the coefficients of the implicit Runge–Kutta method with five stages can be found in Butcher (1964).

### 3.5.4 Radau Method

As in Gauss method, $c_j \in [0, 1]$ $(j = 1, 2, \ldots, s)$ for $t_{k(j)} = t_k + c_j h_{k+1}$ is determined from Radau quadrature formulas, i.e.,

$$\text{Radau I: } \frac{d^{s-1}}{dx^{s-1}}[x^s(1 - x)^{s-1}] = 0 \Rightarrow x_j \quad (j = 1, 2, \ldots, s),$$
$$\quad (3.274)$$
$$\text{Radau II: } \frac{d^{s-1}}{dx^{s-1}}[x^{s-1}(1 - x)^s] = 0 \Rightarrow x_j \quad (j = 1, 2, \ldots, s),$$

and let

$$c_j = x_j \quad (j = 1, 2, \ldots, s). \quad (3.275)$$

For the Radau IA method, we have $p = 2s - 1$, $\eta = s - 1$, and $\xi = s$. For $s = 1$ (order 1), the first one of Eq. (3.274) gives

$$\frac{d^{s-1}}{dx^{s-1}}[x^s(1 - x)^{s-1}] = x = 0 \Rightarrow x_1 = 0. \quad (3.276)$$

Thus, $c_1 = 0$ is selected with $t_{k(1)} = t_k$. From $B(2s - 1)$, $C(s - 1)$, and $D(s)$,

$$\sum_{i=1}^{1} b_i = 1, \quad \sum_{i=1}^{1} b_i a_{i1} = b_1(1 - c_1). \tag{3.277}$$

Thus, we have $b_1 = 1$ and $a_{11} = 1$.

$$\begin{aligned} \mathbf{X}_1 &= \mathbf{x}_k + h_{k+1}\mathbf{f}(\mathbf{X}_1, t_k, \mathbf{p}), \\ \mathbf{x}_{k+1} &= \mathbf{x}_k + h_{k+1}\mathbf{f}(\mathbf{X}_1, t_k, \mathbf{p}). \end{aligned} \tag{3.278}$$

Further, we have

$$\mathbf{X}_1 = \mathbf{x}_{k+1}, \tag{3.279}$$

and

$$\mathbf{x}_{k+1} = \mathbf{x}_k + h_{k+1}\mathbf{f}(\mathbf{x}_{k+1}, t_k, \mathbf{p}). \tag{3.280}$$

The Butcher tableau for the Radau IA method of order 1 ($s = 1$) is given by

$$\begin{array}{c|c} c_1 & a_{11} \\ \hline & b_1 \end{array} \quad \xrightarrow{c_1 = 0} \quad \begin{array}{c|c} 0 & 1 \\ \hline & 1 \end{array} \tag{3.281}$$

For $s = 2$ (order 3), Radau IA method is used, and the first one of Eq. (3.274) gives

$$\frac{\mathrm{d}}{\mathrm{d}x}[x^2(1 - x)] = 2x - 3x^2 = 0 \Rightarrow x_1 = 0, \quad x_2 = \frac{2}{3}. \tag{3.282}$$

Thus, we have $c_1 = 0$ and $c_2 = \frac{2}{3}$ with $t_{k(1)} = t_k$ and $t_{k(2)} = t_k + \frac{2}{3}h_{k+1}$. From $B(2s - 1)$, $C(s - 1)$, and $D(s)$, we have

$$\sum_{i=1}^{2} b_i = 1, \quad \sum_{i=1}^{2} b_i c_i = \frac{1}{2}, \quad \sum_{i=1}^{2} b_i c_i^2 = \frac{1}{3};$$

$$\sum_{j=1}^{2} a_{1j} = c_1; \quad \sum_{j=1}^{2} a_{2j} = c_2;$$

$$\sum_{i=1}^{2} b_i a_{i1} = b_1(1 - c_1), \quad \sum_{i=1}^{2} b_i c_i a_{i1} = \frac{b_1}{2}(1 - c_1^2);$$

$$\sum_{i=1}^{2} b_i a_{i2} = b_2(1 - c_2), \quad \sum_{i=1}^{2} b_i c_i a_{i2} = \frac{b_2}{2}(1 - c_2^2). \tag{3.283}$$

The coefficients for implicit Runge–Kutta method are given by

$$b_1 = \frac{1}{2}, \; b_2 = \frac{3}{4}; \; a_{11} = \frac{1}{4}, \; a_{12} = -\frac{1}{4}; \; a_{21} = \frac{1}{4}, \; a_{22} = \frac{5}{12}; \qquad (3.284)$$

and

$$\mathbf{X}_i = \mathbf{x}_k + \sum_{j=1}^{2} a_{ij} \mathbf{f}(\mathbf{X}_j, t_k + c_j h_{k+1}, \mathbf{p}), \quad i = 1, 2. \qquad (3.285)$$

The implicit Runge–Kutta method ($s = 2$ and order 3) is given by

$$\mathbf{x}_{k+1} = \mathbf{x}_k + h_{k+1} \frac{1}{4} [\mathbf{f}(\mathbf{X}_1, t_{k(1)}, \mathbf{p}) + 3\mathbf{f}(\mathbf{X}_2, t_{k(2)}, \mathbf{p})]. \qquad (3.286)$$

The Butcher tableau for the Radau IA method of order 3 ($s = 2$) is expressed by

$$
\begin{array}{c|cc}
c_1 & a_{11} & a_{12} \\
c_2 & a_{21} & a_{22} \\
\hline
 & b_1 & b_2
\end{array}
\quad \xrightarrow{c_1 = 0, \; c_2 = \frac{2}{3}} \quad
\begin{array}{c|cc}
0 & \frac{1}{4} & -\frac{1}{4} \\
\frac{2}{3} & \frac{1}{4} & \frac{5}{12} \\
\hline
 & \frac{1}{4} & \frac{3}{4}
\end{array}
\qquad (3.287)
$$

For $s = 3$ (order 5), Eq. (3.274) gives

$$\frac{d^2}{dx^2}[x^3(1-x)^2] = 6x - 24x^2 + 20x^3 = 0$$
$$\Rightarrow x_1 = 0, \quad x_2 = \frac{6 - \sqrt{6}}{10}, \quad x_3 = \frac{6 + \sqrt{6}}{10}. \qquad (3.288)$$

Choosing $c_i = x_i (i = 1, 2, 3)$, using $B(2s - 1)$, $C(s - 1)$, and $D(s)$ gives $b_i$ and $a_{ij}$, and the Butcher tableau for the Radau IA method of order 5 ($s = 3$) is

$$
\begin{array}{c|ccc}
c_1 & a_{11} & a_{12} & a_{13} \\
c_2 & a_{21} & a_{22} & a_{23} \\
c_3 & a_{31} & a_{32} & a_{33} \\
\hline
 & b_1 & b_2 & b_3
\end{array}
\quad \xrightarrow[c_3 = \frac{6+\sqrt{6}}{10}]{c_1 = 0, \; c_2 = \frac{6-\sqrt{6}}{10}} \quad
\begin{array}{c|ccc}
0 & \frac{1}{9} & \frac{-1-\sqrt{6}}{18} & \frac{-1+\sqrt{6}}{18} \\
\frac{6-\sqrt{6}}{10} & \frac{1}{9} & \frac{88+7\sqrt{6}}{360} & \frac{88-43\sqrt{6}}{360} \\
\frac{6+\sqrt{6}}{10} & \frac{1}{9} & \frac{88+43\sqrt{6}}{360} & \frac{88-7\sqrt{6}}{360} \\
\hline
 & \frac{1}{9} & \frac{16+\sqrt{6}}{36} & \frac{16-\sqrt{6}}{36}
\end{array}
\qquad (3.289)
$$

For the Radau IIA method, we have $p = 2s - 1$, $\eta = s$, and $\xi = s - 1$. For $s = 1$ (order 1), the second one of Eq. (3.274) gives

$$\frac{d^{s-1}}{dx^{s-1}}[x^{s-1}(1-x)^s] = 1 - x = 0 \Rightarrow x_1 = 1. \qquad (3.290)$$

Thus, $c_1 = 1$ is selected with $t_{k(1)} = t_k$. From $B(2s - 1)$, $C(s)$, and $D(s - 1)$,

$$\sum_{i=1}^{1} b_i = 1, \quad \sum_{i=1}^{1} a_{i1} = c_1. \tag{3.291}$$

Thus, we have $b_1 = 1$ and $a_{11} = 1$.

$$\begin{aligned} \mathbf{X}_1 &= \mathbf{x}_k + h_{k+1}\mathbf{f}(\mathbf{X}_1, t_{k+1}, \mathbf{p}), \\ \mathbf{x}_{k+1} &= \mathbf{x}_k + h_{k+1}\mathbf{f}(\mathbf{X}_1, t_{k+1}, \mathbf{p}). \end{aligned} \tag{3.292}$$

Further, we have

$$\mathbf{X}_1 = \mathbf{x}_{k+1}, \tag{3.293}$$

and

$$\mathbf{x}_{k+1} = \mathbf{x}_k + h_{k+1}\mathbf{f}(\mathbf{x}_{k+1}, t_{k+1}, \mathbf{p}). \tag{3.294}$$

The Butcher tableau for the Radau IIA method of order 1 ($s = 1$) is given by

$$\begin{array}{c|c} c_1 & a_{11} \\ \hline & b_1 \end{array} \quad \xrightarrow{c_1 = 0} \quad \begin{array}{c|c} 1 & 1 \\ \hline & 1 \end{array} \tag{3.295}$$

For $s = 2$ (order 3), the Radau IIA method is used, and the second one of Eq. (3.274) gives

$$\frac{d}{dx}[x(1 - x)^2] = 1 - 4x + 3x^2 = 0 \Rightarrow x_1 = \frac{1}{3}, \quad x_2 = 1. \tag{3.296}$$

Thus, we have $c_1 = \frac{1}{3}$ and $c_2 = 1$ with $t_{k(1)} = t_k + \frac{1}{3}h_{k+1}$ and $t_{k(2)} = t_k + h_{k+1}$. From $B(2s - 1)$, $C(s)$, and $D(s - 1)$, we have

$$\begin{aligned} &\sum_{i=1}^{2} b_i = 1, \quad \sum_{i=1}^{2} b_i c_i = \frac{1}{2}, \quad \sum_{i=1}^{2} b_i c_i^2 = \frac{1}{3}; \\ &\sum_{j=1}^{2} a_{1j} = c_1, \quad \sum_{j=1}^{2} a_{1j}c_j = \frac{1}{2}c_2^2; \\ &\sum_{j=1}^{2} a_{2j} = c_1, \quad \sum_{j=1}^{2} a_{2j}c_j = \frac{1}{2}c_2^2; \\ &\sum_{i=1}^{2} b_i a_{i1} = b_1(1 - c_1); \quad \sum_{i=1}^{2} b_i a_{i2} = b_2(1 - c_2). \end{aligned} \tag{3.297}$$

The coefficients for the implicit Runge–Kutta method are given by

$$b_1 = \frac{3}{4}, b_2 = \frac{1}{4}; \; a_{11} = \frac{5}{12}, a_{12} = -\frac{1}{12}; \; a_{21} = \frac{3}{4}, a_{22} = \frac{1}{4}; \qquad (3.298)$$

and

$$\mathbf{X}_i = \mathbf{x}_k + \sum_{j=1}^{2} a_{ij}\mathbf{f}(\mathbf{X}_j, t_k + c_j h_{k+1}, \mathbf{p}), \quad i = 1, 2. \qquad (3.299)$$

The implicit Runge–Kutta method ($s$ = 2 and order 3) is given by

$$\mathbf{x}_{k+1} = \mathbf{x}_k + h_{k+1}\frac{1}{4}[3\mathbf{f}(\mathbf{X}_1, t_{k(1)}, \mathbf{p}) + \mathbf{f}(\mathbf{X}_2, t_{k(2)}, \mathbf{p})]. \qquad (3.300)$$

The Butcher tableau for the Radau IIA method of order 3 ($s$ = 2) is expressed by

$$\begin{array}{c|cc} c_1 & a_{11} & a_{12} \\ c_2 & a_{21} & a_{22} \\ \hline & b_1 & b_2 \end{array} \xrightarrow{c_1=0, \, c_2=\frac{2}{3}} \begin{array}{c|cc} \frac{1}{3} & \frac{5}{12} & -\frac{1}{12} \\ 1 & \frac{3}{4} & \frac{1}{4} \\ \hline & \frac{3}{4} & \frac{1}{4} \end{array} \qquad (3.301)$$

For $s$ = 3 (order 5), Eq. (3.251) gives

$$\frac{\mathrm{d}^2}{\mathrm{d}x^2}[x^2(1-x)^3] = 2 - 18x + 36x^2 - 20x^3 = 0$$
$$\Rightarrow x_1 = \frac{4 - \sqrt{6}}{10}, x_2 = \frac{4 + \sqrt{6}}{10}, \quad x_3 = 1. \qquad (3.302)$$

Choosing $c_i = x_i (i$ = 1, 2, 3), using $B(2s-1)$, $C(s)$, and $D(s-1)$ gives $b_i$ and $a_{ij}$, and the Butcher tableau for the Radau IIA method of order 5 ($s$ = 3) is

$$\begin{array}{c|ccc} c_1 & a_{11} & a_{12} & a_{13} \\ c_2 & a_{21} & a_{22} & a_{23} \\ c_3 & a_{31} & a_{32} & a_{33} \\ \hline & b_1 & b_2 & b_3 \end{array} \xrightarrow[c_3=1]{c_1=\frac{4-\sqrt{6}}{10}, \, c_2=\frac{4+\sqrt{6}}{10}} \begin{array}{c|ccc} \frac{4-\sqrt{6}}{10} & \frac{88-7\sqrt{6}}{360} & \frac{296-169\sqrt{6}}{1800} & \frac{-2+3\sqrt{6}}{225} \\ \frac{4+\sqrt{6}}{10} & \frac{296+169\sqrt{6}}{1800} & \frac{88+7\sqrt{6}}{360} & \frac{-2-3\sqrt{6}}{225} \\ 1 & \frac{16-\sqrt{6}}{36} & \frac{16+\sqrt{6}}{36} & \frac{1}{9} \\ \hline & \frac{16-\sqrt{6}}{36} & \frac{16+\sqrt{6}}{36} & \frac{1}{9} \end{array} \qquad (3.303)$$

### 3.5.5 Lobatto Method

As in Gauss method, $c_j \in [0, 1]$ ($j$ = 1, 2, ..., $s$) for $t_{k(j)} = t_k + c_j h_{k+1}$ is determined through Lobatto quadrature formulas, i.e.,

$$\text{Lobatto: } \frac{d^{s-1}}{dx^{s-1}}[x^{s-1}(1-x)^{s-1}] = 0 \Rightarrow x_j \quad (j = 1, 2, \ldots, s) \tag{3.304}$$

and let

$$c_j = x_j \quad (j = 1, 2, \ldots, s). \tag{3.305}$$

In Hairer and Wanner (1991), for the Lobatto IIIA method, we have

$$p = 2s - 2, \quad \eta = s, \quad \xi = s - 2. \tag{3.306}$$

The Butcher tableau for the Lobatto IIIA method of order 2 ($s = 2$) is

$$
\begin{array}{c|cc}
c_1 & a_{11} & a_{12} \\
c_2 & a_{21} & a_{22} \\
\hline
& b_1 & b_2
\end{array}
\xrightarrow{c_1=0,\,c_2=1}
\begin{array}{c|cc}
0 & 0 & 0 \\
1 & \frac{1}{2} & \frac{1}{2} \\
\hline
& \frac{1}{2} & \frac{1}{2}
\end{array}
\tag{3.307}
$$

The Butcher tableau for the Lobatto IIIA method of order 4 ($s = 3$) is

$$
\begin{array}{c|ccc}
c_1 & a_{11} & a_{12} & a_{13} \\
c_2 & a_{21} & a_{22} & a_{23} \\
c_3 & a_{31} & a_{32} & a_{33} \\
\hline
& b_1 & b_2 & b_3
\end{array}
\xrightarrow[c_3=1]{c_1=0,\,c_2=\frac{1}{2}}
\begin{array}{c|ccc}
0 & 0 & 0 & 0 \\
\frac{1}{2} & \frac{5}{24} & \frac{1}{3} & \frac{-1}{24} \\
1 & \frac{1}{6} & \frac{2}{3} & \frac{1}{6} \\
\hline
& \frac{1}{6} & \frac{2}{3} & \frac{1}{6}
\end{array}
\tag{3.308}
$$

The Butcher tableau for the Lobatto IIIA method of order 6 ($s = 4$) is

$$
\begin{array}{c|cccc}
c_1 & a_{11} & a_{12} & a_{13} & a_{14} \\
c_2 & a_{21} & a_{22} & a_{23} & a_{24} \\
c_3 & a_{31} & a_{32} & a_{33} & a_{34} \\
c_4 & a_{41} & a_{42} & a_{43} & a_{44} \\
\hline
& b_1 & b_2 & b_3 & b_4
\end{array}
\xrightarrow[c_3=\frac{5+\sqrt{5}}{10},\,c_4=1]{c_1=0,\,c_2=\frac{5-\sqrt{5}}{10}}
\begin{array}{c|cccc}
0 & 0 & 0 & 0 & 0 \\
\frac{5-\sqrt{5}}{10} & \frac{11+\sqrt{5}}{120} & \frac{25-\sqrt{5}}{120} & \frac{25-13\sqrt{5}}{120} & \frac{-1+\sqrt{5}}{120} \\
\frac{5+\sqrt{5}}{10} & \frac{11-\sqrt{5}}{120} & \frac{25+13\sqrt{5}}{120} & \frac{25+\sqrt{5}}{120} & \frac{-1-\sqrt{5}}{120} \\
1 & \frac{1}{12} & \frac{5}{12} & \frac{5}{15} & \frac{1}{12} \\
\hline
& \frac{1}{12} & \frac{5}{12} & \frac{5}{12} & \frac{1}{12}
\end{array}
\tag{3.309}
$$

In Hairer and Wanner (1991), for the Lobatto IIIB method, we have

$$p = 2s - 2, \quad \eta = s - 2, \quad \xi = s. \tag{3.310}$$

The Butcher tableau for the Lobatto IIIB method of order 2 ($s = 2$) is

$$
\begin{array}{c|cc}
c_1 & a_{11} & a_{12} \\
c_2 & a_{21} & a_{22} \\
\hline
& b_1 & b_2
\end{array}
\xrightarrow{c_1=0,\,c_2=1}
\begin{array}{c|cc}
0 & \frac{1}{2} & 0 \\
1 & \frac{1}{2} & 0 \\
\hline
& \frac{1}{2} & \frac{1}{2}
\end{array}
\tag{3.311}
$$

The Butcher tableau for the Lobatto IIIB method of order 4 ($s = 3$) is

$$
\begin{array}{c|ccc}
c_1 & a_{11} & a_{12} & a_{13} \\
c_2 & a_{21} & a_{22} & a_{23} \\
c_3 & a_{31} & a_{32} & a_{33} \\
\hline
 & b_1 & b_2 & b_3
\end{array}
\quad \xrightarrow[\substack{c_1=0,\, c_2=\frac{1}{2}\\ c_3=1}]{} \quad
\begin{array}{c|ccc}
0 & \frac{1}{6} & -\frac{1}{6} & 0 \\
\frac{1}{2} & \frac{1}{6} & \frac{1}{3} & 0 \\
1 & \frac{1}{6} & \frac{5}{6} & 0 \\
\hline
 & \frac{1}{6} & \frac{2}{3} & \frac{1}{6}
\end{array}
\tag{3.312}
$$

The Butcher tableau for the Lobatto IIIB method of order 6 ($s = 4$) is

$$
\begin{array}{c|cccc}
c_1 & a_{11} & a_{12} & a_{13} & a_{14} \\
c_2 & a_{21} & a_{22} & a_{23} & a_{24} \\
c_3 & a_{31} & a_{32} & a_{33} & a_{34} \\
c_4 & a_{41} & a_{42} & a_{43} & a_{44} \\
\hline
 & b_1 & b_2 & b_3 & b_4
\end{array}
\xrightarrow[\substack{c_1=0,\, c_2=\frac{5-\sqrt{5}}{10}\\ c_3=\frac{5+\sqrt{5}}{10},\, c_4=1}]{}
\begin{array}{c|cccc}
0 & \frac{1}{12} & \frac{-1-\sqrt{5}}{120} & \frac{-1+\sqrt{5}}{24} & 0 \\
\frac{5-\sqrt{5}}{10} & \frac{1}{12} & \frac{25+\sqrt{5}}{120} & \frac{25-13\sqrt{5}}{120} & 0 \\
\frac{5+1\sqrt{5}}{10} & \frac{1}{12} & \frac{25+13\sqrt{5}}{120} & \frac{25+\sqrt{5}}{120} & 0 \\
1 & \frac{1}{12} & \frac{11-\sqrt{5}}{24} & \frac{11+\sqrt{5}}{24} & 0 \\
\hline
 & \frac{1}{12} & \frac{5}{12} & \frac{5}{12} & \frac{1}{12}
\end{array}
\tag{3.313}
$$

In Hairer and Wanner (1991), for the Lobatto IIIC method, we have

$$
p = 2s - 2, \quad \eta = s - 1, \quad \xi = s - 1.
\tag{3.314}
$$

The Butcher tableau for the Lobatto IIIC method of order 2 ($s = 2$) is

$$
\begin{array}{c|cc}
c_1 & a_{11} & a_{12} \\
c_2 & a_{21} & a_{22} \\
\hline
 & b_1 & b_2
\end{array}
\quad \xrightarrow[]{c_1=0,\, c_2=1} \quad
\begin{array}{c|cc}
0 & \frac{1}{2} & -\frac{1}{2} \\
1 & \frac{1}{2} & \frac{1}{2} \\
\hline
 & \frac{1}{2} & \frac{1}{2}
\end{array}
\tag{3.315}
$$

The Butcher tableau for the Lobatto IIIC method of order 4 ($s = 3$) is

$$
\begin{array}{c|ccc}
c_1 & a_{11} & a_{12} & a_{13} \\
c_2 & a_{21} & a_{22} & a_{23} \\
c_3 & a_{31} & a_{32} & a_{33} \\
\hline
 & b_1 & b_2 & b_3
\end{array}
\quad \xrightarrow[\substack{c_1=0,\, c_2=\frac{1}{2}\\ c_3=1}]{} \quad
\begin{array}{c|ccc}
0 & \frac{1}{6} & -\frac{1}{6} & 0 \\
\frac{1}{2} & \frac{1}{6} & \frac{1}{3} & 0 \\
1 & \frac{1}{6} & \frac{5}{6} & 0 \\
\hline
 & \frac{1}{6} & \frac{2}{3} & \frac{1}{6}
\end{array}
\tag{3.316}
$$

The Butcher tableau for the Lobatto IIIC method of order 6 ($s = 4$) is

$$
\begin{array}{c|cccc}
c_1 & a_{11} & a_{12} & a_{13} & a_{14} \\
c_2 & a_{21} & a_{22} & a_{23} & a_{24} \\
c_3 & a_{31} & a_{32} & a_{33} & a_{34} \\
c_4 & a_{41} & a_{42} & a_{43} & a_{44} \\
\hline
 & b_1 & b_2 & b_3 & b_4
\end{array}
\xrightarrow[\substack{c_1=0,\, c_2=\frac{5-\sqrt{5}}{10}\\ c_3=\frac{5+\sqrt{5}}{10},\, c_4=1}]{}
\begin{array}{c|cccc}
0 & \frac{1}{12} & -\frac{\sqrt{5}}{12} & \frac{\sqrt{5}}{12} & -\frac{1}{12} \\
\frac{5-\sqrt{5}}{10} & \frac{1}{12} & \frac{1}{4} & \frac{10-7\sqrt{5}}{60} & \frac{\sqrt{5}}{60} \\
\frac{5+1\sqrt{5}}{10} & \frac{1}{12} & \frac{10+7\sqrt{5}}{60} & \frac{1}{4} & -\frac{\sqrt{5}}{60} \\
1 & \frac{1}{12} & \frac{5}{12} & \frac{5}{12} & \frac{1}{12} \\
\hline
 & \frac{1}{12} & \frac{5}{12} & \frac{5}{12} & \frac{1}{12}
\end{array}
\tag{3.317}
$$

## 3.5.6  Diagonally Implicit RK Methods

Consider a semi-diagonally implicit Runge–Kutta (SDIRK) method as

$$
\begin{array}{c|ccccc}
c_1 & \gamma & 0 & \cdots & 0 & 0 \\
c_2 & a_{21} & \gamma & \cdots & 0 & 0 \\
c_3 & a_{31} & a_{32} & \cdots & 0 & 0 \\
\vdots & \vdots & \vdots & \vdots & \vdots & \vdots \\
c_s & a_{s1} & a_{s2} & \cdots & a_{s(s-1)} & \gamma \\
\hline
 & b_1 & b_2 & \cdots & b_{s-1} & b_s
\end{array}
\tag{3.318}
$$

with

$$
\sum_{k=1}^{j-1} a_{jk} = c_j. \tag{3.319}
$$

Consider a fourth-order SDIRK method as

$$
\mathbf{f} : \ \sum_{j=1}^{s} b_j = b_1 + b_2 + b_3 + b_4 = 1,
$$

$$
D\mathbf{f} : \ \sum_{j,k=1}^{s-1,j} b_j a_{jk} = b_2 c_2 + b_3 c_3 + b_4 c_4 = \frac{1}{2} - \gamma,
$$

$$
D^2\mathbf{f} : \ \sum_{j,k,l=1}^{s-1,j,j} b_j a_{jk} a_{jl} = b_2 c_2^2 + b_3 c_3^2 + b_4 c_4^2 = \frac{1}{3} - \gamma + \gamma^2,
$$

$$
\mathbf{f_x} D\mathbf{f} : \ \sum_{j,k,l=1}^{s-1,j,k} b_j a_{jk} a_{kl} = b_3 a_{32} c_2 + b_4 (c_2 a_{42} + c_3 a_{43}) = \frac{1}{6} - \gamma + \gamma^2,
$$

$$
D^3\mathbf{f} : \ \sum_{j,k,l,m=1}^{s-1,j,j,j} b_j a_{jk} a_{jl} a_{jm} = b_3 c_3^3 + b_2 c_2^3 + b_4 c_4^3 = \frac{1}{4} - \gamma + \frac{3}{2}\gamma^2,
$$

$$
\mathbf{f_x} D^2\mathbf{f} : \ \sum_{j,k,l,m=1}^{s-1,j,k,k} b_j a_{jk} a_{kl} a_{km} = b_3 a_{32} c_2^2 + b_4 (c_2^2 a_{42} + c_3^2 a_{43}) a_{32} a_{21}
$$
$$
= \frac{1}{12} - \frac{3}{2}\gamma + \frac{3}{2}\gamma^2 - \gamma^3,
$$

$$
D\mathbf{f_x} \cdot D\mathbf{f} : \ \sum_{j,k,l,m=1}^{s-1,j,k,j} b_j a_{jk} a_{kl} a_{jm} = b_3 a_{32} c_3 c_2 + b_4 c_4 (c_2 a_{42} + c_3 a_{43})
$$
$$
= \frac{1}{8} - \frac{2}{3}\gamma + \frac{3}{2}\gamma^2 - \gamma^3,
$$

$$\mathbf{f_x^2 Df} : \sum_{j,k,l,m=1}^{s-1,j,k,l} b_j a_{jk} a_{kl} a_{lm} = b_4 c_1 a_{32} a_{43} = \frac{1}{24} - \frac{1}{2}\gamma + \frac{3}{2}\gamma^2 - \gamma^3. \qquad (3.320)$$

The general rule can be done

$$\mathbf{f} : \sum_{j=1}^{s} b_j = 1,$$

$$Df : \sum_{j,k=1}^{s-1,j} b_j a_{jk} = \sum_{j,k=1}^{s,j} b_j a_{jk} - \sum_{k=1}^{s} b_s a_{sk},$$

$$D^2\mathbf{f} : \sum_{j,k,l=1}^{s-1,j,j} b_j a_{jk} a_{jl} = \sum_{j,k,l=1}^{s,j,j} b_j a_{jk} a_{jl} - \sum_{k,l=1}^{s,s} b_s a_{sk} a_{sl},$$

$$\mathbf{f_x} Df : \sum_{j,k,l=1}^{s-1,j,k} b_j a_{jk} a_{kl} = \sum_{j,k,l=1}^{s,j,k} b_j a_{jk} a_{kl} - \sum_{k,l=1}^{s,k} b_s a_{sk} a_{kl},$$

$$D^3\mathbf{f} : \sum_{j,k,l,m=1}^{s-1,j,j,j} b_j a_{jk} a_{jl} a_{jm} = \sum_{j,k,l,m=1}^{s,j,j,j} b_j a_{jk} a_{jl} a_{jm} - \sum_{k,l,m=1}^{s,s,s} b_s a_{sk} a_{sl} a_{sm}, \qquad (3.321)$$

$$\mathbf{f_x} D^2\mathbf{f} : \sum_{j,k,l,m=1}^{s-1,j,k,k} b_j a_{jk} a_{kl} a_{km} = \sum_{j,k,l,m=1}^{s,j,k,k} b_j a_{jk} a_{kl} a_{km} - \sum_{k,l,m=1}^{s,k,k} b_s a_{sk} a_{kl} a_{km},$$

$$D\mathbf{f_x} \cdot Df : \sum_{j,k,l,m=1}^{s-1,j,k,j} b_j a_{jk} a_{kl} a_{jm} = \sum_{j,k,l,m=1}^{s,j,k,j} b_j a_{jk} a_{kl} a_{jm} - \sum_{k,l,m=1}^{s,k,s} b_s a_{sk} a_{kl} a_{sm},$$

$$\mathbf{f_x^2} Df : \sum_{j,k,l,m=1}^{s-1,j,k,l} b_j a_{jk} a_{kl} a_{lm} = \sum_{j,k,l,m=1}^{s,j,k,l} b_j a_{jk} a_{kl} a_{lm} - \sum_{k,l,m=1}^{s,k,l} b_s a_{sk} a_{kl} a_{lm},$$

with

$$\mathbf{f} : \sum_{j=1}^{s} b_j = 1,$$

$$Df : \sum_{j,k=1}^{s,j} b_j a_{jk} = \frac{1}{2}$$

$$D^2\mathbf{f} : \sum_{j,k,l=1}^{s,j,j} b_j a_{jk} a_{jl} = \frac{1}{3},$$

$$\mathbf{f_x} Df : \sum_{j,k,l=1}^{s,j,k} b_j a_{jk} a_{kl} = \frac{1}{6},$$

$$D^3\mathbf{f} : \sum_{j,k,l,m=1}^{s,j,j,j} b_j a_{jk} a_{jl} a_{jm} = \frac{1}{4},$$

$$\mathbf{f_x}D^2\mathbf{f} : \sum_{j,k,l,m=1}^{s,j,k,k} b_j a_{jk} a_{kl} a_{km} = \frac{1}{12},$$

$$D\mathbf{f_x} \cdot D\mathbf{f} : \sum_{j,k,l,m=1}^{s-1,j,k,j} b_j a_{jk} a_{kl} a_{jm} = \frac{1}{8},$$

$$\mathbf{f_x^2}D\mathbf{f} : \sum_{j,k,l,m=1}^{s,j,k,l} b_j a_{jk} a_{kl} a_{lm} = \frac{1}{8}.$$

(3.322)

If the following condition is used,

$$a_{sj} = b_j \qquad (j = 1, 2, \ldots, s),$$

(3.323)

the stiffly accurate SDIRK methods are obtained with the Butcher tableau as

$$\begin{array}{c|ccccc}
c_1 & \gamma & 0 & \cdots & 0 & 0 \\
c_2 & a_{21} & \gamma & \cdots & 0 & 0 \\
c_3 & a_{31} & a_{32} & \cdots & 0 & 0 \\
\vdots & \vdots & \vdots & & \vdots & \vdots \\
c_s & b_1 & b_2 & \cdots & b_{s-1} & \gamma \\
\hline
 & b_1 & b_2 & \cdots & b_{s-1} & \gamma
\end{array}$$

(3.324)

The SDIRK method for $s = 5$ with Eq. (3.325) gives

$$\mathbf{f} : \sum_{j=1}^{s-1} b_j = b_1 + b_2 + b_3 + b_4 = 1 - \gamma,$$

$$D\mathbf{f} : \sum_{j,k=1}^{s-1,j} b_j a_{jk} = b_2 c_2 + b_3 c_3 + b_4 c_4 = \frac{1}{2} - 2\gamma + \gamma^2,$$

$$D^2\mathbf{f}: \sum_{j,k,l=1}^{s-1,j,j} b_j a_{jk} a_{jl} = b_2 c_2^2 + b_3 c_3^2 + b_4 c_4^2 = \frac{1}{3} - 2\gamma + 3\gamma^2 - \gamma^3,$$

$$\mathbf{f_x}D\mathbf{f} : \sum_{j,k,l=1}^{s-1,j,k} b_j a_{jk} a_{kl} = b_3 a_{32} c_2 + b_4 (c_2 a_{42} + c_3 a_{43})$$

$$= \frac{1}{6} - \frac{3}{2}\gamma + 3\gamma^2 - \gamma^3,$$

$$D^3\mathbf{f}:\quad \sum_{j,k,l,m=1}^{s-1,j,j,j} b_j a_{jk} a_{jl} a_{jm} = b_3 c_3^3 + b_2 c_2^3 + b_4 c_4^3$$

$$= \frac{1}{4} - 2\gamma + \frac{9}{2}\gamma^2 - 4\gamma^3 + \gamma^4,$$

$$\mathbf{f_x}D^2\mathbf{f}:\quad \sum_{j,k,l,m=1}^{s-1,j,k,k} b_j a_{jk} a_{kl} a_{km} = b_3 a_{32} c_2^2 + b_4(c_2^2 a_{42} + c_3^2 a_{43}) a_{32} a_{21}$$

$$= \frac{1}{12} - \frac{4}{3}\gamma + 4\gamma^2 - 4\gamma^3 + \gamma^4, \qquad\qquad (3.325)$$

$$D\mathbf{f_x}\cdot D\mathbf{f}:\quad \sum_{j,k,l,m=1}^{s-1,j,k,j} b_j a_{jk} a_{kl} a_{jm} = b_3 a_{32} c_3 c_2 + b_4 c_4(c_2 a_{42} + c_3 a_{43})$$

$$= \frac{1}{8} - \frac{5}{6}\gamma + \frac{3}{2}\gamma^2 - \gamma^3,$$

$$\mathbf{f_x^2}D\mathbf{f}:\quad \sum_{j,k,l,m=1}^{s-1,j,k,l} b_j a_{jk} a_{kl} a_{lm} = b_4 c_1 a_{32} a_{43} = \frac{1}{24} - \frac{2}{3}\gamma + 3\gamma^2 - 4\gamma^3 + \gamma^4.$$

### 3.5.7 Stability of Implicit Runge–Kutta Methods

The local error, global error, and stability of the implicit Runge–Kutta method are presented through the following theorem.

**Theorem 3.13** *Suppose the solution* $\mathbf{x} = \mathbf{x}(t)$ *is the unique solution of* Eq. (3.1). *If* $\mathbf{x}(t) \in C^{s+1}[t_0, t_M]$ *and the points* $(t_k, \mathbf{x}_k)$ *for* $k = 0, 1, 2, \ldots, M-1$ *is the sequence of the* s*-stage implicit Runge–Kutta discrete approximation of* Eq. (3.1), *then the corresponding local error satisfies*

$$\|\varepsilon_{k+1}\| = \|\mathbf{x}(t_{k+1}) - \mathbf{x}_k - h_{k+1}\mathbf{\Phi}(\mathbf{x}_k, t_k, \mathbf{p})\| = O(h_{k+1}^{s+1}) \qquad (3.326)$$

where

$$\mathbf{\Phi}(\mathbf{x}_k, t_k, \mathbf{p}) = \sum_{i=1}^{s} b_i \mathbf{f}(\mathbf{X}_i, t_k + c_i h_{k+1}, \mathbf{p}),$$

$$\mathbf{X}_i = \mathbf{x}_k + h_{k+1}\sum_{j=1}^{s} a_{ij}\mathbf{f}(\mathbf{X}_j, t_k + c_j h_{k+1}, \mathbf{p}), \qquad (3.327)$$

and the global error for $h_k = h$ ($k = 1, 2, 3, \ldots$) and $t_k = kh$ is

$$\|\mathbf{e}_k\| \le \frac{h^s L}{(s+1)! K}(e^{t_k K} - 1) \qquad (3.328)$$

with

$$L = \max_{k \in \{1,2,\ldots,M\}} \|\boldsymbol{\delta}^{(s)}(\mathbf{x}_k^c, t_k^c, \mathbf{p})\|,$$

$$K = \max_{k \in \{1,2,\ldots,M\}} \|\mathbf{b}^{\mathrm{T}} \mathbf{J}(\mathbf{I}_{s \times s} - h_{k+1} \mathbf{A} \mathbf{J})^{-1}\|,$$

$$\mathbf{J} = D_{\mathbf{Y}_s} \mathbf{F}_s(\mathbf{Y}_s) \text{ and } \|\boldsymbol{\xi}_s\| \in (\|\bar{\mathbf{Y}}_s\|, \|\mathbf{Y}_s\|),$$

$$\mathbf{Y}_s = (\mathbf{X}_1, \mathbf{X}_2, \ldots, \mathbf{X}_s)^{\mathrm{T}},$$

$$\mathbf{b} = (b_1 \mathbf{I}_{n \times n}, b_2 \mathbf{I}_{n \times n}, \ldots, b_s \mathbf{I}_{n \times n})^{\mathrm{T}}, \tag{3.329}$$

$$\mathbf{A} = (a_{ij} \mathbf{I}_{n \times n})_{s \times s},$$

$$\mathbf{F}_s = (\mathbf{f}_1, \mathbf{f}_2, \ldots, \mathbf{f}_s)^{\mathrm{T}},$$

$$\bar{\mathbf{Y}}_s = (\mathbf{X}(t_1), \mathbf{X}(t_2), \ldots, \mathbf{X}(t_s))^{\mathrm{T}}.$$

For a fixed $t_k = kh$, as $h$ goes to zero,

$$\|\mathbf{e}_k\| = \|\mathbf{x}(t_k) - \mathbf{x}_k\| = O(h^s). \tag{3.330}$$

*Proof* Based on the Taylor series, the Runge–Kutta expansion of $\mathbf{x}(t_{k+1})$ at point $(t_k, \mathbf{x}(t_k))$ as

$$\mathbf{x}(t_{k+1}) = \mathbf{x}(t_k + h_{k+1})$$

$$= \mathbf{x}(t_k) + h_{k+1} \boldsymbol{\Phi}(\mathbf{x}(t_k), t_k, \mathbf{p}) + \frac{1}{(s+1)!} h_{k+1}^{s+1} \boldsymbol{\delta}^{(s+1)}(\mathbf{x}_k^c, t_k^c, \mathbf{p}).$$

The Runge–Kutta approximation of the $s$th-order gives

$$\mathbf{Y}_s = \mathbf{x}_k \mathbf{1} + h_{k+1} \mathbf{A} \mathbf{F}_s(\mathbf{Y}_s),$$

$$\mathbf{x}_{k+1} = \mathbf{x}_k + h_{k+1} \mathbf{b}^{\mathrm{T}} \mathbf{F}_s(\mathbf{Y}_s);$$

where

$$\mathbf{Y}_s = (\mathbf{X}_1, \mathbf{X}_2, \ldots, \mathbf{X}_s)^{\mathrm{T}},$$

$$\mathbf{1} = (\underbrace{\mathbf{I}_{n \times n}, \mathbf{I}_{n \times n} \cdots, \mathbf{I}_{n \times n}}_{s})^{\mathrm{T}},$$

$$\mathbf{b} = (b_1 \mathbf{I}_{n \times n}, b_2 \mathbf{I}_{n \times n}, \ldots, b_s \mathbf{I}_{n \times n})^{\mathrm{T}},$$

$$\mathbf{A} = (a_{ij} \mathbf{I}_{n \times n})_{s \times s},$$

$$\mathbf{F}_s = (\mathbf{f}_1, \mathbf{f}_2, \ldots, \mathbf{f}_s)^{\mathrm{T}}.$$

On the other hand,

$$\bar{\mathbf{Y}}_s = \mathbf{x}(t_k)\mathbf{1} + h_{k+1}\mathbf{A}\mathbf{F}_s(\bar{\mathbf{Y}}_s),$$
$$\mathbf{x}(t_{k+1}) = \mathbf{x}(t_k) + h_{k+1}\mathbf{b}^{\mathsf{T}}\mathbf{F}_s(\bar{\mathbf{Y}}_s)$$

where

$$\bar{\mathbf{Y}}_s = (\mathbf{X}(t_1), \mathbf{X}(t_2), \ldots, \mathbf{X}(t_s))^{\mathsf{T}},$$
$$\mathbf{F}_s(\bar{\mathbf{Y}}_s) = (\mathbf{f}_1(\mathbf{X}(t_1)), \mathbf{f}_2(\mathbf{X}(t_2)), \ldots, \mathbf{f}_s(\mathbf{X}(t_s)))^{\mathsf{T}}.$$

Consider the difference between the true and approximate solutions as

$$\mathbf{x}(t_{k+1}) - \mathbf{x}_{k+1} = \mathbf{x}(t_k) - \mathbf{x}_k + h_{k+1}\mathbf{b}^{\mathsf{T}}[\mathbf{F}_s(\bar{\mathbf{Y}}_s) - \mathbf{F}_s(\mathbf{Y}_s)]$$
$$+ \frac{1}{(s+1)!}h_{k+1}^{(s+1)}\boldsymbol{\delta}^{(s+1)}(\mathbf{x}_k^c, t_k^c, \mathbf{p}),$$
$$\bar{\mathbf{Y}}_s - \mathbf{Y}_s = [\mathbf{x}(t_k) - \mathbf{x}_k]\mathbf{1} + h_{k+1}\mathbf{A}[\mathbf{F}_s(\bar{\mathbf{Y}}_s) - \mathbf{F}_s(\mathbf{Y}_s)].$$

If $\mathbf{x}(t_k) = \mathbf{x}_k$, then $\bar{\mathbf{Y}}_s = \mathbf{Y}_s$. Therefore, the local error is

$$\boldsymbol{\varepsilon}_{k+1} = \mathbf{x}(t_{k+1}) - \mathbf{x}_k - h_{k+1}\boldsymbol{\Phi}(\mathbf{x}_k, t_k, \mathbf{p}) = \frac{1}{(s+1)!}h_{k+1}^{s+1}\boldsymbol{\delta}^{(s+1)}(\mathbf{x}_k^c, t_k^c, \mathbf{p}).$$

Thus, the local error of the Runge–Kutta method of the $s$th-order at the $(k+1)$th step is

$$\|\boldsymbol{\varepsilon}_{k+1}\| = \|\mathbf{x}(t_{k+1}) - \mathbf{x}_k - h_{k+1}\boldsymbol{\Phi}(\mathbf{x}_k, t_k, \mathbf{p})\| = O(h_{k+1}^s).$$

If $\mathbf{x}(t_k) \neq \mathbf{x}_k$, then $\boldsymbol{\Phi}(\mathbf{x}(t_k), t_k, \mathbf{p}) \neq \boldsymbol{\Phi}(\mathbf{x}_k, t_k, \mathbf{p})$, so

$$\mathbf{x}(t_{k+1}) - \mathbf{x}_{k+1} = \mathbf{x}(t_k) - \mathbf{x}_k + h_{k+1}\mathbf{b}^{\mathsf{T}}[\mathbf{F}_s(\bar{\mathbf{Y}}_s) - \mathbf{F}_s(\mathbf{Y}_s)]$$
$$+ \frac{1}{(s+1)!}h_{k+1}^{(s+1)}\boldsymbol{\delta}^{(s+1)}(\mathbf{x}_k^c, t_k^c, \mathbf{p}),$$
$$\bar{\mathbf{Y}}_s - \mathbf{Y}_s = [\mathbf{x}(t_k) - \mathbf{x}_k]\mathbf{1} + h_{k+1}\mathbf{A}[\mathbf{F}_s(\bar{\mathbf{Y}}_s) - \mathbf{F}_s(\mathbf{Y}_s)]$$

and the Lipschitz conditions generate

$$\mathbf{x}(t_{k+1}) - \mathbf{x}_{k+1} = \mathbf{x}(t_k) - \mathbf{x}_k + h_{k+1}\mathbf{b}^{\mathsf{T}}\mathbf{J}_k(\bar{\mathbf{Y}}_s - \mathbf{Y}_s) + \boldsymbol{\varepsilon}_{k+1},$$
$$\bar{\mathbf{Y}}_s - \mathbf{Y}_s = (\mathbf{I}_{s \times s} - h_{k+1}\mathbf{A}\mathbf{J}_k)^{-1}[\mathbf{x}(t_k) - \mathbf{x}_k]\mathbf{1}$$

where

$$\mathbf{J}_k = D_{\mathbf{Y}_s}\mathbf{F}_s(\mathbf{Y}_s) \quad \text{and} \quad \|\boldsymbol{\xi}_s\| \in (\|\bar{\mathbf{Y}}_s\|, \|\mathbf{Y}_s\|).$$

So

$$\mathbf{e}_{k+1} = [\mathbf{I}_{n\times n} + h_{k+1}\mathbf{b}^{\mathrm{T}}\mathbf{J}_k(\mathbf{I}_{s\times s} - h_{k+1}\mathbf{A}\mathbf{J}_k)^{-1}]\mathbf{e}_k + \boldsymbol{\varepsilon}_{k+1}$$

gives

$$\|\mathbf{e}_{k+1}\| \leq (1 + h_{k+1}\|\mathbf{b}^{\mathrm{T}}\mathbf{J}_k(\mathbf{I}_{s\times s} - h_{k+1}\mathbf{A}\mathbf{J}_k)^{-1}\|)\|\mathbf{e}_k\| + \|\boldsymbol{\varepsilon}_{k+1}\|.$$

For $h_k = h$ ($k = 1, 2, 3, \ldots$) and $t_k = kh$, letting

$$L = \max_{k\in\{1,2,\ldots,M\}} \|\boldsymbol{\delta}^{(s)}(\mathbf{x}_k^c, t_k^c, \mathbf{p})\|,$$

$$K = \max_{k\in\{1,2,\ldots,M\}} \|\mathbf{b}^{\mathrm{T}}\mathbf{J}_k(\mathbf{I}_{s\times s} - h_{k+1}\mathbf{A}\mathbf{J}_k)^{-1}\|,$$

consider a simple discrete equation as

$$z_{k+1} = (1 + hK)z_k + \frac{1}{(s+1)!}h^{s+1}L \quad \text{with} \quad z_0 = 0$$

and

$$\begin{aligned}
z_k &= \frac{1}{(s+1)!}h^{s+1}L\left(\sum_{l=0}^{k-1}(1+hK)^l\right) \\
&= \frac{h^{s+1}L}{(s+1)!K}[(1+hK)^k - 1] \quad \text{for} \ k = 1, 2, \ldots, M.
\end{aligned}$$

For $k > 0$, we have $1 + hK > 0$. If $z_k \geq \|\mathbf{e}_k\|$, then $z_{k+1} \geq \|\mathbf{e}_{k+1}\|$. Since $t_k = kh$, for $k \to \infty$ and $h \to 0$, we have

$$\lim_{k\to\infty}\left(1 + \frac{t_k}{k}K\right)^k = \mathrm{e}^{t_kK},$$

so

$$\|\mathbf{e}_k\| \leq \frac{h^sL}{(s+1)!K}(\mathrm{e}^{t_kK} - 1).$$

For a fixed $t_k = kh$, as $h$ goes to zero, we have

$$\|\mathbf{e}_k\| = \|\mathbf{x}(t_k) - \mathbf{x}_k\| = O(h^s) \quad \text{for} \ k = 1, 2, \ldots, M.$$

This theorem is proved.                                                                              □

Consider the global error at the

$$\mathbf{e}_{k+1} = [\mathbf{I}_{n\times n} + h_{k+1}\mathbf{b}^{\mathrm{T}}\mathbf{J}_k(\mathbf{I}_{s\times s} - h_{k+1}\mathbf{A}\mathbf{J}_k)^{-1}]\mathbf{e}_k. \qquad (3.331)$$

If $\mathbf{e}_{k+1} = \lambda\mathbf{e}_k$, then the foregoing equation becomes

$$[(\mathbf{I}_{n\times n} + h_{k+1}\mathbf{b}^{\mathrm{T}}\mathbf{J}_k(\mathbf{I}_{s\times s} - h_{k+1}\mathbf{A}\mathbf{J}_k)^{-1}) - \lambda\mathbf{I}_{n\times n}]\mathbf{e}_k = \mathbf{0}. \qquad (3.332)$$

The corresponding eigenvalues are generated by

$$|(\mathbf{I}_{n\times n} + h_{k+1}\mathbf{b}^{\mathrm{T}}\mathbf{J}_k(\mathbf{I}_{s\times s} - h_{k+1}\mathbf{A}\mathbf{J}_k)^{-1}) - \lambda\mathbf{I}_{n\times n}| = 0. \qquad (3.333)$$

If all eigenvalues $|\lambda_j| < 1$ $(j = 1, 2, \ldots, n)$, then

$$\|\mathbf{e}_{k+1}\| < \|\mathbf{e}_k\|. \qquad (3.334)$$

So, the global error will not be enlarged. Thus, under such a condition, the Runge–Kutta method of order $s$ gives a stable approximation. For one-dimensional systems, we have $\mathbf{J}_k = L_k$ and $\mathbf{I}_{n\times n} = 1$. Equation (3.333) gives the stability interval as

$$R(h_{k+1}L_k) \leq 1 \qquad (3.335)$$

where

$$R(h_{k+1}L_k) = \frac{\det\left(\mathbf{I}_{s\times s} - h_{k+1}L_k\mathbf{A} + h_{k+1}L_k\mathbf{1}\mathbf{b}^{\mathrm{T}}\right)}{\det(\mathbf{I}_{s\times s} - h_{k+1}L_k\mathbf{A})}. \qquad (3.336)$$

## 3.6  Multi-step Methods

If the dynamical system in Eq. (3.1) is converted into an integral equation, the integration of Eq. (3.1) over the interval $[t_k, t_{k+1}]$ gives

$$\mathbf{x}(t_{k+1}) = \mathbf{x}(t_k) + \int_{t_k}^{t_{k+1}} \mathbf{f}(\mathbf{x}, t, \mathbf{p})dt. \qquad (3.337)$$

### 3.6.1  Adams–Bashforth Methods

For a given integer $s > 0$, the Adams–Bashforth method adopts the interpolation polynomial of degree $s$ at the points $(t_k, t_{k-1}, \ldots, t_{k-s})$, as shown in Fig. 3.5.

**Fig. 3.5** The node points at $t_k, t_{k-1}, \ldots, t_{k-s}$ for $[t_k, t_{k+1}]$ for Adams–Bashforth methods (explicit)

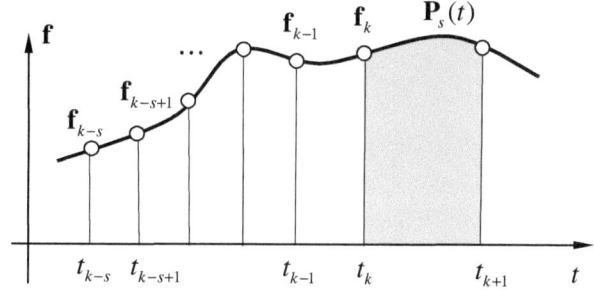

For $s = 1$, the linear interpolation polynomial of $\mathbf{f}(\mathbf{x}, t, \mathbf{p})$ is

$$\mathbf{P}_1(t) = \frac{1}{h}[(t_k - t)\bar{\mathbf{f}}_{k-1} + (t - t_{k-1})\bar{\mathbf{f}}_k]. \tag{3.338}$$

where

$$\bar{\mathbf{f}}_k = \mathbf{f}(\mathbf{x}(t_k), t_k, \mathbf{p}), \quad k = 0, 1, 2, \ldots \tag{3.339}$$

Theory of interpolation polynomial gives

$$\mathbf{f}(\mathbf{x}, t, \mathbf{p}) - \mathbf{P}_1(t) = \frac{1}{2!}(t - t_k)(t - t_{k-1})D^2\mathbf{f}(\mathbf{x}_c, t_c, \mathbf{p}), \tag{3.340}$$

with $t_c \in [t_{k-1}, t_{k+1}]$

$$Df(\mathbf{x}, t, \mathbf{p}) = (\frac{\partial \mathbf{f}}{\partial \mathbf{x}}\dot{\mathbf{x}} + \frac{\partial \mathbf{f}}{\partial t}) \quad \text{and} \quad D^m\mathbf{f}(\mathbf{x}, t, \mathbf{p}) = D(D^{m-1}\mathbf{f}(\mathbf{x}, t, \mathbf{p})). \tag{3.341}$$

The integration of $\mathbf{f}(\mathbf{x}, t, \mathbf{p})$ over $[t_k, t_{k+1}]$ is

$$\int_{t_k}^{t_{k+1}} \mathbf{f}(\mathbf{x}, t, \mathbf{p})\mathrm{d}t = \int_{t_k}^{t_{k+1}} \left[ \mathbf{P}_1(t) + \frac{1}{2!}(t - t_k)(t - t_{k-1})D^2\mathbf{f}(\mathbf{x}_c, t_c, \mathbf{p}) \right]\mathrm{d}t$$

$$= \frac{1}{2}h(3\bar{\mathbf{f}}_k - \bar{\mathbf{f}}_{k-1}) + \frac{5}{12}h^3D^2\mathbf{f}(\mathbf{x}_c, t_c, \mathbf{p}). \tag{3.342}$$

Thus, Eq. (3.337) becomes

$$\mathbf{x}(t_{k+1}) = \mathbf{x}(t_k) + \frac{1}{2}h(3\bar{\mathbf{f}}_k - \bar{\mathbf{f}}_{k-1}) + \frac{5}{12}h^3D^2\mathbf{f}(\mathbf{x}_c, t_c, \mathbf{p}). \tag{3.343}$$

Without truncation error, an approximate discrete map is

$$\mathbf{x}_{k+1} = \mathbf{x}_k + \frac{1}{2}h(3\mathbf{f}_k - \mathbf{f}_{k-1}) \qquad (3.344)$$

where

$$\mathbf{f}_k = \mathbf{f}(\mathbf{x}_k, t_k, \mathbf{p}), \quad k = 0, 1, 2, \ldots. \qquad (3.345)$$

Setting

$$h\mathbf{\Phi}(\mathbf{x}_k, \mathbf{x}_{k-1}) = \frac{1}{2}h(3\mathbf{f}_k - \mathbf{f}_{k-1}), \qquad (3.346)$$

we have a new form

$$\mathbf{x}_{k+1} = \mathbf{x}_k + h\mathbf{\Phi}(\mathbf{x}_k, \mathbf{x}_{k-1}). \qquad (3.347)$$

If $\mathbf{x}(t_\alpha) = \mathbf{x}_\alpha$ ($\alpha = k, k-1$), the local error (or a truncation error) for $t \in [t_k, t_{k+1}]$ is

$$\mathbf{\varepsilon}_{k+1} = \mathbf{x}(t_{k+1}) - \mathbf{x}_k - h\mathbf{\Phi}(\mathbf{x}_k, \mathbf{x}_{k-1}) = \frac{5}{12}h^3 D^2 \mathbf{f}(\mathbf{x}_c, t_c, \mathbf{p}). \qquad (3.348)$$

Let

$$\mathbf{e}_{k+1} = \mathbf{x}(t_{k+1}) - \mathbf{x}_{k+1} \quad \text{and} \quad \mathbf{e}_k = \mathbf{x}(t_k) - \mathbf{x}_k. \qquad (3.349)$$

We have the global error

$$\mathbf{e}_{k+1} = \mathbf{e}_k + \frac{1}{2}h(3\mathbf{A}_k\mathbf{e}_k - \mathbf{A}_{k-1}\mathbf{e}_{k-1}) + \mathbf{\varepsilon}_{k+1}. \qquad (3.350)$$

and

$$\mathbf{A}_k = \left.\frac{\partial \mathbf{f}_k}{\partial \mathbf{x}_k}\right|_{\mathbf{x}_k^c} \quad \text{and} \quad \mathbf{A}_{k-1} = \left.\frac{\partial \mathbf{f}_{k-1}}{\partial \mathbf{x}_{k-1}}\right|_{\mathbf{x}_{k-1}^c} \qquad (3.351)$$

where $\|\mathbf{x}_k^c\| \in (\|\mathbf{x}(t_k)\|, \|\mathbf{x}_k\|)$ and $\mathbf{x}_{k-1}^c\| \in (\|\mathbf{x}(t_{k-1})\|, \|\mathbf{x}_{k-1}\|)$

Consider the stability of discrete mapping through

$$\left\{\begin{array}{c} \mathbf{e}_{k+1} \\ \mathbf{e}_k \end{array}\right\} = \left[\begin{array}{cc} \mathbf{I}_{n\times n} + \frac{3}{2}h\mathbf{A}_k & -\frac{1}{2}h\mathbf{A}_k \\ \mathbf{I}_{n\times n} & \mathbf{0}_{n\times n} \end{array}\right] \left\{\begin{array}{c} \mathbf{e}_k \\ \mathbf{e}_{k-1} \end{array}\right\} = \mathbf{J}_{2n\times 2n}\left\{\begin{array}{c} \mathbf{e}_k \\ \mathbf{e}_{k-1} \end{array}\right\}. \qquad (3.352)$$

Assuming

$$\left\{ \begin{array}{c} \mathbf{e}_{k+1} \\ \mathbf{e}_k \end{array} \right\} = \lambda \left\{ \begin{array}{c} \mathbf{e}_k \\ \mathbf{e}_{k-1} \end{array} \right\}, \tag{3.353}$$

we have

$$|\mathbf{J}_{2n \times 2n} - \lambda \mathbf{I}_{2n \times 2n}| = 0. \tag{3.354}$$

If all eigenvalues $\lambda_j < 1$ $(j = 1, 2, \ldots, 2n)$, then the discrete mapping is stable. In other words,

$$\|\mathbf{e}_{k+1}\| \leq \|\mathbf{e}_k\|. \tag{3.355}$$

The computational error will not be expanded.

For $s = 2$, the linear interpolation polynomial of $\mathbf{f}(\mathbf{x}, t, \mathbf{p})$ is

$$\mathbf{P}_2(t) = l_0(t)\bar{\mathbf{f}}_k + l_1(t)\bar{\mathbf{f}}_{k-1} + l_2(t)\bar{\mathbf{f}}_{k-2} \tag{3.356}$$

where

$$\begin{aligned} l_0(t) &= \frac{1}{2h^2}(t - t_{k-1})(t - t_{k-2}), \\ l_1(t) &= \frac{1}{h^2}(t - t_k)(t - t_{k-2}), \\ l_2(t) &= \frac{1}{2h^2}(t - t_k)(t - t_{k-1}). \end{aligned} \tag{3.357}$$

Theory of interpolation polynomial gives

$$\mathbf{f}(\mathbf{x}, t, \mathbf{p}) - \mathbf{P}_2(t) = \frac{1}{3!}(t - t_k)(t - t_{k-1})(t - t_{k-2})D^3\mathbf{f}(\mathbf{x}_c, t_c, \mathbf{p}), \tag{3.358}$$

with $t_c \in [t_{k-2}, t_{k+1}]$. The integration of $\mathbf{f}(\mathbf{x}, t, \mathbf{p})$ over $[t_k, t_{k+1}]$ is

$$\int_{t_k}^{t_{k+1}} \mathbf{f}(\mathbf{x}, t, \mathbf{p}) dt = \frac{1}{12}h(23\bar{\mathbf{f}}_k - 16\bar{\mathbf{f}}_{k-1} + 5\bar{\mathbf{f}}_{k-2}) + \frac{3}{8}h^4 D^3\mathbf{f}(\mathbf{x}_c, t_c, \mathbf{p}). \tag{3.359}$$

Thus, Eq. (3.337) becomes

$$\mathbf{x}(t_{k+1}) = \mathbf{x}(t_k) + \frac{1}{12}h(23\bar{\mathbf{f}}_k - 16\bar{\mathbf{f}}_{k-1} + 5\bar{\mathbf{f}}_{k-2}) + \frac{3}{8}h^4 D^3\mathbf{f}(\mathbf{x}_c, t_c, \mathbf{p}). \tag{3.360}$$

Without truncation error, an approximate discrete map is

$$\mathbf{x}_{k+1} = \mathbf{x}_k + \frac{1}{12}h(23\mathbf{f}_k - 16\mathbf{f}_{k-1} + 5\mathbf{f}_{k-2}). \tag{3.361}$$

Setting

$$h\mathbf{\Phi}(\mathbf{x}_k, \mathbf{x}_{k-1}, \mathbf{x}_{k-2}) = \frac{1}{12}h(23\mathbf{f}_k - 16\mathbf{f}_{k-1} + 5\mathbf{f}_{k-2}), \tag{3.362}$$

we have a new form

$$\mathbf{x}_{k+1} = \mathbf{x}_k + h\mathbf{\Phi}(\mathbf{x}_k, \mathbf{x}_{k-1}, \mathbf{x}_{k-2}). \tag{3.363}$$

If $\mathbf{x}(t_\alpha) = \mathbf{x}_\alpha$ ($\alpha = k, k-1, k-2$), the local error (or a truncation error) for $t \in [t_k, t_{k+1}]$ is

$$\boldsymbol{\varepsilon}_{k+1} = \mathbf{x}(t_{k+1}) - \mathbf{x}_k - h\mathbf{\Phi}(\mathbf{x}_k, \mathbf{x}_{k-1}, \mathbf{x}_{k-2}) = \frac{3}{8}h^4 D^3 \mathbf{f}(\mathbf{x}_c, t_c, \mathbf{p}). \tag{3.364}$$

Setting

$$\mathbf{e}_j = \mathbf{x}(t_j) - \mathbf{x}_j, \quad j = k, k-1, k-2, \tag{3.365}$$

we have the global error

$$\mathbf{e}_{k+1} = \mathbf{e}_k + \frac{1}{12}h(23\mathbf{A}_k\mathbf{e}_k - 16\mathbf{A}_{k-1}\mathbf{e}_{k-1} + 5\mathbf{A}_{k-2}\mathbf{e}_{k-2}) + \boldsymbol{\varepsilon}_{k+1} \tag{3.366}$$

and

$$\mathbf{A}_j = \left.\frac{\partial \mathbf{f}_j}{\partial \mathbf{x}_j}\right|_{\mathbf{x}_j^c}, \quad \text{for } j = k, k-1, k-2 \tag{3.367}$$

where $\|\mathbf{x}_j^c\| \in (\|\mathbf{x}(t_j)\|, \|\mathbf{x}_j\|)$.

Consider the stability of discrete mapping through

$$\left\{\begin{array}{c} \mathbf{e}_{k+1} \\ \mathbf{e}_k \\ \mathbf{e}_{k-1} \end{array}\right\} = \mathbf{J}_{3n\times 3n} \left\{\begin{array}{c} \mathbf{e}_k \\ \mathbf{e}_{k-1} \\ \mathbf{e}_{k-2} \end{array}\right\} \tag{3.368}$$

where

$$\mathbf{J}_{3n\times 3n} = \begin{bmatrix} \mathbf{I}_{n\times n} + \frac{23}{12}h\mathbf{A}_k & -\frac{16}{12}h\mathbf{A}_{k-1} & \frac{5}{12}h\mathbf{A}_{k-2} \\ \mathbf{I}_{n\times n} & \mathbf{0}_{n\times n} & \mathbf{0}_{n\times n} \\ \mathbf{0}_{n\times n} & \mathbf{I}_{n\times n} & \mathbf{0}_{n\times n} \end{bmatrix}. \tag{3.369}$$

Assuming

$$\left\{ \begin{array}{c} \mathbf{e}_{k+1} \\ \mathbf{e}_k \\ \mathbf{e}_{k-1} \end{array} \right\} = \lambda \left\{ \begin{array}{c} \mathbf{e}_k \\ \mathbf{e}_{k-1} \\ \mathbf{e}_{k-2} \end{array} \right\}, \tag{3.370}$$

we have

$$|\mathbf{J}_{3n \times 3n} - \lambda \mathbf{I}_{3n \times 3n}| = 0. \tag{3.371}$$

If all eigenvalues $\lambda_j < 1$ ($j = 1, 2, \ldots, 3n$), then the discrete mapping is stable. The computational error will not be expanded.

For a general case, the linear interpolation polynomial of $\mathbf{f}(\mathbf{x}, t, \mathbf{p})$ is

$$\mathbf{P}_s(t) = \sum_{j=0}^{s} l_j(t) \bar{\mathbf{f}}_{k-j} \tag{3.372}$$

where

$$l_j(t) = \prod_{\substack{i=0, \\ i \neq j}}^{s} \frac{(t - t_{k-i})}{(t_{k-j} - t_{k-i})}. \tag{3.373}$$

Theory of interpolation polynomial gives

$$\mathbf{f}(\mathbf{x}, t, \mathbf{p}) - \mathbf{P}_s(t) = \frac{1}{(s+1)!} \prod_{j=0}^{s} (t - t_{k-j}) D^{s+1} \mathbf{f}(\mathbf{x}_c, t_c, \mathbf{p}), \tag{3.374}$$

with $t_c \in [t_{k-2}, t_{k+1}]$. The integration of $\mathbf{f}(\mathbf{x}, t, \mathbf{p})$ over $[t_k, t_{k+1}]$ is

$$\int_{t_k}^{t_{k+1}} \mathbf{f}(\mathbf{x}, t, \mathbf{p}) dt = \sum_{j=0}^{s} b_j \bar{\mathbf{f}}_{k-j} + c_s h^{s+2} D^{s+1} \mathbf{f}(\mathbf{x}_c, t_c, \mathbf{p}) \tag{3.375}$$

where

$$b_j = \frac{1}{h_{k+1}} \int_{t_k}^{t_{k+1}} \prod_{\substack{i=0 \\ i \neq j}}^{s} \frac{(t - t_{k-i})}{(t_{k-j} - t_{k-i})} dt,$$

$$c_s = \frac{1}{(s+1)! h_{k+1}^{s+2}} \int_{t_k}^{t_{k+1}} \prod_{j=0}^{s} (t - t_{k-j}) dt. \tag{3.376}$$

Thus, Eq. (3.337) becomes

$$\mathbf{x}(t_{k+1}) = \mathbf{x}(t_k) + h_{k+1} \sum_{j=0}^{s} b_j \bar{\mathbf{f}}_{k-j} + c_s h_{k+1}^{s+2} D^{s+1} \mathbf{f}(\mathbf{x}_c, t_c, \mathbf{p}). \tag{3.377}$$

Without truncation error, an approximate discrete map is

$$\mathbf{x}_{k+1} = \mathbf{x}_k + h_{k+1} \sum_{j=0}^{s} b_j \mathbf{f}_{k-j}. \tag{3.378}$$

Setting

$$\mathbf{\Phi}(\mathbf{x}_k, \mathbf{x}_{k-1}, \ldots, \mathbf{x}_{k-s}) = \sum_{j=0}^{s} b_j \mathbf{f}_{k-j}, \tag{3.379}$$

thus, we have

$$\mathbf{x}_{k+1} = \mathbf{x}_k + h_{k+1} \mathbf{\Phi}(\mathbf{x}_k, \mathbf{x}_{k-1}, \ldots, \mathbf{x}_{k-s}). \tag{3.380}$$

If $\mathbf{x}(t_\alpha) = \mathbf{x}_\alpha$ $(\alpha = k, k-1, \ldots, k-s)$, the local error (or a truncation error) for $t \in [t_k, t_{k+1}]$ is

$$\begin{aligned}
\varepsilon_{k+1} &= \mathbf{x}(t_{k+1}) - \mathbf{x}_k - h_{k+1} \mathbf{\Phi}(\mathbf{x}_k, \mathbf{x}_{k-1}, \ldots, \mathbf{x}_{k-s}) \\
&= c_s h_{k+1}^{s+2} D^{s+1} \mathbf{f}(\mathbf{x}_c, t_c, \mathbf{p}).
\end{aligned} \tag{3.381}$$

Let

$$\mathbf{e}_j = \mathbf{x}(t_j) - \mathbf{x}_j, \, j = k, k-1, k-2. \tag{3.382}$$

We have the global error

$$\mathbf{e}_{k+1} = \mathbf{e}_k + h_{k+1} \sum_{j=0}^{s} b_j \mathbf{A}_{k-j} + \varepsilon_{k+1}, \tag{3.383}$$

and

$$\mathbf{A}_j = \left. \frac{\partial \mathbf{f}_j}{\partial \mathbf{x}_j} \right|_{\mathbf{x}_j^c}, \quad \text{for } j = k, k-1, \ldots, k-s \tag{3.384}$$

where $\|\mathbf{x}_j^c\| \in (\|\mathbf{x}(t_j)\|, \|\mathbf{x}_j\|)$.

Consider the stability of discrete mapping through

$$\left\{ \begin{array}{c} \mathbf{e}_{k+1} \\ \mathbf{e}_k \\ \vdots \\ \mathbf{e}_{k-s+1} \end{array} \right\} = \mathbf{J}_{(s+1)n \times (s+1)n} \left\{ \begin{array}{c} \mathbf{e}_k \\ \mathbf{e}_{k-1} \\ \vdots \\ \mathbf{e}_{k-s} \end{array} \right\} \tag{3.385}$$

where

$$\mathbf{J}_{(s+1)n\times(s+1)n} = \begin{bmatrix} \mathbf{I}_{n\times n} + h_{k+1}b_0\mathbf{A}_k & h_{k+1}b_1\mathbf{A}_{k-1} & h_{k+1}b_2\mathbf{A}_{k-2} & \cdots & h_{k+1}b_s\mathbf{A}_{k-s} \\ \mathbf{I}_{n\times n} & \mathbf{0}_{n\times n} & \mathbf{0}_{n\times n} & \cdots & \mathbf{0}_{n\times n} \\ \mathbf{0}_{n\times n} & \mathbf{I}_{n\times n} & \mathbf{0}_{n\times n} & \cdots & \mathbf{0}_{n\times n} \\ \vdots & \vdots & \vdots & & \vdots \\ \mathbf{0}_{n\times n} & \mathbf{0}_{n\times n} & \mathbf{0}_{n\times n} & \cdots & \mathbf{0}_{n\times n} \end{bmatrix}.$$

$$(3.386)$$

Assuming

$$\left\{ \begin{array}{c} \mathbf{e}_{k+1} \\ \mathbf{e}_k \\ \vdots \\ \mathbf{e}_{k-s+1} \end{array} \right\} = \lambda \left\{ \begin{array}{c} \mathbf{e}_k \\ \mathbf{e}_{k-1} \\ \vdots \\ \mathbf{e}_{k-s} \end{array} \right\}, \tag{3.387}$$

we have

$$\left| \mathbf{J}_{(s+1)n\times(s+1)n} - \lambda \mathbf{I}_{(s+1)n\times(s+1)n} \right| = 0. \tag{3.388}$$

If all eigenvalues $\lambda_j < 1$ $(j = 1, 2, \ldots, (s+1)n)$, then the discrete mapping is stable. In other words,

$$\|\mathbf{e}_{k+1}\| \le \|\mathbf{e}_k\|. \tag{3.389}$$

The computational error will not be expanded.

If $h_j = h$ $(j = k+1, k, k-1, \ldots, k-s+1)$, the interpolation polynomial through the points $(t_{k-j}, \bar{\mathbf{f}}_{k-j})$ $(j = 0, 1, 2, \ldots, s)$ can be expressed by

$$\mathbf{P}_s(t) = \mathbf{P}_s(t_k + rh) = \sum_{j=0}^{s} (-1)^j C_{-r}^j \nabla^j \bar{\mathbf{f}}_k \tag{3.390}$$

with

$$r = \frac{t - t_k}{h}, \quad \text{and} \quad C_r^j = \frac{r(r-1)\cdots(r-j+1)}{1\times 2\times\cdots\times j}, \tag{3.391}$$
$$\nabla^0 \bar{\mathbf{f}}_k = \bar{\mathbf{f}}_k \quad \text{and} \quad \nabla^{j+1}\bar{\mathbf{f}}_k = \nabla^j\bar{\mathbf{f}}_k - \nabla^j\bar{\mathbf{f}}_{k-1}.$$

Thus, Eq. (3.337) becomes

$$\mathbf{x}(t_{k+1}) = \mathbf{x}(t_k) + h\sum_{j=0}^{s}\gamma_j\nabla^j\bar{\mathbf{f}}_k + c_s h^{s+2} D^{s+1}\mathbf{f}(\mathbf{x}_c, t_c, \mathbf{p}) \tag{3.392}$$

**Table 3.1** Coefficients for Adams–Bashforth method

| $j$ | 0 | 1 | 2 | 3 | 4 | 5 | 6 |
|---|---|---|---|---|---|---|---|
| $\gamma_j$ | 1 | $\frac{1}{2}$ | $\frac{5}{12}$ | $\frac{3}{8}$ | $\frac{251}{720}$ | $\frac{95}{285}$ | $\frac{19{,}087}{60{,}480}$ |

where

$$\gamma_j = (-1)^j \frac{1}{h} \int_{t_k}^{t_{k+1}} C^j_{-r}\,dt = (-1)^j \int_0^1 C^j_{-r}\,dr. \tag{3.393}$$

The coefficients are listed in Table 3.1. Without truncation error, an approximate discrete map is given by

$$\mathbf{x}_{k+1} = \mathbf{x}_k + h \sum_{j=0}^{s} \gamma_j \nabla^j \mathbf{f}_k \tag{3.394}$$

where

$$\nabla^0 \mathbf{f}_k = \mathbf{f}_k \quad \text{and} \quad \nabla^{j+1}\mathbf{f}_k = \nabla^j \mathbf{f}_k - \nabla^j \mathbf{f}_{k-1}. \tag{3.395}$$

Equation (3.394) gives the same formulas as Eq. (3.375). The other discussions can be referred to Henrici (1962) and Hairer et al. (1987). The summarization of the Adams–Bashforth methods is in Table 3.2. For $s = 3$, Eq. (3.394) gives the most popular Adams–Bashforth form.

**Table 3.2** Adams–Bashforth method (explicit)

| $s$ | Order | Methods | L. error |
|---|---|---|---|
| 0 | 1 | $\mathbf{x}_{k+1} = \mathbf{x}_k + h\mathbf{f}_k$ (forward Euler) | $\frac{1}{2}h^2 D\mathbf{f}_c$ |
| 1 | 2 | $\mathbf{x}_{k+1} = \mathbf{x}_k + \frac{1}{2}h(3\mathbf{f}_k - \mathbf{f}_{k-1})$ | $\frac{5}{12}h^3 D^2\mathbf{f}_c$ |
| 2 | 3 | $\mathbf{x}_{k+1} = \mathbf{x}_k + \frac{1}{12}h(23\mathbf{f}_k - 16\mathbf{f}_{k-1} + 5\mathbf{f}_{k-2})$ | $\frac{3}{8}h^4 D^3\mathbf{f}_c$ |
| 3 | 4 | $\mathbf{x}_{k+1} = \mathbf{x}_k + \frac{1}{24}h(55\mathbf{f}_k - 59\mathbf{f}_{k-1} + 37\mathbf{f}_{k-2} - 9\mathbf{f}_{k-3})$ | $\frac{251}{720}h^5 D^4\mathbf{f}_c$ |
| 4 | 5 | $\mathbf{x}_{k+1} = \mathbf{x}_k + \dfrac{1}{720}h(1901\mathbf{f}_k - 2774\mathbf{f}_{k-1} + 2616\mathbf{f}_{k-2}$ $- 1274\mathbf{f}_{k-3} + 251\mathbf{f}_{k-4})$ | $\frac{95}{288}h^6 D^5\mathbf{f}_c$ |
| 5 | 6 | $\mathbf{x}_{k+1} = \mathbf{x}_k + \dfrac{1}{1440}h(4277\mathbf{f}_k - 7923\mathbf{f}_{k-1} + 9982\mathbf{f}_{k-2}$ $- 7298\mathbf{f}_{k-3} + 2877\mathbf{f}_{k-4} - 475\mathbf{f}_{k-5})$ | $\frac{19{,}087}{60{,}480}h^7 D^6\mathbf{f}_c$ |

Note that $D^j\mathbf{f}_c = D^j\mathbf{f}(\mathbf{x}(t_c), t_c, p)$ $(j = 0, 1, 2, 3, \ldots)$

### 3.6.2 Adams–Moulton Methods

For a given integer $s > 0$, the Adams–Moulton method uses the interpolation polynomial of degree $s$ at the points $(t_{k+1}, t_k, \ldots, t_{k-s+1})$, as shown in Fig. 3.6.

For $s = 1$, the linear interpolation polynomial of $\mathbf{f}(\mathbf{x}, t, \mathbf{p})$ is

$$\mathbf{P}_1^*(t) = \frac{1}{h}[(t_{k+1} - t)\bar{\mathbf{f}}_k + (t - t_k)\bar{\mathbf{f}}_{k+1}]. \tag{3.396}$$

Theory of interpolation polynomial gives

$$\mathbf{f}(\mathbf{x}, t, \mathbf{p}) - \mathbf{P}_1^*(t) = \frac{1}{2!}(t - t_{k+1})(t - t_k)D^2\mathbf{f}(\mathbf{x}_c, t_c, \mathbf{p}), \tag{3.397}$$

with $t_c \in [t_{k-1}, t_{k+1}]$. The integration of $\mathbf{f}(\mathbf{x}, t, \mathbf{p})$ over $[t_k, t_{k+1}]$ is

$$\int_{t_k}^{t_{k+1}} \mathbf{f}(\mathbf{x}, t, \mathbf{p})dt = \int_{t_k}^{t_{k+1}} [\mathbf{P}_1^*(t) + \frac{1}{2!}(t - t_{k+1})(t - t_k)D^2\mathbf{f}(\mathbf{x}_c, t_c, \mathbf{p})]dt$$
$$= \frac{1}{2}h(\bar{\mathbf{f}}_{k+1} + \bar{\mathbf{f}}_{k-1}) - \frac{1}{12}h^3 D^2\mathbf{f}(\mathbf{x}_c, t_c, \mathbf{p}). \tag{3.398}$$

Thus, Eq. (3.337) becomes

$$\mathbf{x}(t_{k+1}) = \mathbf{x}(t_k) + \frac{1}{2}h(\bar{\mathbf{f}}_{k+1} + \bar{\mathbf{f}}_k) - \frac{1}{12}h^3 D^2\mathbf{f}(\mathbf{x}_c, t_c, \mathbf{p}). \tag{3.399}$$

Without truncation error, an approximate discrete map is

$$\mathbf{x}_{k+1} = \mathbf{x}_k + \frac{1}{2}h(\mathbf{f}_{k+1} + \mathbf{f}_k). \tag{3.400}$$

This gives the trapezoidal method.

**Fig. 3.6** The node points at $t_k, t_{k-1}, \ldots, t_{k-s}$ for $[t_k, t_{k+1}]$ for Adams–Moulton methods (implicit)

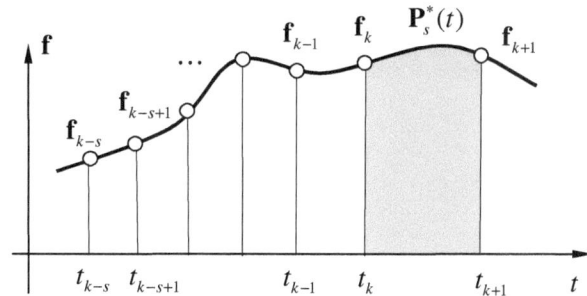

For $s = 2$, the linear interpolation polynomial of $\mathbf{f}(\mathbf{x}, t, \mathbf{p})$ is

$$\mathbf{P}_1^*(t) = l_0(t)\bar{\mathbf{f}}_{k+1} + l_1(t)\bar{\mathbf{f}}_k + l_2(t)\bar{\mathbf{f}}_{k-1} \tag{3.401}$$

where

$$
\begin{aligned}
l_0(t) &= \frac{1}{2h^2}(t - t_k)(t - t_{k-1}), \\
l_1(t) &= \frac{1}{h^2}(t - t_{k+1})(t - t_{k-1}), \\
l_2(t) &= \frac{1}{2h^2}(t - t_{k+1})(t - t_k).
\end{aligned}
\tag{3.402}
$$

Theory of interpolation polynomial gives

$$\mathbf{f}(\mathbf{x}, t, \mathbf{p}) - \mathbf{P}_2(t) = \frac{1}{3!}(t - t_{k+1})(t - t_k)(t - t_{k-1})D^3\mathbf{f}(\mathbf{x}_c, t_c, \mathbf{p}), \tag{3.403}$$

with $t_c \in [t_{k-1}, t_{k+1}]$. The integration of $\mathbf{f}(\mathbf{x}, t, \mathbf{p})$ over $[t_k, t_{k+1}]$ is

$$\int_{t_k}^{t_{k+1}} \mathbf{f}(\mathbf{x}, t, \mathbf{p})dt = \frac{1}{12}h(5\bar{\mathbf{f}}_{k+1} + 8\bar{\mathbf{f}}_k - \bar{\mathbf{f}}_{k-1}) - \frac{1}{24}h^4 D^3\mathbf{f}(\mathbf{x}_c, t_c, \mathbf{p}). \tag{3.404}$$

Thus, Eq. (3.337) becomes

$$\mathbf{x}(t_{k+1}) = \mathbf{x}(t_k) + \frac{1}{12}h(5\bar{\mathbf{f}}_{k+1} + 8\bar{\mathbf{f}}_k - \bar{\mathbf{f}}_{k-1}) - \frac{1}{24}h^4 D^3\mathbf{f}(\mathbf{x}_c, t_c, \mathbf{p}). \tag{3.405}$$

Without truncation error, an approximate discrete map is

$$\mathbf{x}_{k+1} = \mathbf{x}_k + \frac{1}{12}h(5\mathbf{f}_{k+1} + 8\mathbf{f}_k - \mathbf{f}_{k-1}). \tag{3.406}$$

Setting

$$h\mathbf{\Phi}(\mathbf{x}_k, \mathbf{x}_{k-1}, \mathbf{x}_{k-2}) = \frac{1}{12}h(5\mathbf{f}_{k+1} + 8\mathbf{f}_k - \mathbf{f}_{k-1}), \tag{3.407}$$

we have a new form

$$\mathbf{x}_{k+1} = \mathbf{x}_k + h\mathbf{\Phi}(\mathbf{x}_{k+1}, \mathbf{x}_k, \mathbf{x}_{k-1}). \tag{3.408}$$

If $\mathbf{x}(t_\alpha) = \mathbf{x}_\alpha$ ($\alpha = k, k - 1$), a truncation error) for $t \in [t_k, t_{k+1}]$ is

$$\mathbf{T}_e = -\frac{1}{24}h^4 D^3\mathbf{f}(\mathbf{x}_c, t_c, \mathbf{p}), \tag{3.409}$$

and the local error can be expressed by

$$\boldsymbol{\varepsilon}_{k+1} = \mathbf{x}(t_{k+1}) - \mathbf{x}_k - \frac{1}{12}h(5\mathbf{f}_{k+1} + 8\mathbf{f}_k - 5\mathbf{f}_{k-1}) + \mathbf{T}_e. \tag{3.410}$$

Thus

$$\boldsymbol{\varepsilon}_{k+1} = \left(\mathbf{I}_{n \times n} - \frac{5}{12}h\mathbf{A}_{k+1}\right)^{-1}\mathbf{T}_e = -\frac{1}{24}\left(\mathbf{I}_{n \times n} - \frac{5}{12}h\mathbf{A}_{k+1}\right)^{-1}h^4 D^3\mathbf{f}(\mathbf{x}_c, t_c, \mathbf{p}) \tag{3.411}$$

where

$$\mathbf{A}_{k+1-j} = \left.\frac{\partial \mathbf{f}}{\partial \mathbf{x}}\right|_{\mathbf{x}_{k+1-j}^c}, \quad j = 0, 1, 2, \tag{3.412}$$

with $\|\mathbf{x}_{k+1-j}^c\| \in (\|\mathbf{x}(t_{k+1-j})\|, \|\mathbf{x}_{k+1-j}\|)$. Let

$$\mathbf{e}_j = \mathbf{x}(t_j) - \mathbf{x}_j, \quad j = k, k-1, k-2. \tag{3.413}$$

We have the global error

$$\mathbf{e}_{k+1} = \mathbf{e}_k + \frac{1}{12}h(5\mathbf{A}_{k+1}\mathbf{e}_{k+1} + 8\mathbf{A}_k\mathbf{e}_k - \mathbf{A}_{k-1}\mathbf{e}_{k-1}) + \mathbf{T}_e. \tag{3.414}$$

Consider the stability of discrete mapping through

$$\left\{\begin{array}{c}\mathbf{e}_{k+1} \\ \mathbf{e}_k\end{array}\right\} = \mathbf{J}_{2n \times 2n}\left\{\begin{array}{c}\mathbf{e}_k \\ \mathbf{e}_{k-1}\end{array}\right\} \tag{3.415}$$

where

$$\mathbf{J}_{2n \times 2n} = \left[\begin{array}{cc}\frac{8}{12}\left(1 - \frac{5}{12}h\mathbf{A}_{k+1}\right)^{-1}\mathbf{A}_k & -\frac{1}{12}\left(1 - \frac{5}{12}h\mathbf{A}_{k+1}\right)^{-1}\mathbf{A}_{k-1} \\ \mathbf{I}_{n \times n} & \mathbf{0}_{n \times n}\end{array}\right]. \tag{3.416}$$

Assuming

$$\left\{\begin{array}{c}\mathbf{e}_{k+1} \\ \mathbf{e}_k\end{array}\right\} = \lambda\left\{\begin{array}{c}\mathbf{e}_k \\ \mathbf{e}_{k-1}\end{array}\right\}, \tag{3.417}$$

we have

$$|\mathbf{J}_{2n \times 2n} - \lambda\mathbf{I}_{2n \times 2n}| = 0. \tag{3.418}$$

If all eigenvalues $\lambda_j < 1$ $(j = 1, 2, \ldots, 2n)$, then the discrete mapping is stable. The computational error will not be expanded.

For a general case, the linear interpolation polynomial of $\mathbf{f}(\mathbf{x}, t, \mathbf{p})$ is

$$\mathbf{P}_s^*(t) = \sum_{j=0}^{s} l_j(t) \bar{\mathbf{f}}_{k-j+1} \tag{3.419}$$

where

$$l_j(t) = \prod_{\substack{i=0 \\ i \neq j}}^{s} \frac{(t - t_{k+1-i})}{(t_{k+1-j} - t_{k+1-i})}. \tag{3.420}$$

From the theory of interpolation polynomial, we have

$$\mathbf{f}(\mathbf{x}, t, \mathbf{p}) - \mathbf{P}_s(t) = \frac{1}{(s+1)!} \prod_{j=0}^{s} (t - t_{k+1-j}) D^{s+1} \mathbf{f}(\mathbf{x}_c, t_c, \mathbf{p}), \tag{3.421}$$

with $t_c \in [t_{k+1-s}, t_{k+1}]$. The integration of $\mathbf{f}(\mathbf{x}, t, \mathbf{p})$ over $[t_k, t_{k+1}]$ is

$$\int_{t_k}^{t_{k+1}} \mathbf{f}(\mathbf{x}, t, \mathbf{p}) dt = \sum_{j=0}^{s} b_j^* \bar{\mathbf{f}}_{k-j} + c_s^* h_{k+1}^{s+2} D^{s+1} \mathbf{f}(\mathbf{x}_c, t_c, \mathbf{p}) \tag{3.422}$$

where

$$b_j^* = \frac{1}{h_{k+1}} \int_{t_k}^{t_{k+1}} \prod_{\substack{i=0 \\ i \neq j}}^{s} \frac{(t - t_{k+1-i})}{(t_{k+1-j} - t_{k+1-i})} dt,$$

$$c_s^* = \frac{1}{(s+1)! h_{k+1}^{s+2}} \int_{t_k}^{t_{k+1}} \prod_{i=0}^{s} (t - t_{k+1-i}) dt. \tag{3.423}$$

Thus, Eq. (3.337) becomes

$$\mathbf{x}(t_{k+1}) = \mathbf{x}(t_k) + h_{k+1} \sum_{j=0}^{s} b_j^* \bar{\mathbf{f}}_{k+1-j} + c_s^* h_{k+1}^{s+2} D^{s+1} \mathbf{f}(\mathbf{x}_c, t_c, \mathbf{p}). \tag{3.424}$$

Without truncation error, an approximate discrete map is

$$\mathbf{x}_{k+1} = \mathbf{x}_k + h_{k+1} \sum_{j=0}^{s} b_j^* \mathbf{f}_{k+1-j}. \tag{3.425}$$

Setting

$$\mathbf{\Phi}(\mathbf{x}_{k+1}, \mathbf{x}_k, \ldots, \mathbf{x}_{k+1-s}) = \sum_{j=0}^{s} b_j^* \mathbf{f}_{k+1-j}, \qquad (3.426)$$

we have a new form

$$\mathbf{x}_{k+1} = \mathbf{x}_k + h_{k+1} \mathbf{\Phi}(\mathbf{x}_{k+1}, \mathbf{x}_k, \ldots, \mathbf{x}_{k+1-s}). \qquad (3.427)$$

The truncation error is

$$\mathbf{T}_e = c_s^* h_{k+1}^{s+2} D^{s+1} \mathbf{f}(\mathbf{x}_c, t_c, \mathbf{p}). \qquad (3.428)$$

If $\mathbf{x}(t_\alpha) = \mathbf{x}_\alpha$ ($\alpha = k, k-1, \ldots, k+1-s$), the local error for $t \in [t_k, t_{k+1}]$ is

$$\boldsymbol{\varepsilon}_{k+1} = h_{k+1} \mathbf{A}_{k+1} \boldsymbol{\varepsilon}_{k+1} + c_s^* h_{k+1}^{s+2} D^{s+1} \mathbf{f}(\mathbf{x}_c, t_c, \mathbf{p}). \qquad (3.429)$$

where

$$\mathbf{A}_{k+1-j} = \left. \frac{\partial \mathbf{f}_{k+1-j}}{\partial \mathbf{x}_{k+1-j}} \right|_{\mathbf{x}_{k+1-j}^c}, \quad j = 0, 1, 2, \ldots, s+1. \qquad (3.430)$$

Thus, the local error is

$$\begin{aligned}
\boldsymbol{\varepsilon}_{k+1} &= (\mathbf{I}_{n \times n} - h_{k+1} b_j^* \mathbf{A}_{k+1})^{-1} \mathbf{T}_e \\
&= c_s^* h_{k+1}^{s+2} (\mathbf{I}_{n \times n} - h_{k+1} b_j^* \mathbf{A}_{k+1})^{-1} D^{s+2} \mathbf{f}(\mathbf{x}_c, t_c, \mathbf{p})
\end{aligned} \qquad (3.431)$$

where $\|\mathbf{x}_{k+1-j}^c\| \in (\|\mathbf{x}(t_{k+1-j})\|, \|\mathbf{x}_{k+1-j}\|)$. As usual, let

$$\mathbf{e}_j = \mathbf{x}(t_j) - \mathbf{x}_j, \quad j = k, k-1, k-2. \qquad (3.432)$$

We have the global error

$$\mathbf{e}_{k+1} = \mathbf{e}_k + h_{k+1} \sum_{j=0}^{s} b_j^* \mathbf{A}_{k+1-j} \mathbf{e}_{k+1-j} + \mathbf{T}_e. \qquad (3.433)$$

Consider the stability of discrete mapping through

$$\left\{ \begin{array}{c} \mathbf{e}_{k+1} \\ \mathbf{e}_k \\ \vdots \\ \mathbf{e}_{k+2-s} \end{array} \right\} = \mathbf{J}_{sn \times sn} \left\{ \begin{array}{c} \mathbf{e}_k \\ \mathbf{e}_{k-1} \\ \vdots \\ \mathbf{e}_{k+1-s} \end{array} \right\} \qquad (3.434)$$

where

$$\mathbf{J}_{(s+1)n\times(s+1)n} = \begin{bmatrix} h_{k+1}b_1^*\bar{\mathbf{A}}_k & h_{k+1}b_2^*\bar{\mathbf{A}}_{k-1} & h_{k+1}b_3^*\bar{\mathbf{A}}_{k-2} & \cdots & h_{k+1}b_s^*\bar{\mathbf{A}}_{k+1-s} \\ \mathbf{I}_{n\times n} & \mathbf{0}_{n\times n} & \mathbf{0}_{n\times n} & \cdots & \mathbf{0}_{n\times n} \\ \mathbf{0}_{n\times n} & \mathbf{I}_{n\times n} & \mathbf{0}_{n\times n} & \cdots & \mathbf{0}_{n\times n} \\ \vdots & \vdots & \vdots & \vdots & \vdots \\ \mathbf{0}_{n\times n} & \mathbf{0}_{n\times n} & \mathbf{0}_{n\times n} & \cdots & \mathbf{0}_{n\times n} \end{bmatrix}$$

(3.435)

and

$$\bar{\mathbf{A}}_{k+1-j} = (\mathbf{I}_{n\times n} - h_{k+1}b_1^*\mathbf{A}_{k+1})^{-1}\mathbf{A}_{k+1-j}, \quad j = 1, 2, \ldots, s-1. \quad (3.436)$$

Assuming

$$\begin{Bmatrix} \mathbf{e}_{k+1} \\ \mathbf{e}_k \\ \vdots \\ \mathbf{e}_{k-s+2} \end{Bmatrix} = \lambda \begin{Bmatrix} \mathbf{e}_k \\ \mathbf{e}_{k-1} \\ \vdots \\ \mathbf{e}_{k-s+1} \end{Bmatrix}, \quad (3.437)$$

we have

$$|\mathbf{J}_{sn\times sn} - \lambda\mathbf{I}_{sn\times sn}| = 0. \quad (3.438)$$

If all eigenvalues $\lambda_j < 1$ $(j = 1, 2, \ldots, sn)$, then the discrete mapping is stable. The computational error will not be expanded.

If $h_k = h$ $(j = k+1, k, k-1, \ldots, k-s+1)$, the interpolation polynomial through the points $(t_{k-j}, \bar{\mathbf{f}}_{k-j})$ $(j = 0, 1, 2, \ldots, s)$ can be expressed by

$$\mathbf{P}_s^*(t) = \mathbf{P}_s^*(t_k + rh) = \sum_{j=0}^{s} (-1)^j C_{-r+1}^j \nabla^j \bar{\mathbf{f}}_{k+1}. \quad (3.439)$$

Thus, Eq. (3.337) becomes

$$\mathbf{x}(t_{k+1}) = \mathbf{x}(t_k) + h \sum_{j=0}^{s} \gamma_j^* \nabla^j \bar{\mathbf{f}}_{k+1} + c_s^* h^{s+2} D^{s+1} \mathbf{f}(\mathbf{x}_c, t_c, \mathbf{p}) \quad (3.440)$$

where

$$\gamma_s^* = (-1)^j \frac{1}{h} \int_{t_k}^{t_{k+1}} C_{-r+1}^j dt = (-1)^j \int_0^1 C_{-r+1}^j dr. \quad (3.441)$$

**Table 3.3** Coefficients for Adams–Moulton methods

| $j$ | 0 | 1 | 2 | 3 | 4 | 5 | 6 |
|-----|---|---|---|---|---|---|---|
| $\gamma_j$ | 1 | $-\frac{1}{2}$ | $-\frac{1}{12}$ | $-\frac{1}{24}$ | $-\frac{19}{720}$ | $-\frac{3}{160}$ | $-\frac{863}{60,480}$ |

Without truncation error, an approximate discrete map is

$$\mathbf{x}_{k+1} = \mathbf{x}_k + h\sum_{j=0}^{s}\gamma_s^*\nabla^j\mathbf{f}_{k+1}. \tag{3.442}$$

The corresponding coefficients are listed in Table 3.3. Equation (3.443) gives the same formulas as Eq. (3.425). The other discussions can also be referred to Henrici (1962) and Hairer et al. (1987). The summarization of the Adams–Moulton methods is in Table 3.4. For $s = 3$, Eq. (3.442) gives the most popular Adams–Moulton form. In numerical iterations, the Adams–Bashforth form is called the Adams–Bashforth predictor, and the Adams–Moulton form provides the Adams–Moulton corrector. Consider $s = 3$ as an example for the Adams–Bashforth–Moulton method. The predictor is

$$\mathbf{p}_{k+1} = \mathbf{x}_k + \frac{1}{24}h(55\mathbf{f}_k - 59\mathbf{f}_{k-1} + 37\mathbf{f}_{k-2} - 9\mathbf{f}_{k-3}) \tag{3.443}$$

and the corrector is

$$\mathbf{x}_{k+1} = \mathbf{x}_k + \frac{1}{24}h(9\mathbf{f}_{k+1} + 19\mathbf{f}_k - 5\mathbf{f}_{k-1} + \mathbf{f}_{k-2}) \tag{3.444}$$

where

$$\mathbf{f}_{k+1} \approx \mathbf{f}(\mathbf{p}_{k+1}, t_{k+1}, \mathbf{p}). \tag{3.445}$$

**Table 3.4** Adams–Moulton method (implicit)

| $s$ | Order | Methods | T. error |
|-----|-------|---------|----------|
| 0 | 1 | $\mathbf{x}_{k+1} = \mathbf{x}_k + h\mathbf{f}_{k+1}$ (backward Euler) | $-\frac{1}{2}h^2 D\mathbf{f}_c$ |
| 1 | 2 | $\mathbf{x}_{k+1} = \mathbf{x}_k + \frac{1}{2}h(\mathbf{f}_{k+1} + \mathbf{f}_k)$ | $-\frac{1}{12}h^3 D^2\mathbf{f}_c$ |
| 2 | 3 | $\mathbf{x}_{k+1} = \mathbf{x}_k + \frac{1}{12}h(5\mathbf{f}_{k+1} + 8\mathbf{f}_k - \mathbf{f}_{k-1})$ | $-\frac{1}{24}h^4 D^3\mathbf{f}_c$ |
| 3 | 4 | $\mathbf{x}_{k+1} = \mathbf{x}_k + \frac{1}{24}h(9\mathbf{f}_{k+1} + 19\mathbf{f}_k - 5\mathbf{f}_{k-1} + \mathbf{f}_{k-2})$ | $-\frac{19}{720}h^5 D^4\mathbf{f}_c$ |
| 4 | 5 | $\mathbf{x}_{k+1} = \mathbf{x}_k + \dfrac{1}{720}h(251\mathbf{f}_{k+1} + 464\mathbf{f}_k - 264\mathbf{f}_{k-1}$ $+ 106\mathbf{f}_{k-2} - 19\mathbf{f}_{k-3})$ | $-\frac{3}{160}h^6 D^5\mathbf{f}_c$ |
| 5 | 6 | $\mathbf{x}_{k+1} = \mathbf{x}_k + \dfrac{1}{1440}h(475\mathbf{f}_{k+1} + 1427\mathbf{f}_k - 789\mathbf{f}_{k-1}$ $+ 482\mathbf{f}_{k-2} - 173\mathbf{f}_{k-3} + 27\mathbf{f}_{k-4})$ | $-\frac{863}{60,480}h^7 D^6\mathbf{f}_c$ |

Note that $D^j\mathbf{f}_c = D^j\mathbf{f}(\mathbf{x}(t_c), t_c, p)$ $(j = 0, 1, 2, 3, \ldots)$

From the truncation errors of the predictor and corrector, we have

$$\mathbf{x}(t_{k+1}) - \mathbf{p}_{k+1} = \frac{251}{720} h^5 D^4 \mathbf{f}_d \quad \text{and} \quad \mathbf{x}(t_{k+1}) - \mathbf{x}_{k+1} = -\frac{19}{720} h^5 D^4 \mathbf{f}_c. \quad (3.446)$$

If $D^4 \mathbf{f}_d \approx D^4 \mathbf{f}_c$, the forgoing equation gives

$$\mathbf{x}(t_{k+1}) - \mathbf{x}_{k+1} = -\frac{19}{270}(\mathbf{x}_{k+1} - \mathbf{p}_{k+1}). \quad (3.447)$$

Thus, the error estimate can be done by Eq. (3.447). Let

$$\|\mathbf{x}(t_{k+1}) - \mathbf{x}_{k+1}\| = \varepsilon \Rightarrow \|\mathbf{x}_{k+1}\| - \varepsilon \leq \|\mathbf{x}(t_{k+1})\| \leq \|\mathbf{x}_{k+1}\| + \varepsilon. \quad (3.448)$$

The relative error can be computed by

$$\varepsilon_r = \frac{\|\mathbf{x}(t_{k+1}) - \mathbf{x}_{k+1}\|}{\|\mathbf{x}(t_{k+1})\|}. \quad (3.449)$$

If the following condition is satisfied,

$$\frac{19}{270} \frac{\|\mathbf{x}_{k+1} - \mathbf{p}_{k+1}\|}{\|\mathbf{x}_{k+1}\| + \varepsilon} \leq \varepsilon_r, \quad (3.450)$$

then, we have $\mathbf{x}_{k+1} \approx \mathbf{x}(t_{k+1})$.

### 3.6.3  Explicit Adams Methods

If the dynamical system in Eq. (3.1) is converted into an integral equation, the integration of Eq. (3.1) over the interval $[t_{k-l}, t_{k+1}]$ gives

$$\mathbf{x}(t_{k+1}) = \mathbf{x}(t_{k-l}) + \int_{t_{k-l}}^{t_{k+1}} \mathbf{f}(\mathbf{x}, t, \mathbf{p}) dt. \quad (3.451)$$

For $l = 0$, the Adams–Bashforth and Adams–Moulton methods were presented before. The other methods for $l \neq 0$ will be discussed.

Using $\mathbf{P}(t)$ in Eqs. (3.390) and (3.391) to approximate $\mathbf{f}(\mathbf{x}, t, \mathbf{p})$ in Eq. (3.451) gives

$$\mathbf{x}(t_{k+1}) = \mathbf{x}(t_{k-l}) + h \sum_{j=0}^{s} \kappa_{j,l} \nabla^j \bar{\mathbf{f}}_k + c_{s,l} h^{s+2} D^{s+1} \mathbf{f}(\mathbf{x}_c, t_c, \mathbf{p}) \quad (3.452)$$

**Fig. 3.7** The node points at $t_k, t_{k-1}, \ldots, t_{k-s}$ for $[t_{k-1}, t_{k+1}]$ for the Nyström methods (explicit)

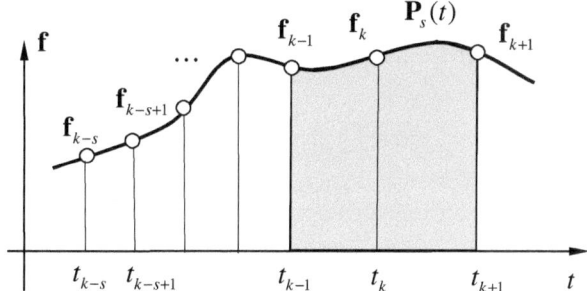

where

$$\kappa_{j,l} = (-1)^j \frac{1}{h} \int\limits_{t_{k-l}}^{t_{k+1}} C^j_{-r} \mathrm{d}t = (-1)^j \int\limits_{-l}^{1} C^j_{-r} \mathrm{d}r,$$

$$c_{s,l} = \frac{1}{(s+1)! h^{s+2}} \int\limits_{t_{k-l}}^{t_{k+1}} \prod_{j=0}^{s} (t - t_{k-j}) \mathrm{d}t = (-1)^{s+1} \int\limits_{-l}^{1} C^{s+1}_{-r} \mathrm{d}r.$$

(3.453)

Without any truncation error, an approximate discrete map is

$$\mathbf{x}_{k+1} = \mathbf{x}_{k-l} + h \sum_{j=0}^{s} \kappa_{j,l} \nabla^j \mathbf{f}_k.$$

(3.454)

For $l = 1$, the above method is called the Nyström methods. The node points at $t_k, t_{k-1}, \ldots, t_{k-s}$ for $[t_{k-1}, t_{k+1}]$ for the Nyström methods are sketched in Fig. 3.7. The corresponding coefficients are listed in Table 3.5. A few Nyström schemes are summarized in Table 3.6. For $s = 3$, Eq. (3.454) gives the popular Nyström method. The Nyström method uses the polynomial vector function based on the mesh points $t_k, t_{k-1}, \ldots, t_{k-s}$, similar to the Adams–Bashforth method.

For $l = 2$, we have

$$\mathbf{x}_{k+1} = \mathbf{x}_{k-2} + h \sum_{j=0}^{s-1} \kappa_{j,2} \nabla^j \mathbf{f}_k.$$

(3.455)

**Table 3.5** Coefficients for the Nyström methods

| $j$ | 0 | 1 | 2 | 3 | 4 | 5 | 6 |
|---|---|---|---|---|---|---|---|
| $\kappa_{j,1}$ | 2 | 0 | $\frac{1}{3}$ | $\frac{1}{3}$ | $\frac{29}{90}$ | $\frac{14}{45}$ | $\frac{1139}{3780}$ |

**Table 3.6** The Nyström methods (explicit)

| $s$ | Order | Methods | T. error |
|---|---|---|---|
| 0 | 1 | $\mathbf{x}_{k+1} = \mathbf{x}_{k-1} + 2h\mathbf{f}_k$ | $\frac{1}{3}h^3 D^2 \mathbf{f}_c$ |
| 1 | 2 | $\mathbf{x}_{k+1} = \mathbf{x}_{k-1} + 2h\mathbf{f}_k$ | $\frac{1}{3}h^3 D^2 \mathbf{f}_c$ |
| 2 | 3 | $\mathbf{x}_{k+1} = \mathbf{x}_{k-1} + \frac{1}{3}h(7\mathbf{f}_k - 2\mathbf{f}_{k-1} + \mathbf{f}_{k-2})$ | $\frac{1}{3}h^4 D^3 \mathbf{f}_c$ |
| 3 | 4 | $\mathbf{x}_{k+1} = \mathbf{x}_{k-1} + \frac{1}{3}h(8\mathbf{f}_k - 5\mathbf{f}_{k-1} + 4\mathbf{f}_{k-2} - \mathbf{f}_{k-3})$ | $\frac{29}{90}h^5 D^4 \mathbf{f}_c$ |
| 4 | 5 | $\mathbf{x}_{k+1} = \mathbf{x}_k + \dfrac{1}{90}h(269\mathbf{f}_k - 266\mathbf{f}_{k-1} + 294\mathbf{f}_{k-2}$ $- 146\mathbf{f}_{k-3} + 29\mathbf{f}_{k-5})$ | $\frac{14}{45}h^6 D^5 \mathbf{f}_c$ |

**Table 3.7** Coefficients for the explicit Adams methods ($l = 2$)

| $j$ | 0 | 1 | 2 | 3 | 4 | 5 | 6 |
|---|---|---|---|---|---|---|---|
| $\kappa_{j,2}$ | 3 | $-\frac{3}{2}$ | $\frac{3}{4}$ | $\frac{3}{8}$ | $\frac{27}{80}$ | $\frac{51}{160}$ | $\frac{411}{1124}$ |

**Table 3.8** The explicit Adams methods ($l = 2$)

| $s$ | Order | Methods | T. error |
|---|---|---|---|
| 0 | 1 | $\mathbf{x}_{k+1} = \mathbf{x}_{k-2} + 3h\mathbf{f}_k$ | $-\frac{3}{2}h^2 D\mathbf{f}_c$ |
| 1 | 2 | $\mathbf{x}_{k+1} = \mathbf{x}_{k-2} + \frac{1}{2}h(3\mathbf{f}_k + 3\mathbf{f}_{k-1})$ | $\frac{3}{4}h^3 D^2 \mathbf{f}_c$ |
| 2 | 3 | $\mathbf{x}_{k+1} = \mathbf{x}_{k-2} + \frac{1}{4}h(9\mathbf{f}_k + 3\mathbf{f}_{k-2})$ | $\frac{3}{8}h^4 D^3 \mathbf{f}_c$ |
| 3 | 4 | $\mathbf{x}_{k+1} = \mathbf{x}_{k-2} + \frac{1}{24}h(63\mathbf{f}_k - 27\mathbf{f}_{k-1} + 45\mathbf{f}_{k-2} - 9\mathbf{f}_{k-3})$ | $\frac{27}{80}h^5 D^4 \mathbf{f}_c$ |
| 4 | 5 | $\mathbf{x}_{k+1} = \mathbf{x}_{k-2} + \frac{1}{80}h(237\mathbf{f}_k - 198\mathbf{f}_{k-1} + 312\mathbf{f}_{k-2}$ $- 138\mathbf{f}_{k-3} + 27\mathbf{f}_{k-5})$ | $\frac{51}{160}h^6 D^5 \mathbf{f}_c$ |

The coefficients $\kappa_{j,2}$ are listed in Table 3.7. A few schemes are summarized in Table 3.8, which is also to the alike Adams–Bashforth method ($l = 2$). As in Ceschino and Kuntzmann (1966), the coefficients of $\kappa_{j,l}$ is listed in Table 3.9.

### 3.6.4 Implicit Adams Methods

Using $\mathbf{P}^*(t)$ in Eq. (3.439) to approximate $\mathbf{f}(\mathbf{x}, t, \mathbf{p})$ in Eq. (3.451) gives

$$\mathbf{x}(t_{k+1}) = \mathbf{x}(t_{k-l}) + h \sum_{j=0}^{s-1} \kappa_{j,l}^* \nabla^j \bar{\mathbf{f}}_k + c_s^* h^{s+2} D^{s+1} \mathbf{f}(\mathbf{x}_c, t_c, \mathbf{p}) \qquad (3.456)$$

**Table 3.9** Coefficients $\kappa_{j,l}$ for the explicit Adams methods

| $\frac{1}{\eta}\kappa_{j,l}\ \underline{\frac{|j}{|}}$ | 0 | $1(\frac{1}{2})$ | $2(\frac{1}{12})$ | $3(\frac{1}{24})$ | $4(\frac{1}{720})$ |
|---|---|---|---|---|---|
| 0 | 1 | 1 | 5 | 9 | 251 |
| 1 | 2 | 0 | 4 | 8 | 232 |
| 2 | 3 | −3 | 9 | 9 | 243 |
| 3 | 4 | −8 | 32 | 0 | 224 |
| 4 | 5 | −15 | 85 | −55 | 475 |
| 5 | 6 | −24 | 180 | −216 | 376 |
| 6 | 7 | −35 | 329 | −567 | 9,107 |
| 7 | 8 | −48 | 544 | −1216 | 26,368 |
| $\frac{1}{\eta}\kappa_{j,l}\ \underline{\frac{|j}{|}}$ | $5(\frac{1}{1440})$ | $6(\frac{1}{64,480})$ | $7(\frac{1}{120,960})$ | $8(\frac{1}{3,628,800})$ | $9(\frac{1}{7,257,600})$ |
| 0 | 375 | 19,087 | 36,799 | 1,070,017 | 2,082,753 |
| 1 | 448 | 18,233 | 35,424 | 1,036,064 | 2,025,472 |
| 2 | 459 | 18,495 | 35,775 | 1,043,361 | 2,036,097 |
| 3 | 448 | 18,304 | 35,584 | 1,040,128 | 2,032,128 |
| 4 | 475 | 18,575 | 35,775 | 1,042,625 | 2,034,625 |
| 5 | 0 | 17,712 | 35,424 | 1,039,392 | 2,032,128 |
| 6 | −4277 | 36,799 | 36,799 | 1,046,689 | 2,036,097 |
| 7 | −22,016 | 235,520 | 0 | 1,012,736 | 2,025,472 |

where

$$\kappa_{j,l}^{*} = (-1)^{j}\frac{1}{h}\int_{t_{k-l}}^{t_{k+1}} C_{-r+1}^{j}\,\mathrm{d}t = (-1)^{j}\int_{-l}^{1} C_{-r+1}^{j}\,\mathrm{d}r,$$

$$c_{s,l}^{*} = \frac{1}{(s+1)!h^{s+2}}\int_{t_{k-l}}^{t_{k+1}}\prod_{j=0}^{s}(t - t_{k+1-j})\,\mathrm{d}t = (-1)^{s+1}\int_{-l}^{1} C_{-r+1}^{s+1}\,\mathrm{d}r.$$

(3.457)

Without truncation error, an approximate discrete map is

$$\mathbf{x}_{k+1} = \mathbf{x}_{k-l} + h\sum_{j=0}^{s-1}\kappa_{j,l}^{*}\nabla^{j}\mathbf{f}_{k}.$$

(3.458)

For $l = 1$, the above method is called the Milne–Simpson method. The node points at $t_{k+1}, t_k, \ldots, t_{k-s+1}$ for $[t_{k-1}, t_{k+1}]$ for the Milne–Simpson methods are sketched in Fig. 3.8. The corresponding coefficients are listed in Table 3.10. A few Milne–Simpson schemes are summarized in Table 3.11. For $s = 3$, Eq. (3.446) gives the popular Milne–Simpson method.

**Fig. 3.8** The node points at $t_{k+1}, t_{k-1}, \ldots, t_{k-s+1}$ for $[t_{k-1}, t_{k+1}]$ for the Milne–Simpson methods (implicit)

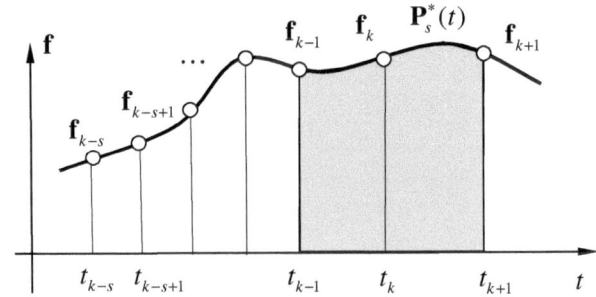

**Table 3.10** Coefficients for the Milne–Simpson methods

| $j$ | 0 | 1 | 2 | 3 | 4 | 5 | 6 |
|---|---|---|---|---|---|---|---|
| $\kappa_{j,1}^*$ | 2 | $-2$ | $\frac{1}{3}$ | 0 | $-\frac{1}{90}$ | $-\frac{1}{90}$ | $-\frac{37}{3780}$ |

As in Ceschino and Kuntzmann (1966), the coefficients of $\kappa_{j,l}$ is summarized in Table 3.12.

In numerical iterations, the explicit Adams form is called the explicit Adams predictor, and the implicit Adams form provides the implicit Adams corrector. Consider $l = 3$ and $s = 2$ as an example for the Milne–Simpson method. If $\mathbf{p}_{k+1} \approx \mathbf{x}_{k+1}$, the Milne predictor is

$$\mathbf{p}_{k+1} = \mathbf{x}_{k-3} + \frac{1}{3}h(8\mathbf{f}_k - 4\mathbf{f}_{k-1} + 2\mathbf{f}_{k-2}) \tag{3.459}$$

and the Simpson corrector is

$$\mathbf{x}_{k+1} = \mathbf{x}_{k-1} + \frac{1}{3}h(\mathbf{f}_{k+1} + 4\mathbf{f}_k + \mathbf{f}_{k-1}) \tag{3.460}$$

where

$$\mathbf{f}_{k+1} \approx \mathbf{f}(\mathbf{p}_{k+1}, t_{k+1}, \mathbf{p}). \tag{3.461}$$

**Table 3.11** Milne–Simpson methods (implicit)

| $s$ | Order | Methods | T. error |
|---|---|---|---|
| 0 | 1 | $\mathbf{x}_{k+1} = \mathbf{x}_{k-1} + 2h\mathbf{f}_{k+1}$ | $-2h^2 D\mathbf{f}_c$ |
| 1 | 2 | $\mathbf{x}_{k+1} = \mathbf{x}_{k-1} + 2h\mathbf{f}_k$ | $\frac{1}{3}h^3 D^2\mathbf{f}_c$ |
| 2 | 3 | $\mathbf{x}_{k+1} = \mathbf{x}_{k-1} + \frac{1}{3}h(\mathbf{f}_{k+1} + 4\mathbf{f}_k + \mathbf{f}_{k-1})$ | $-\frac{1}{90}h^5 D^4\mathbf{f}_c$ |
| 3 | 4 | $\mathbf{x}_{k+1} = \mathbf{x}_{k-1} + \frac{1}{3}h(\mathbf{f}_{k+1} + 4\mathbf{f}_k + \mathbf{f}_{k-1})$ | $-\frac{1}{90}h^5 D^4\mathbf{f}_c$ |
| 4 | 5 | $\mathbf{x}_{k+1} = \mathbf{x}_{k-1} + \frac{1}{90}h(29\mathbf{f}_{k+1} + 124\mathbf{f}_k + 24\mathbf{f}_{k-1}$ $+ 4\mathbf{f}_{k-2} - \mathbf{f}_{k-3})$ | $-\frac{37}{3790}h^6 D^5\mathbf{f}_c$ |

**Table 3.12** Coefficients $\kappa_{j,l}^*$ for the implicit Adams methods

| $\boxed{\dfrac{l}{i}}\kappa_{j,l}^* \boxed{\dfrac{j}{}}$ | 0 | $1(\frac{1}{2})$ | $2(\frac{1}{12})$ | $3(\frac{1}{24})$ | $4(\frac{1}{720})$ |
|---|---|---|---|---|---|
| 0 | 1 | −1 | −1 | −1 | −19 |
| 1 | 2 | −4 | 4 | 0 | −8 |
| 2 | 3 | −9 | 27 | −9 | −27 |
| 3 | 4 | −16 | 80 | −64 | 224 |
| 4 | 5 | −25 | 175 | −225 | 2,125 |
| 5 | 6 | −36 | 324 | −576 | 8,856 |
| 6 | 7 | −49 | 539 | −1255 | 25,117 |
| 7 | 8 | −64 | 832 | −2304 | 26,848 |
| $\boxed{\dfrac{l}{i}}\kappa_{j,l}^* \boxed{\dfrac{j}{}}$ | $5(\frac{1}{1440})$ | $6(\frac{1}{64,480})$ | $7(\frac{1}{120,960})$ | $8(\frac{1}{3,628,800})$ | $9(\frac{1}{7,257,600})$ |
| 0 | −27 | −863 | −1375 | −33,953 | −57,281 |
| 1 | −16 | −592 | −1024 | −26,656 | −46,656 |
| 2 | −27 | −783 | −1215 | −29,889 | −51,138 |
| 3 | 0 | −512 | −1024 | −27,392 | −48,128 |
| 4 | −475 | −1375 | −1375 | −30,625 | −51,138 |
| 5 | −4752 | 17,712 | 0 | −23,328 | −46,656 |
| 6 | −22,491 | 316,433 | −36,799 | −57,281 | −57,281 |
| 7 | −74,752 | 1,160,192 | −471,040 | 1,012,736 | 0 |

From the truncation errors of the predictor and corrector, we have

$$\mathbf{x}(t_{k+1}) - \mathbf{p}_{k+1} = \frac{28}{90} h^5 D^4 \mathbf{f}_d \quad \text{and} \quad \mathbf{x}(t_{k+1}) - \mathbf{x}_{k+1} = -\frac{1}{90} h^5 D^4 \mathbf{f}_c. \qquad (3.462)$$

If $D^4 \mathbf{f}_d \approx D^4 \mathbf{f}_c$, the forgoing equation gives

$$\mathbf{x}(t_{k+1}) - \mathbf{p}_{k+1} = \frac{28}{29} (\mathbf{x}_{k+1} - \mathbf{p}_{k+1}). \qquad (3.463)$$

If $\mathbf{x}_{k+1} - \mathbf{p}_{k+1} \approx \mathbf{x}_k - \mathbf{p}_k$, a modifier $\mathbf{m}_{k+1}$ is introduced to replace $\mathbf{x}(t_{k+1})$. The forgoing equation becomes

$$\mathbf{m}_{k+1} \approx \mathbf{p}_{k+1} + \frac{28}{29} (\mathbf{x}_k - \mathbf{p}_k). \qquad (3.464)$$

Thus, Eq. (3.461) becomes

$$\mathbf{f}_{k+1} \approx \mathbf{f}(\mathbf{m}_{k+1}, t_{k+1}, \mathbf{p}). \qquad (3.465)$$

In summary, the modified Milne–Simpson method is given by

$$\mathbf{p}_{k+1} = \mathbf{x}_{k-3} + \frac{1}{3}h(8\mathbf{f}_k - 4\mathbf{f}_{k-1} + 2\mathbf{f}_{k-2}),$$

$$\mathbf{m}_{k+1} \approx \mathbf{p}_{k+1} + \frac{28}{29}(\mathbf{x}_k - \mathbf{p}_k),$$

$$\mathbf{f}_{k+1} \approx \mathbf{f}(\mathbf{m}_{k+1}, t_{k+1}, \mathbf{p}),$$

$$\mathbf{x}_{k+1} = \mathbf{x}_{k-1} + \frac{1}{3}h(\mathbf{f}_{k+1} + 4\mathbf{f}_k + \mathbf{f}_{k-1}).$$

(3.466)

### 3.6.5  General Forms

From the explicit and implicit Adams methods, a general form of multi-step methods can be developed, which can be referred to Henrici (1962). Consider a general difference equation as

$$\mathbf{x}_{k+1} = \sum_{i=0}^{s} a_i \mathbf{x}_{k-i} + h \sum_{i=-1}^{s} b_i \mathbf{f}_{k-i} \qquad (3.467)$$

where the coefficients $a_i$ $(i = 0, 1, 2, \ldots, s)$ and $b_i$ $(i = -1, 0, 1, \ldots, s)$ are constant, and

$$\mathbf{f}_{k-i} = \mathbf{f}(\mathbf{x}_{k-i}, t_{k-i}, \mathbf{p}), \qquad (3.468)$$

For $b_{-1} \neq 0$, Eq. (3.467) gives an implicit method. For $b_{-1} = 0$, Eq. (3.467) gives an explicit method. The predictor given by the explicit method can be

$$\mathbf{x}_{k+1}^{(i+1)} = \sum_{i=0}^{s} a_i \mathbf{x}_{k-i} + h \sum_{i=0}^{s} b_i \mathbf{f}_{k-i} \qquad (3.469)$$

and the corrector given by the implicit method can be

$$\mathbf{x}_{k+1} = \sum_{i=0}^{s} a_i \mathbf{x}_{k-i} + b_{-1} h \mathbf{f}_{k+1}^{(i+1)} + h \sum_{i=0}^{s} b_i \mathbf{f}_{k-i} \qquad (3.470)$$

where

$$\mathbf{f}_{k+1}^{(i+1)} = \mathbf{f}(\mathbf{x}_{k+1}^{(i+1)}, t_{k+1}, \mathbf{p}). \qquad (3.471)$$

From the above scheme, one can complete the numerical computation.

Consider the exact expression as

$$\mathbf{x}(t_{k+1}) = \sum_{i=0}^{s} a_i \mathbf{x}(t_{k-i}) + \sum_{i=-1}^{s} b_i \mathbf{f}(\mathbf{x}(t_{k-i}), t_{k-i}, p) + \mathbf{T_e} \tag{3.472}$$

where $\mathbf{T_e}$ is a truncated error. Thus, using the notation of $\mathbf{e}_{k-i} = \mathbf{x}(t_{k-i}) - \mathbf{x}_{k-i}$, Eq. (3.472) minus Eq. (3.467) gives

$$\mathbf{e}_{k+1} = \sum_{i=0}^{s} a_i \mathbf{e}_{k-i} + h \sum_{i=0}^{s} b_i \mathbf{A}_{k-i} \mathbf{e}_{k-i} + b_{-1} h \mathbf{A}_{k+1} \mathbf{e}_{k+1} + \mathbf{T_e} \tag{3.473}$$

where

$$\mathbf{A}_{k-i} = \left. \frac{\partial \mathbf{f}}{\partial \mathbf{x}} \right|_{\mathbf{x}_{k-i}^c}. \tag{3.474}$$

If

$$\mathbf{x}(t_j) = \mathbf{x}_j, \quad j = k, k-1, \ldots, k-s, \tag{3.475}$$

then

$$\mathbf{f}(\mathbf{x}(t_j), t_j, \mathbf{p}) = \mathbf{f}_j, \quad j = k, k-1, \ldots, k-s. \tag{3.476}$$

The local error is

$$\boldsymbol{\varepsilon}_{k+1} = \mathbf{x}(t_{k+1}) - \mathbf{x}_{k+1} = (\mathbf{I}_{n \times n} - h b_{-1} \mathbf{A}_{k+1})^{-1} \mathbf{T_e}. \tag{3.477}$$

If $\mathbf{x}(t_j) \neq \mathbf{x}_j (j = k, k-1, \ldots, k-s)$, we have

$$\mathbf{e}_{k+1} = (\mathbf{I}_{n \times n} - b_{-1} h \mathbf{A}_{k+1})^{-1} \sum_{i=0}^{s} (a_i \mathbf{I}_{n \times n} + h b_i \mathbf{A}_{k-i})$$
$$+ (\mathbf{I}_{n \times n} - b_{-1} h \mathbf{A}_{k+1})^{-1} \mathbf{T_e}. \tag{3.478}$$

For $\mathbf{T_e} = \mathbf{0}$, Eq. (3.473) becomes

$$\mathbf{e}_{k+1} = (\mathbf{I}_{n \times n} + h b_{-1} \mathbf{A}_{k+1})^{-1} \sum_{i=0}^{s} (a_i \mathbf{I}_{n \times n} + h b_i \mathbf{A}_{k-i}) \mathbf{e}_{k-i}. \tag{3.479}$$

Consider the stability of discrete mapping through

$$\left\{ \begin{array}{c} \mathbf{e}_{k+1} \\ \mathbf{e}_k \\ \vdots \\ \mathbf{e}_{k+2-s} \end{array} \right\} = \mathbf{J}_{sn \times sn} \left\{ \begin{array}{c} \mathbf{e}_k \\ \mathbf{e}_{k-1} \\ \vdots \\ \mathbf{e}_{k+1-s} \end{array} \right\} \tag{3.480}$$

where

$$\mathbf{J}_{(s+1)n \times (s+1)n} = \begin{bmatrix} hb_1^* \bar{\mathbf{A}}_k & hb_2^* \bar{\mathbf{A}}_{k-1} & hb_3^* \bar{\mathbf{A}}_{k-2} & \cdots & hb_s^* \bar{\mathbf{A}}_{k+1-s} \\ \mathbf{I}_{n \times n} & \mathbf{0}_{n \times n} & \mathbf{0}_{n \times n} & \cdots & \mathbf{0}_{n \times n} \\ \mathbf{0}_{n \times n} & \mathbf{I}_{n \times n} & \mathbf{0}_{n \times n} & \cdots & \mathbf{0}_{n \times n} \\ \vdots & \vdots & \vdots & & \vdots \\ \mathbf{0}_{n \times n} & \mathbf{0}_{n \times n} & \mathbf{0}_{n \times n} & \cdots & \mathbf{0}_{n \times n} \end{bmatrix} \tag{3.481}$$

and

$$\bar{\mathbf{A}}_{k+1-j} = (\mathbf{I}_{n \times n} - hb_1^* \mathbf{A}_{k+1})^{-1}(a_i \mathbf{I}_{n \times n} + b_i \mathbf{A}_{k+1-j}), \quad j = 1, 2, \ldots, s-1. \tag{3.482}$$

Assuming

$$\left\{ \begin{array}{c} \mathbf{e}_{k+1} \\ \mathbf{e}_k \\ \vdots \\ \mathbf{e}_{k-s+2} \end{array} \right\} = \lambda \left\{ \begin{array}{c} \mathbf{e}_k \\ \mathbf{e}_{k-1} \\ \vdots \\ \mathbf{e}_{k+1-s} \end{array} \right\}, \tag{3.483}$$

we have

$$|\mathbf{J}_{sn \times sn} - \lambda \mathbf{I}_{sn \times sn}| = 0. \tag{3.484}$$

If all eigenvalues $\lambda_j < 1$ $(j = 1, 2, \ldots, sn)$, then the discrete mapping is stable. The computational error will not be expanded.

## 3.7 Generalized Implicit Multi-step Methods

Consider the mesh node points with $t_{k+s_2}, t_{k+s_2-1}, \ldots, t_k, \ldots, t_{k-s_1}$ for an time interval $[t_{k-s_1}, t_{k+s_2}]$, as shown in Fig. 3.9. Integration of Eq. (3.1) during $t \in [t_{k+s}, t_{k+s+1}]$ gives

$$\mathbf{x}(t_{k+s+1}) = \mathbf{x}(t_{k+s}) + \int_{t_{k+s}}^{t_{k+s+1}} \mathbf{f}(\mathbf{x}, t, \mathbf{p}) dt \tag{3.485}$$

$$\text{for} \quad s = -s_1, -s_1 + 1, \ldots, s_2 - 2, s_2 - 1.$$

From the $s_2 + s_1 + 1$ points, the linear interpolation polynomial of $\mathbf{f}(\mathbf{x}, t, \mathbf{p})$ is

$$\mathbf{P}_{s_1+s_2}(t) = \sum_{j=-s_2}^{s_1} l_j(t) \bar{\mathbf{f}}_{k-j} \tag{3.486}$$

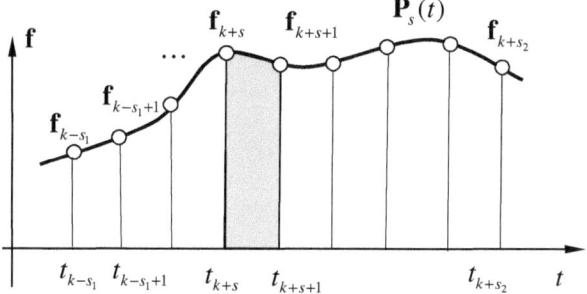

**Fig. 3.9** The node points at $t_{k+s_2}, t_{k+s_2-1}, \ldots, t_k, \ldots, t_{k-s_1}$ for the time interval $[t_{k-s_1}, t_{k+s_2}]$ for the generalized implicit methods

where

$$l_j(t) = \prod_{\substack{i=-s_2 \\ i \neq j}}^{s_1} \frac{(t - t_{k-i})}{(t_{k-j} - t_{k-i})}. \tag{3.487}$$

Using the theory of interpolation polynomial, we have

$$\mathbf{f}(\mathbf{x}, t, \mathbf{p}) - \mathbf{P}_{s_1+s_2}(t) = \frac{1}{(s_1 + s_2 + 1)!} \prod_{j=-s_2}^{s_1} (t - t_{k-j}) D^{s_1+s_2+1} \mathbf{f}(\mathbf{x}_c, t_c, \mathbf{p}), \tag{3.488}$$

with $t_c \in [t_{k-s_1}, t_{k+s_2}]$. The integration of $\mathbf{f}(\mathbf{x}, t, \mathbf{p})$ over $[t_{k+s}, t_{k+s+1}]$ is

$$\int_{t_{k+s}}^{t_{k+s+1}} \mathbf{f}(\mathbf{x}, t, \mathbf{p}) \mathrm{d}t = \sum_{j=-s_2}^{s_1} b_j^{(s)} \bar{\mathbf{f}}_{k-j} + c_{s_1+s_2}^{(s)} h^{s_1+s_2+2} D^{s_1+s_2+1} \mathbf{f}(\mathbf{x}_c, t_c, \mathbf{p}) \tag{3.489}$$

where

$$b_j^{(s)} = \frac{1}{h_{k+s+1}} \int_{t_{k+s}}^{t_{k+s+1}} \prod_{\substack{i=-s_2 \\ i \neq j}}^{s_1} \frac{(t - t_{k-i})}{(t_{k-j} - t_{k-i})} \mathrm{d}t, \quad \text{and}$$

$$\tag{3.490}$$

$$c_{s_1+s_2}^{(s)} = \frac{1}{(s_1 + s_2 + 1)! h_{k+1}^{s_1+s_2+2}} \int_{t_k}^{t_{k+1}} \prod_{j=-s_2}^{s_1} (t - t_{k-j}) \mathrm{d}t.$$

Thus, Eq. (3.485) becomes

$$
\begin{aligned}
\mathbf{x}(t_{k+s+1}) = \mathbf{x}(t_{k+s}) + h_{k+s+1} \sum_{j=-s_2}^{s_1} b_j^{(s)} \overline{\mathbf{f}}_{k-j} \\
+ c_{s_1+s_2}^{(s)} h_{k+1}^{s_1+s_2} D^{s_1+s_2+1} \mathbf{f}(\mathbf{x}_c, t_c, \mathbf{p}) \\
\text{for } s = -s_1, -s_1 + 1, \ldots, s_2 - 2, s_2 - 1.
\end{aligned}
\tag{3.491}
$$

Without truncation error, an approximate discrete map is

$$
\begin{aligned}
\mathbf{x}_{k+s+1} = \mathbf{x}_{k+s} + h_{k+s+1} \sum_{j=-s_2}^{s_1} b_j^{(s)} \mathbf{f}_{k-j} \\
\text{for } s = -s_1, -s_1 + 1, \ldots, s_2 - 2, s_2 - 1.
\end{aligned}
\tag{3.492}
$$

Setting

$$
\boldsymbol{\Phi}^{(s)}(\mathbf{x}_{k+s_2}, \mathbf{x}_{k+s_2-1}, \ldots, \mathbf{x}_{k-s_1}) = \sum_{j=-s_2}^{s_1} b_j^{(s)} \mathbf{f}_{k-j},
\tag{3.493}
$$

we have

$$
\mathbf{x}_{k+s+1} = \mathbf{x}_{k+s} + h_{k+s+1} \boldsymbol{\Phi}^{(s)}(\mathbf{x}_{k+s_2}, \mathbf{x}_{k+s_2-1}, \ldots, \mathbf{x}_{k-s_1}).
\tag{3.494}
$$

The truncation error is

$$
\mathbf{T}_e = c_{s_1+s_2}^{(s)} h_{k+1}^{s_1+s_2} D^{s_1+s_2+1} \mathbf{f}(\mathbf{x}_c, t_c, \mathbf{p}).
\tag{3.495}
$$

If $\mathbf{x}(t_{k+j}) = \mathbf{x}_{k+j}$ $(j \neq s + 1)$, the local error for $t \in [t_{k+s}, t_{k+s+1}]$ is

$$
\begin{aligned}
\boldsymbol{\varepsilon}_{k+s+1} &= \mathbf{x}(t_{k+s+1}) - \mathbf{x}_{k+s} - h_{k+s+1} \boldsymbol{\Phi}^{(s)}(\mathbf{x}_{k+s_2}, \mathbf{x}_{k+s_2-1}, \ldots, \mathbf{x}_{k-s_1}) \\
&= (\mathbf{I}_{n \times n} - b_{s+1}^{(s)} \mathbf{A}_{k+s+1}^{(s)})^{-1} c_{s_1+s_2}^{(s)} h_{k+1}^{s_1+s_2} D^{s_1+s_2+1} \mathbf{f}(\mathbf{x}_c, t_c, \mathbf{p}),
\end{aligned}
\tag{3.396}
$$

where

$$
\mathbf{A}_{k+s+1}^{(s)} = \left. \frac{\partial \mathbf{f}_{k+s+1}}{\partial \mathbf{x}_{k+s+1}} \right|_{\mathbf{x}_{k+s+1}^c}
\tag{3.397}
$$

with $\|\mathbf{x}_{k+s+1}^c\| \in (\|\mathbf{x}(t_{k+s+1})\|, \|\mathbf{x}_{k+s+1}\|)$. Let

$$
\mathbf{e}_{k-j} = \mathbf{x}(t_{k-j}) - \mathbf{x}_{k-j}, \quad j = -s_1, -s_1 + 1, \ldots, s_2.
\tag{3.498}
$$

If $\mathbf{x}(t_{k+j}) \neq \mathbf{x}_{k+j}$, the global error for each step is

$$
\mathbf{e}_{k+s+1} = (\mathbf{I}_{n\times n} - b_{k+s+1}^{(s)}\mathbf{A}_{k+s+1}^{(s)})^{-1}[\mathbf{e}_{k+s} + h_{k+s+1}\sum_{\substack{j=-s_2\\j\neq-(s+1)}}^{s_2} b_j^{(s)}\mathbf{A}_{k-j}^{(s)}\mathbf{e}_{k-j}]
$$

$$
+ (\mathbf{I}_{n\times n} - b_{k+s+1}^{(s)}\mathbf{A}_{k+s+1}^{(s)})^{-1}c_{s_1+s_2}^{(s)}h_{k+1}^{s_1+s_2}D^{s_1+s_2+1}\mathbf{f}(\mathbf{x}_c, t_c, \mathbf{p}).
$$

(3.499)

where

$$
\mathbf{A}_{k-j}^{(s)} = \left.\frac{\partial \mathbf{f}_{k-j}}{\partial \mathbf{x}_{k-j}}\right|_{\mathbf{x}_{k-j}^c}, \quad \text{for } j = -s_2, -s_2+1, \ldots, s_1
$$

(3.500)

where $\|\mathbf{x}_{k-j}^c\| \in (\|\mathbf{x}(t_{k-j})\|, \|\mathbf{x}_{k-j}\|)$. The global error for the time interval $[t_{k-s_1}, t_{k+s_2}]$ is computed by

$$
\mathbf{E}_{k-s_1}^{k+s_2} = \sum_{s=-s_1+1}^{s_2-1} \mathbf{e}_{k+s+1}.
$$

(3.501)

Consider the stability of discrete mapping in Eq. (3.492) through

$$
\begin{Bmatrix} \mathbf{e}_{k+s_2} \\ \mathbf{e}_{k+s_2-1} \\ \vdots \\ \mathbf{e}_{k-s_1+1} \end{Bmatrix} = \mathbf{J}_{(s_1+s_2)n\times(s_1+s_2)n} \begin{Bmatrix} \mathbf{e}_{k+s_2-1} \\ \mathbf{e}_{k+s_2-2} \\ \vdots \\ \mathbf{e}_{k-s_1} \end{Bmatrix}
$$

(3.502)

where

$$
\mathbf{J}_{(s_1+s_2)n\times(s_1+s_2)n} = \begin{bmatrix} \bar{\mathbf{A}}_{k+s_2-1}^{(s_2-1)} & \bar{\mathbf{A}}_{k+s_2-2}^{(s_2-1)} & \bar{\mathbf{A}}_{k+s_2-3}^{(s_2-1)} & \cdots & \bar{\mathbf{A}}_{k-s_1}^{(s_2-1)} \\ \bar{\mathbf{A}}_{k+s_2-1}^{(s_2-2)} & \bar{\mathbf{A}}_{k+s_2-2}^{(s_2-2)} & \bar{\mathbf{A}}_{k+s_2-3}^{(s_2-2)} & \cdots & \bar{\mathbf{A}}_{k-s_1}^{(s_2-2)} \\ \bar{\mathbf{A}}_{k+s_2-1}^{(s_2-3)} & \bar{\mathbf{A}}_{k+s_2-2}^{(s_2-3)} & \bar{\mathbf{A}}_{k+s_2-3}^{(s_2-3)} & \cdots & \bar{\mathbf{A}}_{k-s_1}^{(s_2-3)} \\ \vdots & \vdots & \vdots & & \vdots \\ \bar{\mathbf{A}}_{k-s_1}^{(-s_1)} & \bar{\mathbf{A}}_{k-s_1}^{(-s_1)} & \bar{\mathbf{A}}_{k-s_1}^{(-s_1)} & \cdots & \bar{\mathbf{A}}_{k-s_1}^{(-s_1)} \end{bmatrix}
$$

(3.503)

where

$$
\bar{\mathbf{A}}_{k-j}^{(s)} = (\mathbf{I}_{n\times n} - h_{k+s+1}b_{k+s+1}^{(s)}\mathbf{A}_{k+s+1}^{(s)})^{-1}(\mathbf{I}_{n\times n}\delta_j^{s+1} + h_{k+s+1}b_{k-j}^{(s)}\mathbf{A}_{k-j}^{(s)}).
$$

(3.504)

Assuming

$$\left\{\begin{array}{c} \mathbf{e}_{k+s_2} \\ \mathbf{e}_{k+s_2-1} \\ \vdots \\ \mathbf{e}_{k-s_1+1} \end{array}\right\} = \lambda \left\{\begin{array}{c} \mathbf{e}_{k+s_2-1} \\ \mathbf{e}_{k+s_2-2} \\ \vdots \\ \mathbf{e}_{k-s_1} \end{array}\right\}, \tag{3.505}$$

we have

$$\left| \mathbf{J}_{(s_1+s_2)n \times (s_1+s_2)n} - \lambda \mathbf{I}_{(s_1+s_2)n \times (s_1+s_2)n} \right| = 0. \tag{3.506}$$

If all eigenvalues $\lambda_j < 1$ $(j = 1, 2, \ldots, (s_1 + s_2)n)$, then the discrete mapping is stable. In other words,

$$\|\mathbf{e}_{k+s+1}\| \leq \|\mathbf{e}_{k+s}\| \quad \text{for } s = -s_1, -s_1 + 1, \ldots, s_2 - 2, s_2 - 1. \tag{3.507}$$

The computational error will not be expanded.

The implicit Runge–Kutta method presented before cannot be used for multistep method. However, the accuracy may not be better than the single-step implicit Runge–Kutta methods. Thus, herein, the multi-step implicit Runge–Kutta method will not be discussed herein.

For $s = 2$, the linear interpolation polynomial of $\mathbf{f}(\mathbf{x}, t, \mathbf{p})$ is

$$\mathbf{P}_2(t) = l_0(t)\bar{\mathbf{f}}_{k+1} + l_1(t)\bar{\mathbf{f}}_k + l_2(t)\bar{\mathbf{f}}_{k-1} \tag{3.508}$$

where

$$\begin{aligned} l_0(t) &= \frac{1}{2h^2}(t - t_k)(t - t_{k-1}), \\ l_1(t) &= \frac{1}{h^2}(t - t_{k+1})(t - t_{k-1}), \\ l_2(t) &= \frac{1}{2h^2}(t - t_{k+1})(t - t_k). \end{aligned} \tag{3.509}$$

Theory of interpolation polynomial gives

$$\mathbf{f}(\mathbf{x}, t, \mathbf{p}) - \mathbf{P}_2(t) = \frac{1}{3!}(t - t_{k+1})(t - t_k)(t - t_{k-1})D^3\mathbf{f}(\mathbf{x}_c, t_c, \mathbf{p}), \tag{3.510}$$

with $t_c \in [t_{k-1}, t_{k+1}]$. The integration of $\mathbf{f}(\mathbf{x}, t, \mathbf{p})$ over $[t_{k-1}, t_k]$ is

$$\int_{t_{k-1}}^{t_k} \mathbf{f}(\mathbf{x}, t, \mathbf{p})dt = \frac{1}{12}h(\bar{\mathbf{f}}_{k+1} - 4\bar{\mathbf{f}}_k + 5\bar{\mathbf{f}}_{k-1}) + \frac{1}{24}h^4 D^3\mathbf{f}(\mathbf{x}_c, t_c, \mathbf{p}). \tag{3.511}$$

Thus, Eq. (3.485) becomes

$$\mathbf{x}(t_{k+1}) = \mathbf{x}(t_k) + \frac{1}{12}h(\bar{\mathbf{f}}_{k+1} - 4\bar{\mathbf{f}}_k + 5\bar{\mathbf{f}}_{k-1}) + \frac{1}{24}h^4 D^3 \mathbf{f}(\mathbf{x}_c, t_c, \mathbf{p}). \qquad (3.512)$$

Without truncation error, an approximate discrete map is

$$\mathbf{x}_k = \mathbf{x}_{k-1} + \frac{1}{12}h(\mathbf{f}_{k+1} - 4\mathbf{f}_k + 5\mathbf{f}_{k-1}) \qquad (3.513)$$

The integration of $\mathbf{f}(\mathbf{x}, t, \mathbf{p})$ over $[t_k, t_{k+1}]$ is

$$\int_{t_k}^{t_{k+1}} \mathbf{f}(\mathbf{x}, t, \mathbf{p})\mathrm{d}t = \frac{1}{12}h(5\bar{\mathbf{f}}_{k+1} + 8\bar{\mathbf{f}}_k - \bar{\mathbf{f}}_{k-1}) - \frac{1}{24}h^4 D^3 \mathbf{f}(\mathbf{x}_c, t_c, \mathbf{p}) \qquad (3.514)$$

Thus, Eq. (3.485) becomes

$$\mathbf{x}(t_{k+1}) = \mathbf{x}(t_k) + \frac{1}{12}h(5\bar{\mathbf{f}}_{k+1} + 8\bar{\mathbf{f}}_k - \bar{\mathbf{f}}_{k-1}) - \frac{1}{24}h^4 D^3 \mathbf{f}(\mathbf{x}_c, t_c, \mathbf{p}). \qquad (3.515)$$

Without truncation error, an approximate discrete map is

$$\mathbf{x}_{k+1} = \mathbf{x}_k + \frac{1}{12}h(5\mathbf{f}_{k+1} + 8\mathbf{f}_k - \mathbf{f}_{k-1}). \qquad (3.516)$$

In summary, we have

$$\begin{aligned} \mathbf{x}_k &= \mathbf{x}_{k-1} + \frac{1}{12}h(\mathbf{f}_{k+1} - 4\mathbf{f}_k + 5\mathbf{f}_{k-1}), \\ \mathbf{x}_{k+1} &= \mathbf{x}_k + \frac{1}{12}h(5\mathbf{f}_{k+1} + 8\mathbf{f}_k - \mathbf{f}_{k-1}). \end{aligned} \qquad (3.517)$$

# References

Butcher, J. C. (1964). Implicit Runge-Kutta process. *Mathematical Computation, 18,* 50–64.

Ceschino, F., & Kuntzmann, J. (1966). *Numerical solution of initial value problems* (D. Boyanovitch, Trans.). Englwood Cliffs, New Jersey: Prentice-Hall.

Coddington, E. A., & Levinson, N. (1955). *Theory of ordinary differential equations.* New York: McGraw-Hill.

Hairer, E., Norsett, S. P., & Wanner, G. (1987). *Solving ordinary differential equations I: Nonstiff problems.* Berlin: Springer.

Hairer, E., & Wanner, G. (1991). *Solving ordinary differential equations II: Stiff problems and differential-algebraic equations.* Springer: Berlin.

Henrici, P. (1962). *Discrete variable methods in ordinary differential equations.* New York: Wiley.

Kahaner, D., Moler, C., & Nash, S. (1989). *Numerical methods and software*. New Jersey: Prentice Hall.

Lapidus, L., & Seinfeld, J. H. (1971). *Numerical solutions of ordinary differential equations*. New York and London: Academic Press.

Luo, A. C. J. (2012). *Regularity and complexity in nonlinear systems*. New York: Springer.

# Chapter 4
# Implicit Mapping Dynamics

This chapter presents a Yin–Yang theory for implicit, nonlinear, discrete dynamical systems with consideration of positive and negative iterations of discrete iterative maps. In existing analysis, the solutions relative to "Yang" in nonlinear dynamical systems are extensively investigated. However, the solutions pertaining to "Yin" in nonlinear dynamical systems are not investigated yet. A set of concepts on "Yin" and "Yang" in implicit, nonlinear, discrete dynamical systems are introduced. Based on the Yin–Yang theory, the complete dynamics of implicit discrete dynamical systems can be discussed. A discrete dynamical system with the Henon map is investigated as an example. Period-$m$ solutions, stability, and bifurcations for multi-step, implicit discrete systems will be discussed.

## 4.1 Single-Step Implicit Maps

**Definition 4.1** Consider an implicit vector function $\mathbf{f} : D \to D$ on an open set $D \subset \mathscr{R}^n$ in an $n$-dimensional discrete dynamical system. For $\mathbf{x}_k, \mathbf{x}_{k+1} \in D$, there is a discrete relation as

$$\mathbf{f}(\mathbf{x}_k, \mathbf{x}_{k+1}, \mathbf{p}) = \mathbf{0} \tag{4.1}$$

where the vector function is $\mathbf{f} = (f_1, f_2, \ldots, f_n)^{\mathrm{T}} \in \mathscr{R}^n$ and discrete variable vector is $\mathbf{x}_k = (x_{k1}, x_{k2}, \ldots, x_{kn})^{\mathrm{T}} \in D$ with a parameter vector $\mathbf{p} = (p_1, p_2, \ldots, p_m)^{\mathrm{T}} \in \mathscr{R}^m$.

As in Luo (2010), to symbolically describe the discrete dynamical systems, introduce two discrete sets.

**Definition 4.2** For a discrete dynamical system in Eq. (4.1), the positive and negative discrete sets are defined by

$$\left. \begin{aligned} \Sigma_+ &= \{\mathbf{x}_{k+i} | \mathbf{x}_{k+i} \in \mathscr{R}^n, i \in \mathbb{Z}_+\} \subset D \text{ and} \\ \Sigma_- &= \{\mathbf{x}_{k-i} | \mathbf{x}_{k-i} \in \mathscr{R}^n, i \in \mathbb{Z}_+\} \subset D \end{aligned} \right\} \tag{4.2}$$

© Higher Education Press, Beijing and Springer-Verlag Berlin Heidelberg 2015     159
A.C.J. Luo, *Discretization and Implicit Mapping Dynamics*,
Nonlinear Physical Science, DOI 10.1007/978-3-662-47275-0_4

respectively. The discrete set is

$$\Sigma = \Sigma_+ \cup \Sigma_-. \tag{4.3}$$

A positive mapping is defined as

$$P_+ : \Sigma \to \Sigma_+ \Rightarrow P_+ : \mathbf{x}_k \to \mathbf{x}_{k+1} \tag{4.4}$$

and a negative mapping is defined by

$$P_- : \Sigma \to \Sigma_- \Rightarrow P_- : \mathbf{x}_k \to \mathbf{x}_{k-1}. \tag{4.5}$$

**Definition 4.3** For a discrete dynamical system in Eq. (4.1), consider two points $\mathbf{x}_k \in D$ and $\mathbf{x}_{k+1} \in D$, and there is a specific, differentiable, vector function $\mathbf{g} \in \mathscr{R}^n$ to make $\mathbf{g}(\mathbf{x}_k, \mathbf{x}_{k+1}, \lambda) = \mathbf{0}$.

 (i)  The stable solution based on $\mathbf{x}_{k+1} = P_+ \mathbf{x}_k$ for the positive mapping $P_+$ is called the "Yang" of the discrete dynamical system in Eq. (4.1) in the sense of $\mathbf{g}(\mathbf{x}_k, \mathbf{x}_{k+1}, \lambda) = \mathbf{0}$ if solutions $(\mathbf{x}_k^*, \mathbf{x}_{k+1}^*)$ of $\mathbf{f}(\mathbf{x}_k, \mathbf{x}_{k+1}, \mathbf{p}) = \mathbf{0}$ and $\mathbf{g}(\mathbf{x}_k, \mathbf{x}_{k+1}, \lambda) = \mathbf{0}$ exist.
 (ii)  The stable solution based on $\mathbf{x}_k = P_- \mathbf{x}_{k+1}$ for the negative mapping $P_-$ is called the "Yin" of the discrete dynamical system in Eq. (4.1) in the sense of $\mathbf{g}(\mathbf{x}_k, \mathbf{x}_{k+1}, \lambda) = \mathbf{0}$ if solutions $(\mathbf{x}_k^*, \mathbf{x}_{k+1}^*)$ of $\mathbf{f}(\mathbf{x}_k, \mathbf{x}_{k+1}, \mathbf{p}) = \mathbf{0}$ and $\mathbf{g}(\mathbf{x}_k, \mathbf{x}_{k+1}, \lambda) = \mathbf{0}$ exist.
 (iii)  The solution based on $\mathbf{x}_{k+1} = P_+ \mathbf{x}_k$ is called "Yin–Yang" for the positive mapping $P_+$ of the discrete dynamical system in Eq. (4.1) in the sense of $\mathbf{g}(\mathbf{x}_k, \mathbf{x}_{k+1}, \lambda) = \mathbf{0}$ if solutions $(\mathbf{x}_k^*, \mathbf{x}_{k+1}^*)$ of $\mathbf{f}(\mathbf{x}_k, \mathbf{x}_{k+1}, \mathbf{p}) = \mathbf{0}$ and $\mathbf{g}(\mathbf{x}_k, \mathbf{x}_{k+1}, \lambda) = \mathbf{0}$ exist and the eigenvalues of $DP_+(\mathbf{x}_k^*)$ are distributed inside and outside the unit cycle.
 (iv)  The solution based on $\mathbf{x}_k = P_- \mathbf{x}_{k+1}$ is called the "Yin–Yang" for the negative mapping $P_-$ of the discrete dynamical system in Eq. (4.1) in the sense of $\mathbf{g}(\mathbf{x}_k, \mathbf{x}_{k+1}, \lambda) = \mathbf{0}$ if solutions $(\mathbf{x}_k^*, \mathbf{x}_{k+1}^*)$ of $\mathbf{f}(\mathbf{x}_k, \mathbf{x}_{k+1}, \mathbf{p}) = \mathbf{0}$ and $\mathbf{g}(\mathbf{x}_k, \mathbf{x}_{k+1}, \lambda) = \mathbf{0}$ exist and the eigenvalues of $DP_-(\mathbf{x}_{k+1}^*)$ are distributed inside and outside unit cycle.

Consider that the positive and negative mappings are

$$\mathbf{x}_{k+1} = P_+ \mathbf{x}_k \quad \text{and} \quad \mathbf{x}_k = P_- \mathbf{x}_{k+1}. \tag{4.6}$$

For the simplest case, consider the constraint condition of $\mathbf{g}(\mathbf{x}_k, \mathbf{x}_{k+1}, \lambda) = \mathbf{x}_{k+1} - \mathbf{x}_k = \mathbf{0}$. Thus, the positive and negative mappings have, respectively, the constraints

$$\mathbf{x}_{k+1} = \mathbf{x}_k \quad \text{and} \quad \mathbf{x}_k = \mathbf{x}_{k+1}. \tag{4.7}$$

Both positive and negative mappings are governed by the discrete relation in Eq. (4.1). In other words, Eq. (4.6) gives

$$\mathbf{f}(\mathbf{x}_k, \mathbf{x}_{k+1}, \mathbf{p}) = \mathbf{0} \quad \text{and} \quad \mathbf{f}(\mathbf{x}_k, \mathbf{x}_{k+1}, \mathbf{p}) = \mathbf{0}. \tag{4.8}$$

Setting the period-1 solution $\mathbf{x}_k^*$ and substitution of Eq. (4.7) into Eq. (4.8) gives

$$\mathbf{f}(\mathbf{x}_k^*, \mathbf{x}_k^*, \mathbf{p}) = \mathbf{0} \quad \text{and} \quad \mathbf{f}(\mathbf{x}_k^*, \mathbf{x}_k^*, \mathbf{p}) = \mathbf{0}. \tag{4.9}$$

It is observed that the period-1 solutions for the positive and negative mappings are identical. The two relations for positive and negative mappings are illustrated in Fig. 4.1a, b, respectively. To determine the period-1 solution, the fixed points of Eq. (4.7) exist under constraints in Eq. (4.8), which are also shown in Fig. 4.1. The two thick lines on the axis are two sets for the mappings from the starting to final states. The relation in Eq. (4.7) is presented by a solid curve. The intersection points of the curves and straight lines for relations in Eqs. (4.7) and (4.8) give the fixed points of Eq. (4.9), which are period-1 solutions, labeled by the circular symbols. However, their stability and bifurcation for the period-1 solutions is different. The stability and bifurcation of fixed points for the positive and negative mappings can be stated as follows.

**Fig. 4.1** Period-1 solution for **a** positive mapping and **b** negative mapping. The two *thick lines* on the axis are two sets for the mappings from the starting to final states. The mapping relation is presented by a *solid curve*. The *circular symbols* give period-1 solutions for the positive and negative mappings

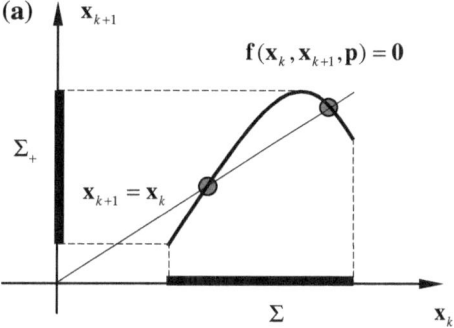

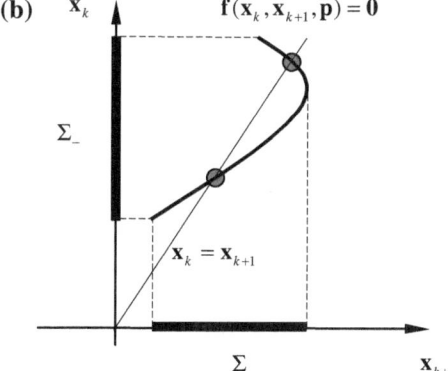

**Theorem 4.1** *For a discrete dynamical system in* Eq. (4.1), *there are two points* $\mathbf{x}_k \in D$ *and* $\mathbf{x}_{k+1} \in D$, *and two positive and negative mappings are*

$$\mathbf{x}_{k+1} = P_+\mathbf{x}_k \quad and \quad \mathbf{x}_k = P_-\mathbf{x}_{k+1} \tag{4.10}$$

*with*

$$\mathbf{f}(\mathbf{x}_k, \mathbf{x}_{k+1}, \mathbf{p}) = \mathbf{0} \quad and \quad \mathbf{f}(\mathbf{x}_k, \mathbf{x}_{k+1}, \mathbf{p}) = \mathbf{0}. \tag{4.11}$$

*Suppose a specific, differentiable, vector function* $\mathbf{g} \in \mathscr{R}^n$ *makes* $\mathbf{g}(\mathbf{x}_k, \mathbf{x}_{k+1}, \lambda) = \mathbf{0}$ *hold. If the solutions* $(\mathbf{x}_k^*, \mathbf{x}_{k+1}^*)$ *of both* $\mathbf{f}(\mathbf{x}_k, \mathbf{x}_{k+1}, \mathbf{p}) = \mathbf{0}$ *and* $\mathbf{g}(\mathbf{x}_k, \mathbf{x}_{k+1}, \lambda) = \mathbf{0}$ *exist, then the following conclusions in the sense of* $\mathbf{g}(\mathbf{x}_k, \mathbf{x}_{k+1}, \lambda) = \mathbf{0}$ *hold.*

(i) *The stable* $P_+$*-1 solutions are the unstable* $P_-$*-1 solutions with all eigenvalues of* $DP_-(\mathbf{x}_{k+1}^*)$ *outside the unit cycle, vice versa.*

(ii) *The unstable* $P_+$*-1 solutions with all eigenvalues of* $DP_+(\mathbf{x}_k^*)$ *outside the unit cycle are the stable* $P_-$*-1 solutions, vice versa.*

(iii) *For the unstable* $P_+$*-1 solutions with eigenvalue distribution of* $DP_+(\mathbf{x}_k^*)$ *inside and outside the unit cycle, the corresponding* $P_-$*-1 solution is also unstable with switching the eigenvalue distribution of* $DP_-(\mathbf{x}_{k-1}^*)$ *inside and outside the unit cycle, vice versa.*

(iv) *All the bifurcations of the* stable *and* unstable *$P_+$-1 solutions are all the bifurcations of the* unstable *and* stable *$P_-$-1 solutions, respectively.*

*Proof* From Luo (2012), the proof is given as follows. Consider the positive and negative mappings with relations in Eq. (4.9). The periodic solution in the sense of $\mathbf{g}(\mathbf{x}_k, \mathbf{x}_{k+1}, \lambda) = \mathbf{0}$ is given by

$$\mathbf{f}(\mathbf{x}_k, \mathbf{x}_{k+1}, \mathbf{p}) = \mathbf{0} \quad and \quad \mathbf{g}(\mathbf{x}_k, \mathbf{x}_{k+1}, \lambda) = \mathbf{0}$$

from which the fixed points $(\mathbf{x}_k^*, \mathbf{x}_{k+1}^*)$ can be determined. Consider a small perturbation

$$\mathbf{x}_{k+1} = \mathbf{x}_{k+1}^* + \delta\mathbf{x}_{k+1} \quad and \quad \mathbf{x}_k = \mathbf{x}_k^* + \delta\mathbf{x}_k.$$

The linearization of mappings in Eq. (4.9) gives

$$\delta\mathbf{x}_{k+1} = DP_+(\mathbf{x}_k^*)\delta\mathbf{x}_k \quad and \quad \delta\mathbf{x}_k = DP_-(\mathbf{x}_{k+1}^*)\delta\mathbf{x}_{k+1}$$

where

$$DP_+(\mathbf{x}_k^*) = \left[\frac{\partial\mathbf{x}_{k+1}}{\partial\mathbf{x}_k}\right]_{\mathbf{x}_k^*} \quad and \quad DP_-(\mathbf{x}_{k+1}^*) = \left[\frac{\partial\mathbf{x}_k}{\partial\mathbf{x}_{k+1}}\right]_{\mathbf{x}_{k+1}^*}.$$

From Eq. (4.10), one obtains

$$\left(\left[\frac{\partial \mathbf{f}}{\partial \mathbf{x}_k}\right] + \left[\frac{\partial \mathbf{f}}{\partial \mathbf{x}_{k+1}}\right]\left[\frac{\partial \mathbf{x}_{k+1}}{\partial \mathbf{x}_k}\right]\right)_{(\mathbf{x}_k^*, \mathbf{x}_{k+1}^*)} = \mathbf{0},$$

$$\mathbf{g}(\mathbf{x}_k^*, \mathbf{x}_{k+1}^*, \lambda) = \mathbf{g}(\mathbf{x}_k, \mathbf{x}_{k+1}, \lambda) = \mathbf{0};$$

$$\left(\left[\frac{\partial \mathbf{f}}{\partial \mathbf{x}_{k+1}}\right] + \left[\frac{\partial \mathbf{f}}{\partial \mathbf{x}_k}\right]\left[\frac{\partial \mathbf{x}_k}{\partial \mathbf{x}_{k+1}}\right]\right)_{(\mathbf{x}_k^*, \mathbf{x}_{k+1}^*)} = \mathbf{0},$$

$$\mathbf{g}(\mathbf{x}_k^*, \mathbf{x}_{k+1}^*, \lambda) = \mathbf{g}(\mathbf{x}_k, \mathbf{x}_{k+1}, \lambda) = \mathbf{0}.$$

That is,

$$DP_+(\mathbf{x}_k^*) = \left[\frac{\partial \mathbf{x}_{k+1}}{\partial \mathbf{x}_k}\right]\Big|\mathbf{x}_k^* = -\left(\left[\frac{\partial \mathbf{f}}{\partial \mathbf{x}_{k+1}}\right]^{-1}\left[\frac{\partial \mathbf{f}}{\partial \mathbf{x}_k}\right]\right)_{(\mathbf{x}_k^*, \mathbf{x}_{k+1}^*)},$$

$$DP_-(\mathbf{x}_{k+1}^*) = \left[\frac{\partial \mathbf{x}_k}{\partial \mathbf{x}_{k+1}}\right]_{\mathbf{x}_{k+1}^*} = -\left(\left[\frac{\partial \mathbf{f}}{\partial \mathbf{x}_k}\right]^{-1}\left[\frac{\partial \mathbf{f}}{\partial \mathbf{x}_{k+1}}\right]\right)_{(\mathbf{x}_k^*, \mathbf{x}_{k+1}^*)}.$$

Taking the inverse of the second equation in the foregoing equation gives

$$DP_-^{-1}(\mathbf{x}_{k+1}^*) = \left[\frac{\partial \mathbf{x}_k}{\partial \mathbf{x}_{k+1}}\right]^{-1}\Big|\mathbf{x}_{k+1}^* = -\left[\frac{\partial \mathbf{f}}{\partial \mathbf{x}_{k+1}}\right]^{-1}\left[\frac{\partial \mathbf{f}}{\partial \mathbf{x}_k}\right]_{(\mathbf{x}_{k+1}^*, \mathbf{x}_k^*)}$$

which is identical to $DP_+(\mathbf{x}_k^*)$. Therefore, one obtains

$$DP_-^{-1}(\mathbf{x}_{k+1}^*) = DP_+(\mathbf{x}_k^*).$$

In other words, $DP_+(\mathbf{x}_k^*)$ is the inverse of $DP_-(\mathbf{x}_{k+1}^*)$.

Consider the eigenvalues $\lambda_-$ and $\lambda_+$ of $DP_-(\mathbf{x}_{k+1}^*)$ and $DP_+(\mathbf{x}_k^*)$, accordingly. The following relations hold

$$(DP_-(\mathbf{x}_{k+1}^*) - \lambda_-\mathbf{I})\delta\mathbf{x}_k = \mathbf{0},$$
$$(DP_+(\mathbf{x}_k^*) - \lambda_+\mathbf{I})\delta\mathbf{x}_k = \mathbf{0}.$$

Left multiplication of $DP_+(\mathbf{x}_k^*)$ in the first equation of the foregoing equation, division of $\lambda_-$ on both sides, and application of $DP_-^{-1}(\mathbf{x}_{k+1}^*) = DP_+(\mathbf{x}_k^*)$ give

$$[DP_+(\mathbf{x}_k^*) - \lambda_-^{-1}\mathbf{I}]\delta\mathbf{x}_k = \mathbf{0}.$$

Thus, one can obtain

$$\lambda_+ = \lambda_-^{-1}.$$

From the stability and bifurcation theory for $P_+$-1 and $P_-$-1 solutions for discrete dynamical system in Eq. (4.1), the following conclusions can be given as follows:

(i)   The stable $P_+$-1 solutions are the unstable $P_-$-1 solutions with all eigenvalues of $DP_-(\mathbf{x}_{k+1}^*)$ outside the unit cycle, vice versa.

(ii)   The unstable $P_+$-1 solutions with all eigenvalues of $DP_+(\mathbf{x}_k^*)$ outside the unit cycle are the stable $P_-$-1 solutions, vice versa.

(iii)   For the unstable $P_+$-1 solutions with eigenvalue distribution of $DP_+(\mathbf{x}_k^*)$ inside and outside the unit cycle, the corresponding $P_-$-1 solution is also unstable with switching the eigenvalue distribution of $DP_-(\mathbf{x}_{k+1}^*)$ inside and outside the unit cycle, vice versa.

(iv)   All the bifurcations of the *stable* and *unstable* $P_+$-1 solutions are all the bifurcations of the *unstable* and *stable* $P_-$-1 solutions, respectively.

This theorem is proved.                                                                    □

From the foregoing theorem, the *Yin*, *Yang*, and *Yin–Yang* states in discrete dynamical systems exist. To generate the above ideas to $P_+^{(N)}$-1 and $P_-^{(N)}$-1 solutions in discrete dynamical systems in the sense of $\mathbf{g}(\mathbf{x}_k, \mathbf{x}_{k+N}, \lambda) = 0$, the mapping structure consisting of $N$-positive or negative mappings is considered.

**Definition 4.4** For a discrete dynamical system in Eq. (4.1), the mapping structures of $N$-mappings for the positive and negative mappings are defined as

$$\mathbf{x}_{k+N} = \underbrace{P_+ \circ P_+ \circ \cdots \circ P_+}_{N} \mathbf{x}_k = P_+^{(N)} \mathbf{x}_k, \tag{4.12}$$

$$\mathbf{x}_k = \underbrace{P_- \circ P_- \circ \cdots \circ P_-}_{N} \mathbf{x}_{k+N} = P_-^{(N)} \mathbf{x}_{k+N} \tag{4.13}$$

with

$$\mathbf{f}(\mathbf{x}_{k+i-1}, \mathbf{x}_{k+i}, \mathbf{p}) = 0 \quad \text{for } i = 1, 2, \ldots, N \tag{4.14}$$

where $P_+^{(0)} = \mathbf{I}_{n \times n}$ and $P_-^{(0)} = \mathbf{I}_{n \times n}$ for $N = 0$.

**Definition 4.5** For a discrete dynamical system in Eq. (4.1), consider two points $\mathbf{x}_{k+i-1} \in D$ ($i = 1, 2, \ldots, N$) and $\mathbf{x}_{k+N} \in D$, and there is a specific, differentiable, vector function $\mathbf{g} \in \mathbb{R}^n$ to make $\mathbf{g}(\mathbf{x}_k, \mathbf{x}_{k+N}, \lambda) = 0$.

(i)   The stable solution based on $\mathbf{x}_{k+N} = P_+^{(N)} \mathbf{x}_k$ for the positive mapping $P_+$ is called the "Yang" of the discrete dynamical system in Eq. (4.1) in the sense of

$\mathbf{g}(\mathbf{x}_k, \mathbf{x}_{k+N}, \lambda) = \mathbf{0}$ if the solutions $(\mathbf{x}_k^*, \mathbf{x}_{k+1}^*, \ldots, \mathbf{x}_{k+N}^*)$ of Eq. (4.14) with $\mathbf{g}(\mathbf{x}_k, \mathbf{x}_{k+N}, \lambda) = \mathbf{0}$ exist.

(ii) The stable solution based on $\mathbf{x}_k = P_-^{(N)} \mathbf{x}_{k+N}$ for the negative mapping $P_-$ is called "Yin" of the discrete dynamical system in Eq. (4.1) in the sense of $\mathbf{g}(\mathbf{x}_k, \mathbf{x}_{k+N}, \lambda) = \mathbf{0}$ if the solutions $(\mathbf{x}_k^*, \mathbf{x}_{k+1}^*, \ldots, \mathbf{x}_{k+N}^*)$ of Eq. (4.14) with $\mathbf{g}(\mathbf{x}_k, \mathbf{x}_{k+N}, \lambda) = \mathbf{0}$ exist.

(iii) The solution based on $\mathbf{x}_{k+N} = P_+^{(N)} \mathbf{x}_k$ is called "Yin–Yang" for the positive mapping $P_+$ of the discrete dynamical system in Eq. (4.1) in the sense of $\mathbf{g}(\mathbf{x}_k, \mathbf{x}_{k+N}, \lambda) = \mathbf{0}$ if the solutions $(\mathbf{x}_k^*, \mathbf{x}_{k+1}^*, \ldots, \mathbf{x}_{k+N}^*)$ of Eq. (4.14) with $\mathbf{g}(\mathbf{x}_k, \mathbf{x}_{k+N}, \lambda) = \mathbf{0}$ exist and the eigenvalues of $DP_+^{(N)}(\mathbf{x}_k^*)$ are distributed inside and outside the unit cycle.

(iv) The solution based on $\mathbf{x}_k = P_-^{(N)} \mathbf{x}_{k+N}$ is called "Yin–Yang" for the negative mapping $P_-$ of the discrete dynamical system in Eq. (4.1) in the sense of $\mathbf{g}(\mathbf{x}_k, \mathbf{x}_{k+N}, \lambda) = \mathbf{0}$ if the solutions $(\mathbf{x}_k^*, \mathbf{x}_{k+1}^*, \ldots, \mathbf{x}_{k+N}^*)$ of Eq. (4.14) with $\mathbf{g}(\mathbf{x}_k, \mathbf{x}_{k+N}, \lambda) = \mathbf{0}$ exist and the eigenvalues of $DP_-^{(N)}(\mathbf{x}_{k+N}^*)$ are distributed inside and outside unit cycle.

To determine the Yin–Yang properties of $P_+^{(N)}$-1 and $P_-^{(N)}$-1 in the discrete mapping system in Eq. (4.1), the corresponding theorem is presented as follows.

**Theorem 4.2** *For a discrete dynamical system in* Eq. (4.1), *there are two points* $\mathbf{x}_k \in D$ *and* $\mathbf{x}_{k+N} \in D$, *and two positive and negative mappings are*

$$\mathbf{x}_{k+N} = P_+^{(N)} \mathbf{x}_k \text{ and } \mathbf{x}_k = P_-^{(N)} \mathbf{x}_{k+N} \tag{4.15}$$

*and* $\mathbf{x}_{k+i} = P_+ \mathbf{x}_{k+i-1}$ *and* $\mathbf{x}_{k+i-1} = P_- \mathbf{x}_{k+i}$ *can be governed by*

$$\mathbf{f}(\mathbf{x}_{k+i-1}, \mathbf{x}_{k+i}, \mathbf{p}) = \mathbf{0} \quad \text{for } i = 1, 2, \ldots, N. \tag{4.16}$$

*Suppose a specific, differentiable, vector function of* $\mathbf{g} \in \mathcal{R}^n$ *makes* $\mathbf{g}(\mathbf{x}_k, \mathbf{x}_{k+N}, \lambda) = \mathbf{0}$ *hold. If the solutions* $(\mathbf{x}_k^*, \ldots, \mathbf{x}_{k+i}^*)$ *of* Eq. (4.16) *with* $\mathbf{g}(\mathbf{x}_k, \mathbf{x}_{k+N}, \lambda) = \mathbf{0}$ *exist, then the following conclusions in the sense of* $\mathbf{g}(\mathbf{x}_k, \mathbf{x}_{k+N}, \lambda) = \mathbf{0}$ *hold.*

(i) *The stable* $P_+^{(N)}$-1 *solution is the unstable* $P_-^{(N)}$-1 *solutions with all eigenvalues of* $DP_-^{(N)}(\mathbf{x}_{k+N}^*)$ *outside the unit cycle, vice versa.*

(ii) *The unstable* $P_+^{(N)}$-1 *solutions with all eigenvalues of* $DP_+^{(N)}(\mathbf{x}_k^*)$ *outside the unit cycle are the stable* $P_-^{(N)}$-1 *solutions, vice versa.*

(iii) *For the unstable* $P_+^{(N)}$-1 *solution with eigenvalue distribution of* $DP_+^{(N)}(\mathbf{x}_k^*)$ *inside and outside the unit cycle, the corresponding* $P_-^{(N)}$-1 *solution is also unstable with switching the eigenvalue distribution of* $DP_-^{(N)}(\mathbf{x}_{k+N}^*)$ *inside and outside the unit cycle, vice versa.*

(iv) *All the bifurcations of the* stable *and* unstable $P_+^{(N)}$-1 *solution are all the bifurcations of the* unstable *and* stable $P_-^{(N)}$-1 *solution, respectively.*

*Proof* From Luo (2012), the proof is given as follows. Consider positive and negative mappings with relations in Eq. (4.15), i.e.,

$$\mathbf{f}(\mathbf{x}_{k+i-1}, \mathbf{x}_{k+i}, \mathbf{p}) = \mathbf{0} \quad \text{for } i = 1, 2, \ldots, N$$

from which $\mathbf{x}_{k+i}$ is a function of $\mathbf{x}_{k+i-1}$ in the positive mapping iteration and $\mathbf{x}_{k+i-1}$ is a function of $\mathbf{x}_{k+i}$ in the negative mapping iteration. The periodic solution in the sense of $\mathbf{g}(\mathbf{x}_k, \mathbf{x}_{k+N}, \boldsymbol{\lambda}) = 0$ is given by

$$\mathbf{f}(\mathbf{x}_{k+i-1}, \mathbf{x}_{k+i}, \mathbf{p}) = \mathbf{0} \quad \text{for } i = 1, 2, \ldots, N,$$
$$\mathbf{g}(\mathbf{x}_k, \mathbf{x}_{k+N}, \boldsymbol{\lambda}) = \mathbf{0}.$$

Setting the period-1 solution be $\mathbf{x}^*_{k+i-1}$ or $\mathbf{x}^*_{k+i}$ $(i = 1, 2, \ldots, N)$ and the foregoing equation give

$$\mathbf{f}(\mathbf{x}^*_{k+i-1}, \mathbf{x}^*_{k+i}, \mathbf{p}) = \mathbf{0} \quad \text{for } i = 0, 1, \ldots, N,$$
$$\mathbf{g}(\mathbf{x}^*_k, \mathbf{x}^*_{k+N}, \boldsymbol{\lambda}) = \mathbf{0}.$$

for both the positive and negative mapping iterations. The existence condition of the foregoing equation requires

$$\det[(D_{ij})_{N \times N}] \neq 0$$

where

$$D_{N1} = -\left( \left[ \frac{\partial \mathbf{f}(\mathbf{x}_{k+N-1}, \mathbf{x}_{k+N}, \mathbf{p})}{\partial \mathbf{x}_{k+N}} \right]_{n \times n} \left[ \frac{\partial \mathbf{x}_k}{\partial \mathbf{x}_{k+N}} \right]^{-1}_{n \times n} \right)_{(\mathbf{x}^*_{k+N-1}, \mathbf{x}^*_k)}$$

$$= -\left( \left[ \frac{\partial \mathbf{f}(\mathbf{x}_{k+N-1}, \mathbf{x}_{k+N}, \mathbf{p})}{\partial \mathbf{x}_{k+N}} \right]_{n \times n} \left[ \frac{\partial \mathbf{g}(\mathbf{x}_{k+N}, \mathbf{x}_k, \mathbf{p})}{\partial \mathbf{x}_{k+N}} \right]^{-1}_{n \times n} \times \left[ \frac{\partial \mathbf{g}(\mathbf{x}_{k+N}, \mathbf{x}_k, \mathbf{p})}{\partial \mathbf{x}_k} \right]_{n \times n} \right)\Big|_{(\mathbf{x}^*_{k+N-1}, \mathbf{x}^*_k)},$$

$$D_{NN} = \left[ \frac{\partial \mathbf{f}(\mathbf{x}_{k+N-1}, \mathbf{x}_k, \mathbf{p})}{\partial \mathbf{x}_{k+N-1}} \right]_{n \times n} \Big|_{(\mathbf{x}^*_{k+N-1}, \mathbf{x}^*_k)},$$

$$D_{Nj} = [\mathbf{0}]_{n \times n} \quad \text{for } j = 2, 3, \ldots, N-1;$$

$$D_{ii} = \left[ \frac{\partial \mathbf{f}(\mathbf{x}_{k+i-1}, \mathbf{x}_{k+i}, \mathbf{p})}{\partial \mathbf{x}_{k+i-1}} \right]_{n \times n} \Big|_{(\mathbf{x}^*_{k+i-1}, \mathbf{x}^*_{k+i})},$$

$$D_{i(i+1)} = \left[ \frac{\partial \mathbf{f}(\mathbf{x}_{k+i-1}, \mathbf{x}_{k+i}, \mathbf{p})}{\partial \mathbf{x}_{k+i}} \right]_{n \times n} \Big|_{(\mathbf{x}^*_{k+i-1}, \mathbf{x}^*_{k+i})},$$

$$D_{ij} = [\mathbf{0}]_{n \times n} \quad \text{for } i = 1, 2, \ldots, N-1;$$
$$j = 1, 2, \ldots, i-1; \ i+2, \ i+3, \ldots, N.$$

Once $\mathbf{x}^*_{k+i-1}$ or $\mathbf{x}^*_{k+i}$ $(i = 1, 2, \ldots, N)$ is obtained in the sense of $\mathbf{g}(\mathbf{x}_k, \mathbf{x}_{k+N}, \boldsymbol{\lambda}) \equiv 0$, the corresponding stability and bifurcation of the periodic solutions can be determined. However, the stability and bifurcation of $P^{(N)}_+$-1 and $P^{(N)}_-$-1 solutions will be different. Herein, consider a small perturbation from the periodic solution

$$\left.\begin{array}{l} \mathbf{x}_{k+i} = \mathbf{x}^*_{k+i} + \delta\mathbf{x}_{k+i} \\ \mathbf{x}_{k+i+1} = \mathbf{x}^*_{k+i+1} + \delta\mathbf{x}_{k+i+1} \end{array}\right\} \quad \text{for} \;\; i = 0, 1, \ldots, N.$$

With the foregoing equation, linearization of Eq. (4.15) gives

$$\delta\mathbf{x}_{k+N} = \underbrace{DP_+ \cdot DP_+ \cdots DP_+}_{N}(\mathbf{x}^*_k)\delta\mathbf{x}_k$$

$$= DP^{(N)}_+(\mathbf{x}^*_k)\delta\mathbf{x}_k,$$

$$\delta\mathbf{x}_k = \underbrace{DP_- \cdot DP_- \cdots DP_-}_{N}(\mathbf{x}^*_{k+N})\delta\mathbf{x}_{k+N}$$

$$= DP^{(N)}_-(\mathbf{x}^*_{k+N})\delta\mathbf{x}_{k+N}.$$

On the other hand, each single positive and negative mappings gives

$$\delta\mathbf{x}_{k+i} = DP_+(\mathbf{x}^*_{k+i-1})\delta\mathbf{x}_{k+i-1} \quad \text{for} \;\; i = 1, 2, \ldots, N,$$

$$\delta\mathbf{x}_{k+i-1} = DP_-(\mathbf{x}^*_{k+i})\delta\mathbf{x}_{k+i} \quad \text{for} \;\; i = 1, 2, \ldots, N$$

where

$$DP_+(\mathbf{x}^*_{k+i-1}) = \left[\frac{\partial\mathbf{x}_{k+i}}{\partial\mathbf{x}_{k+i-1}}\right]_{\mathbf{x}^*_{k+i-1}} \quad \text{for} \;\; i = 1, 2, \ldots, N,$$

$$DP_-(\mathbf{x}^*_{k+i}) = \left[\frac{\partial\mathbf{x}_{k+i-1}}{\partial\mathbf{x}_{k+i}}\right]_{\mathbf{x}^*_{k+i}} \quad \text{for} \;\; i = 1, 2, \ldots, N$$

and for $i = 1, 2, \ldots, N$, linearization of Eq. (4.15) gives

$$DP_+(\mathbf{x}^*_{k+i-1}) = \left[\frac{\partial\mathbf{x}_{k+i}}{\partial\mathbf{x}_{k+i-1}}\right]_{\mathbf{x}^*_{k+i-1}} = -\left(\left[\frac{\partial\mathbf{f}}{\partial\mathbf{x}_{k+i}}\right]^{-1}\left[\frac{\partial\mathbf{f}}{\partial\mathbf{x}_{k+i-1}}\right]\right)_{(\mathbf{x}^*_{k+i},\mathbf{x}^*_{k+i-1})},$$

$$DP_-(\mathbf{x}^*_{k+i}) = \left[\frac{\partial\mathbf{x}_{k+i-1}}{\partial\mathbf{x}_{k+i}}\right]_{\mathbf{x}^*_{k+i}} = -\left(\left[\frac{\partial\mathbf{f}}{\partial\mathbf{x}_{k+i-1}}\right]^{-1}\left[\frac{\partial\mathbf{f}}{\partial\mathbf{x}_{k+i}}\right]\right)_{(\mathbf{x}^*_{k+i},\mathbf{x}^*_{k+i-1})}.$$

Therefore, the resultant Jacobian matrices for $P^{(N)}_+$-1 and $P^{(N)}_-$-1 are

$$DP^{(N)}_+(\mathbf{x}^*_k) = DP_+(\mathbf{x}^*_{k+N-1}) \cdot DP_+(\mathbf{x}^*_{k+N-2}) \cdots \cdot DP_+(\mathbf{x}^*_{k+1}) \cdot DP_+(\mathbf{x}^*_k)$$

$$= \left[\frac{\partial\mathbf{x}_{k+N}}{\partial\mathbf{x}_{k+N-1}}\right]_{\mathbf{x}^*_{k+N-1}} \cdot \left[\frac{\partial\mathbf{x}_{k+N}}{\partial\mathbf{x}_{k+N-1}}\right]_{\mathbf{x}^*_{k+N-2}} \cdots \cdot \left[\frac{\partial\mathbf{x}_{k+2}}{\partial\mathbf{x}_{k+1}}\right]_{\mathbf{x}^*_{k+1}} \cdot \left[\frac{\partial\mathbf{x}_{k+1}}{\partial\mathbf{x}_k}\right]_{\mathbf{x}^*_k}$$

$$= (-1)^N \left( \left[ \frac{\partial \mathbf{f}}{\partial \mathbf{x}_{k+N}} \right]^{-1} \left[ \frac{\partial \mathbf{f}}{\partial \mathbf{x}_{k+N-1}} \right] \right)_{(\mathbf{x}^*_{k+N}, \mathbf{x}^*_{k+N-1})} \cdots$$

$$\cdot \left( \left[ \frac{\partial \mathbf{f}}{\partial \mathbf{x}_{k+1}} \right]^{-1} \left[ \frac{\partial \mathbf{f}}{\partial \mathbf{x}_k} \right] \right)_{(\mathbf{x}^*_{k+1}, \mathbf{x}^*_k)},$$

$$DP_-^{(N)}(\mathbf{x}^*_{k+N}) = DP_-(\mathbf{x}^*_{k+1}) \cdot DP_-(\mathbf{x}^*_{k+2}) \cdots DP_-(\mathbf{x}^*_{k+N-1}) \cdot DP_-(\mathbf{x}^*_{k+N})$$

$$= \left[ \frac{\partial \mathbf{x}_k}{\partial \mathbf{x}_{k+1}} \right]_{\mathbf{x}^*_{k+1}} \cdot \left[ \frac{\partial \mathbf{x}_{k+1}}{\partial \mathbf{x}_{k+2}} \right]_{\mathbf{x}^*_{k+2}} \cdots \left[ \frac{\partial \mathbf{x}_{k+N-2}}{\partial \mathbf{x}_{k+N-1}} \right]_{\mathbf{x}^*_{k+N-1}} \cdot \left[ \frac{\partial \mathbf{x}_{k+N-1}}{\partial \mathbf{x}_{k+N}} \right]_{\mathbf{x}^*_{k+N}}$$

$$= (-1)^N \left( \left[ \frac{\partial \mathbf{f}}{\partial \mathbf{x}_k} \right]^{-1} \left[ \frac{\partial \mathbf{f}}{\partial \mathbf{x}_{k+1}} \right] \right)_{(\mathbf{x}^*_{k+1}, \mathbf{x}^*_k)} \cdots$$

$$\cdot \left( \left[ \frac{\partial \mathbf{f}}{\partial \mathbf{x}_{k+N-1}} \right]^{-1} \left[ \frac{\partial \mathbf{f}}{\partial \mathbf{x}_{k+N}} \right] \right)_{(\mathbf{x}^*_{k+N}, \mathbf{x}^*_{k+N-1})}.$$

From the two equations, it is very easily proved that the two resultant Jacobian matrices are inverse of each other, i.e.,

$$DP_+^{(N)}(\mathbf{x}^*_k) \cdot DP_-^{(N)}(\mathbf{x}^*_{k+N}) = \mathbf{I}_{n \times n}.$$

Similarly, consider eigenvalues $\lambda_-$ and $\lambda_+$ of $DP_-^{(N)}(\mathbf{x}^*_{k+N})$ and $DP_+^{(N)}(\mathbf{x}^*_k)$, accordingly. The following relations hold

$$(DP_-^{(N)}(\mathbf{x}^*_{k+N}) - \lambda_- \mathbf{I}) \delta \mathbf{x}_{k+N} = \mathbf{0},$$
$$(DP_+^{(N)}(\mathbf{x}^*_k) - \lambda_+ \mathbf{I}) \delta \mathbf{x}_k = \mathbf{0}.$$

Left multiplication of $DP_+^{(N)}(\mathbf{x}^*_k)$ in the first equation of the foregoing equation, division of $\lambda_-$ on both sides give

$$[DP_+^{(N)}(\mathbf{x}^*_k) - \lambda_-^{-1} \mathbf{I}] \delta \mathbf{x}_{k+N} = \mathbf{0}.$$

Since $\delta \mathbf{x}_{k+N}$ is arbitrarily selected, compared to $(DP_+^{(N)}(\mathbf{x}^*_k) - \lambda_+ \mathbf{I}) \delta \mathbf{x}_k = \mathbf{0}$, one obtains

$$\lambda_+ = \lambda_-^{-1}$$

in the sense of $\mathbf{g}(\mathbf{x}_k, \mathbf{x}_{k+N}, \lambda) = \mathbf{0}$ hold. From the stability and bifurcation theory for discrete dynamical systems, the following conclusions can be summarized as

(i)  The stable $P_+^{(N)}$-1 solution is the unstable $P^{(N)}$-1 solutions with all eigenvalues of $DP_-^{(N)}(\mathbf{x}^*_{k+N})$ outside the unit cycle, vice versa.

(ii)  The unstable $P_+^{(N)}$-1 solutions with all eigenvalues of $DP_+^{(N)}(\mathbf{x}_k^*)$ outside the unit cycle are the stable $P_-^{(N)}$-1 solutions, vice versa.

(iii) For the unstable $P_+^{(N)}$-1 solution with eigenvalue distribution of $DP_+^{(N)}(\mathbf{x}_k^*)$ inside and outside the unit cycle, the corresponding $P_-^{(N)}$-1 solution is also unstable with switching eigenvalue distribution of $DP_-^{(N)}(\mathbf{x}_{k+N}^*)$ *inside* and *outside* the unit cycle, vice versa.

(iv) All the bifurcations of the *stable* and *unstable* $P_+^{(N)}$-1 solution are all the bifurcations of the *unstable* and *stable* $P_-^{(N)}$-1 solution, respectively.

This theorem is proved.                                                            □

Notice that the number $N$ for the $P_+^{(N)}$-1 and $P_-^{(N)}$-1 solutions in the discrete dynamical system can be any integer if such a solution exists in the sense of $\mathbf{g}(\mathbf{x}_k, \mathbf{x}_{k+N}, \lambda) = \mathbf{0}$.

**Theorem 4.3** *For a discrete dynamical system in* Eq. (4.1), *there are two points* $\mathbf{x}_k \in D$ *and* $\mathbf{x}_{k+N} \in D$. *If the period-doubling cascade of the* $P_+^{(N)}$-1 *and* $P_-^{(N)}$-1 *solution occurs, the corresponding mapping structures are given by*

$$\mathbf{x}_{k+2N} = P_+^{(N)} \circ P_+^{(N)} \mathbf{x}_k = P_+^{(2N)} \mathbf{x}_k \quad and \quad \mathbf{g}(\mathbf{x}_k, \mathbf{x}_{k+2N}, \lambda) = \mathbf{0};$$

$$\mathbf{x}_{k+2^2N} = P_+^{(2N)} \circ P_+^{(2N)} \mathbf{x}_k = P_+^{(2^2N)} \mathbf{x}_k \quad and \quad \mathbf{g}(\mathbf{x}_k, \mathbf{x}_{k+2^2N}, \lambda) = \mathbf{0};$$

$$\vdots \qquad\qquad\qquad\qquad\qquad\qquad\qquad\qquad (4.17)$$

$$\mathbf{x}_{k+2^lN} = P_+^{(2^{l-1}N)} \circ P_+^{(2^{l-1}N)} \mathbf{x}_k = P_+^{(2^lN)} \mathbf{x}_k \quad and \quad \mathbf{g}(\mathbf{x}_k, \mathbf{x}_{k+2^lN}, \lambda) = \mathbf{0};$$

*for positive mappings and*

$$\mathbf{x}_k = P_-^{(N)} \circ P_-^{(N)} \mathbf{x}_{k+2N} = P_-^{(2N)} \mathbf{x}_{k+2N} \quad and \quad \mathbf{g}(\mathbf{x}_k, \mathbf{x}_{k+2N}, \lambda) = \mathbf{0};$$

$$\mathbf{x}_k = P_-^{(2N)} \circ P_-^{(2N)} \mathbf{x}_{k+2^2N} = P_-^{(2^2N)} \mathbf{x}_{k+2^2N} \quad and \quad \mathbf{g}(\mathbf{x}_k, \mathbf{x}_{k+2^2N}, \lambda) = \mathbf{0};$$

$$\vdots \qquad\qquad\qquad\qquad\qquad\qquad\qquad\qquad (4.18)$$

$$\mathbf{x}_k = P_-^{(2^{l-1}N)} \circ P_-^{(2^{l-1}N)} \mathbf{x}_{k+2^lN} = P_-^{(2^lN)} \mathbf{x}_{k+2^lN} \quad and \quad \mathbf{g}(\mathbf{x}_k, \mathbf{x}_{k+2^lN}, \lambda) = \mathbf{0}$$

*for negative mapping, then the following statements hold, i.e.,*

(i)  *The stable chaos generated by the limit state of the stable* $P_+^{(2^lN)}$-1 *solutions* $(l \to \infty)$ *in the sense of* $\mathbf{g}(\mathbf{x}_k, \mathbf{x}_{k+2^lN}, \lambda) = \mathbf{0}$ *is the unstable chaos generated by the limit state of the unstable–stable* $P_-^{(2^lN)}$-1 *solutions* $(l \to \infty)$ *in the sense of* $\mathbf{g}(\mathbf{x}_k, \mathbf{x}_{k+2^lN}, \lambda) = \mathbf{0}$ *with all eigenvalue distribution of* $DP_-^{(2^lN)}$ *outside unit cycle, vice versa. Such a chaos is the "Yang" chaos in nonlinear discrete dynamical systems.*

(ii) *The unstable chaos generated by the limit state of the unstable $P_+^{(2^l N)}$-1 solutions $(l \to \infty)$ in the sense of $\mathbf{g}(\mathbf{x}_k, \mathbf{x}_{k+2^l N}, \boldsymbol{\lambda}) = \mathbf{0}$ with all eigenvalue distribution of $DP_+^{(2^l N)}$ outside the unit cycle is the stable chaos generated by the limit state of the stable $P_+^{(2^l N)} - 1$ solution $(l \to \infty)$ in the sense of $\mathbf{g}(\mathbf{x}_k, \mathbf{x}_{k+2^l N}, \boldsymbol{\lambda}) = \mathbf{0}$, vice versa. Such a chaos is the "Yin" chaos in nonlinear discrete dynamical systems.*

(iii) *The unstable chaos generated by the limit state of the unstable $P_+^{(2^l N)} - 1$ solutions $(l \to \infty)$ in the sense of $\mathbf{g}(\mathbf{x}_k, \mathbf{x}_{k+2^l N}, \boldsymbol{\lambda}) = \mathbf{0}$ with all eigenvalue distribution of $DP_+^{(2^l N)}$ inside and outside the unit cycle is the unstable chaos generated by the limit state of the unstable $P_-^{(2^l N)} - 1$ solution $(l \to \infty)$ in the sense of $\mathbf{g}(\mathbf{x}_k, \mathbf{x}_{k+2^l N}, \boldsymbol{\lambda}) = \mathbf{0}$ with switching all eigenvalue distribution of $DP_+^{(2^l N)}$ inside and outside the unit cycle, vice versa. Such a chaos is the "Yin–Yang" chaos in nonlinear discrete dynamical systems.*

**Proof** The proof is similar to the proof of Theorem 4.2, and the chaos is obtained by $l \to \infty$. This theorem is proved.                                                                            $\square$

## 4.2  Discrete Systems with Multiple Maps

**Definition 4.5** Consider implicit vector functions $\mathbf{f}^{(j)} : D \to D$ $(j = 1, 2, \ldots)$ on an open set $D \subset \mathscr{R}^n$ in an $n$-dimensional discrete dynamical system. For $\mathbf{x}_k, \mathbf{x}_{k+1} \in D$, there is a discrete relation as

$$\mathbf{f}^{(j)}(\mathbf{x}_k, \mathbf{x}_{k+1}, \mathbf{p}^{(j)}) = \mathbf{0} \quad \text{for} \quad j = 1, 2, \ldots \tag{4.19}$$

where the vector function is $\mathbf{f}^{(j)} = (f_1^{(j)}, f_2^{(j)}, \ldots, f_n^{(j)})^{\mathrm{T}} \in \mathscr{R}^n$ and discrete variable vector is $\mathbf{x}_k = (x_{k1}, x_{k2}, \ldots, x_{kn})^{\mathrm{T}} \in \Omega$ with a parameter vector $\mathbf{p}^{(j)} = (p_1^{(j)}, p_2^{(j)}, \ldots, p_{m_j}^{(j)})^{\mathrm{T}} \in \mathscr{R}^{m_j}$.

**Definition 4.6** Consider a set of implicit vector functions $\mathbf{f}^{(j)} : D \to D$ $(j = 1, 2, \ldots)$ on an open set $D \subset \mathscr{R}^n$ in an $n$-dimensional discrete dynamical system.

(i)  A set for discrete relations is defined as

$$\Phi = \left\{ \mathbf{f}^{(j)} | \mathbf{f}^{(j)}(\mathbf{x}_k, \mathbf{x}_{k+1}, \mathbf{p}^{(j)}) = \mathbf{0}, \ j \in \mathbb{Z}_+; \ k \in \mathbb{Z} \right\}. \tag{4.20}$$

(ii)  The positive and negative discrete sets are defined as

$$\begin{aligned}
\Sigma_+ &= \{ \mathbf{x}_{k+i} | \mathbf{x}_{k+i} \in \mathscr{R}^n, i \in \mathbb{Z}_+ \} \subset D \quad \text{and} \\
\Sigma_- &= \{ \mathbf{x}_{k-i} | \mathbf{x}_{k-i} \in \mathscr{R}^n, i \in \mathbb{Z}_+ \} \subset D
\end{aligned} \tag{4.21}$$

respectively, and the total set of the discrete states is

$$\Sigma = \Sigma_+ \cup \Sigma_-. \tag{4.22}$$

(iii) A positive mapping for $\mathbf{f}^{(j)} \in \Phi$ is defined as

$$P_j^+ : \Sigma \to \Sigma_+ \Rightarrow P_j^+ : \mathbf{x}_k \to \mathbf{x}_{k+1} \tag{4.23}$$

and a negative mapping is defined by

$$P_j^- : \Sigma \to \Sigma_- \Rightarrow P_j^- : \mathbf{x}_k \to \mathbf{x}_{k-1}. \tag{4.24}$$

(iv) Two sets for positive and negative mappings are defined as

$$
\begin{aligned}
\Theta_+ &= \left\{ P_j^+ | P_j^+ : \mathbf{x}_k \to \mathbf{x}_{k+1} \text{ with } \mathbf{f}^{(j)}(\mathbf{x}_k, \mathbf{x}_{k+1}, \mathbf{p}^{(j)}) = \mathbf{0}, \quad j \in \mathbb{Z}_+; \ k \in \mathbb{Z} \right\}, \\
\Theta_- &= \left\{ P_j^- | P_j^- : \mathbf{x}_{k+1} \to \mathbf{x}_k \text{ with } \mathbf{f}^{(j)}(\mathbf{x}_k, \mathbf{x}_{k+1}, \mathbf{p}^{(j)}) = \mathbf{0}, \quad j \in \mathbb{Z}_+; \ k \in \mathbb{Z} \right\}
\end{aligned} \tag{4.25}
$$

with the total mapping sets are

$$\Theta = \Theta_+ \cup \Theta_-. \tag{4.26}$$

**Definition 4.7** Consider a discrete dynamical system with a set of implicit vector functions $\mathbf{f}^{(j)} : D \to D$ $(j = 1, 2, \ldots)$. For a mapping $P_j^+ \in \Theta_+$ with $N_j$-actions and $P_j^- \in \Theta_-$ with $N_j$-actions, the resultant mapping is defined as

$$P_{jN}^+ = \underbrace{P_j^+ \circ P_j^+ \circ \cdots \circ P_j^+}_{N} \quad \text{and} \quad P_{jN}^- = \underbrace{P_j^- \circ P_j^- \circ \cdots \circ P_j^-}_{N}. \tag{4.27}$$

**Definition 4.8** Consider a discrete dynamical system with a set of implicit vector functions $\mathbf{f}^{(j)} : D \to D$ $(j = 1, 2, \ldots)$. For the $m$-positive mappings of $P_{j_i}^+ \in \Theta_+$ $(i = 1, 2, \ldots, m)$ with $N_{j_i}$-actions $(N_{j_i} \in \{0, \mathbb{Z}_+\})$ and the corresponding $m$-negative mappings of $P_{j_i}^- \in \Theta_-$ $(i = 1, 2, \ldots, m)$ with $N_{j_i}$-actions, the resultant nonlinear mapping cluster with pure positive or negative mappings is defined as

$$
\begin{aligned}
P_{(N_{j_m} \cdots N_{j_2} N_{j_1})}^+ &= \underbrace{P_{j_m}^{+N_{j_m}} \circ \cdots \circ P_{j_2}^{+N_{j_2}} \circ P_{j_1}^{+N_{j_1}}}_{m-terms}; \\
P_{(N_{j_1} N_{j_2} \cdots N_{j_m})}^- &= \underbrace{P_{j_1}^{-N_{j_1}} \circ P_{j_2}^{-N_{j_2}} \circ \cdots \circ P_{j_m}^{-N_{j_m}}}_{m-terms}.
\end{aligned} \tag{4.28}
$$

in which at least one of the mappings ($P_{j_i}^+$ and $P_{j_i}^-$) with $N_{j_i} \in \mathbb{Z}_+$ possesses a nonlinear iterative relation.

**Theorem 4.4** *Consider a discrete dynamical system with a set of implicit vector functions* $\mathbf{f}^{(j)} : D \to D$ ($j = 1, 2, \ldots$). *For the m-positive mappings of* $P_{j_i}^+ \in \Theta_+$ ($i = 1, 2, \ldots, m$) *with* $N_{j_i}$-*actions* ($N_{j_i} \in \{0, \mathbb{Z}_+\}$) *and the corresponding m-negative mappings of* $P_{j_i}^- \in \Theta_-$ ($i = 1, 2, \ldots, m$) *with* $N_{j_i}$-*actions, the resultant nonlinear mapping with pure positive and negative mappings is defined as*

$$\mathbf{x}_{k+\Sigma_{s=1}^m N_{j_s}} = P_{(N_{j_m}\ldots N_{j_2} N_{j_1})}^+ \mathbf{x}_k \quad and \quad \mathbf{x}_k = P_{(N_{j_1} N_{j_2}\ldots N_{j_m})}^- \mathbf{x}_{k+\Sigma_{s=1}^m N_{j_s}} \tag{4.29}$$

*and* $\mathbf{x}_{k+i} = P_{j_s}^+ \mathbf{x}_{k+i-1}$ *and* $\mathbf{x}_{k+i-1} = P_{j_s}^- \mathbf{x}_{k+i}$ *can be governed by*

$$\mathbf{f}(\mathbf{x}_{k+i-1}, \mathbf{x}_{k+i}, \mathbf{p}) = \mathbf{0} \quad for \ \ i = 1, 2, \ldots, \sum_{s=1}^m N_{j_s}. \tag{4.30}$$

*Suppose there is a specific, differentiable, vector function* $\mathbf{g} \in \mathscr{R}^n$ *to make* $\mathbf{g}(\mathbf{x}_k, \mathbf{x}_{k+\Sigma_{s=1}^m N_{j_s}}, \boldsymbol{\lambda}) = \mathbf{0}$ *hold. If the solutions* $(\mathbf{x}_k^*, \ldots, \mathbf{x}_{k+\Sigma_{s=1}^m N_{j_s}}^*)$ *of Eq. (4.29) with* $\mathbf{g}(\mathbf{x}_k, \mathbf{x}_{k+\Sigma_{s=1}^m N_{j_s}}, \boldsymbol{\lambda}) = \mathbf{0}$ *exist, then the following conclusions in the sense of* $\mathbf{g}(\mathbf{x}_k, \mathbf{x}_{k+\Sigma_{s=1}^m N_{j_s}}, \boldsymbol{\lambda}) = \mathbf{0}$ *hold.*

(i)   *The stable* $P_{(N_{j_m}\ldots N_{j_2} N_{j_1})}^+$-1 *solution is the unstable* $P_{(N_{j_1} N_{j_2}\ldots N_{j_m})}^-$-1 *solutions with all eigenvalues of* $DP_{(N_{j_1} N_{j_2}\ldots N_{j_m})}^- (\mathbf{x}_{k+\Sigma_{s=1}^m N_{j_s}}^*)$ *outside the unit cycle, vice versa.*

(ii)  *The unstable* $P_{(N_{j_m}\ldots N_{j_2} N_{j_1})}^+$-1 *solutions with all eigenvalues of* $DP_{(N_{j_m}\ldots N_{j_2} N_{j_1})}^+ (\mathbf{x}_k^*)$ *outside the unit cycle are the stable* $P_{(N_{j_1} N_{j_2}\ldots N_{j_m})}^-$-1 *solutions, vice versa.*

(iii) *For the unstable* $P_{(N_{j_m}\ldots N_{j_2} N_{j_1})}^+$-1 *solution with eigenvalue distribution of* $DP_{(N_{j_m}\ldots N_{j_2} N_{j_1})}^+ (\mathbf{x}_k^*)$ *inside and outside the unit cycle, the corresponding* $P_{(N_{j_1} N_{j_2}\ldots N_{j_m})}^-$-1 *solution is also unstable with switching eigenvalue distribution of* $DP_{(N_{j_1} N_{j_2}\ldots N_{j_m})}^- (\mathbf{x}_{k+\Sigma_{s=1}^m N_{j_s}}^*)$ *inside and outside the unit cycle, vice versa.*

(iv)  *All the bifurcations of the stable and unstable* $P_{(N_{j_m}\ldots N_{j_2} N_{j_1})}^+$-1 *solution are all the bifurcations of the unstable and stable* $P_{(N_{j_1} N_{j_2}\ldots N_{j_m})}^-$-1 *solution, respectively.*

*Proof* The proof is similar to the proof of Theorem 4.2. This theorem is proved.□

The chaos generated by the period-doubling of the $P_{(N_{j_m}\ldots N_{j_2} N_{j_1})}^+$-1 and $P_{(N_{j_1} N_{j_2}\ldots N_{j_m})}^-$-1 solutions can be described through the following theorem.

**Theorem 4.5** *Consider a discrete dynamical system with a set of implicit vector functions* $\mathbf{f}^{(j)} : D \to D$ ($j = 1, 2, \ldots$). *For the m-positive mappings of* $P_{j_i}^+ \in \Theta_+$ ($i = 1, 2, \ldots, m$) *with* $N_{j_i}$-*actions* ($N_{j_i} \in \{0, \mathbb{Z}_+\}$) *and the corresponding m-negative mappings of* $P_{j_i}^- \in \Theta_-$ ($i = 1, 2, \ldots, m$) *with* $N_{j_i}$-*actions, the resultant nonlinear mapping with pure positive and negative mappings is defined as*

$$\mathbf{x}_{k+\Sigma_{s=1}^{m}N_{j_s}} = P^{+}_{(N_{j_m}...N_{j_2}N_{j_1})}\mathbf{x}_k \quad \text{and} \quad \mathbf{x}_k = P^{-}_{(N_{j_1}N_{j_2}...N_{j_m})}\mathbf{x}_{k+\Sigma_{s=1}^{m}N_{j_s}}; \qquad (4.31)$$

*and* $\mathbf{x}_{k+i} = P^{+}_{j_s}\mathbf{x}_{k+i-1}$ *and* $\mathbf{x}_{k+i-1} = P^{-}_{j_s}\mathbf{x}_{k+i}$ *can be governed by*

$$\mathbf{f}^{(j)}\left(\mathbf{x}_{k+i-1}, \mathbf{x}_{k+i}, \mathbf{p}\right) = \mathbf{0} \quad \text{for } i = 1, 2, \ldots, \sum_{s=1}^{m}N_{j_s};$$

$$\mathbf{f}^{(j)}\left(\mathbf{x}_{k+i-1}, \mathbf{x}_{k+i}, \mathbf{p}\right) = \mathbf{0} \quad \text{for } i = \sum_{s=1}^{m}N_{j_s}, \ldots, 2, 1.$$

$$(4.32)$$

*Suppose a specific, differentiable, vectorfunction* $\mathbf{g} \in \mathscr{R}^n$ *makes* $\mathbf{g}(\mathbf{x}_k, \mathbf{x}_{k+\Sigma_{s=1}^{m}N_{j_s}}, \boldsymbol{\lambda})$ $= \mathbf{0}$ *hold. If the period-doubling cascade of the* $P^{+}_{(N_{j_m}...N_{j_2}N_{j_1})}$-1 *and* $P^{-}_{(N_{j_1}N_{j_2}...N_{j_m})}$-1 *solution occurs, the corresponding mapping structures are given by*

$$\left.\begin{aligned}\mathbf{x}_{k+2\Sigma_{s=1}^{m}N_{j_s}} &= P^{+}_{(N_{j_m}...N_{j_2}N_{j_1})} \circ P^{+}_{(N_{j_m}...N_{j_2}N_{j_1})}\mathbf{x}_k = P^{+}_{2(N_{j_m}...N_{j_2}N_{j_1})}\mathbf{x}_k \\ & \qquad\qquad \mathbf{g}(\mathbf{x}_k, \mathbf{x}_{k+2\Sigma_{s=1}^{m}N_{j_s}}, \boldsymbol{\lambda}) = \mathbf{0};\end{aligned}\right\}$$

$$\left.\begin{aligned}\mathbf{x}_{k+2^2\Sigma_{s=1}^{m}N_{j_s}} &= P^{+}_{2(N_{j_m}...N_{j_2}N_{j_1})} \circ P^{+}_{2(N_{j_m}...N_{j_2}N_{j_1})}\mathbf{x}_k = P^{+}_{2^2(N_{j_m}...N_{j_2}N_{j_1})}\mathbf{x}_k \\ & \qquad\qquad \mathbf{g}(\mathbf{x}_k, \mathbf{x}_{k+2^2\Sigma_{s=1}^{m}N_{j_s}}, \boldsymbol{\lambda}) = \mathbf{0};\end{aligned}\right\} \quad (4.33)$$

$$\vdots$$

$$\left.\begin{aligned}\mathbf{x}_{k+2^l\Sigma_{s=1}^{m}N_{j_s}} &= P^{+}_{2^{l-1}(N_{j_m}...N_{j_2}N_{j_1})} \circ P^{+}_{2^{l-1}(N_{j_m}...N_{j_2}N_{j_1})}\mathbf{x}_k = P^{+}_{2^l(N_{j_m}...N_{j_2}N_{j_1})}\mathbf{x}_k \\ & \qquad\qquad \mathbf{g}(\mathbf{x}_k, \mathbf{x}_{k+2^l\Sigma_{s=1}^{m}N_{j_s}}, \boldsymbol{\lambda}) = \mathbf{0};\end{aligned}\right\}$$

*for positive mappings and*

$$\left.\begin{aligned}\mathbf{x}_k &= P^{-}_{(N_{j_1}N_{j_2}...N_{j_m})} \circ P^{-}_{(N_{j_1}N_{j_2}...N_{j_m})}\mathbf{x}_{k+2\Sigma_{s=1}^{m}N_{j_s}} = P^{-}_{2(N_{j_1}N_{j_2}...N_{j_m})}\mathbf{x}_{k+2\Sigma_{s=1}^{m}N_{j_s}} \\ & \qquad\qquad \mathbf{g}(\mathbf{x}_k, \mathbf{x}_{k+2\Sigma_{s=1}^{m}N_{j_s}}, \boldsymbol{\lambda}) = \mathbf{0};\end{aligned}\right\}$$

$$\left.\begin{aligned}\mathbf{x}_k &= P^{-}_{2(N_{j_1}N_{j_2}...N_{j_m})} \circ P^{-}_{2(N_{j_1}N_{j_2}...N_{j_m})}\mathbf{x}_{k+2^2\Sigma_{s=1}^{m}N_{j_s}} = P^{-}_{2^2(N_{j_1}N_{j_2}...N_{j_m})}\mathbf{x}_{k+2^2\Sigma_{s=1}^{m}N_{j_s}} \\ & \qquad\qquad \mathbf{g}(\mathbf{x}_k, \mathbf{x}_{k+2^2\Sigma_{s=1}^{m}N_{j_s}}, \boldsymbol{\lambda}) = \mathbf{0};\end{aligned}\right\} \quad (4.34)$$

$$\vdots$$

$$\left.\begin{aligned}\mathbf{x}_k &= P^{-}_{2^{l-1}(N_{j_1}N_{j_2}...N_{j_m})} \circ P^{-}_{2^{l-1}(N_{j_1}N_{j_2}...N_{j_m})}\mathbf{x}_{k+2^l\Sigma_{s=1}^{m}N_{j_s}} = P^{-}_{2^l(N_{j_1}N_{j_2}...N_{j_m})}\mathbf{x}_{k+2^l\Sigma_{s=1}^{m}N_{j_s}} \\ & \qquad\qquad \mathbf{g}(\mathbf{x}_k, \mathbf{x}_{k+2^l\Sigma_{s=1}^{m}N_{j_s}}, \boldsymbol{\lambda}) = \mathbf{0};\end{aligned}\right\}$$

*for negative mapping, then the following statements hold, i.e.,*

(i)  *The stable chaos generated by the limit state of the stable $P^+_{2^l(N_{jm}...N_{j_2}N_{j_1})}$-1*
     *solutions $(l \to \infty)$ in the sense of $\mathbf{g}(\mathbf{x}_k, \mathbf{x}_{k+2^l\Sigma^m_{s=1}N_{js}}, \lambda) = \mathbf{0}$ is the unstable*
     *chaos generated by the limit state of the unstable–stable $P^-_{2^l(N_{j_1}N_{j_2}...N_{jm})}$-1*
     *solutions $(l \to \infty)$ in the sense of $\mathbf{g}(\mathbf{x}_k, \mathbf{x}_{k+2^l\Sigma^m_{s=1}N_{js}}, \lambda) = \mathbf{0}$ with all eigenvalue*
     *distribution of $DP^-_{2^l(N_{jm}...N_{j_2}N_{j_1})}$ outside unit cycle, vice versa. Such a chaos is*
     *the "Yang" chaos in nonlinear discrete dynamical systems.*

(ii) *The unstable chaos generated by the limit state of the unstable $P^+_{2^l(N_{jm}...N_{j_2}N_{j_1})}$-1*
     *solutions $(l \to \infty)$ in the sense of $\mathbf{g}(\mathbf{x}_k, \mathbf{x}_{k+2^l\Sigma^m_{s=1}N_{js}}, \lambda) = \mathbf{0}$ with all eigenvalue*
     *distribution of $P^+_{2^l(N_{jm}...N_{j_2}N_{j_1})}$-1 outside the unit cycle is the stable chaos gen-*
     *erated by the limit state of the stable $P^-_{2^l(N_{j_1}N_{j_2}...N_{jm})}$-1 solution $(l \to \infty)$ in the*
     *sense of $\mathbf{g}(\mathbf{x}_k, \mathbf{x}_{k+2^l\Sigma^m_{s=1}N_{js}}, \lambda) = \mathbf{0}$, vice versa. Such a chaos is the "Yin" chaos*
     *in nonlinear discrete dynamical systems.*

(iii) *The unstable chaos generated by the limit state of the unstable $P^+_{2^l(N_{jm}...N_{j_2}N_{j_1})}$-1*
     *solutions $(l \to \infty)$ in the sense of $\mathbf{g}(\mathbf{x}_k, \mathbf{x}_{k+2^l\Sigma^m_{s=1}N_{js}}, \lambda) = \mathbf{0}$ with all eigenvalue*
     *distribution of $DP^+_{2^l(N_{jm}...N_{j_2}N_{j_1})}$ inside and outside the unit cycle is the unstable*
     *chaos generated by the limit state of the unstable $P^-_{2^l(N_{j_1}N_{j_2}...N_{jm})}$-1 solution*
     *$(l \to \infty)$ in the sense of $\mathbf{g}(\mathbf{x}_k, \mathbf{x}_{k+2^l\Sigma^m_{s=1}N_{js}}, \lambda) = \mathbf{0}$ with switching all eigenvalue*
     *distribution of $DP^-_{2^l(N_{j_1}N_{j_2}...N_{jm})}$ inside and outside the unit cycle, vice versa.*
     *Such a chaos is the "Yin–Yang" chaos in nonlinear discrete dynamical*
     *systems.*

*Proof* The proof is similar to the proof of Theorem 4.2, and the chaos is obtained
by $l \to \infty$. This theorem is proved.                                              □

## 4.3  Complete Dynamics of a Henon Map System

As in Luo and Guo (2010), consider the Henon map

$$f_1(\mathbf{x}_k, \mathbf{x}_{k+1}, \mathbf{p}) = x_{k+1} - y_k - 1 + ax_k^2 = 0,$$
$$f_2(\mathbf{x}_k, \mathbf{x}_{k+1}, \mathbf{p}) = y_{k+1} - bx_k = 0 \tag{4.35}$$

where $\mathbf{x}_k = (x_k, y_k)^T$, $\mathbf{f} = (f_1, f_2)^T$, and $\mathbf{p} = (a, b)^T$. The positive and negative
mappings are

$$\mathbf{x}_{k+1} = P_+\mathbf{x}_k \quad \text{and} \quad \mathbf{x}_k = P_-\mathbf{x}_{k+1}. \tag{4.36}$$

Consider two positive and negative mapping structures as

$$\mathbf{x}_{k+N} = P_+^{(N)}\mathbf{x}_k = \underbrace{P_+ \circ \cdots P_+ \circ P_+}_{N-\text{terms}} \mathbf{x}_k,$$

$$\mathbf{x}_k = P_-^{(N)}\mathbf{x}_{k+N} = \underbrace{P_- \circ \cdots P_- \circ P_-}_{N-\text{terms}} \mathbf{x}_{k+N}. \qquad (4.37)$$

Equations (4.36) and (4.37) give

$$\left.\begin{array}{l} \mathbf{f}(\mathbf{x}_k, \mathbf{x}_{k+1}, \mathbf{p}) = \mathbf{0}, \\ \mathbf{f}(\mathbf{x}_{k+1}, \mathbf{x}_{k+2}, \mathbf{p}) = \mathbf{0}, \\ \vdots \\ \mathbf{f}(\mathbf{x}_{k+N-1}, \mathbf{x}_{k+N}, \mathbf{p}) = \mathbf{0}; \end{array}\right\} \qquad (4.38)$$

and

$$\left.\begin{array}{l} \mathbf{f}(\mathbf{x}_{k+N-1}, \mathbf{x}_{k+N}, \mathbf{p}) = \mathbf{0}, \\ \mathbf{f}(\mathbf{x}_{k+N-2}, \mathbf{x}_{k+N-1}, \mathbf{p}) = \mathbf{0}, \\ \vdots \\ \mathbf{f}(\mathbf{x}_k, \mathbf{x}_{k+1}, \mathbf{p}) = \mathbf{0}. \end{array}\right\} \qquad (4.39)$$

The switching of equation order in Eq. (4.38) shows that Eqs. (4.38) and (4.39) are identical. For periodic solutions for the positive and negative maps, the periodicity of the positive and negative mapping structures of the Henon map requires

$$\mathbf{x}_{k+N} = \mathbf{x}_k \quad \text{or} \quad \mathbf{x}_k = \mathbf{x}_{k+N}. \qquad (4.40)$$

So the periodic solutions $\mathbf{x}_{k+j}^*$ $(j = 0, 1, \ldots, N)$ for the negative and positive mapping structures are the same, which are given by solving Eqs. (4.39) and (4.40). Thus, the resultant perturbation of the mapping structure in Eq. (4.37) gives

$$\delta\mathbf{x}_{k+N} = DP_+^{(N)}\mathbf{x}_k = \underbrace{DP_+ \cdot \ldots \cdot DP_+ \cdot DP_+}_{N-\text{terms}} \delta\mathbf{x}_k,$$

$$\delta\mathbf{x}_k = DP_-^{(N)}\delta\mathbf{x}_{k+N} = \underbrace{DP_- \cdot \ldots \cdot DP_- \cdot DP_-}_{N-\text{terms}} \delta\mathbf{x}_{k+N} \qquad (4.41)$$

where

$$DP_+^{(N)} = \prod_{j=1}^{N} DP_+(\mathbf{x}_{k+N-j}^*),$$

$$DP_-^{(N)} = \prod_{j=1}^{N} DP_-(\mathbf{x}_{k+N-j+1}^*), \qquad (4.42)$$

$$DP_+(\mathbf{x}^*_{k+j-1}) = \begin{bmatrix} 2ax^*_{k+j-1} & -1 \\ -b & 0 \end{bmatrix}, \tag{4.43}$$

$$DP_-(\mathbf{x}^*_{k+j}) = -\frac{1}{b}\begin{bmatrix} 0 & 1 \\ b & 2ax^*_{k+j-1} \end{bmatrix}. \tag{4.44}$$

From the resultant Jacobian matrix, the eigenvalue analysis can be completed. Before analytical prediction of periodic motion, a numerical prediction of the periodic solutions of the Henon map is presented with varying parameter $b$ for $a = 0.2$, as shown in Fig. 4.2. The dashed vertical lines give the bifurcation points. The acronyms "PD," "SN," and "NB" are presented the period-doubling bifurcation, saddle–stable node bifurcation, and Neimark bifurcation, respectively.

From the numerical prediction, the stable periodic solutions of the Henon map are obtained. Herein, through the corresponding mapping structures, the stable and unstable periodic solutions for positive and negative mappings of the Henon maps are presented in Fig. 4.3. The acronyms "PD," "SN," and "NB" are also presented the period-doubling bifurcation, saddle–stable node bifurcation, and Neimark bifurcation, respectively. The acronyms "UPD" and "USN" are presented the period-doubling bifurcation relative to unstable nodes and saddle–unstable node bifurcation, respectively. From eigenvalue analysis, the stable periodic solutions for positive mapping $P_+$ lie in $b \in (-1.0, 1.0)$, which is the same as the numerical prediction. In other words, the stable period-1 solution of $P_+$ is in $b \in (-1, 0.4805)$. For $b \in (0.4805, +\infty)$, the unstable period-1 solution of $P_+$ is saddle. The unstable period-1 solution of $P_+$ is unstable focus for $b \in (-\infty, -1.0)$. The corresponding bifurcations are Neimark bifurcation (NB) and period-doubling bifurcation (PD). However, another unstable period-1 solution of $P_+$ exists. For $b \in (1.5215, +\infty)$, the unstable periodic solution of $P_+$ is unstable node. However, the unstable periodic solution of $P_+$ is saddle for $b \in (-\infty, 1.5215)$. Thus, the unstable period-doubling bifurcation (UPD) of the period-1 solution of $P_+$ occurs at

**Fig. 4.2** Numerical predictions of periodic solutions of the Henon mapping with negative and positive mappings ($a = 0.2$)

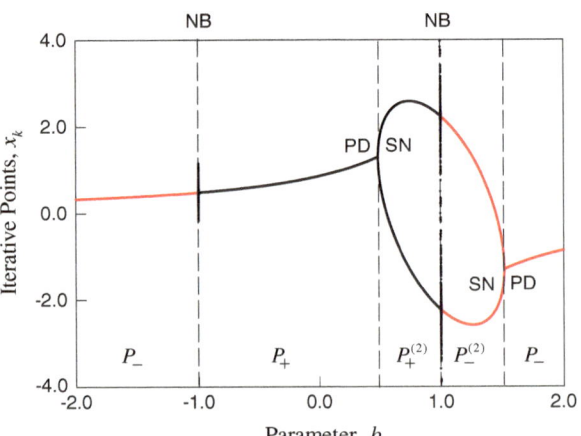

**Fig. 4.3** Analytical predictions of stable and unstable periodic solutions of the Henon map: **a** positive mapping $(P_+)$ and **b** positive mapping $(P_-)$ $(a = 0.2$ and $b \in (-\infty, +\infty))$

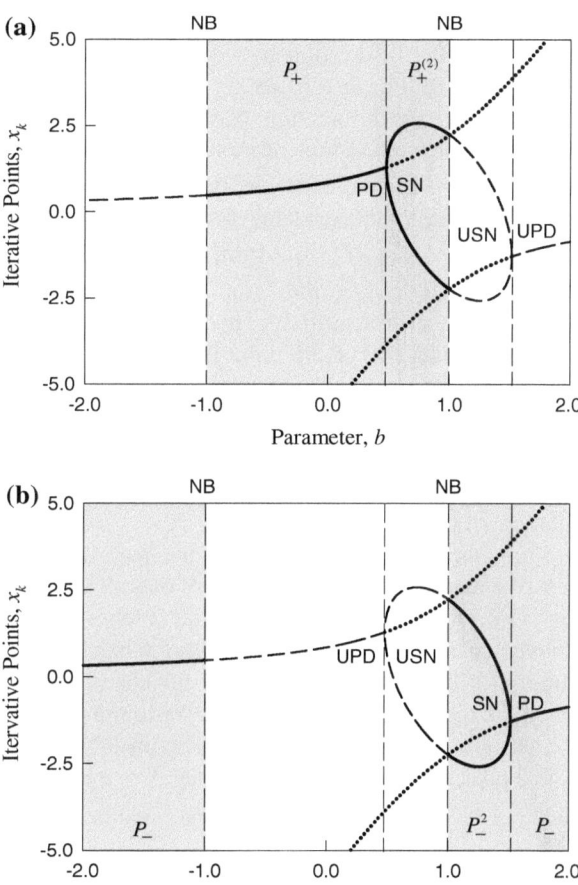

$b \approx 1.5215$. At this point, the unstable periodic solution is from an unstable node to saddle. Because of the unstable period-doubling bifurcation, the unstable periodic solution of $P_+^{(2)}$ is obtained for $b \in (1.0, 1.5215)$. This unstable periodic solution is from unstable focus to unstable node during the parameter of $b \in (1.0, 1.5215)$. At $b \approx 1.5215$, the bifurcation of the unstable periodic solution of $P_+^{(2)}$ occurs between the saddle and unstable node. This bifurcation is called the saddle–unstable node bifurcation. At $b = 1.0$, the NB between the periodic solutions of $P_+^{(2)}$ pertaining to the unstable and stable focuses occurs. The stable periodic solution of $P_+^{(2)}$ is from the stable node to the stable focus for $b \in (0.4805, 1.0)$.

Again, from the eigenvalue analysis, the stable periodic solutions for positive mapping $P_-$ lie in $b \in (-\infty, -1.0)$ and $b \in (1.0, +\infty)$, which is the same as in numerical prediction. The stable period-1 solution of $P_-$ is stable focuses in $b \in (-\infty, -1.0)$ and stable nodes in $b \in (1.5215, +\infty)$. For $b \in (-1.0, 0.4805)$, the unstable period-1 solution of $P_-$ is from the unstable focus to unstable node. At $b = -1$, the bifurcation between the stable and unstable period-1 solution of $P_-$ is the NB. For $b \in (0.4805, +\infty)$, the unstable period-1 solution of $P_-$ is saddle. Thus, the bifurcation between the period-1 solution of $P_-$ between the unstable node and saddle occurs at $b = 0.4805$, which is called the UPD. For $b \in (0.4805, +1)$, the unstable period-2 solution of $P_-$ (i.e., $P_-^{(2)}$) is from the unstable node to the unstable focus. For $b \in (1.0, 1.5215)$, the stable period-2 solution of $P_-$ (i.e., $P_-^{(2)}$) is from the stable focus to the stable nodes. Thus, the point at $b \approx 0.4805$ is the bifurcation of the unstable periodic solution of $P_-^{(2)}$ which is the saddle–unstable node bifurcation between the unstable node and saddle (i.e., USN). For the point at $b = 1$, the NB between the periodic solutions of $P_-^{(2)}$ relative to the unstable and stable focuses occurs. The point at $b \approx 1.5215$ is the bifurcation of the stable periodic solution of $P_-^{(2)}$ which is the saddle bifurcation between the stable node and saddle (SN). For $b \in (-\infty, 1.5215)$, the unstable period-1 solution of $P_-$ is saddle. At $b \approx 1.5215$, the PD of the period-1 solution of $P_-$ takes place.

From the analytical prediction, the parameter maps of both the positive and negative mappings are developed. An overall view of the parameter map is given in Fig. 4.4a. The corresponding periodic solutions are labeled by mapping structures. "None" represents no periodic solutions exist, which means the solution goes to infinity. "Chaos" gives the regions for chaotic solutions. The existing theory can only give the periodic solutions relative to the positive mapping. The coexistence of the periodic solutions is observed. The unstable periodic solutions with saddle will not be presented. The positive and negative mappings are separated by the two NBs at $b = \pm 1$. The zoomed views of the parameter map for periodic solutions of $P_+^{(5)}$ and $P_+^{(7)}$ are presented in Fig. 4.4b, c for better illustration, respectively. The NB of the periodic solution is relative to the unstable and stable focuses, which is presented for a better understanding of the solution switching from positive to negative mappings.

The Poincare mapping relative to the NB of positive (or negative) mapping at $a = 1.1$ and $b = -1$ is presented in Fig. 4.5. Two NBs coexist with different initial conditions. The NB of period-1 solution is presented in Fig. 4.5a, and the initial values of $(x_k, y_k)$ are tabulated in Table 4.1. The most inside point $(x_k^*, y_k^*) \approx (0.4083, -0.4083)$ is the point for the period-1 solution of $P_+$ or $P_-$ relative to the NB. The most outside curve with the initial condition $(x_k^*, y_k^*) \approx (0.5131, -0.4083)$ is the separatrix for the strange attractors around the period-1 solutions with the NB. The NB of period-3 solution is presented in Fig. 4.5b. The initial conditions are listed in Table 4.2. For this case, there are three portions of the strange attractor. The most inside points are $(x_k^*, y_k^*) \approx (-0.2877, -1.1967)$, $(-0.2877, 0.2877)$, and $(1.1966, 0.2877)$ for the period-3 solution of $P_+$ or $P_-$ relative to the NB. The initial condition for three portions of the strange attractor is $(x_k^*, y_k^*) \approx (1.2067, 0.2877)$.

**Fig. 4.4** Parameter map of $(a, b)$: **a** global view, **b** zoomed view for periodic solution of $P_+^{(5)}$, **c** zoomed view for periodic solution of $P_+^{(7)}$

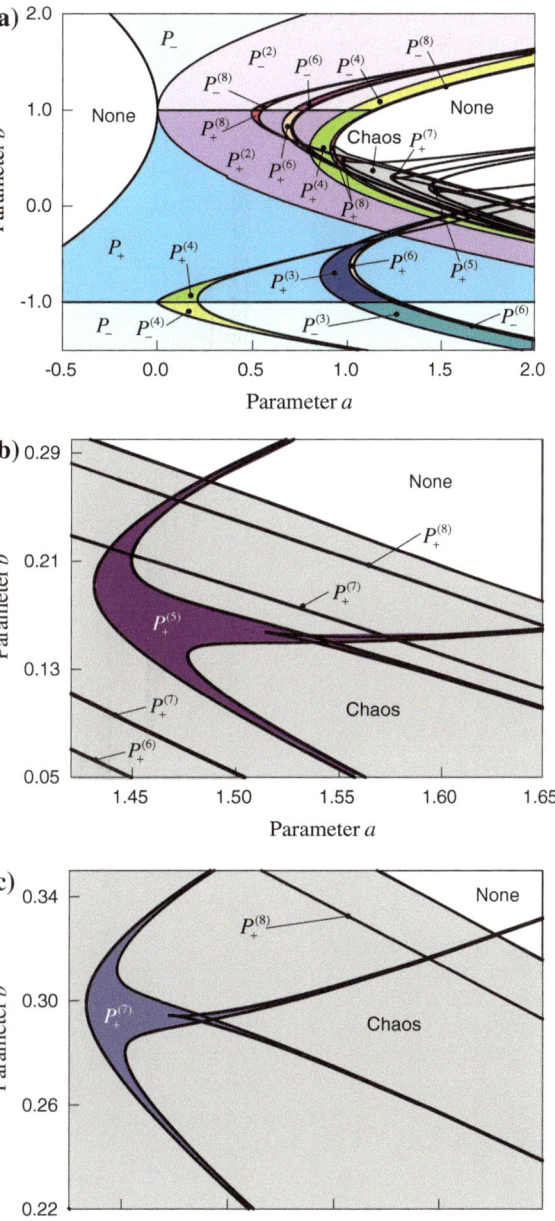

**Fig. 4.5** Poincare mappings at the Neimark bifurcation of period-1 and period-3 solution of the Henon map (i.e., $P_+$-1 or $P_-$-1, $P_+^{(3)}$-1 or $P_-^{(3)}$-1). **a** Neimark bifurcation of period-1 solution, **b** Neimark bifurcation of period-3 solution ($a = 1.1$ and $b = -1$)

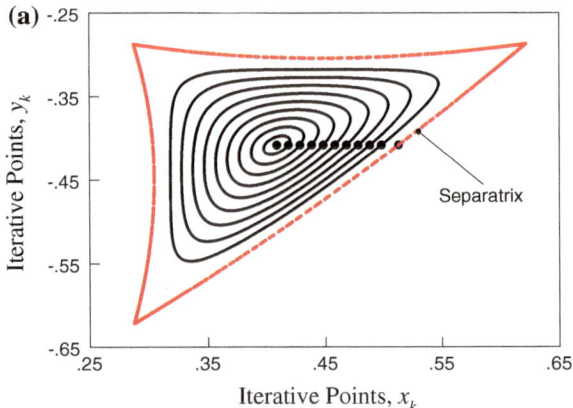

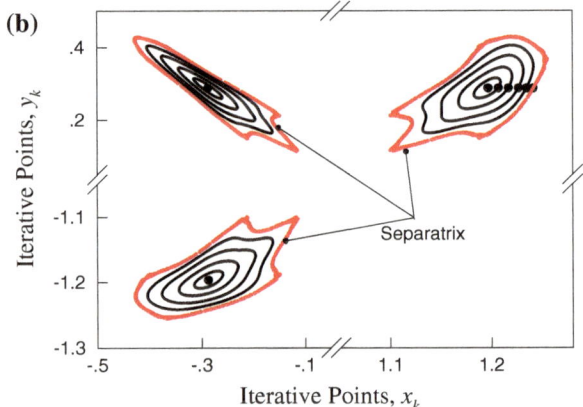

**Table 4.1** Input data for Poincare mappings of period-1 at the Neimark bifurcation ($a = 1.1$ and $b = -1.0$)

| $x_k, y_k$ | $x_k, y_k$ |
|---|---|
| $(0.4083, -0.4083)$ | $(0.4683, -0.4083)$ |
| $(0.4283, -0.4083)$ | $(0.4783, -0.4083)$ |
| $(0.4383, -0.4083)$ | $(0.4883, -0.4083)$ |
| $(0.4483, -0.4083)$ | $(0.4983, -0.4083)$ |
| $(0.4583, -0.4083)$ | $(0.5131, -0.4083)$ |

**Table 4.2** Input data for Poincare mappings of period-3 at the Neimark bifurcation ($a = 1.1$ and $b = -1$)

| $x_k, y_k$ | $x_k, y_k$ |
|---|---|
| $(1.1966, 0.2877)$ | $(1.2267, 0.2877)$ |
| $(1.2067, 0.2877)$ | $(1.2367, 0.2877)$ |
| $(1.2167, 0.2877)$ | $(1.2413, 0.2877)$ |

## 4.4  Multi-step Implicit Maps

**Definition 4.9** Consider an implicit vector functions $\mathbf{f}_j : D \to D$ on an open set $D \subset \mathscr{R}^n$ in an $n$-dimensional discrete dynamical system. For $\mathbf{x}_j \in D$ ($j = k+1, k, \ldots, k-l+1$), there is a discrete relation as

$$\mathbf{f}(\mathbf{x}_{k-l+1}, \ldots, \mathbf{x}_k, \mathbf{x}_{k+1}, \mathbf{p}) = \mathbf{0} \tag{4.45}$$

where the vector function is $\mathbf{f} = (f_1, f_2, \ldots, f_n)^{\mathrm{T}} \in \mathscr{R}^n$ and discrete variable vector is $\mathbf{x}_j = (x_{j1}, x_{j2}, \ldots, x_{jn})^{\mathrm{T}} \in D$ ($j = k+1, k, \ldots, k-l+1$) with a set of parameter vectors $\mathbf{p} = (p_1, p_2, \ldots, p_m)^{\mathrm{T}} \in \mathscr{R}^m$.

**Definition 4.10** For a discrete dynamical system in Eq. (4.45), the positive and negative discrete sets based on a single algebraic relation are defined by

$$
\begin{aligned}
\Sigma_+ &= \left\{ \mathbf{x}_{k+1} \left| \begin{array}{l} \forall \mathbf{x}_{k+1-r} \in \mathscr{R}^n, \quad k \in \mathbb{Z}_+, \quad r = 1, 2, \ldots, l; \\ \mathbf{f}(\mathbf{x}_{k-l+1}, \ldots, \mathbf{x}_k, \mathbf{x}_{k+1}, \mathbf{p}) = \mathbf{0} \end{array} \right. \right\} \subset D, \\
\Sigma_- &= \left\{ \mathbf{x}_{k-r+1} \left| \begin{array}{l} \forall \mathbf{x}_{k-r+2} \in \mathscr{R}^n, \quad k \in \mathbb{Z}_+, \quad r = 1, 2, \ldots, l; \\ \mathbf{f}(\mathbf{x}_{k-l+1}, \ldots, \mathbf{x}_k, \mathbf{x}_{k+1}, \mathbf{p}) = \mathbf{0} \end{array} \right. \right\} \subset D, \\
\Sigma_{r\pm} &= \left\{ \mathbf{x}_{k-r} \left| \begin{array}{l} \forall \mathbf{x}_{k-j+1} \in \mathscr{R}^n, \quad k \in \mathbb{Z}_+, \quad j = 0, 1, \ldots, r-1, r+1, \ldots, l; \\ \mathbf{f}(\mathbf{x}_{k-l+1}, \ldots, \mathbf{x}_k, \mathbf{x}_{k+1}, \mathbf{p}) = \mathbf{0} \end{array} \right. \right\} \subset D
\end{aligned}
\tag{4.46}
$$

respectively. The discrete set is

$$\Sigma = \{ \mathbf{x}_{k+1-r} | \mathbf{x}_{k+1-r} \in \mathscr{R}^n, \quad k \in \mathbb{Z}_+, \quad r = 0, 1, 2, \ldots, l \} \subset D. \tag{4.47}$$

A positive mapping based on the single mapping relation is defined as

$$P_+ : \Sigma \to \Sigma_+ \Rightarrow P_+ : (\mathbf{x}_{k-l+1}, \ldots, \mathbf{x}_{k-1}, \mathbf{x}_k) \to \mathbf{x}_{k+1} \tag{4.48}$$

and a negative mapping on the single mapping relation is defined by

$$P_- : \Sigma \to \Sigma_- \Rightarrow P_- : (\mathbf{x}_{k-l+2}, \ldots, \mathbf{x}_k, \mathbf{x}_{k+1}) \to \mathbf{x}_{k-l+1} \tag{4.49}$$

and a general map on the single mapping relation is defined by

$$P_{r\pm} : \Sigma \to \Sigma_{r\pm} \Rightarrow P_{r\pm} : (\mathbf{x}_{k-l+2}, \ldots, \mathbf{x}_{k-r}, \mathbf{x}_{k+r}, \ldots, \mathbf{x}_{k+1}) \to \mathbf{x}_{k-r}. \tag{4.50}$$

For all given points $(\mathbf{x}_{k-l+1}, \ldots, \mathbf{x}_{k-1}, \mathbf{x}_k)$, the next point $\mathbf{x}_{k+1}$ can be determined by Eq. (4.45). The positive mapping $P_+$ can be defined. For all given points

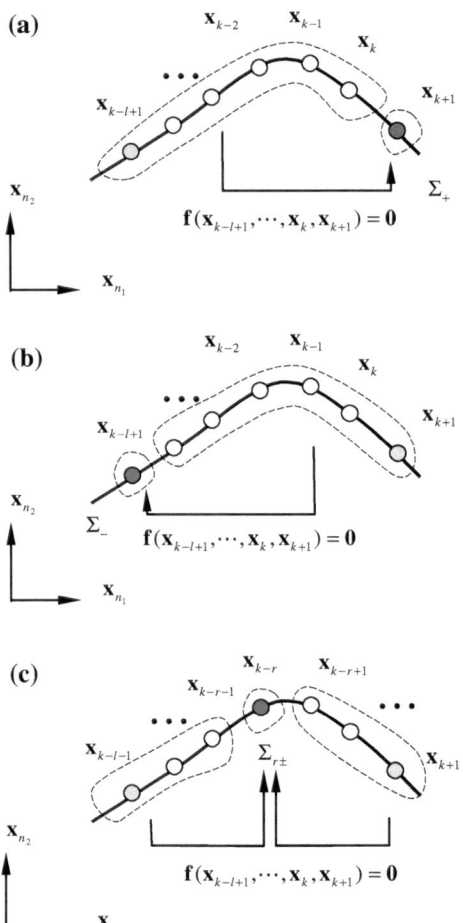

**Fig. 4.6 a** Positive mapping from $(\mathbf{x}_{k-l+1}, \ldots, \mathbf{x}_k)$ to $\mathbf{x}_{k+1}$, and **b** negative mapping from $(\mathbf{x}_{k-l+2}, \ldots, \mathbf{x}_{k+1})$ to $\mathbf{x}_{k-l+1}$, **c** general mapping from $(\mathbf{x}_{k-l+1}, \ldots, \mathbf{x}_{k-r-1}, \mathbf{x}_{k-r+1}, \ldots \mathbf{x}_{k+1})$ to $\mathbf{x}_{k-r}$

$(\mathbf{x}_{k-l+2}, \ldots, \mathbf{x}_k, \mathbf{x}_{k+2})$, the next point $\mathbf{x}_{k-l+1}$ can be determined by Eq. (4.45). The negative mapping $P_-$ can be defined. For a general case, for given points $(\mathbf{x}_{k-l+1}, \ldots, \mathbf{x}_{k-r-1})$ and $(\mathbf{x}_{k-r+1}, \ldots, \mathbf{x}_{k-1}, \mathbf{x}_k)$, any point $\mathbf{x}_r$ can be determined by Eq. (4.45). The mapping $P_{r\pm}$ can be defined. The three mappings are sketched in Fig. 4.6.

For given points, other mapping relations should be added. In a positive way of mappings, one likes to use the following relations for direct iterations,

$$\mathbf{f}_1(\mathbf{x}_{k-l+1}, \mathbf{x}_{k-l+2}, \mathbf{p}) = \mathbf{0},$$
$$\mathbf{f}_2(\mathbf{x}_{k-l+1}, \mathbf{x}_{k-l+2}, \mathbf{x}_{k-l+3}, \mathbf{p}) = \mathbf{0},$$
$$\vdots \qquad\qquad (4.51)$$
$$\mathbf{f}_{l-1}(\mathbf{x}_{k-l+1}, \ldots, \mathbf{x}_{k-1}, \mathbf{x}_k, \mathbf{p}) = \mathbf{0},$$
$$\mathbf{f}_l(\mathbf{x}_{k-l+1}, \ldots, \mathbf{x}_k, \mathbf{x}_{k+1}, \mathbf{p}) = \mathbf{0}.$$

From the above equations, once $\mathbf{x}_{k-l+1}$ is given, one can compute all discrete points, $(\mathbf{x}_{k-l+2}, \ldots, \mathbf{x}_k, \mathbf{x}_{k+1})$, numerically in a positive way. That is, the first equation of Eq. (4.51) computes $\mathbf{x}_{k-l+2}$ and the second equation computes $\mathbf{x}_{k-l+3}$. Continuously, the last equation computes $\mathbf{x}_{k+1}$. In the negative way of mapping relations, one adopts the following relations for direct iteration,

$$\mathbf{f}_1(\mathbf{x}_k, \mathbf{x}_{k+1}, \mathbf{p}) = \mathbf{0},$$
$$\mathbf{f}_2(\mathbf{x}_{k-1}, \mathbf{x}_k, \mathbf{x}_{k+1}, \mathbf{p}) = \mathbf{0},$$
$$\vdots \qquad\qquad (4.52)$$
$$\mathbf{f}_{l-1}(\mathbf{x}_{k-l+2}, \ldots, \mathbf{x}_k, \mathbf{x}_{k+1}, \mathbf{p}) = \mathbf{0},$$
$$\mathbf{f}_l(\mathbf{x}_{k-l+1}, \mathbf{x}_{k-l+2}, \ldots, \mathbf{x}_k, \mathbf{x}_{k+1}, \mathbf{p}) = \mathbf{0}.$$

In the negative way, the first equation of Eq. (4.52) gives $\mathbf{x}_k$ for a given $\mathbf{x}_{k+1}$. The second equation computes $\mathbf{x}_{k-1}$ from the achieved $\mathbf{x}_k$ for a given $\mathbf{x}_{k+1}$. Continuously, the last equation of Eq. (4.52) can compute $\mathbf{x}_{k-l+1}$ from the achieved $\mathbf{x}_{k-l+2}, \ldots, \mathbf{x}_k$ for a given $\mathbf{x}_{k+1}$. All equations in Eqs. (4.51) or (4.52) can be generalized.

$$\mathbf{f}_1(\mathbf{x}_{k-l+1}, \ldots, \mathbf{x}_k, \mathbf{x}_{k+1}, \mathbf{p}) = \mathbf{0},$$
$$\mathbf{f}_2(\mathbf{x}_{k-l+1}, \ldots, \mathbf{x}_k, \mathbf{x}_{k+1}, \mathbf{p}) = \mathbf{0},$$
$$\vdots \qquad\qquad (4.53)$$
$$\mathbf{f}_{l-1}(\mathbf{x}_{k-l+1}, \ldots, \mathbf{x}_k, \mathbf{x}_{k+1}, \mathbf{p}) = \mathbf{0},$$
$$\mathbf{f}_l(\mathbf{x}_{k-l+1}, \ldots, \mathbf{x}_k, \mathbf{x}_{k+1}, \mathbf{p}) = \mathbf{0}.$$

In Eq. (4.53), for given one of $\mathbf{x}_j$ ($j = k-l+1, k-l+2, \ldots, k, k+1$), the rest of them can be computed from $l$-mappings. From the above discussion, we have the following definitions.

**Definition 4.11** Consider implicit vector functions $\mathbf{f}_r : D \to D$ ($r = 1, 2, \ldots, l$) on an open set $D \subset \mathcal{R}^n$ in an $n$-dimensional discrete dynamical system. For $\mathbf{x}_j \in D$ ($j = k+1, k, \ldots, k-l+1$), there is a set of discrete relations as

$$\mathbf{f}_r(\mathbf{x}_{k-l+1}, \ldots, \mathbf{x}_k, \mathbf{x}_{k+1}, \mathbf{p}_r) = \mathbf{0} \quad \text{for } r = 1, 2, \ldots, l \qquad (4.54)$$

**Fig. 4.7 a** positive mapping
and **b** negative mapping. The
mapping relation is presented
by a solid curve. The *circular
symbols* are mapping points
for the positive and negative
mappings through *r*-algebraic
relations

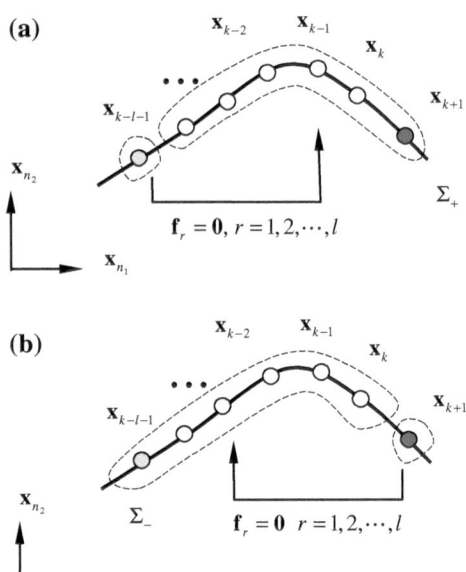

where the vector function is $\mathbf{f}_r = (f_{r1}, f_{r2}, \ldots, f_{rm})^{\mathrm{T}} \in \mathscr{R}^n$ $(r = 1, 2, \ldots, l)$ and discrete variable vector is $\mathbf{x}_j = (x_{j1}, x_{j2}, \ldots, x_{jn})^{\mathrm{T}} \in D$ $(j = k+1, k, \ldots, k-l+1)$ with a set of parameter vectors $\mathbf{p}_r = (p_{r1}, p_{r2}, \ldots, p_{rm})^{\mathrm{T}} \in \mathscr{R}^m$.

The positive and negative mappings based on Eq. (4.54) can be illustrated in Fig. 4.7. As similar to Sect. 4.1, to symbolically describe the discrete dynamical systems, introduce two discrete sets for multi-step mappings.

**Definition 4.12** For a discrete dynamical system in Eq. (4.54), the positive and negative discrete sets are defined by

$$
\begin{aligned}
\Sigma_{+r} &= \left\{ \mathbf{x}_{k-r+2} \left| \begin{array}{l} \mathbf{x}_{k+2-r} \in \mathscr{R}^n, \quad k \in \mathbb{Z}_+, \\ \mathbf{f}_r(\mathbf{x}_{k+1}, \mathbf{x}_k, \ldots, \mathbf{x}_{k-l+1}, \mathbf{p}_r) = \mathbf{0}, \quad r = 1, 2, \ldots, l \end{array} \right. \right\} \subset D, \\
\Sigma_{-r} &= \left\{ \mathbf{x}_{k-r+1} \left| \begin{array}{l} \mathbf{x}_{k-r+1} \in \mathscr{R}^n, \quad k \in \mathbb{Z}_+, \\ \mathbf{f}_r(\mathbf{x}_{k+1}, \mathbf{x}_k, \ldots, \mathbf{x}_{k-l+1}, \mathbf{p}_r) = \mathbf{0}, \quad r = 1, 2, \ldots, l \end{array} \right. \right\} \subset D.
\end{aligned}
\tag{4.55}
$$

respectively. The discrete set is

$$
\Sigma = \cup_{r=0}^{l} \Sigma_{+r} \cup \Sigma_{-r}.
\tag{4.56}
$$

A positive mapping with *l*-algebraic relations is defined as

$$
P_+ : \Sigma \to \Sigma_+ \Rightarrow P_+ : \mathbf{x}_{k-l+1} \to (\mathbf{x}_{k-l+2}, \ldots, \mathbf{x}_k, \mathbf{x}_{k+1})
\tag{4.57}
$$

and a negative mapping with $l$-algebraic relations is defined by

$$P_- : \Sigma \to \Sigma_- \Rightarrow P_- : \mathbf{x}_{k+1} \to (\mathbf{x}_{k-l+1}, \mathbf{x}_{k-l+2}, \ldots, \mathbf{x}_k). \qquad (4.58)$$

From Eq. (4.54), the fixed point based on the positive and negative mappings can be stated as follows.

**Definition 4.13** For a discrete dynamical system in Eq. (4.54), consider two points $\mathbf{x}_{k-l+1} \in D$ and $\mathbf{x}_{k+1} \in D$, and there is a specific, differentiable, vector function $\mathbf{g} \in \mathcal{R}^n$ to make $\mathbf{g}(\mathbf{x}_{k-l+1}, \mathbf{x}_{k+1}, \boldsymbol{\lambda}) = \mathbf{0}$. That is,

$$\begin{aligned}
\mathbf{f}_r(\mathbf{x}_{k-l+1}, \ldots, \mathbf{x}_k, \mathbf{x}_{k+1}, \mathbf{p}_r) &= \mathbf{0} \quad \text{for } r = 1, 2, \ldots, l, \\
\mathbf{g}(\mathbf{x}_{k-l+1}, \mathbf{x}_{k+1}, \boldsymbol{\lambda}) &= \mathbf{0}.
\end{aligned} \qquad (4.59)$$

If a solution $(\mathbf{x}_{k-l+1}^*, \ldots, \mathbf{x}_k^*, \mathbf{x}_{k+1}^*)$ of Eq. (4.59) exists, the corresponding definitions are given as follows:

 (i)   The stable solution based on $\mathbf{x}_{k+1} = P_+ \mathbf{x}_{k-l+1}$ for the positive mapping $P_+$
        with $l$-algebraic relations is called the "Yang" of the discrete dynamical system
        in Eq. (4.54) in the sense of $\mathbf{g}(\mathbf{x}_{k-l+1}, \mathbf{x}_{k+1}, \boldsymbol{\lambda}) = \mathbf{0}$
 (ii)  The stable solution based on $\mathbf{x}_{k-l+1} = P_- \mathbf{x}_{k+1}$ for the negative mapping $P_-$
        with $l$-algebraic relations is called the "Yin" of the discrete dynamical system
        in Eq. (4.54) in the sense of $\mathbf{g}(\mathbf{x}_{k-l+1}, \mathbf{x}_{k+1}, \boldsymbol{\lambda}) = \mathbf{0}$.
 (iii) The solution based on $\mathbf{x}_{k+1} = P_+ \mathbf{x}_{k-l+1}$ is called "Yin–Yang" for the positive
        mapping $P_+$ with $l$-algebraic relations in the discrete dynamical system of
        Eq. (4.54) in the sense of $\mathbf{g}(\mathbf{x}_{k-l+1}, \mathbf{x}_{k+1}, \boldsymbol{\lambda}) = \mathbf{0}$ once the eigenvalues of
        $DP_+(\mathbf{x}_{k-l+1}^*) = [\partial \mathbf{x}_{k+1}/\partial \mathbf{x}_{k-l+1}]_{(\mathbf{x}_{k-l+1}^*, \ldots, \mathbf{x}_k^*, \mathbf{x}_{k+1}^*)}$, are distributed inside and out-
        side the unit cycle.
 (iv)  The solution based on $\mathbf{x}_{k-l+1} = P_- \mathbf{x}_{k+1}$ is called the "Yin–Yang" for the
        negative mapping $P_-$ with $l$-algebraic relations in the discrete dynamical
        system of Eq. (4.54) in the sense of $\mathbf{g}(\mathbf{x}_{k-l+1}, \mathbf{x}_{k+1}, \boldsymbol{\lambda}) = \mathbf{0}$ once the eigen-
        values of $DP_-(\mathbf{x}_{k+1}^*) = [\partial \mathbf{x}_{k-l+1}/\partial \mathbf{x}_{k+1}]_{(\mathbf{x}_{k-l+1}^*, \ldots, \mathbf{x}_k^*, \mathbf{x}_{k+1}^*)}$, are distributed inside
        and outside unit cycle.

Consider that the positive and negative mappings are

$$\mathbf{x}_{k+1} = P_+ \mathbf{x}_{k-l+1} \quad \text{and} \quad \mathbf{x}_{k-l+1} = P_- \mathbf{x}_{k+1}. \qquad (4.60)$$

For the simplest case, consider the constraint condition of $\mathbf{g}(\mathbf{x}_{k-l+1}, \mathbf{x}_{k+1}, \boldsymbol{\lambda}) = \mathbf{x}_{k+1} - \mathbf{x}_{k-l+1} = \mathbf{0}$. Thus, the positive and negative mappings have, respectively, the constraints

$$\mathbf{x}_{k+1} = \mathbf{x}_{k-l+1}. \qquad (4.61)$$

Both positive and negative mappings are governed by the discrete relation in Eq. (4.54). In other words, Eq. (4.54) gives

$$\mathbf{f}_1(\mathbf{x}_{k-l+1}, \ldots, \mathbf{x}_k, \mathbf{x}_{k+1}, \mathbf{p}_1) = \mathbf{0},$$
$$\mathbf{f}_2(\mathbf{x}_{k-l+1}, \ldots, \mathbf{x}_k, \mathbf{x}_{k+1}, \mathbf{p}_2) = \mathbf{0},$$
$$\vdots \qquad\qquad (4.62)$$
$$\mathbf{f}_l(\mathbf{x}_{k-l+1}, \ldots, \mathbf{x}_k, \mathbf{x}_{k+1}, \mathbf{p}_l) = \mathbf{0}$$

and

$$\mathbf{f}_l(\mathbf{x}_{k-l+1}, \ldots, \mathbf{x}_k, \mathbf{x}_{k+1}, \mathbf{p}_l) = \mathbf{0},$$
$$\vdots \qquad\qquad (4.63)$$
$$\mathbf{f}_2(\mathbf{x}_{k-l+1}, \ldots, \mathbf{x}_k, \mathbf{x}_{k+1}, \mathbf{p}_2) = \mathbf{0},$$
$$\mathbf{f}_1(\mathbf{x}_{k-l+1}, \ldots, \mathbf{x}_k, \mathbf{x}_{k+1}, \mathbf{p}_1) = \mathbf{0}.$$

Setting the period-1 solution $\mathbf{x}_k^*, \mathbf{x}_{k-1}^*, \ldots, \mathbf{x}_{k-l+1}^*$ and substitution of Eq. (4.61) into Eqs. (4.62) and (4.63) give

$$\mathbf{f}_1(\mathbf{x}_{k-l+1}^*, \ldots, \mathbf{x}_k^*, \mathbf{x}_{k-l+1}^*, \mathbf{p}_1) = \mathbf{0},$$
$$\mathbf{f}_2(\mathbf{x}_{k-l+1}^*, \ldots, \mathbf{x}_k^*, \mathbf{x}_{k-l+1}^*, \mathbf{p}_2) = \mathbf{0},$$
$$\vdots \qquad\qquad (4.64)$$
$$\mathbf{f}_l(\mathbf{x}_{k-l+1}^*, \ldots, \mathbf{x}_k^*, \mathbf{x}_{k-l+1}^*, \mathbf{p}_l) = \mathbf{0}$$

and

$$\mathbf{f}_l(\mathbf{x}_{k-l+1}^*, \ldots, \mathbf{x}_k^*, \mathbf{x}_{k-l+1}^*, \mathbf{p}_l) = \mathbf{0},$$
$$\vdots \qquad\qquad (4.65)$$
$$\mathbf{f}_2(\mathbf{x}_{k-l+1}^*, \ldots, \mathbf{x}_k^*, \mathbf{x}_{k-l+1}^*, \mathbf{p}_2) = \mathbf{0},$$
$$\mathbf{f}_1(\mathbf{x}_{k-l+1}^*, \ldots, \mathbf{x}_k^*, \mathbf{x}_{k-l+1}^*, \mathbf{p}_1) = \mathbf{0}.$$

The stability and bifurcation of fixed points for the positive and negative mappings can be stated as follows:

**Theorem 4.6** *For a discrete dynamical system in Eq. (4.54), there are two points* $\mathbf{x}_{k-l+1} \in D$ *and* $\mathbf{x}_{k+1} \in D$, *and two positive and negative mappings are*

$$\mathbf{x}_{k+1} = P_+\mathbf{x}_{k-l+1} \quad and \quad \mathbf{x}_{k-l+1} = P_-\mathbf{x}_{k+1} \qquad (4.66)$$

*with*

$$\mathbf{f}_r(\mathbf{x}_{k-l+1}, \ldots, \mathbf{x}_k, \mathbf{x}_{k+1}, \mathbf{p}_r) = \mathbf{0}, \quad r = 1, 2, \ldots, l, \qquad (4.67)$$

$$\mathbf{f}_r(\mathbf{x}_{k-l+1}, \ldots, \mathbf{x}_k, \mathbf{x}_{k+1}, \mathbf{p}_r) = \mathbf{0}, \quad r = l, l-1, \ldots, 1. \tag{4.68}$$

*Suppose there is a specific, differentiable, vector function* $\mathbf{g} \in \mathcal{R}^n$ *to make* $\mathbf{g}(\mathbf{x}_{k-l+1}, \mathbf{x}_{k+1}, \lambda) = \mathbf{0}$. *If the solutions* $(\mathbf{x}_{k-l+1}^*, \ldots, \mathbf{x}_k^*, \mathbf{x}_{k+1}^*)$ *of* $\mathbf{g}(\mathbf{x}_{k-l+1}, \mathbf{x}_{k+1}, \lambda) = \mathbf{0}$ *and* $\mathbf{f}_r(\mathbf{x}_{k-l+1}, \ldots, \mathbf{x}_k, \mathbf{x}_{k+1}, \mathbf{p}_r) = \mathbf{0}$ $(r = 1, 2, \ldots, l)$ *exist, then the following conclusions in the sense of* $\mathbf{g}(\mathbf{x}_{k-l+1}, \mathbf{x}_{k+1}, \lambda) = \mathbf{0}$ *hold.*

(i) *The stable* $P_+$-1 *solutions are the unstable* $P_-$-1 *solutions with all eigenvalues of* $DP_-(\mathbf{x}_{k+1}^*) = [\partial\mathbf{x}_{k-l+1}/\partial\mathbf{x}_{k+1}]_{(\mathbf{x}_{k-l+1}^*, \ldots, \mathbf{x}_k^*, \mathbf{x}_{k+1}^*)}$, *outside the unit cycle, vice versa.*

(ii) *The unstable* $P_+$-1 *solutions with all eigenvalues of* $DP_+(\mathbf{x}_{k-l+1}^*) = [\partial\mathbf{x}_{k+1}/\partial\mathbf{x}_{k-l+1}]_{(\mathbf{x}_{k-l+1}^*, \ldots, \mathbf{x}_k^*, \mathbf{x}_{k+1}^*)}$, *outside the unit cycle are the stable* $P_-$-1 *solutions, vice versa.*

(iii) *For the unstable* $P_+$-1 *solutions with eigenvalue distribution of* $DP_+(\mathbf{x}_{k-l+1}^*) = [\partial\mathbf{x}_{k+1}/\partial\mathbf{x}_{k-l+1}]_{(\mathbf{x}_{k-l+1}^*, \ldots, \mathbf{x}_k^*, \mathbf{x}_{k+1}^*)}$, *inside and outside the unit cycle, the corresponding* $P_-$-1 *solution is also unstable with switching the eigenvalue distribution of* $DP_-(\mathbf{x}_{k+1}^*) = [\partial\mathbf{x}_{k-l+1}/\partial\mathbf{x}_{k+1}]_{(\mathbf{x}_{k-l+1}^*, \ldots, \mathbf{x}_k^*, \mathbf{x}_{k+1}^*)}$, *outside and inside the unit cycle, vice versa.*

(iv) *All the bifurcations of the* stable *and* unstable $P_+$-1 *solutions are all the bifurcations of the* unstable *and* stable $P_-$-1 *solutions, respectively.*

*Proof* Consider the positive and negative mappings with relations in Eq. (4.54). The periodic solution in the sense of $\mathbf{g}(\mathbf{x}_{k-l+1}, \mathbf{x}_{k+1}, \lambda) = \mathbf{0}$ is given by

$$\mathbf{f}_1(\mathbf{x}_{k-l+1}, \ldots, \mathbf{x}_k, \mathbf{x}_{k+1}, \mathbf{p}_1) = \mathbf{0},$$
$$\mathbf{f}_2(\mathbf{x}_{k-l+1}, \ldots, \mathbf{x}_k, \mathbf{x}_{k+1}, \mathbf{p}_2) = \mathbf{0},$$
$$\vdots$$
$$\mathbf{f}_l(\mathbf{x}_{k-l+1}, \ldots, \mathbf{x}_k, \mathbf{x}_{k+1}, \mathbf{p}_l) = \mathbf{0}$$

and

$$\mathbf{f}_l(\mathbf{x}_{k-l+1}, \ldots, \mathbf{x}_k, \mathbf{x}_{k+1}, \mathbf{p}_l) = \mathbf{0},$$
$$\vdots$$
$$\mathbf{f}_2(\mathbf{x}_{k-l+1}, \ldots, \mathbf{x}_k, \mathbf{x}_{k+1}, \mathbf{p}_2) = \mathbf{0},$$
$$\mathbf{f}_1(\mathbf{x}_{k-l+1}, \ldots, \mathbf{x}_k, \mathbf{x}_{k+1}, \mathbf{p}_1) = \mathbf{0}$$

from which the fixed points $(\mathbf{x}_{k-l+1}^*, \ldots, \mathbf{x}_k^*, \mathbf{x}_{k+1}^*)$ can be determined. Consider a small perturbation

$$\mathbf{x}_{k-l+1} = \mathbf{x}^*_{k-l+1} + \delta\mathbf{x}_{k-l+1}, \ldots, \ \mathbf{x}_k = \mathbf{x}^*_k + \delta\mathbf{x}_k, \quad \text{and} \quad \mathbf{x}_{k+1} = \mathbf{x}^*_{k+1} + \delta\mathbf{x}_{k+1}.$$

The linearization of mappings in Eq. (4.66) gives

$$\delta\mathbf{x}_{k+1} = DP_+(\mathbf{x}^*_{k-l+1})\delta\mathbf{x}_{k-l+1} \quad \text{and} \quad \delta\mathbf{x}_{k-l+1} = DP_-(\mathbf{x}^*_{k+1})\delta\mathbf{x}_{k+1}$$

where

$$DP_+(\mathbf{x}^*_{k-l+1}) = \left[\frac{\partial\mathbf{x}_{k+1}}{\partial\mathbf{x}_{k-l+1}}\right]_{(\mathbf{x}^*_{k-l+1},\ldots,\mathbf{x}^*_k,\mathbf{x}^*_{k+1})},$$

$$DP_-(\mathbf{x}^*_{k+1}) = \left[\frac{\partial\mathbf{x}_{k-l+1}}{\partial\mathbf{x}_{k+1}}\right]_{(\mathbf{x}^*_{k-l+1},\ldots,\mathbf{x}^*_k,\mathbf{x}^*_{k+1})}.$$

From Eqs. (4.67) and (4.68), one obtains

$$\left(\left[\frac{\partial\mathbf{f}_r}{\partial\mathbf{x}_{k-l+1}}\right] + \sum_{s=1}^{l}\left[\frac{\partial\mathbf{f}_r}{\partial\mathbf{x}_{k-s+2}}\right]\left[\frac{\partial\mathbf{x}_{k-s+2}}{\partial\mathbf{x}_{k-l+1}}\right]\right)_{(\mathbf{x}^*_{k-l+1},\ldots,\mathbf{x}^*_k,\mathbf{x}^*_{k+1})} = \mathbf{0}, \quad r = 1, 2, \ldots, l,$$

$$\mathbf{g}(\mathbf{x}^*_{k-l+1}, \mathbf{x}^*_{k+1}, \boldsymbol{\lambda}) = \mathbf{g}(\mathbf{x}_{k-l+1}, \mathbf{x}_{k+1}, \boldsymbol{\lambda}) = \mathbf{0};$$

$$\left(\left[\frac{\partial\mathbf{f}_r}{\partial\mathbf{x}_{k+1}}\right] + \sum_{s=1}^{l}\left[\frac{\partial\mathbf{f}_r}{\partial\mathbf{x}_{k-s+1}}\right]\left[\frac{\partial\mathbf{x}_{k-s+1}}{\partial\mathbf{x}_{k+1}}\right]\right)_{(\mathbf{x}^*_{k-l+1},\ldots,\mathbf{x}^*_k,\mathbf{x}^*_{k+1})} = \mathbf{0}, \quad r = l, l-1, \ldots, 1,$$

$$\mathbf{g}(\mathbf{x}^*_{k-l+1}, \mathbf{x}^*_{k+1}, \boldsymbol{\lambda}) = \mathbf{g}(\mathbf{x}_{k-l+1}, \mathbf{x}_{k+1}, \boldsymbol{\lambda}) = \mathbf{0}.$$

That is,

$$\left\{\begin{array}{c} \dfrac{\partial\mathbf{x}_{k-l+2}}{\partial\mathbf{x}_{k-l+1}} \\ \vdots \\ \dfrac{\partial\mathbf{x}_k}{\partial\mathbf{x}_{k-l+1}} \\ \dfrac{\partial\mathbf{x}_{k+1}}{\partial\mathbf{x}_{k-l+1}} \end{array}\right\} = -\left[\begin{array}{cccc} \dfrac{\partial\mathbf{f}_1}{\partial\mathbf{x}_{k-l+2}} & \cdots & \dfrac{\partial\mathbf{f}_1}{\partial\mathbf{x}_k} & \dfrac{\partial\mathbf{f}_1}{\partial\mathbf{x}_{k+1}} \\ \vdots & & \vdots & \vdots \\ \dfrac{\partial\mathbf{f}_{l-1}}{\partial\mathbf{x}_{k-l+2}} & \cdots & \dfrac{\partial\mathbf{f}_{l-1}}{\partial\mathbf{x}_k} & \dfrac{\partial\mathbf{f}_{l-1}}{\partial\mathbf{x}_{k+1}} \\ \dfrac{\partial\mathbf{f}_l}{\partial\mathbf{x}_{k-l+2}} & \cdots & \dfrac{\partial\mathbf{f}_l}{\partial\mathbf{x}_k} & \dfrac{\partial\mathbf{f}_l}{\partial\mathbf{x}_{k+1}} \end{array}\right]^{-1}\left\{\begin{array}{c} \dfrac{\partial\mathbf{f}_1}{\partial\mathbf{x}_{k-l+1}} \\ \vdots \\ \dfrac{\partial\mathbf{f}_{l-1}}{\partial\mathbf{x}_{k-l+1}} \\ \dfrac{\partial\mathbf{f}_1}{\partial\mathbf{x}_{k-l+2}} \end{array}\right\},$$

$$
\left\{
\begin{array}{c}
\dfrac{\partial \mathbf{x}_{k-l+2}}{\partial \mathbf{x}_{k+1}} \\[2ex]
\vdots \\[1ex]
\dfrac{\partial \mathbf{x}_{k}}{\partial \mathbf{x}_{k+1}} \\[2ex]
\dfrac{\partial \mathbf{x}_{k-l+1}}{\partial \mathbf{x}_{k+1}}
\end{array}
\right\}
= -
\begin{bmatrix}
\dfrac{\partial \mathbf{f}_{1}}{\partial \mathbf{x}_{k-l+2}} & \cdots & \dfrac{\partial \mathbf{f}_{1}}{\partial \mathbf{x}_{k}} & \dfrac{\partial \mathbf{f}_{1}}{\partial \mathbf{x}_{k-l+1}} \\[2ex]
\vdots & & \vdots & \vdots \\[1ex]
\dfrac{\partial \mathbf{f}_{l-1}}{\partial \mathbf{x}_{k-l+2}} & \cdots & \dfrac{\partial \mathbf{f}_{l-1}}{\partial \mathbf{x}_{k}} & \dfrac{\partial \mathbf{f}_{l-1}}{\partial \mathbf{x}_{k-l+1}} \\[2ex]
\dfrac{\partial \mathbf{f}_{l}}{\partial \mathbf{x}_{k-l+2}} & \cdots & \dfrac{\partial \mathbf{f}_{l}}{\partial \mathbf{x}_{k}} & \dfrac{\partial \mathbf{f}_{l}}{\partial \mathbf{x}_{k-l+1}}
\end{bmatrix}^{-1}
\left(
\begin{array}{c}
\dfrac{\partial \mathbf{f}_{1}}{\partial \mathbf{x}_{k+1}} \\[2ex]
\vdots \\[1ex]
\dfrac{\partial \mathbf{f}_{l-1}}{\partial \mathbf{x}_{k+1}} \\[2ex]
\dfrac{\partial \mathbf{f}_{1}}{\partial \mathbf{x}_{k+1}}
\end{array}
\right).
$$

From two equations, we obtain

$$
DP_{+}(\mathbf{x}_{k-l+1}^{*}) = \left[ \frac{\partial \mathbf{x}_{k+1}}{\partial \mathbf{x}_{k-l+1}} \right]\Big|_{(\mathbf{x}_{k-l+1}^{*}, \cdots, \mathbf{x}_{k}^{*}, \mathbf{x}_{k+1}^{*})},
$$

$$
DP_{-}(\mathbf{x}_{k+1}^{*}) = \left[ \frac{\partial \mathbf{x}_{k-l+1}}{\partial \mathbf{x}_{k+1}} \right]\Big|_{(\mathbf{x}_{k-l+1}^{*}, \cdots, \mathbf{x}_{k}^{*}, \mathbf{x}_{k+1}^{*})}.
$$

Since

$$
\left[ \frac{\partial \mathbf{x}_{k+1}}{\partial \mathbf{x}_{k-l+1}} \right] \cdot \left[ \frac{\partial \mathbf{x}_{k-l+1}}{\partial \mathbf{x}_{k+1}} \right]\Big|_{(\mathbf{x}_{k-l+1}^{*}, \cdots, \mathbf{x}_{k}^{*}, \mathbf{x}_{k+1}^{*})} = \mathbf{I}_{n \times n},
$$

we have

$$
DP_{+}(\mathbf{x}_{k-l+1}^{*}) \cdot DP_{-}(\mathbf{x}_{k+1}^{*}) = \mathbf{I}_{n \times n}.
$$

Therefore, one obtains

$$
DP_{-}^{-1}(\mathbf{x}_{k+1}^{*}) = DP_{+}(\mathbf{x}_{k-l+1}^{*}).
$$

In other words, $DP_{+}(\mathbf{x}_{k-l+1}^{*})$ is the inverse of $DP_{-}(\mathbf{x}_{k+!}^{*})$.

For eigenvalues $\lambda_{-}$ and $\lambda_{+}$ of $DP_{-}(\mathbf{x}_{k+1}^{*})$ and $DP_{+}(\mathbf{x}_{k-l+1}^{*})$ accordingly, we have

$$
(DP_{-}(\mathbf{x}_{k+1}^{*}) - \lambda_{-}\mathbf{I})\delta \mathbf{x}_{k+1} = \mathbf{0},
$$
$$
(DP_{+}(\mathbf{x}_{k-l+1}^{*}) - \lambda_{+}\mathbf{I})\delta \mathbf{x}_{k-l+1} = \mathbf{0}.
$$

Left multiplication of $DP_{+}(\mathbf{x}_{k-l+1}^{*})$ in the first equation of the foregoing equation, division of $\lambda_{-}$ on both sides, and application of $DP_{-}^{-1}(\mathbf{x}_{k+1}^{*}) = DP_{+}(\mathbf{x}_{k-l+1}^{*})$ give

$$
[DP_{+}(\mathbf{x}_{k-l+1}^{*}) - \lambda_{-}^{-1}\mathbf{I}]\delta \mathbf{x}_{k+1} = \mathbf{0}.
$$

Thus, one can obtain

$$\lambda_+ = \lambda_-^{-1}.$$

From the stability and bifurcation theory for $P_+$−1 and $P_-$−1 solutions for discrete dynamical system in Eq. (4.54), the following conclusions can be given as follows:

(i)  The stable $P_+$-1 solutions are the unstable $P_-$-1 solutions with all eigenvalues of $DP_-(\mathbf{x}_{k+1}^*)$ outside the unit cycle, vice versa.

(ii)  The unstable $P_+$-1 solutions with all eigenvalues of $DP_+(\mathbf{x}_{k-l+1}^*)$ outside the unit cycle are the stable $P_-$-1 solutions, vice versa.

(iii)  For the unstable $P_+$-1 solutions with eigenvalue distribution of $DP_+(\mathbf{x}_{k-l+1}^*)$ inside and outside the unit cycle, the corresponding $P_-$-1 solution is also unstable with switching the eigenvalue distribution of $DP_-(\mathbf{x}_{k+1}^*)$ inside and outside the unit cycle, vice versa.

(iv)  All the bifurcations of the *stable* and *unstable* $P_+$-1 solutions are all the bifurcations of the *unstable* and *stable* $P_-$-1 solutions, respectively.

This theorem is proved.                                                                    $\square$

From the foregoing theorem, the *Yin*, *Yang*, and *Yin–Yang* states in discrete dynamical systems exist. To generate the above ideas to $P_+^{(N)}$-1 and $P_-^{(N)}$-1 solutions in discrete dynamical systems in the sense of $\mathbf{g}(\mathbf{x}_k, \mathbf{x}_{k+lN}, \lambda) = \mathbf{0}$, the mapping structure consisting of $N$-positive or negative mappings is considered.

**Definition 4.14** For a discrete dynamical system in Eq. (4.54), the mapping structures of $N$-mappings for the positive and negative mappings are defined as

$$\mathbf{x}_{k+l(N-1)+1} = \underbrace{P_+ \circ P_+ \circ \cdots \circ P_+}_{N} \mathbf{x}_{k-l+1} = P_+^{(N)} \mathbf{x}_{k-l+1}, \qquad (4.69)$$

$$\mathbf{x}_{k-l+1} = \underbrace{P_- \circ P_- \circ \cdots \circ P_-}_{N} \mathbf{x}_{k+lN} = P_-^{(N)} \mathbf{x}_{k+(N-1)l+1}. \qquad (4.70)$$

with

$$\mathbf{f}_r(\mathbf{x}_{k+(i-1)l+1}, \ldots, \mathbf{x}_{k+il}, \mathbf{x}_{k+il+1}, \mathbf{p}_r) = \mathbf{0} \quad (r = 1, 2, \ldots, l),$$
$$\text{for } i = 0, 1, 2, \ldots, N-1, \qquad (4.71)$$

where $P_+^{(0)} = \mathbf{I}_{n \times n}$ and $P_-^{(0)} = \mathbf{I}_{n \times n}$ for $N = 0$.

**Definition 4.15** For a discrete dynamical system in Eq. (4.54), consider two points $\mathbf{x}_{k-l+1} \in D$ and $\mathbf{x}_{k+(N-1)l+1} \in D$, and there is a specific, differentiable, vector function $\mathbf{g} \in \mathscr{R}^n$ to make $\mathbf{g}(\mathbf{x}_{k-l+1}, \mathbf{x}_{k+(N-1)l+1}, \boldsymbol{\lambda}) = \mathbf{0}$.

(i) The stable solution of $\mathbf{x}_{k+(N-1)l+1} = P_+^{(N)} \mathbf{x}_{k-l+1}$ for the positive mapping $P_+$ is called the "Yang" of the discrete dynamical system in Eq. (4.54) in the sense of $\mathbf{g}(\mathbf{x}_{k-l+1}, \mathbf{x}_{k+(N-1)l+1}, \boldsymbol{\lambda}) = \mathbf{0}$ if the solutions $(\mathbf{x}_{k-l+1}^*, \mathbf{x}_{k-l+2}^*, \ldots, \mathbf{x}_{k+(N-1)l+1}^*)$ of Eq. (4.71) with $\mathbf{g}(\mathbf{x}_{k-l+1}, \mathbf{x}_{k+(N-1)l+1}, \boldsymbol{\lambda}) = \mathbf{0}$ exist.

(ii) The stable solution of $\mathbf{x}_{k-l+1} = P_-^{(N)} \mathbf{x}_{k+(N-1)l+1}$ for the negative mapping $P_-$ is called "Yin" of the discrete dynamical system in Eq. (4.54) in the sense of $\mathbf{g}(\mathbf{x}_{k-l+1}, \mathbf{x}_{k+(N-1)l+1}, \boldsymbol{\lambda}) = \mathbf{0}$ if the solutions $(\mathbf{x}_{k-l+1}^*, \mathbf{x}_{k-l+2}^*, \ldots, \mathbf{x}_{k+(N-1)l+1}^*)$ of Eq. (4.71) with $\mathbf{g}(\mathbf{x}_{k-l+1}, \mathbf{x}_{k+(N-1)l+1}, \boldsymbol{\lambda}) = \mathbf{0}$ exist.

(iii) The solution of $\mathbf{x}_{k+(N-1)l+1} = P_+^{(N)} \mathbf{x}_{k-l+1}$ is called "Yin–Yang" for the negative mapping $P_+$ of the discrete dynamical system in Eq. (4.54) in the sense of $\mathbf{g}(\mathbf{x}_{k-l+1}, \mathbf{x}_{k+(N-1)l+1}, \boldsymbol{\lambda}) = \mathbf{0}$ if the solutions $(\mathbf{x}_{k-l+1}^*, \mathbf{x}_{k-l+2}^*, \ldots, \mathbf{x}_{k+(N-1)l+1}^*)$ of Eq. (4.71) with $\mathbf{g}(\mathbf{x}_{k-l+1}, \mathbf{x}_{k+(N-1)l+1}, \boldsymbol{\lambda}) = \mathbf{0}$ exist and the eigenvalues of $DP_+^{(N)}(\mathbf{x}_{k-l+1}^*)$ are distributed inside and outside the unit cycle.

(iv) The solution of $\mathbf{x}_{k-l+1} = P_-^{(N)} \mathbf{x}_{k+(N-1)l+1}$ is called "Yin–Yang" for the negative mapping $P_-$ of the discrete dynamical system in Eq. (4.54) in the sense of $\mathbf{g}(\mathbf{x}_{k-l+1}, \mathbf{x}_{k+(N-1)l+1}, \boldsymbol{\lambda}) = \mathbf{0}$ if the solutions $(\mathbf{x}_{k-l+1}^*, \mathbf{x}_{k-l+2}^*, \ldots, \mathbf{x}_{k+(N-1)l+1}^*)$ of Eq. (4.71) with $\mathbf{g}(\mathbf{x}_{k-l+1}, \mathbf{x}_{k+(N-1)l+1}, \boldsymbol{\lambda}) = \mathbf{0}$ exist and the eigenvalues of $DP_-^{(N)}(\mathbf{x}_{k+(N-1)l+1}^*)$ are distributed inside and outside unit cycle.

Similarly, to determine the Yin–Yang properties of $P_+^{(N)}$-1 and $P_-^{(N)}$-1 in the discrete mapping system with multi-step mappings in Eq. (4.54), the corresponding theorem is stated as follows.

**Theorem 4.7** *For a discrete dynamical system in Eq. (4.54), there are two points $\mathbf{x}_{k-l+1} \in D$ and $\mathbf{x}_{k+(N-1)l+1} \in D$, and two positive and negative mappings are*

$$\mathbf{x}_{k+(N-1)l+1} = P_+^{(N)} \mathbf{x}_{k-l+1} \quad and \quad \mathbf{x}_{k-l+1} = P_-^{(N)} \mathbf{x}_{k+(N-1)l+1}. \tag{4.72}$$

*and $\mathbf{x}_{k+il+1} = P_+ \mathbf{x}_{k+(i-1)l+1}$ and $\mathbf{x}_{k+(i-1)l+1} = P_- \mathbf{x}_{k+il+1}$ can be governed by*

$$\mathbf{f}_r(\mathbf{x}_{k+(i-1)l+1}, \ldots, \mathbf{x}_{k+il}, \mathbf{x}_{k+il+1}, \mathbf{p}_r) = \mathbf{0} \quad (r = 1, 2, \ldots, l)$$
$$for \; i = 0, 1, 2, \ldots, N-1; \tag{4.73}$$

$$\mathbf{f}_r(\mathbf{x}_{k+(i-1)l+1}, \ldots, \mathbf{x}_{k+il}, \mathbf{x}_{k+il+1}, \mathbf{p}_r) = \mathbf{0} \quad (r = l, l-1, \ldots, 1)$$
$$for \; i = N-1, N-2, \ldots, 0. \tag{4.74}$$

*Suppose there is a specific, differentiable, vector function of* $\mathbf{g} \in \mathscr{R}^n$ *to make* $\mathbf{g}(\mathbf{x}_{k-l+1}, \mathbf{x}_{k+(N-1)l+1}, \boldsymbol{\lambda}) = \mathbf{0}$ *hold. If the solutions* $(\mathbf{x}^*_{k-l+1}, \mathbf{x}^*_{k-l+2}, \cdots, \mathbf{x}^*_{k+(N-1)l+1})$ *of* Eq. (4.73) *or* (4.74) *with* $\mathbf{g}(\mathbf{x}_{k-l+1}, \mathbf{x}_{k+(N-1)l+1}, \boldsymbol{\lambda}) = \mathbf{0}$ *exist, then the following conclusions in the sense of* $\mathbf{g}(\mathbf{x}_{k-l+1}, \mathbf{x}_{k+(N-1)l+1}, \boldsymbol{\lambda}) = \mathbf{0}$ *hold.*

(i) *The stable* $P^{(N)}_+$-1 *solution is the unstable* $P^{(N)}_-$-1 *solutions with all eigenvalues of* $DP^{(N)}_+(\mathbf{x}^*_{k+(N-1)l+1})$ *outside the unit cycle, vice versa.*

(ii) *The unstable* $P^{(N)}_+$-1 *solutions with all eigenvalues of* $DP^{(N)}_+(\mathbf{x}^*_{k-l+1})$ *outside the unit cycle are the stable* $P^{(N)}_-$-1 *solutions with all eigenvalues of* $DP^{(N)}_+(\mathbf{x}^*_{k+(N-1)l+1})$ *inside the unit cycle, vice versa.*

(iii) *For the unstable* $P^{(N)}_+$-1 *solution with eigenvalue distribution of* $DP^{(N)}_+(\mathbf{x}^*_{k-l+1})$ *inside and outside the unit cycle, the corresponding* $P^{(N)}_-$-1 *solution is also unstable with switching eigenvalue distribution of* $DP^{(N)}_-(\mathbf{x}^*_{k+(N-1)l+1})$ *inside and outside the unit cycle, vice versa.*

(iv) *All the bifurcations of the* stable *and* unstable $P^{(N)}_+$-1 *solution are all the bifurcations of the* unstable *and* stable $P^{(N)}_-$-1 *solution, respectively.*

*Proof* Consider positive and negative mappings with relations in Eq. (4.76), i.e.,

$$\mathbf{f}_r(\mathbf{x}_{k+(i-1)l+1}, \cdots, \mathbf{x}_{k+il}, \mathbf{x}_{k+il+1}, \mathbf{p}_r) = \mathbf{0} \quad (r = 1, 2, \cdots, l)$$
$$\text{for } i = 0, 1, 2, \ldots, N - 1.$$

from which $\mathbf{x}_{k+il+1}$ is a function of $\mathbf{x}_{k+(i-1)l+1}$ in the positive mapping and $\mathbf{x}_{k+(i-1)l+1}$ is a function of $\mathbf{x}_{k+il+1}$ in the negative mapping. The periodic solution in the sense of $\mathbf{g}(\mathbf{x}_{k-l+1}, \mathbf{x}_{k+(N-1)l+1}, \boldsymbol{\lambda}) = 0$ is given by

$$\mathbf{f}_r(\mathbf{x}_{k+(i-1)l+1}, \cdots, \mathbf{x}_{k+il}, \mathbf{x}_{k+il+1}, \mathbf{p}_r) = \mathbf{0} \quad (r = 1, 2, \ldots, l)$$
$$\text{for } i = 0, 1, 2, \cdots, N - 1;$$
$$\mathbf{g}(\mathbf{x}_{k-l+1}, \mathbf{x}_{k+(N-1)l+1}, \boldsymbol{\lambda}) = \mathbf{0}.$$

Letting the period-$N$ solution be $\{\mathbf{x}^*_{k-l+1}, \ldots, \mathbf{x}^*_{k-(N-1)l}, \mathbf{x}^*_{k-(N-1)l+1}\}$, the foregoing equation gives

$$\mathbf{f}_r(\mathbf{x}^*_{k+(i-1)l+1}, \cdots, \mathbf{x}^*_{k+il}, \mathbf{x}^*_{k+il+1}, \mathbf{p}_r) = \mathbf{0} \quad (r = 1, 2, \ldots, l)$$
$$\text{for } i = 0, 1, 2, \cdots, N - 1;$$
$$\mathbf{g}(\mathbf{x}^*_{k-l+1}, \mathbf{x}^*_{k+(N-1)l+1}, \boldsymbol{\lambda}) = \mathbf{0},$$

for both the positive and negative mappings.

In the sense of $\mathbf{g}(\mathbf{x}_{k-l+1}, \mathbf{x}_{k+(N-1)l+1}, \boldsymbol{\lambda}) \equiv \mathbf{0}$, once $\mathbf{x}_{k-l+1}^{*}, \cdots, \mathbf{x}_{k-(N-1)l}^{*}$, $\mathbf{x}_{k-(N-1)l+1}^{*}$ are obtained, the corresponding stability and bifurcation of the periodic solutions can be determined. Herein, consider a small perturbation

$$\mathbf{x}_{k-(i-1)l+1} = \mathbf{x}_{k-l+1}^{*} + \delta\mathbf{x}_{k-l+1},$$

$$\vdots$$

$$\mathbf{x}_{k+(N-1)l} = \mathbf{x}_{k+(N-1)l}^{*} + \delta\mathbf{x}_{k+(N-1)l},$$

$$\mathbf{x}_{k+(N-1)l+1} = \mathbf{x}_{k+(N-1)l+1}^{*} + \delta\mathbf{x}_{k+(N-1)l+1}$$

The linearization of mappings in Eqs. (4.73) and (4.74) gives

$$\delta\mathbf{x}_{k+(N-1)l+1} = DP_{+}(\mathbf{x}_{k-l+1}^{*})\delta\mathbf{x}_{k-l+1},$$

$$\delta\mathbf{x}_{k-l+1} = DP_{-}(\mathbf{x}_{k+(N-1)l+1}^{*})\delta\mathbf{x}_{k+(N-1)l+1}$$

where

$$DP_{+}(\mathbf{x}_{k-l+1}^{*}) = \left[\frac{\partial\mathbf{x}_{k+(N-1)l+1}}{\partial\mathbf{x}_{k-l+1}}\right]_{(\mathbf{x}_{k-l+1}^{*}, \cdots, \mathbf{x}_{k+(N-1)l}^{*}, \mathbf{x}_{k+(N-1)l+1}^{*})},$$

$$DP_{-}(\mathbf{x}_{k+(N-1)l+1}^{*}) = \left[\frac{\partial\mathbf{x}_{k-l+1}}{\partial\mathbf{x}_{k+(N-1)l+1}}\right]_{(\mathbf{x}_{k-l+1}^{*}, \cdots, \mathbf{x}_{k+(N-1)l}^{*}, \mathbf{x}_{k+(N-1)l+1}^{*})}.$$

From Eqs. (4.67) and (4.68), one obtains

$$\left(\left[\frac{\partial\mathbf{f}_{r}}{\partial\mathbf{x}_{k+(i-1)l+1}}\right] + \sum_{s=1}^{l}\left[\frac{\partial\mathbf{f}_{r}}{\partial\mathbf{x}_{k+(i-1)l+s+1}}\right]\left[\frac{\partial\mathbf{x}_{k+(i-1)l+s+1}}{\partial\mathbf{x}_{k+(i-1)l+1}}\right]\right)_{(\mathbf{x}_{k+(i-1)l+1}^{*}, \cdots, \mathbf{x}_{k+il}^{*}, \mathbf{x}_{k+il+1}^{*})} = \mathbf{0},$$

$$r = 1, 2, \cdots, l; \quad i = 0, 1, 2, \cdots, N-1;$$

$$\mathbf{g}(\mathbf{x}_{k-l+1}^{*}, \mathbf{x}_{k+(N-1)l+1}^{*}, \boldsymbol{\lambda}) = \mathbf{g}(\mathbf{x}_{k-l+1}, \mathbf{x}_{k+(N-1)l+1}, \boldsymbol{\lambda}) = \mathbf{0};$$

$$\left(\left[\frac{\partial\mathbf{f}_{r}}{\partial\mathbf{x}_{k+il+1}}\right] + \sum_{s=1}^{l}\left[\frac{\partial\mathbf{f}_{r}}{\partial\mathbf{x}_{k+il-s+1}}\right]\left[\frac{\partial\mathbf{x}_{k+il-s+1}}{\partial\mathbf{x}_{k+il+1}}\right]\right)_{(\mathbf{x}_{k+(i-1)l+1}^{*}, \cdots, \mathbf{x}_{k+il}^{*}, \mathbf{x}_{k+il+1}^{*})} = \mathbf{0},$$

$$r = l, l-1, \cdots, 1; \quad i = N-1, \cdots, 2, 1, 0;$$

$$\mathbf{g}(\mathbf{x}_{k-l+1}^{*}, \mathbf{x}_{k+(N-1)l+1}^{*}, \boldsymbol{\lambda}) = \mathbf{g}(\mathbf{x}_{k-l+1}, \mathbf{x}_{k+(N-1)l+1}, \boldsymbol{\lambda}) = \mathbf{0}.$$

That is, for $r = 1, 2, \cdots, l; \; i = 0, 1, 2, \cdots, N-1$

$$
\left\{
\begin{array}{c}
\dfrac{\partial \mathbf{x}_{k+(i-1)l+2}}{\partial \mathbf{x}_{k+(i-1)l+1}} \\
\vdots \\
\dfrac{\partial \mathbf{x}_{k+il}}{\partial \mathbf{x}_{k+(i-1)l+1}} \\
\dfrac{\partial \mathbf{x}_{k+il+1}}{\partial \mathbf{x}_{k+(i-1)l+1}}
\end{array}
\right\}
= -
\left[
\begin{array}{cccc}
\dfrac{\partial \mathbf{f}_1}{\partial \mathbf{x}_{k+(i-1)l+2}} & \cdots & \dfrac{\partial \mathbf{f}_1}{\partial \mathbf{x}_{k+il}} & \dfrac{\partial \mathbf{f}_1}{\partial \mathbf{x}_{k+il+1}} \\
\vdots & & \vdots & \vdots \\
\dfrac{\partial \mathbf{f}_{l-1}}{\partial \mathbf{x}_{k+(i-1)l+2}} & \cdots & \dfrac{\partial \mathbf{f}_{l-1}}{\partial \mathbf{x}_{k+il}} & \dfrac{\partial \mathbf{f}_{l-1}}{\partial \mathbf{x}_{k+il+1}} \\
\dfrac{\partial \mathbf{f}_l}{\partial \mathbf{x}_{k+(i-1)l+2}} & \cdots & \dfrac{\partial \mathbf{f}_l}{\partial \mathbf{x}_{k+il}} & \dfrac{\partial \mathbf{f}_l}{\partial \mathbf{x}_{k+il+1}}
\end{array}
\right]^{-1}
\left\{
\begin{array}{c}
\dfrac{\partial \mathbf{f}_1}{\partial \mathbf{x}_{k+(i-1)l+1}} \\
\vdots \\
\dfrac{\partial \mathbf{f}_{l-1}}{\partial \mathbf{x}_{k+(i-1)l+1}} \\
\dfrac{\partial \mathbf{f}_l}{\partial \mathbf{x}_{k+(i-1)l+1}}
\end{array}
\right\},
$$

$$
\left\{
\begin{array}{c}
\dfrac{\partial \mathbf{x}_{k+(i-1)l+2}}{\partial \mathbf{x}_{k+il+1}} \\
\vdots \\
\dfrac{\partial \mathbf{x}_{k+il}}{\partial \mathbf{x}_{k+il+1}} \\
\dfrac{\partial \mathbf{x}_{k-l+1}}{\partial \mathbf{x}_{k+(i-1)l+1}}
\end{array}
\right\}
= -
\left[
\begin{array}{cccc}
\dfrac{\partial \mathbf{f}_1}{\partial \mathbf{x}_{k+(i-1)l+2}} & \cdots & \dfrac{\partial \mathbf{f}_1}{\partial \mathbf{x}_{k+il}} & \dfrac{\partial \mathbf{f}_1}{\partial \mathbf{x}_{k+(i-1)l+1}} \\
\vdots & & \vdots & \vdots \\
\dfrac{\partial \mathbf{f}_{l-1}}{\partial \mathbf{x}_{k+(i-1)l+2}} & \cdots & \dfrac{\partial \mathbf{f}_{l-1}}{\partial \mathbf{x}_{k+il}} & \dfrac{\partial \mathbf{f}_{l-1}}{\partial \mathbf{x}_{k+(i-1)l+1}} \\
\dfrac{\partial \mathbf{f}_l}{\partial \mathbf{x}_{k+(i-1)l+2}} & \cdots & \dfrac{\partial \mathbf{f}_l}{\partial \mathbf{x}_k} & \dfrac{\partial \mathbf{f}_l}{\partial \mathbf{x}_{k+(i-1)l+1}}
\end{array}
\right]^{-1}
\left\{
\begin{array}{c}
\dfrac{\partial \mathbf{f}_1}{\partial \mathbf{x}_{k+il++1}} \\
\vdots \\
\dfrac{\partial \mathbf{f}_{l-1}}{\partial \mathbf{x}_{k+il+1}} \\
\dfrac{\partial \mathbf{f}_l}{\partial \mathbf{x}_{k+il+1}}
\end{array}
\right\}.
$$

From two equations, we obtain

$$
DP_+(\mathbf{x}^*_{k+(i-1)l+1}) = \left[\dfrac{\partial \mathbf{x}_{k+il+1}}{\partial \mathbf{x}_{k+(i-1)l+1}}\right]\Big|_{(\mathbf{x}^*_{k+(i-1)l+1},\cdots,\mathbf{x}^*_{k+il},\mathbf{x}^*_{k+il+1})},
$$

$$
DP_-(\mathbf{x}^*_{k+il+1}) = \left[\dfrac{\partial \mathbf{x}_{k+(i-1)l+1}}{\partial \mathbf{x}_{k+il+1}}\right]\Big|_{(\mathbf{x}^*_{k+(i-1)l+1},\cdots,\mathbf{x}^*_{k+il},\mathbf{x}^*_{k+il+1})}.
$$

Since

$$
\left[\dfrac{\partial \mathbf{x}_{k+(N-1)l+1}}{\partial \mathbf{x}_{k-l+1}}\right] \cdot \left[\dfrac{\partial \mathbf{x}_{k-l+1}}{\partial \mathbf{x}_{k+(N-1)l+1}}\right]\Big|_{(\mathbf{x}^*_{k-l+1},\cdots,\mathbf{x}^*_{k+(N-1)l},\mathbf{x}^*_{k+(N-1)l+1})} = \mathbf{I}_{n\times n}.
$$

we have

$$
DP^{(N)}_+(\mathbf{x}^*_{k-l+1}) \cdot DP^{(N)}_-(\mathbf{x}^*_{k+(N-1)l+1}) = \mathbf{I}_{n\times n}.
$$

where

$$
DP^{(N)}_+(\mathbf{x}^*_{k-l+1}) = DP_+(\mathbf{x}^*_{k+(N-2)l+1}) \cdot DP_+(\mathbf{x}^*_{k+(N-3)l+1}) \cdots \cdots DP_+(\mathbf{x}^*_{k-l+1})
$$
$$
= \left[\dfrac{\partial \mathbf{x}_{k+(N-1)l+1}}{\partial \mathbf{x}_{k+(N-2)l+1}}\right] \cdot \left[\dfrac{\partial \mathbf{x}_{k+(N-2)l+1}}{\partial \mathbf{x}_{k+(N-3)l+1}}\right] \cdots \cdots \left[\dfrac{\partial \mathbf{x}_{k+1}}{\partial \mathbf{x}_{k-l+1}}\right]
$$

and

$$DP^{(N)}_-(\mathbf{x}^*_{k+(N-1)l+1}) = DP_+(\mathbf{x}^*_{k+1}) \cdot \cdots \cdot DP_+(\mathbf{x}^*_{k+(N-2)l+1}) \cdot DP_+(\mathbf{x}^*_{k+(N-1)l+1})$$

$$= \left[\frac{\partial \mathbf{x}_{k-l+1}}{\partial \mathbf{x}_{k+1}}\right] \cdot \cdots \cdot \left[\frac{\partial \mathbf{x}_{k+(N-3)l+1}}{\partial \mathbf{x}_{k+(N-2)l+1}}\right] \cdot \left[\frac{\partial \mathbf{x}_{k+(N-2)l+1}}{\partial \mathbf{x}_{k+(N-1)l+1}}\right].$$

Therefore, one obtains

$$\left[DP^{(N)}_-(\mathbf{x}^*_{k+(N-1)l+1})\right]^{-1} = DP^{(N)}_+(\mathbf{x}^*_{k-l+1}).$$

In other words, $DP^{(N)}_+(\mathbf{x}^*_{k-l+1})$ is the inverse of $DP^{(N)}_-(\mathbf{x}^*_{k+(N-1)l+1})$, vice versa.

For eigenvalues $\lambda_-$ and $\lambda_+$ of $DP^{(N)}_-(\mathbf{x}^*_{k+(N-1)l+1})$ and $DP^{(N)}_+(\mathbf{x}^*_{k-l+1})$ accordingly, we have

$$(DP^{(N)}_-(\mathbf{x}^*_{k+(N-1)l+1}) - \lambda_-\mathbf{I})\delta\mathbf{x}_{k+(N-1)l+1} = \mathbf{0},$$

$$(DP^{(N)}_+(\mathbf{x}^*_{k-l+1}) - \lambda_+\mathbf{I})\delta\mathbf{x}_{k-l+1} = \mathbf{0}.$$

Left multiplication of $DP^{(N)}_+(\mathbf{x}^*_{k-l+1})$ in the first equation of the foregoing equation, division of $\lambda_-$ on both sides, and application of $[DP^{(N)}_-(\mathbf{x}^*_{k+(N-1)l+1})]^{-1} = DP^{(N)}_+(\mathbf{x}^*_{k-l+1})$ give

$$[DP^{(N)}_+(\mathbf{x}^*_{k-l+1}) - \lambda_-^{-1}\mathbf{I}]\delta\mathbf{x}_{k+(N-1)l+1} = \mathbf{0}.$$

Thus, one can obtain

$$\lambda_+ = \lambda_-^{-1}.$$

From the stability and bifurcation theory for discrete dynamical systems, the following conclusions can be summarized as

(i)   The stable $P^{(N)}_+$-1 solution is the unstable $P^{(N)}_-$-1 solutions with all eigenvalues of $DP^{(N)}_-(\mathbf{x}^*_{k+(N-1)l+1})$ outside the unit cycle, vice versa.

(ii)  The unstable $P^{(N)}_+$-1 solutions with all eigenvalues of $DP^{(N)}_+(\mathbf{x}^*_{k-l+1})$ outside the unit cycle are the stable $P^{(N)}_-$-1 solutions, vice versa.

(iii) For the unstable $P^{(N)}_+$-1 solution with eigenvalue distribution of $DP^{(N)}_+(\mathbf{x}^*_{k-l+1})$ inside and outside the unit cycle, the corresponding $P^{(N)}_-$-1 solution is also unstable with switching eigenvalue distribution of $DP^{(N)}_-(\mathbf{x}^*_{k+(N-1)l+1})$ *inside* and *outside* the unit cycle, vice versa.

(iv)  All the bifurcations of the *stable* and *unstable* $P_+^{(N)}$-1 solution are all the bifurcations of the *unstable* and *stable* $P_-^{(N)}$-1 solution, respectively.

This theorem is proved.                                                                    $\square$

As before, the number $N$ for the $P_+^{(N)}$-1 and $P_-^{(N)}$-1 solutions in the discrete dynamical system can be any integer if such a solution exists in the sense of $\mathbf{g}(\mathbf{x}_{k-l+1}, \mathbf{x}_{k+(N-1)l+1}, \boldsymbol{\lambda}) = \mathbf{0}$.

**Theorem 4.8** *For a discrete dynamical system in* Eq. (4.54), *there are two points* $\mathbf{x}_{k-l+1} \in D$ *and* $\mathbf{x}_{k+(N-1)l+1} \in D$. *If the period-doubling cascade of the* $P_+^{(N)}$-1 *and* $P_-^{(N)}$-1 *solution occurs, the corresponding mapping structures are given by*

$$
\left.
\begin{array}{l}
\mathbf{x}_{k+(2N-1)l+1} = P_+^{(N)} \circ P_+^{(N)} \mathbf{x}_{k-l+1} = P_+^{(2N)} \mathbf{x}_{k-l+1} \\
\mathbf{g}(\mathbf{x}_{k-l+1}, \mathbf{x}_{k+(2N-1)l+1}, \boldsymbol{\lambda}) = \mathbf{0}
\end{array}
\right\};
$$

$$
\left.
\begin{array}{l}
\mathbf{x}_{k+(2^2N-1)l+1} = P_+^{(2N)} \circ P_+^{(2N)} \mathbf{x}_{k-l+1} = P_+^{(2^2N)} \mathbf{x}_{k-l+1} \\
\mathbf{g}(\mathbf{x}_{k-l+1}, \mathbf{x}_{k+(2^2N-1)l+1}, \boldsymbol{\lambda}) = \mathbf{0}
\end{array}
\right\}; \qquad (4.75)
$$

$$
\vdots
$$

$$
\left.
\begin{array}{l}
\mathbf{x}_{k+(2^rN-1)l+1} = P_+^{(2^{r-1}N)} \circ P_+^{(2^{r-1}N)} \mathbf{x}_{k-l+1} = P_+^{(2^rN)} \mathbf{x}_{k-l+1} \\
\mathbf{g}(\mathbf{x}_{k-l+1}, \mathbf{x}_{k+(2^rN-1)l+1}, \boldsymbol{\lambda}) = \mathbf{0}
\end{array}
\right\};
$$

*for positive mappings and*

$$
\left.
\begin{array}{l}
\mathbf{x}_{k-l+1} = P_-^{(N)} \circ P_-^{(N)} \mathbf{x}_{k+(2N-1)l+1} = P_-^{(2N)} \mathbf{x}_{k+(2N-1)l+1} \\
\mathbf{g}(\mathbf{x}_{k-l+1}, \mathbf{x}_{k+(2N-1)l+1}, \boldsymbol{\lambda}) = \mathbf{0}
\end{array}
\right\};
$$

$$
\left.
\begin{array}{l}
\mathbf{x}_{k-l+1} = P_-^{(2N)} \circ P_-^{(2N)} \mathbf{x}_{k+(2^2N-1)l+1} = P_-^{(2^2N)} \mathbf{x}_{k+(2^2N-1)l+1} \\
\mathbf{g}(\mathbf{x}_k, \mathbf{x}_{k+2^2N}, \boldsymbol{\lambda}) = \mathbf{0}
\end{array}
\right\}; \qquad (4.76)
$$

$$
\vdots
$$

$$
\left.
\begin{array}{l}
\mathbf{x}_{k-l+1} = P_-^{(2^{r-1}N)} \circ P_-^{(2^{r-1}N)} \mathbf{x}_{k+(2^rN-1)l+1} = P_-^{(2^rN)} \mathbf{x}_{k+(2^rN-1)l+1} \\
\mathbf{g}(\mathbf{x}_{k-l+1}, \mathbf{x}_{k+(2^rN-1)l+1}, \boldsymbol{\lambda}) = \mathbf{0}
\end{array}
\right\}
$$

*for negative mapping, then the following statements hold*, i.e.,

(i)  *The stable chaos generated by the limit state of the stable* $P_+^{(2^rN)}$-1 *solutions* $(r \to \infty)$ *in the sense of* $\mathbf{g}(\mathbf{x}_{k-l+1}, \mathbf{x}_{k+(2^rN-1)l+1}, \boldsymbol{\lambda}) = \mathbf{0}$ *is the unstable chaos generated by the limit state of the unstable stable* $P_-^{(2^rN)}$-1 *solution* $(r \to \infty)$

*in the sense of* $\mathbf{g}(\mathbf{x}_{k-l+1}, \mathbf{x}_{k+(2^r N-1)l+1}, \boldsymbol{\lambda}) = 0$ *with all eigenvalue distribution of* $DP_{-}^{(2^r N)}$ *outside unit cycle, vice versa. Such a chaos is the "Yang" chaos in nonlinear discrete dynamical systems.*

(ii) *The unstable chaos generated by the limit state of the unstable* $P_{+}^{(2^r N)}$-1 *solutions ($r \to \infty$) in the sense of* $\mathbf{g}(\mathbf{x}_{k-l+1}, \mathbf{x}_{k+(2^r N-1)l+1}, \boldsymbol{\lambda}) = 0$ *with all eigenvalue distribution of* $DP_{+}^{(2^r N)}$ *outside the unit cycle is the stable chaos generated by the limit state of the stable* $P_{-}^{(2^r N)}$-1 *solution ($r \to \infty$) in the sense of* $\mathbf{g}(\mathbf{x}_{k-l+1}, \mathbf{x}_{k+(2^r N-1)l+1}, \boldsymbol{\lambda}) = 0$, *vice versa. Such a chaos is the "Yin" chaos in nonlinear discrete dynamical systems.*

(iii) *The unstable chaos generated by the limit state of the unstable* $P_{+}^{(2^r N)}$-1 *solutions ($r \to \infty$) in the sense of* $\mathbf{g}(\mathbf{x}_{k-l+1}, \mathbf{x}_{k+(2^r N-1)l+1}, \boldsymbol{\lambda}) = 0$ *with all eigenvalue distribution of* $DP_{+}^{(2^r N)}$ *inside and outside the unit cycle is the unstable chaos generated by the limit state of the unstable* $P_{-}^{(2^r N)}$-1 *solution ($r \to \infty$) in the sense of* $\mathbf{g}(\mathbf{x}_{k-l+1}, \mathbf{x}_{k+(2^r N-1)l+1}, \boldsymbol{\lambda}) = 0$ *with all eigenvalue distribution of* $DP_{-}^{(2^r N)}$ *inside and outside the unit cycle, vice versa. Such a chaos is the "Yin–Yang" chaos in nonlinear discrete dynamical systems.*

Proof The proof is similar to the proof of Theorem 4.7, and the chaos is obtained by $r \to \infty$. This theorem is proved. □

# References

Luo, A. C. J. (2010). A Ying-Yang theory in nonlinear discrete dynamical systems. *International Journal of Bifurcation and Chaos, 20*, 1085–1098.

Luo, A. C. J., & Guo, Y. (2010). Parameter characteristics for stable and unstable solutions in nonlinear discrete dynamical systems. *International Journal of Bifurcation and Chaos, 20*, 3173–3191.

Luo, A. C. J. (2012). *Regularity and Complexity in Dynamical Systems*. New York: Springer.

# Chapter 5
# Periodic Flows in Continuous Systems

This chapter will present periodic flows in nonlinear dynamical systems through the discrete implicit mappings. The period-1 flows in nonlinear dynamical systems will be discussed first by the one-step discrete maps, and then, the period-m flows in nonlinear dynamical systems will also be discussed through the one-step discrete maps. Multi-step, implicit discrete maps will be used to discuss the period-1 and period-m motions in nonlinear dynamical systems. Periodic flows in nonlinear time-delay dynamical systems will be discussed with time-delay discrete nodes interpolated by two non-delay discrete nodes. In addition, periodic flows in time-delay nonlinear dynamical systems will be also discussed through the delay nodes determined by integration. Through the discrete nodes in periodic flows, the periodic flows will be approximated by the discrete Fourier series and the frequency space of the periodic flows can be determined through amplitude spectrums.

## 5.1 Continuous Nonlinear Systems

As in Luo (2014), periodic flows in continuous dynamical systems will be presented. If a nonlinear system has a periodic flow with a period of $T = 2\pi/\Omega$, then such a periodic flow can be determined by discrete points through discrete mappings of the continuous system. The method is stated as follows.

**Theorem 5.1** *Consider a nonlinear dynamical system as*

$$\dot{\mathbf{x}} = \mathbf{f}(\mathbf{x}, t, \mathbf{p}) \in \mathscr{R}^n \qquad (5.1)$$

*where $\mathbf{f}(\mathbf{x}, t, \mathbf{p})$ is a $C^r$-continuous nonlinear vector function $r \geq 1$. If such a system has a periodic flow $\mathbf{x}(t)$ with finite norm $\|\mathbf{x}\|$ and period $T = 2\pi/\Omega$, there is a set of discrete time $t_k$ $(k = 0, 1, \ldots, N)$ with $(N \to \infty)$ during one period $T$, and the corresponding solution $\mathbf{x}(t_k)$ and vector field $\mathbf{f}(\mathbf{x}(t_k), t_k, \mathbf{p})$ are exact. Suppose a discrete node $\mathbf{x}_k$ is on the approximate solution of the periodic flow under $\|\mathbf{x}(t_k) - \mathbf{x}_k\| \leq \varepsilon_k$ with a small $\varepsilon_k \geq 0$ and*

© Higher Education Press, Beijing and Springer-Verlag Berlin Heidelberg 2015
A.C.J. Luo, *Discretization and Implicit Mapping Dynamics*,
Nonlinear Physical Science, DOI 10.1007/978-3-662-47275-0_5

$$||\mathbf{f}(\mathbf{x}(t_k), t_k, \mathbf{p}) - \mathbf{f}(\mathbf{x}_k, t_k, \mathbf{p})|| \le \delta_k \tag{5.2}$$

with a small $\delta_k \ge 0$. During a time interval $t \in [t_k, t_{k+1}]$, there is a mapping $P_k : \mathbf{x}_{k-1} \rightarrow \mathbf{x}_k$ $(k = 1, 2, \ldots, N)$, i.e.,

$$\mathbf{x}_k = P_k \mathbf{x}_{k-1} \quad \text{with } \mathbf{g}_k(\mathbf{x}_{k-1}, \mathbf{x}_k, \mathbf{p}) = \mathbf{0}, \quad k = 1, 2, \ldots, N \tag{5.3}$$

where $\mathbf{g}_k$ is an implicit vector function. Consider a mapping structure as

$$\begin{aligned}
&P = P_N \circ P_{N-1} \circ \cdots \circ P_2 \circ P_1 : \quad \mathbf{x}_0 \rightarrow \mathbf{x}_N; \\
&\text{with } P_k : \quad \mathbf{x}_{k-1} \rightarrow \mathbf{x}_k \quad (k = 1, 2, \ldots, N).
\end{aligned} \tag{5.4}$$

For $\mathbf{x}_N = P\mathbf{x}_0$, if there is a set of points $\mathbf{x}_k^*$ $(k = 0, 1, \ldots, N)$ computed by

$$\begin{aligned}
&\mathbf{g}_k(\mathbf{x}_{k-1}^*, \mathbf{x}_k^*, \mathbf{p}) = \mathbf{0}, \quad (k = 1, 2, \ldots, N) \\
&\mathbf{x}_0^* = \mathbf{x}_N^*,
\end{aligned} \tag{5.5}$$

then the points $\mathbf{x}_k^*$ $(k = 0, 1, \ldots, N)$ are approximations of points $\mathbf{x}(t_k)$ of the periodic solution. In the neighborhood of $\mathbf{x}_k^*$, with $\mathbf{x}_k = \mathbf{x}_k^* + \Delta\mathbf{x}_k$, the linearized equation is given by

$$\begin{aligned}
&\Delta\mathbf{x}_k = DP_k \cdot \Delta\mathbf{x}_{k-1} \\
&\text{with } \mathbf{g}_k(\mathbf{x}_{k-1}^* + \Delta\mathbf{x}_{k-1}, \mathbf{x}_k^* + \Delta\mathbf{x}_k, \mathbf{p}) = \mathbf{0} \\
&(k = 1, 2, \ldots, N).
\end{aligned} \tag{5.6}$$

The resultant Jacobian matrices of the periodic flow are

$$\begin{aligned}
&DP_{k(k-1)\cdots 1} = DP_N \cdot DP_{N-1} \cdot \cdots \cdot DP_1, \quad (k = 1, 2, \ldots, N) \\
&DP \equiv DP_{N(N-1)\cdots 1} = DP_N \cdot DP_{N-1} \cdot \cdots \cdot DP_1
\end{aligned} \tag{5.7}$$

where

$$DP_k = \left[ \frac{\partial\mathbf{x}_k}{\partial\mathbf{x}_{k-1}} \right]_{(\mathbf{x}_{k-1}^*, \mathbf{x}_k^*)} = -\left[ \frac{\partial\mathbf{g}_k}{\partial\mathbf{x}_k} \right]_{(\mathbf{x}_{k-1}^*, \mathbf{x}_k^*)}^{-1} \left[ \frac{\partial\mathbf{g}_k}{\partial\mathbf{x}_{k-1}} \right]_{(\mathbf{x}_{k-1}^*, \mathbf{x}_k^*)} \tag{5.8}$$

$(k = 1, 2, \ldots, N)$.

The eigenvalues of $DP$ and $DP_{k(k-1)\cdots 1}$ for such a periodic flow are determined by

$$\begin{aligned}
&|DP_{k(k-1)\cdots 1} - \bar{\lambda}\mathbf{I}_{n\times n}| = 0, \quad (k = 1, 2, \ldots, N); \\
&|DP - \lambda\mathbf{I}_{n\times n}| = 0.
\end{aligned} \tag{5.9}$$

*Thus, the eigenvalues of* $\mathrm{DP}_{k(k-1)\cdots 1}$ *give the properties of* $\mathbf{x}_k$ *varying with* $\mathbf{x}_0$ *for the periodic flow. The stability and bifurcation of the periodic flow can be classified by the eigenvalues of* $\mathrm{DP}(\mathbf{x}_0^*)$ *with*

$$([n_1^m, n_1^o] : [n_2^m, n_2^o] : [n_3, \kappa_3] : [n_4, \kappa_4]|n_5 : n_6 : [n_7, l, \kappa_7]) \qquad (5.10)$$

*where* $n_1$ *is the total number of real eigenvalues with magnitudes less than one* $(n_1 = n_1^m + n_1^o)$; $n_2$ *is the total number of real eigenvalues with magnitude greater than one* $n_2 = n_2^m + n_2^o$; $n_3$ *is the total number of real eigenvalues equal to* $+1$; $n_4$ *is the total number of real eigenvalues equal to* $-1$; $n_5$ *is the total pair number of complex eigenvalues with magnitudes less than one,* $n_6$ *is the total pair number of complex eigenvalues with magnitudes greater than one; and* $n_7$ *is the total pair number of complex eigenvalues with magnitudes equal to one.*

(i) *If the magnitudes of all eigenvalues of DP are less than one (i.e.,* $|\lambda_i| < 1$, $i = 1, 2, \ldots, n$), *the approximate periodic solution is stable.*

(ii) *If at least the magnitude of one eigenvalue of DP is greater than one (i.e.,* $|\lambda_i| > 1$, $i \in \{1, 2, \ldots, n\}$), *the approximate periodic solution is unstable.*

(iii) *The boundaries between stable and unstable periodic flow with higher-order singularity give bifurcation and stability conditions.*

*Proof* If $\mathbf{f}(\mathbf{x}, \mathbf{p}, t)$ is a $C^r$-continuous nonlinear vector function $(r \geq 1)$, then the velocity $\dot{\mathbf{x}}$ should be $C^r$-continuous $(r \geq 1)$. If such a dynamical system has a periodic flow $\mathbf{x}(t)$ with finite norm $\|\mathbf{x}\|$ and period $T = 2\pi/\Omega$, there is a set of discrete time $t_k$ $(k = 0, 1, \ldots, N)$ with $(N \to \infty)$ during one period $T$. The corresponding solution $\mathbf{x}(t_k)$ and vector fields $\mathbf{f}(\mathbf{x}(t_k), t_k, \mathbf{p})$ are exact. For $t \in [t_{k-1}, t_k]$,

$$\mathbf{x}(t) = \mathbf{x}(t_{k-1}) + \int_{t_{k-1}}^{t} \mathbf{f}(\mathbf{x}, t, \mathbf{p})dt. \qquad (5.11)$$

For the time interval $[t_k, t_{k+1}]$ divided into $s$-nodes $t_{k(i)} = t_{k-1} + c_i h_k$ with $c_i \in [0, 1]$ and $\mathbf{f}(\mathbf{x}(t_{k(i)}), t_{k(i)}, \mathbf{p})$ $(i = 1, \ldots, s)$, there is an approximate function $\mathbf{P}(t, \mathbf{C})$ with unknown $\mathbf{C} = (C_1, \ldots, C_s)^T$ and $C_i$ $(i = 1, \ldots, s)$, and the following condition is

$$\mathbf{f}(\mathbf{x}(t_{k(i)}), t_{k(i)}, \mathbf{p}) = \mathbf{P}(t_{k(i)}, \mathbf{C}), \quad i = 1, 2, \ldots, s;$$
$$|\frac{\partial \mathbf{P}}{\partial \mathbf{C}}| \neq 0. \qquad (5.12)$$

The unknowns $\mathbf{C}(\mathbf{t}_k) = (C_1, \ldots, C_s)^T$ with $\mathbf{t}_k = (t_{k(1)}, \ldots, t_{k(s)})^T = t_k(1, 1, \ldots, 1)^T + h_k(c_1, \ldots, c_s)^T$ are determined. For a small $\delta > 0$, if there is a relation

$$|\mathbf{P}(t, \mathbf{C}(\mathbf{t}_k)) - \mathbf{f}(\mathbf{x}, t, \mathbf{p})| < \delta \qquad (5.13)$$

for $t \in [t_{k-1}, t_k]$, Eq. (5.11) can be approximated as

$$\mathbf{x}(t) = \mathbf{x}(t_{k-1}) + \int_{t_{k-1}}^{t} [\mathbf{P}(t, \mathbf{C}(\mathbf{t}_k)) + O(\delta)]dt;$$

$$\bar{\mathbf{x}}(t) = \bar{\mathbf{x}}(t_{k-1}) + \int_{t_{k-1}}^{t} \mathbf{P}(t, \mathbf{C}(\mathbf{t}_k))dt \tag{5.14}$$

and

$$\bar{\mathbf{x}}(t_{k+1}) = \bar{\mathbf{x}}(t_k) + \int_{t_k}^{t_{k+1}} \mathbf{P}(t, \mathbf{C}(\mathbf{t}_k))dt. \tag{5.15}$$

Let $\bar{\mathbf{x}}(t_{k-1}) = \mathbf{x}_{k-1}$ and $\bar{\mathbf{x}}(t_k) = \mathbf{x}_k$. For any small $\varepsilon_{k-1} > 0$ and $\varepsilon_k > 0$, under $||\mathbf{x}(t_{k-1}) - \mathbf{x}_{k-1}|| \le \varepsilon_{k-1}$ and $||\mathbf{x}(t_k) - \mathbf{x}_k|| \le \varepsilon_k$, Eq. (5.15) gives

$$\mathbf{x}_k = \mathbf{x}_{k-1} + \bar{\mathbf{g}}_k(\mathbf{x}_{k-1}, \mathbf{x}_k, \mathbf{p}),$$

$$\bar{\mathbf{g}}_k(\mathbf{x}_{k-1}, \mathbf{x}_k, \mathbf{p}) = \int_{t_{k-1}}^{t_k} \mathbf{P}(t, \mathbf{C}(\mathbf{t}_k))dt. \tag{5.16}$$

Thus, a discrete mapping relation is obtained by

$$\mathbf{g}_k(\mathbf{x}_{k-1}, \mathbf{x}_k, \mathbf{p}) \equiv \mathbf{x}_k - \mathbf{x}_{k-1} - \bar{\mathbf{g}}_k(\mathbf{x}_{k-1}, \mathbf{x}_k, \mathbf{p}) = \mathbf{0}. \tag{5.17}$$

From the discrete mapping, two points $\mathbf{x}(t_{k-1})$ and $\mathbf{x}(t_k)$ for the time interval $t \in [t_{k-1}, t_k]$ $(k = 0, 1, \ldots, N)$ can be approximated by $\mathbf{x}_{k-1}$ and $\mathbf{x}_k$, respectively. If $\mathbf{f}(\mathbf{x}, t, \mathbf{p})$ is a $C^r$-continuous nonlinear vector function, we have $||\mathbf{f}|| \le L$ ($L$ constant). Thus

$$||\mathbf{f}(\mathbf{x}(t_{k-1}), t_{k-1}, \mathbf{p}) - \mathbf{f}(\mathbf{x}_{k-1}, t_{k-1}, \mathbf{p})|| \le L||\mathbf{x}(t_{k-1}) - \mathbf{x}_{k-1}|| \le L\varepsilon_{k-1} = \delta_{k-1},$$

$$||\mathbf{f}(\mathbf{x}(t_k), t_k, \mathbf{p}) - \mathbf{f}(\mathbf{x}_k, t_k, \mathbf{p})|| \le L||\mathbf{x}(t_k) - \mathbf{x}_k|| \le L\varepsilon_k = \delta_k. \tag{5.18}$$

Once the mapping $P_k : \mathbf{x}_{k-1} \to \mathbf{x}_k (k = 1, 2, \ldots, N)$ with $\mathbf{g}_k(\mathbf{x}_{k-1}, \mathbf{x}_k, \mathbf{p}) = \mathbf{0}$ exists, then the periodic flow can be formed by $P : \mathbf{x}_0 \to \mathbf{x}_N$ with $P = P_N \circ \cdots \circ P_2 \circ P_1$, i.e.,

$$\begin{aligned} P_1 : \; & \mathbf{x}_0 \to \mathbf{x}_1 \Rightarrow \mathbf{g}_1(\mathbf{x}_0, \mathbf{x}_1, \mathbf{p}) = \mathbf{0}, \\ P_2 : \; & \mathbf{x}_1 \to \mathbf{x}_2 \Rightarrow \mathbf{g}_2(\mathbf{x}_1, \mathbf{x}_2, \mathbf{p}) = \mathbf{0}, \\ & \vdots \\ P_k : \; & \mathbf{x}_{k-1} \to \mathbf{x}_k \Rightarrow \mathbf{g}_k(\mathbf{x}_{k-1}, \mathbf{x}_k, \mathbf{p}) = \mathbf{0}, \\ & \vdots \\ P_N : \; & \mathbf{x}_{N-1} \to \mathbf{x}_N \Rightarrow \mathbf{g}_N(\mathbf{x}_{N-1}, \mathbf{x}_N, \mathbf{p}) = \mathbf{0}. \end{aligned} \tag{5.19}$$

With the periodicity condition, we have

$$\mathbf{x}_0 = \mathbf{x}_N. \tag{5.20}$$

Solving Eqs. (5.19) and (5.20) gives $\mathbf{x}_k^*$ ($k = 1, 2, \ldots, N$) to get the period-1 flow. For the stability of such a periodic flow, $\mathbf{x}_k = \mathbf{x}_k^* + \Delta\mathbf{x}_k$ ($k = 0, 1, 2, \ldots, N$) is considered, and Eq. (5.19) becomes

$$
\begin{aligned}
&\mathbf{g}_1(\mathbf{x}_0^* + \Delta\mathbf{x}_0, \mathbf{x}_1^* + \Delta\mathbf{x}_1, \mathbf{p}) = \mathbf{0}, \\
&\mathbf{g}_2(\mathbf{x}_1^* + \Delta\mathbf{x}_1, \mathbf{x}_2^* + \Delta\mathbf{x}_2, \mathbf{p}) = \mathbf{0}, \\
&\quad\vdots \\
&\mathbf{g}_k(\mathbf{x}_{k-1}^* + \Delta\mathbf{x}_{k-1}, \mathbf{x}_k^* + \Delta\mathbf{x}_k, \mathbf{p}) = \mathbf{0}, \\
&\quad\vdots \\
&\mathbf{g}_N(\mathbf{x}_{N-1}^* + \Delta\mathbf{x}_{N-1}, \mathbf{x}_N^* + \Delta\mathbf{x}_N, \mathbf{p}) = \mathbf{0}.
\end{aligned}
\tag{5.21}
$$

Thus, derivatives of $\mathbf{g}_k(\mathbf{x}_{k-1}, \mathbf{x}_k, \mathbf{p}) = \mathbf{0}$ with respect to $\mathbf{x}_{k-1}$ gives

$$\left[\frac{\partial\mathbf{g}_k}{\partial\mathbf{x}_{k-1}}\right]_{(\mathbf{x}_{k-1}^*, \mathbf{x}_k^*)} + \left[\frac{\partial\mathbf{g}_k}{\partial\mathbf{x}_k}\right]_{(\mathbf{x}_{k-1}^*, \mathbf{x}_k^*)}\left[\frac{\partial\mathbf{x}_k}{\partial\mathbf{x}_{k-1}}\right] = \mathbf{0} \quad (k = 1, 2, \ldots, N). \tag{5.22}$$

The deformation of the forgoing equation is

$$\left[\frac{\partial\mathbf{x}_k}{\partial\mathbf{x}_{k-1}}\right]_{(\mathbf{x}_{k-1}^*, \mathbf{x}_k^*)} = -\left[\frac{\partial\mathbf{g}_k}{\partial\mathbf{x}_k}\right]^{-1}\left[\frac{\partial\mathbf{g}_k}{\partial\mathbf{x}_{k-1}}\right]_{(\mathbf{x}_{k-1}^*, \mathbf{x}_k^*)} \quad (k = 1, 2, \ldots, N) \tag{5.23}$$

and the linearization of the forgoing equation gives

$$\Delta\mathbf{x}_k = DP_k \cdot \Delta\mathbf{x}_{k-1} \quad \text{with } DP_k = \left[\frac{\partial\mathbf{x}_k}{\partial\mathbf{x}_{k-1}}\right]_{(\mathbf{x}_{k-1}^*, \mathbf{x}_k^*)} \quad (k = 1, 2, \ldots, N). \tag{5.24}$$

In other words,

$$
\begin{aligned}
&\Delta\mathbf{x}_k = DP_{k(k-1)\cdots 1} \cdot \Delta\mathbf{x}_0 \quad \text{with} \\
&DP_{k(k-1)\cdots 1} = DP_k \cdot DP_{k-1} \cdot \cdots \cdot DP_1 = \prod_{j=k}^{1}\left[\frac{\partial\mathbf{x}_j}{\partial\mathbf{x}_{j-1}}\right]_{(\mathbf{x}_{j-1}^*, \mathbf{x}_j^*)}; \\
&\Delta\mathbf{x}_N = DP \cdot \Delta\mathbf{x}_0 \quad \text{with} \\
&DP \equiv DP_{N(N-1)\cdots 1} = DP_N \cdot DP_{N-1} \cdot \cdots \cdot DP_1 = \prod_{k=N}^{1}\left[\frac{\partial\mathbf{x}_k}{\partial\mathbf{x}_{k-1}}\right]_{(\mathbf{x}_{k-1}^*, \mathbf{x}_k^*)}.
\end{aligned}
\tag{5.25}
$$

Setting $\Delta \mathbf{x}_N = \lambda \Delta \mathbf{x}_0$ and $\Delta \mathbf{x}_k = \bar{\lambda} \Delta \mathbf{x}_0$, the forgoing equation becomes

$$
\begin{aligned}
(\mathbf{DP}_{k(k-1)\cdots 1} - \bar{\lambda} \mathbf{I}_{n \times n}) \Delta \mathbf{x}_0 = \mathbf{0}, \\
(\mathbf{DP} - \lambda \mathbf{I}_{n \times n}) \Delta \mathbf{x}_0 = \mathbf{0}.
\end{aligned} \tag{5.26}
$$

For any non-trivial solution $(\|\Delta \mathbf{x}_0\| \neq 0)$, we have

$$
\begin{aligned}
|\mathbf{DP}_{k(k-1)\cdots 1} - \bar{\lambda} \mathbf{I}_{n \times n}| = 0, \quad (k = 1, 2, \ldots, N); \\
|\mathbf{DP} - \lambda \mathbf{I}_{n \times n}| = 0.
\end{aligned} \tag{5.27}
$$

Thus, the eigenvalues of DP and $\mathbf{DP}_{k(k-1)\cdots 1}$ are computed for the periodic solution. The eigenvalues of $\mathbf{DP}_{k(k-1)\cdots 1}$ give the properties of $\mathbf{x}_k$ varying with $\mathbf{x}_0$ for the periodic flow. From the stability and bifurcation theory of dynamical systems at fixed points in discrete nonlinear systems, the stability and bifurcation of the periodic solution can be classified as in Luo (2012a, b). This theorem is proved.  □

To explain how to approximate the periodic flow in an $n$-dimensional nonlinear dynamical system, consider an $n_1 \times n_2$ plane $(n_1 + n_2 = n)$, as shown in Fig. 5.1. $N$-nodes of the periodic flow are chosen for an approximate solution with a certain accuracy  $\|\mathbf{x}(t_k) - \mathbf{x}_k\| \leq \varepsilon_k$   $(\varepsilon_k > 0)$   and   $\|\mathbf{f}(\mathbf{x}(t_k), t_k, \mathbf{p}) - \mathbf{f}(\mathbf{x}_k, t_k, \mathbf{p})\| \leq \delta_k$ $(\delta_k > 0)$. Letting  $\delta = \max\{\delta_k\}_{k \in \{1,2,\ldots,N\}}$  and  $\varepsilon = \max\{\varepsilon_k\}_{k \in \{1,2,\ldots,N\}}$  be small positive quantities prescribed, the periodic flow can be approximately described by a set of mappings $P_k$ with $\mathbf{g}_k(\mathbf{x}_{k-1}, \mathbf{x}_k, \mathbf{p}) = \mathbf{0}$ $(k = 1, 2, \ldots, N)$ with periodicity condition $\mathbf{x}_N = \mathbf{x}_0$. Based on the approximate mapping functions, the nodes of periodic motions are computed approximately, which is depicted by a solid curve. The exact solution of the periodic flow is described by a dashed curve. The node points on the periodic flows are depicted with short lines. The red symbols are node points on the exact solution of the periodic flow. The discrete mapping $P_k$ is developed from the differential equation. With the control of computational accuracy, the nodes of the periodic flow can be obtained with a good approximation.

From the previous methodology, a set of nonlinear discrete mappings $P_k$ with $\mathbf{g}_k(\mathbf{x}_{k-1}, \mathbf{x}_k, \mathbf{p}) = \mathbf{0}$ $(k = 1, 2, \ldots, N)$ are developed for periodic flows. Such mapping can be used for numerical simulations. For given $\mathbf{x}_{k-1}$, one can compute $\mathbf{x}_k$ through $\mathbf{g}_k(\mathbf{x}_{k-1}, \mathbf{x}_k, \mathbf{p}) = \mathbf{0}$. For the explicit form, the mapping is directly used for computation of $\mathbf{x}_k$. For the implicit form, the mapping iteration or Newton–Raphson method can be adopted to compute $\mathbf{x}_k$. In addition to a one-step mapping of $P_k$ with $\mathbf{g}_k(\mathbf{x}_{k-1}, \mathbf{x}_k, \mathbf{p}) = \mathbf{0}$, one can develop a multi-step (or $l$-steps) mapping of $P_k$ with

$$
\begin{aligned}
\mathbf{g}_k(\mathbf{x}_{k-l}, \ldots, \mathbf{x}_{k-1}, \mathbf{x}_k, \mathbf{p}) = \mathbf{0}, \quad (k = 1, 2, \ldots, N) \\
l \in \{1, 2, \ldots, k\}.
\end{aligned} \tag{5.28}
$$

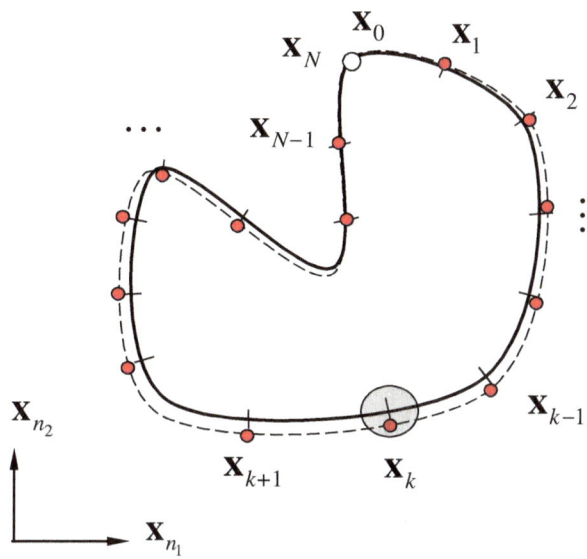

**Fig. 5.1** Period-1 flow with $N$-nodes with short lines. *Solid curve* expected exact results, and *dashed curve* expected numerical results. The local *shaded* area is a small neighborhood of the exact solution at the $k$th node. The *red* symbols are for node points on the numerical solution of the periodic flow

(i)   If $l = 1$, the one-step mapping is recovered from the multi-step mapping.
(ii)  If $l = 2$, the two-step mapping is obtained from the multi-step mapping as

$$\mathbf{g}_k(\mathbf{x}_{k-2}, \mathbf{x}_{k-1}, \mathbf{x}_k, \mathbf{p}) = \mathbf{0}, \quad (k = 1, 2, \ldots, N) \tag{5.29}$$

which can be expanded as

$$\mathbf{g}_1(\mathbf{x}_0, \mathbf{x}_1, \mathbf{p}) = \mathbf{0},$$
$$\vdots \tag{5.30}$$
$$\mathbf{g}_k(\mathbf{x}_{k-2}, \mathbf{x}_{k-1}, \mathbf{x}_k, \mathbf{p}) = \mathbf{0}, \quad (k = 1, 2, \ldots, N).$$

(iii)  If $l = k$, the $k$-steps mapping is obtained, i.e.,

$$\mathbf{g}_k(\mathbf{x}_0, \mathbf{x}_1 \ldots, \mathbf{x}_{k-1}, \mathbf{x}_k, \mathbf{p}) = \mathbf{0}, \quad (k = 1, 2, \ldots, N). \tag{5.31}$$

and the forgoing equation can be expanded as

$$\mathbf{g}_1(\mathbf{x}_0, \mathbf{x}_1, \mathbf{p}) = \mathbf{0},$$
$$\vdots \tag{5.32}$$
$$\mathbf{g}_k(\mathbf{x}_0, \mathbf{x}_1, \ldots, \mathbf{x}_{k-1}, \mathbf{x}_k, \mathbf{p}) = \mathbf{0}, \quad (k = 1, 2, \ldots, N).$$

From the multi-step (or $l$-steps) mapping of $P_k$, with the periodicity condition $(\mathbf{x}_0 = \mathbf{x}_N)$, the periodic flow can be obtained via

$$\mathbf{g}_k(\mathbf{x}_{k-l}, \ldots, \mathbf{x}_{k-1}, \mathbf{x}_k, \mathbf{p}) = \mathbf{0},$$
$$(k = 1, 2, \ldots, N; l \in \{1, 2, \ldots, k\}) \tag{5.33}$$
$$\mathbf{x}_0 = \mathbf{x}_N.$$

Suppose node points $\mathbf{x}_k^*$ $(k = 0, 1, 2, \ldots, N)$ of periodic flows are obtained, the stability and bifurcation can be in the vicinity of $\mathbf{x}_k^*$ with $\mathbf{x}_k = \mathbf{x}_k^* + \Delta\mathbf{x}_k$, i.e.,

$$\frac{\partial \mathbf{g}_k}{\partial \mathbf{x}_{k-l}} \frac{\partial \mathbf{x}_{k-l}}{\partial \mathbf{x}_0} + \cdots + \frac{\partial \mathbf{g}_k}{\partial \mathbf{x}_{k-1}} \frac{\partial \mathbf{x}_{k-1}}{\partial \mathbf{x}_0} + \frac{\partial \mathbf{g}_k}{\partial \mathbf{x}_k} \frac{\partial \mathbf{x}_k}{\partial \mathbf{x}_0} = \mathbf{0}_{n \times n} \tag{5.34}$$
$$(k = 1, 2, \ldots, N; l \in \{1, 2, \ldots, k\}).$$

In other words, we have

$$
\begin{bmatrix} \frac{\partial \mathbf{x}_1}{\partial \mathbf{x}_0} \\ \frac{\partial \mathbf{x}_2}{\partial \mathbf{x}_0} \\ \vdots \\ \frac{\partial \mathbf{x}_l}{\partial \mathbf{x}_0} \\ \vdots \\ \frac{\partial \mathbf{x}_{N-1}}{\partial \mathbf{x}_0} \\ \vdots \\ \frac{\partial \mathbf{x}_N}{\partial \mathbf{x}_0} \end{bmatrix} = -\begin{bmatrix} \frac{\partial \mathbf{g}_1}{\partial \mathbf{x}_1} & \mathbf{0}_{n \times n} & \cdots & \mathbf{0}_{n \times n} & \cdots & \mathbf{0}_{n \times n} & \cdots & \mathbf{0}_{n \times n} \\ \frac{\partial \mathbf{g}_2}{\partial \mathbf{x}_1} & \frac{\partial \mathbf{g}_2}{\partial \mathbf{x}_2} & \cdots & \mathbf{0}_{n \times n} & \cdots & \mathbf{0}_{n \times n} & \cdots & \mathbf{0}_{n \times n} \\ \vdots & \vdots & & \vdots & & \vdots & & \vdots \\ \frac{\partial \mathbf{g}_l}{\partial \mathbf{x}_1} & \frac{\partial \mathbf{g}_l}{\partial \mathbf{x}_2} & \cdots & \frac{\partial \mathbf{g}_l}{\partial \mathbf{x}_l} & \cdots & \mathbf{0}_{n \times n} & \cdots & \mathbf{0}_{n \times n} \\ \vdots & \vdots & & \vdots & & \vdots & & \vdots \\ \mathbf{0}_{n \times n} & \mathbf{0}_{n \times n} & \cdots & \mathbf{0}_{n \times n} & \cdots & \frac{\partial \mathbf{g}_{N-1}}{\partial \mathbf{x}_{N-l}} & \cdots & \mathbf{0}_{n \times n} \\ \vdots & \vdots & & \vdots & & \vdots & & \vdots \\ \mathbf{0}_{n \times n} & \mathbf{0}_{n \times n} & \cdots & \mathbf{0}_{n \times n} & \cdots & \frac{\partial \mathbf{g}_N}{\partial \mathbf{x}_{N-l}} & \cdots & \frac{\partial \mathbf{g}_N}{\partial \mathbf{x}_N} \end{bmatrix}^{-1} \begin{bmatrix} \frac{\partial \mathbf{g}_1}{\partial \mathbf{x}_0} \\ \frac{\partial \mathbf{g}_2}{\partial \mathbf{x}_0} \\ \vdots \\ \frac{\partial \mathbf{g}_l}{\partial \mathbf{x}_0} \\ \vdots \\ \mathbf{0}_{n \times n} \\ \vdots \\ \mathbf{0}_{n \times n} \end{bmatrix}
\tag{5.35}
$$

From the mapping structure, we have

$$\Delta\mathbf{x}_N = \mathrm{DP} \cdot \Delta\mathbf{x}_0 \quad \text{and} \quad \mathrm{DP} = \left[\frac{\partial \mathbf{x}_N}{\partial \mathbf{x}_0}\right]. \tag{5.36}$$

Letting $\Delta\mathbf{x}_N = \lambda\Delta\mathbf{x}_0$, we have

$$(\mathrm{DP} - \lambda\mathbf{I}_{n \times n})\Delta\mathbf{x}_0 = \mathbf{0}. \tag{5.37}$$

The eigenvalue of DP is given by $|\mathrm{DP} - \lambda\mathbf{I}_{n \times n}| = 0$. In addition, we have

$$\Delta\mathbf{x}_k = \mathrm{DP}_{k(k-1)\cdots 1} \cdot \Delta\mathbf{x}_0 \quad \text{and} \quad \mathrm{DP}_{k(k-1)\cdots 1} = \left[\frac{\partial \mathbf{x}_k}{\partial \mathbf{x}_0}\right] \tag{5.38}$$
$$(k = 1, 2, \ldots, N).$$

Letting $\Delta \mathbf{x}_k = \lambda \Delta \mathbf{x}_0$, we have

$$(\mathrm{DP}_{k(k-1)\cdots 1} - \lambda \mathbf{I}_{n\times n})\Delta \mathbf{x}_0 = \mathbf{0}. \tag{5.39}$$

The eigenvalues of $\mathrm{DP}_{k(k-1)\cdots 1}$ are given by $|\mathrm{DP}_{k(k-1)\cdots 1} - \lambda \mathbf{I}_{n\times n}| = 0$. Such eigenvalues tell effects of variation of $\mathbf{x}_0$ on node $\mathbf{x}_k$ in its vicinity. The neighborhood of $\mathbf{x}_k^*$ (i.e., $U_k(\mathbf{x}_k^*)$) is presented in Fig. 5.2 as large circles. In such a neighborhood, the eigenvalues are used to measure the effects $\Delta \mathbf{x}_k$ of $\mathbf{x}_k^*$ varying with $\Delta \mathbf{x}_0$ at $\mathbf{x}_0^*$.

(i)  If $l = 1$, Eq. (5.34) becomes

$$\frac{\partial \mathbf{g}_k}{\partial \mathbf{x}_{k-1}} \frac{\partial \mathbf{x}_{k-1}}{\partial \mathbf{x}_0} + \frac{\partial \mathbf{g}_k}{\partial \mathbf{x}_k} \frac{\partial \mathbf{x}_k}{\partial \mathbf{x}_0} = \mathbf{0} \quad (k = 1, 2, \ldots, N). \tag{5.40}$$

The deformation of the forgoing equation yields

$$\frac{\partial \mathbf{g}_k}{\partial \mathbf{x}_{k-1}} + \frac{\partial \mathbf{g}_k}{\partial \mathbf{x}_k} \frac{\partial \mathbf{x}_k}{\partial \mathbf{x}_{k-1}} = \mathbf{0} \quad (k = 1, 2, \ldots, N). \tag{5.41}$$

That is,

$$\frac{\partial \mathbf{x}_k}{\partial \mathbf{x}_{k-1}} = -\left[\frac{\partial \mathbf{g}_k}{\partial \mathbf{x}_k}\right]^{-1} \cdot \frac{\partial \mathbf{g}_k}{\partial \mathbf{x}_{k-1}} \quad (k = 1, 2, \ldots, N). \tag{5.42}$$

**Fig. 5.2** Neighborhoods of $N$-nodes with short lines for a period-1 flow. *Solid curve* expected numerical results, and *dashed curve* expected exact results. The local *shaded area* is a small neighborhood of the exact solution at the $k$th node. The *red symbols* are node points on the exact solution of the periodic flow

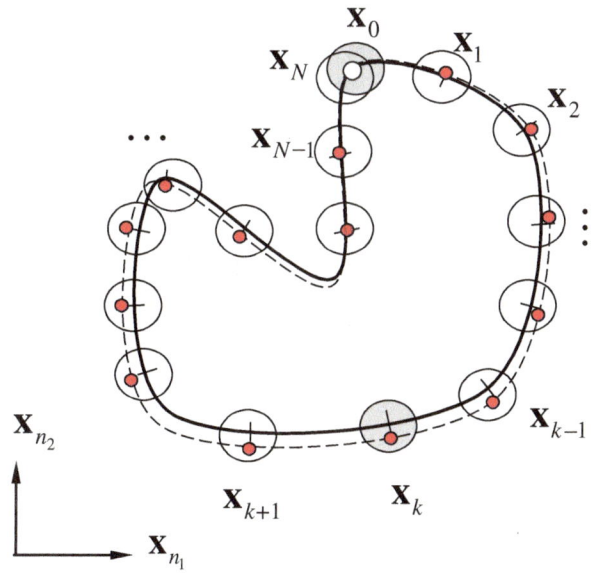

From Eq. (5.40), the following matrix form can be formed.

$$
\begin{bmatrix}
\dfrac{\partial \mathbf{g}_1}{\partial \mathbf{x}_1} & \mathbf{0}_{n\times n} & \cdots & \mathbf{0}_{n\times n} & \mathbf{0}_{n\times n} & \cdots & \mathbf{0}_{n\times n} & \mathbf{0}_{n\times n} \\[2mm]
\dfrac{\partial \mathbf{g}_2}{\partial \mathbf{x}_1} & \dfrac{\partial \mathbf{g}_2}{\partial \mathbf{x}_2} & \cdots & \mathbf{0}_{n\times n} & \mathbf{0}_{n\times n} & \cdots & \mathbf{0}_{n\times n} & \mathbf{0}_{n\times n} \\[2mm]
\vdots & \vdots & & \mathbf{0}_{n\times n} & \mathbf{0}_{n\times n} & & \vdots & \vdots \\[2mm]
\mathbf{0}_{n\times n} & \mathbf{0}_{n\times n} & \cdots & \dfrac{\partial \mathbf{g}_{k-1}}{\partial \mathbf{x}_{k-1}} & \mathbf{0}_{n\times n} & \cdots & \mathbf{0}_{n\times n} & \mathbf{0}_{n\times n} \\[2mm]
\mathbf{0}_{n\times n} & \mathbf{0}_{n\times n} & \cdots & \dfrac{\partial \mathbf{g}_k}{\partial \mathbf{x}_{k-1}} & \dfrac{\partial \mathbf{g}_k}{\partial \mathbf{x}_k} & \cdots & \mathbf{0}_{n\times n} & \mathbf{0}_{n\times n} \\[2mm]
\vdots & \vdots & & \vdots & \vdots & & \vdots & \vdots \\[2mm]
\mathbf{0}_{n\times n} & \mathbf{0}_{n\times n} & \cdots & \mathbf{0}_{n\times n} & \mathbf{0}_{n\times n} & \cdots & \dfrac{\partial \mathbf{g}_{N-1}}{\partial \mathbf{x}_{N-1}} & \mathbf{0}_{n\times n} \\[2mm]
\mathbf{0}_{n\times n} & \mathbf{0}_{n\times n} & \cdots & \mathbf{0}_{n\times n} & \mathbf{0}_{n\times n} & \cdots & \dfrac{\partial \mathbf{g}_N}{\partial \mathbf{x}_{N-1}} & \dfrac{\partial \mathbf{g}_N}{\partial \mathbf{x}_N}
\end{bmatrix}
\begin{bmatrix}
\dfrac{\partial \mathbf{x}_1}{\partial \mathbf{x}_0} \\[2mm]
\dfrac{\partial \mathbf{x}_2}{\partial \mathbf{x}_0} \\[2mm]
\vdots \\[2mm]
\dfrac{\partial \mathbf{x}_{k-1}}{\partial \mathbf{x}_0} \\[2mm]
\dfrac{\partial \mathbf{x}_k}{\partial \mathbf{x}_0} \\[2mm]
\vdots \\[2mm]
\dfrac{\partial \mathbf{x}_{N-1}}{\partial \mathbf{x}_0} \\[2mm]
\dfrac{\partial \mathbf{x}_N}{\partial \mathbf{x}_0}
\end{bmatrix}
= -
\begin{bmatrix}
\dfrac{\partial \mathbf{g}_1}{\partial \mathbf{x}_0} \\[2mm]
\mathbf{0}_{n\times n} \\[2mm]
\vdots \\[2mm]
\mathbf{0}_{n\times n} \\[2mm]
\mathbf{0}_{n\times n} \\[2mm]
\vdots \\[2mm]
\mathbf{0}_{n\times n} \\[2mm]
\mathbf{0}_{n\times n}
\end{bmatrix}.
$$

$$(5.43)$$

So we have

$$
\mathrm{DP} = \left[\frac{\partial \mathbf{x}_N}{\partial \mathbf{x}_0}\right] = \left[\frac{\partial \mathbf{x}_N}{\partial \mathbf{x}_{N-1}}\right] \cdots \left[\frac{\partial \mathbf{x}_1}{\partial \mathbf{x}_0}\right]. \tag{5.44}
$$

(i) For $l = k$, Eq. (5.33) with periodicity condition ($\mathbf{x}_0 = \mathbf{x}_N$) gives node points $\mathbf{x}_k^*$ ($k = 0, 1, 2, \ldots, N$). The stability and bifurcation can be analyzed in the vicinity of $\mathbf{x}_k^*$ with $\mathbf{x}_k = \mathbf{x}_k^* + \Delta\mathbf{x}_k$. Equation (5.34) becomes

$$
\frac{\partial \mathbf{g}_k}{\partial \mathbf{x}_0} + \frac{\partial \mathbf{g}_k}{\partial \mathbf{x}_1}\frac{\partial \mathbf{x}_1}{\partial \mathbf{x}_0} + \cdots + \frac{\partial \mathbf{g}_k}{\partial \mathbf{x}_{k-1}}\frac{\partial \mathbf{x}_{k-1}}{\partial \mathbf{x}_0} + \frac{\partial \mathbf{g}_k}{\partial \mathbf{x}_k}\frac{\partial \mathbf{x}_k}{\partial \mathbf{x}_0} = \mathbf{0}_{n\times n}.
$$
$$(k = 1, 2, \ldots, N) \tag{5.45}$$

In other words,

$$
\begin{bmatrix}
\dfrac{\partial \mathbf{g}_1}{\partial \mathbf{x}_1} & \cdots & \mathbf{0}_{n\times n} & \mathbf{0}_{n\times n} \\[2mm]
\vdots & & \vdots & \vdots \\[2mm]
\dfrac{\partial \mathbf{g}_{N-1}}{\partial \mathbf{x}_1} & \cdots & \dfrac{\partial \mathbf{g}_{N-1}}{\partial \mathbf{x}_{N-1}} & \mathbf{0}_{n\times n} \\[2mm]
\dfrac{\partial \mathbf{g}_N}{\partial \mathbf{x}_1} & \cdots & \dfrac{\partial \mathbf{g}_N}{\partial \mathbf{x}_{N-1}} & \dfrac{\partial \mathbf{g}_N}{\partial \mathbf{x}_N}
\end{bmatrix}
\begin{bmatrix}
\dfrac{\partial \mathbf{x}_1}{\partial \mathbf{x}_0} \\[2mm]
\vdots \\[2mm]
\dfrac{\partial \mathbf{x}_{N-1}}{\partial \mathbf{x}_0} \\[2mm]
\dfrac{\partial \mathbf{x}_N}{\partial \mathbf{x}_0}
\end{bmatrix}
= -
\begin{bmatrix}
\dfrac{\partial \mathbf{g}_1}{\partial \mathbf{x}_0} \\[2mm]
\vdots \\[2mm]
\dfrac{\partial \mathbf{g}_{N-1}}{\partial \mathbf{x}_0} \\[2mm]
\dfrac{\partial \mathbf{g}_N}{\partial \mathbf{x}_0}
\end{bmatrix}
\tag{5.46}
$$

and

$$
\begin{bmatrix}
\frac{\partial \mathbf{x}_1}{\partial \mathbf{x}_0} \\
\vdots \\
\frac{\partial \mathbf{x}_{N-1}}{\partial \mathbf{x}_0} \\
\frac{\partial \mathbf{x}_N}{\partial \mathbf{x}_0}
\end{bmatrix}
= -
\begin{bmatrix}
\frac{\partial \mathbf{g}_1}{\partial \mathbf{x}_1} & \cdots & \mathbf{0}_{n\times n} & \mathbf{0}_{n\times n} \\
\vdots & & \vdots & \vdots \\
\frac{\partial \mathbf{g}_{N-1}}{\partial \mathbf{x}_1} & \cdots & \frac{\partial \mathbf{g}_{N-1}}{\partial \mathbf{x}_{N-1}} & \mathbf{0}_{n\times n} \\
\frac{\partial \mathbf{g}_N}{\partial \mathbf{x}_1} & \cdots & \frac{\partial \mathbf{g}_N}{\partial \mathbf{x}_{N-1}} & \frac{\partial \mathbf{g}_N}{\partial \mathbf{x}_N}
\end{bmatrix}^{-1}
\begin{bmatrix}
\frac{\partial \mathbf{g}_1}{\partial \mathbf{x}_0} \\
\vdots \\
\frac{\partial \mathbf{g}_{N-1}}{\partial \mathbf{x}_0} \\
\frac{\partial \mathbf{g}_N}{\partial \mathbf{x}_0}
\end{bmatrix}.
\tag{5.47}
$$

Using $\partial \mathbf{x}_k / \partial \mathbf{x}_0$, the eigenvalues are determined by

$$
|DP_{k(k-1)\cdots 1} - \bar{\lambda} \mathbf{I}_{n\times n}| = 0 \quad \text{with } DP_{k(k-1)\cdots 1} = \left[ \frac{\partial \mathbf{x}_k}{\partial \mathbf{x}_0} \right]
\tag{5.48}
$$

which is used to measure the properties of node points on the periodic flow.

The multi-step mappings are developed from the previously determined nodes of periodic motion. During time interval $t \in [t_0, t_0 + T]$, the periodic flow can be determined by

$$
\mathbf{x}(t) = \mathbf{x}(t_l) + \int_{t_l}^{t} \mathbf{f}(\mathbf{x}, t, \mathbf{p}) dt, \quad l \in \{0, 1, \ldots, k-1\}.
\tag{5.49}
$$

For such a periodic flow, at most, all of $N$-nodes during the time interval $t \in [t_0, t_0 + T]$ are selected, and the corresponding points $\mathbf{x}(t_k) (k = 1, 2, \ldots, N)$. Under $||\mathbf{x}(t_k) - \mathbf{x}_k|| \le \varepsilon_k$ with $\varepsilon_k \ge 0$,

$$
||\mathbf{f}(\mathbf{x}(t_k), t_k, \mathbf{p}) - \mathbf{f}(\mathbf{x}_k, t_k, \mathbf{p})|| \le \delta_k.
\tag{5.50}
$$

Suppose that $\mathbf{x}_0, \ldots, \mathbf{x}_N$ are given, $\mathbf{f}(\mathbf{x}_k, t_k, \mathbf{p})$ $(k = 0, 1, \ldots, N)$ can be determined. An interpolation polynomial $\mathbf{P}(t, \mathbf{x}_0, \ldots, \mathbf{x}_N, t_0, \ldots, t_N, \mathbf{p})$ is determined, which can be used to approximate $\mathbf{f}(\mathbf{x}, t, \mathbf{p})$. That is,

$$
\mathbf{f}(\mathbf{x}, t, \mathbf{p}) \approx \mathbf{P}(t, \mathbf{x}_0, \ldots, \mathbf{x}_N, t_0, \ldots, t_N, \mathbf{p})
\tag{5.51}
$$

and $\mathbf{x}(t_k) \approx \mathbf{x}_k$ can be computed by

$$
\mathbf{x}_k = \mathbf{x}_l + \int_{t_l}^{t_k} \mathbf{P}(t, \mathbf{x}_0, \ldots, \mathbf{x}_N, t_0, \ldots, t_N, \mathbf{p}) dt \quad (l \in \{0, 1, \ldots, k-1\}).
\tag{5.52}
$$

Therefore, we have

$$
\mathbf{x}_k = \mathbf{x}_l + \bar{\mathbf{g}}_k(\mathbf{x}_0, \ldots, \mathbf{x}_N, \mathbf{p}) \quad (l \in \{0, 1, \ldots, k-1\}).
\tag{5.53}
$$

The mapping $P_k$ $(k \in \{1, 2, \ldots, N\})$ for a specific $l$ is

$$\mathbf{g}_k(\mathbf{x}_0, \ldots, \mathbf{x}_N, \mathbf{p}) = \mathbf{0}. \tag{5.54}$$

The periodic motion is determined by mapping $P_k$ $(k = 1, 2, \ldots, N)$ and periodicity conditions

$$\mathbf{g}_k(\mathbf{x}_0^*, \ldots, \mathbf{x}_N^*, \mathbf{p}) = \mathbf{0} \quad \text{for } k = 1, 2, \ldots, N,$$
$$\mathbf{x}_0^* = \mathbf{x}_N^*. \tag{5.55}$$

From the forgoing equation, node points $\mathbf{x}_k^*$ $(k = 0, 1, 2, \ldots, N)$ can be determined. The corresponding stability and bifurcation can be discussed in the neighborhood of $\mathbf{x}_k^*$ with $\mathbf{x}_k = \mathbf{x}_k^* + \Delta\mathbf{x}_k$. The derivative of Eq. (5.55) with respect to $\mathbf{x}_0$ gives

$$\frac{\partial \mathbf{g}_k}{\partial \mathbf{x}_0} + \frac{\partial \mathbf{g}_k}{\partial \mathbf{x}_1}\frac{\partial \mathbf{x}_1}{\partial \mathbf{x}_0} + \cdots + \frac{\partial \mathbf{g}_k}{\partial \mathbf{x}_{k-1}}\frac{\partial \mathbf{x}_{k-1}}{\partial \mathbf{x}_0} + \frac{\partial \mathbf{g}_k}{\partial \mathbf{x}_k}\frac{\partial \mathbf{x}_k}{\partial \mathbf{x}_0} = \mathbf{0}_{n \times n},$$
$$(k = 1, 2, \ldots, N) \tag{5.56}$$

In other words, we have

$$\begin{bmatrix} \dfrac{\partial \mathbf{g}_1}{\partial \mathbf{x}_1} & \cdots & \dfrac{\partial \mathbf{g}_1}{\partial \mathbf{x}_{N-1}} & \dfrac{\partial \mathbf{g}_1}{\partial \mathbf{x}_N} \\ \vdots & & \vdots & \vdots \\ \dfrac{\partial \mathbf{g}_{N-1}}{\partial \mathbf{x}_1} & \cdots & \dfrac{\partial \mathbf{g}_{N-1}}{\partial \mathbf{x}_{N-1}} & \dfrac{\partial \mathbf{g}_{N-1}}{\partial \mathbf{x}_N} \\ \dfrac{\partial \mathbf{g}_N}{\partial \mathbf{x}_1} & \cdots & \dfrac{\partial \mathbf{g}_N}{\partial \mathbf{x}_{N-1}} & \dfrac{\partial \mathbf{g}_N}{\partial \mathbf{x}_N} \end{bmatrix} \begin{bmatrix} \dfrac{\partial \mathbf{x}_1}{\partial \mathbf{x}_0} \\ \vdots \\ \dfrac{\partial \mathbf{x}_{N-1}}{\partial \mathbf{x}_0} \\ \dfrac{\partial \mathbf{x}_N}{\partial \mathbf{x}_0} \end{bmatrix} = - \begin{bmatrix} \dfrac{\partial \mathbf{g}_1}{\partial \mathbf{x}_0} \\ \vdots \\ \dfrac{\partial \mathbf{g}_{N-1}}{\partial \mathbf{x}_0} \\ \dfrac{\partial \mathbf{g}_N}{\partial \mathbf{x}_0} \end{bmatrix} \tag{5.57}$$

and

$$\begin{bmatrix} \dfrac{\partial \mathbf{x}_1}{\partial \mathbf{x}_0} \\ \vdots \\ \dfrac{\partial \mathbf{x}_{N-1}}{\partial \mathbf{x}_0} \\ \dfrac{\partial \mathbf{x}_N}{\partial \mathbf{x}_0} \end{bmatrix} = - \begin{bmatrix} \dfrac{\partial \mathbf{g}_1}{\partial \mathbf{x}_1} & \cdots & \dfrac{\partial \mathbf{g}_1}{\partial \mathbf{x}_{N-1}} & \dfrac{\partial \mathbf{g}_1}{\partial \mathbf{x}_N} \\ \vdots & & \vdots & \vdots \\ \dfrac{\partial \mathbf{g}_{N-1}}{\partial \mathbf{x}_1} & \cdots & \dfrac{\partial \mathbf{g}_{N-1}}{\partial \mathbf{x}_{N-1}} & \dfrac{\partial \mathbf{g}_{N-1}}{\partial \mathbf{x}_N} \\ \dfrac{\partial \mathbf{g}_N}{\partial \mathbf{x}_1} & \cdots & \dfrac{\partial \mathbf{g}_N}{\partial \mathbf{x}_{N-1}} & \dfrac{\partial \mathbf{g}_N}{\partial \mathbf{x}_N} \end{bmatrix}^{-1} \begin{bmatrix} \dfrac{\partial \mathbf{g}_1}{\partial \mathbf{x}_0} \\ \vdots \\ \dfrac{\partial \mathbf{g}_{N-1}}{\partial \mathbf{x}_0} \\ \dfrac{\partial \mathbf{g}_N}{\partial \mathbf{x}_0} \end{bmatrix}. \tag{5.58}$$

From the above discussion, the discrete mapping can be developed through many forward and backward nodes. The periodic flow in a nonlinear dynamical system can be determined through the following theorem.

**Theorem 5.2** *Consider a nonlinear dynamical system in Eq. (5.1). If such a dynamical system has a periodic flow $\mathbf{x}(t)$ with finite norm $\|\mathbf{x}\|$ and period*

$T = 2\pi/\Omega$, there is a set of discrete time $t_k$ $(k = 0, 1, \ldots, N)$ with $(N \to \infty)$ during one period $T$, and the corresponding solution $\mathbf{x}(t_k)$ and vector field $\mathbf{f}(\mathbf{x}(t_k), t_k, \mathbf{p})$ are exact. Suppose a discrete node $\mathbf{x}_k$ is on the approximate solutions of the periodic flow under $\|\mathbf{x}(t_k) - \mathbf{x}_k\| \le \varepsilon_k$ with a small $\varepsilon_k \ge 0$ and

$$\|\mathbf{f}(\mathbf{x}(t_k), t_k, \mathbf{p}) - \mathbf{f}(\mathbf{x}_k, t_k, \mathbf{p})\| \le \delta_k \tag{5.59}$$

with a small $\delta_k \ge 0$. During a time interval $t \in [t_{k-1}, t_k]$, there is a mapping $P_k :$ $\mathbf{x}_{k-1} \to \mathbf{x}_k$ $(k = 1, 2, \ldots, N)$ as

$$
\begin{aligned}
&\mathbf{x}_k = P_k \mathbf{x}_{k-1} \quad \text{with } \mathbf{g}_k(\mathbf{x}_{s_{kl_1}}, \ldots, \mathbf{x}_{s_{k1}}, \mathbf{x}_{s_{k0}}, \mathbf{x}_{s_{k(-1)}} \cdots, \mathbf{x}_{s_{k(-l_2)}}, \mathbf{p}) = \mathbf{0}, \\
&s_{kj} = \mathrm{mod}(k - j + N, N), j = -l_2, -l_2 + 1, \ldots, l_1 - 1, l_1; \\
&l_1, l_2 \in \{0, 1, 2, \ldots, N\}, 1 \le l_1 + l_2 \le N; l_1 \ge 1; (k = 1, 2, \ldots, N)
\end{aligned}
\tag{5.60}
$$

where $\mathbf{g}_k$ is an implicit vector function. Consider a mapping structure as

$$
\begin{aligned}
&P = P_N \circ \cdots \circ P_2 \circ P_1 : \ \mathbf{x}_0 \to \mathbf{x}_N; \\
&\text{with } P_k : \ \mathbf{x}_{k-1} \to \mathbf{x}_k (k = 1, 2, \ldots, N).
\end{aligned}
\tag{5.61}
$$

For $\mathbf{x}_N = P\mathbf{x}_0$, if there is a set of points $\mathbf{x}_k^*$ $(k = 0, 1, \ldots, N)$ computed by

$$
\begin{aligned}
&\mathbf{g}_k(\mathbf{x}_{s_{kl_1}}^*, \ldots, \mathbf{x}_{s_{k1}}^*, \mathbf{x}_{s_{k0}}^*, \mathbf{x}_{s_{k(-1)}}^* \cdots, \mathbf{x}_{s_{k(-l_2)}}^*, \mathbf{p}) = \mathbf{0}, \\
&(k = 1, 2, \ldots, N) \\
&\mathbf{x}_0^* = \mathbf{x}_N^*,
\end{aligned}
\tag{5.62}
$$

then the points $\mathbf{x}_k^* (k = 0, 1, \ldots, N)$ are approximations of points $\mathbf{x}(t_k)$ of the periodic solution. In the neighborhood of $\mathbf{x}_k^*$, with $\mathbf{x}_k = \mathbf{x}_k^* + \Delta\mathbf{x}_k$, the linearized equation is given by

$$\frac{\partial \mathbf{g}_k}{\partial \mathbf{x}_0} + \cdots + \frac{\partial \mathbf{g}_k}{\partial \mathbf{x}_{k-1}} \frac{\partial \mathbf{x}_{k-1}}{\partial \mathbf{x}_0} + \frac{\partial \mathbf{g}_k}{\partial \mathbf{x}_k} \frac{\partial \mathbf{x}_k}{\partial \mathbf{x}_0} + \frac{\partial \mathbf{g}_k}{\partial \mathbf{x}_{k+1}} \frac{\partial \mathbf{x}_{k+1}}{\partial \mathbf{x}_0} + \cdots + \frac{\partial \mathbf{g}_k}{\partial \mathbf{x}_N} \frac{\partial \mathbf{x}_N}{\partial \mathbf{x}_0} = \mathbf{0}$$

with $\dfrac{\partial \mathbf{g}_k}{\partial \mathbf{x}_\alpha} = \mathbf{0}(\alpha \ne s_{kj})$, $j = -l_2, -l_2 + 1, \ldots, l_1 - 1, l_1; (k = 1, 2, \ldots, N)$. $\quad$ (5.63)

The resultant Jacobian matrices of the periodic flow are

$$
\begin{aligned}
&DP_{k(k-1)\cdots 1} = \left[ \frac{\partial \mathbf{x}_k}{\partial \mathbf{x}_0} \right]_{(\mathbf{x}_0^*, \mathbf{x}_1^*, \ldots, \mathbf{x}_N^*)} \qquad (k = 1, 2, \ldots, N), \\
&\text{and} \quad DP = DP_{N(N-1)\cdots 1} = \left[ \frac{\partial \mathbf{x}_N}{\partial \mathbf{x}_0} \right]_{(\mathbf{x}_0^*, \mathbf{x}_1^*, \ldots, \mathbf{x}_N^*)}
\end{aligned}
\tag{5.64}
$$

*where*

$$
\begin{bmatrix} \dfrac{\partial \mathbf{x}_1}{\partial \mathbf{x}_0} \\[2mm] \vdots \\[2mm] \dfrac{\partial \mathbf{x}_{N-1}}{\partial \mathbf{x}_0} \\[2mm] \dfrac{\partial \mathbf{x}_N}{\partial \mathbf{x}_0} \end{bmatrix} = - \begin{bmatrix} \dfrac{\partial \mathbf{g}_1}{\partial \mathbf{x}_1} & \cdots & \dfrac{\partial \mathbf{g}_1}{\partial \mathbf{x}_{N-1}} & \dfrac{\partial \mathbf{g}_1}{\partial \mathbf{x}_N} \\[2mm] \vdots & & \vdots & \vdots \\[2mm] \dfrac{\partial \mathbf{g}_{N-1}}{\partial \mathbf{x}_1} & \cdots & \dfrac{\partial \mathbf{g}_{N-1}}{\partial \mathbf{x}_{N-1}} & \dfrac{\partial \mathbf{g}_{N-1}}{\partial \mathbf{x}_N} \\[2mm] \dfrac{\partial \mathbf{g}_N}{\partial \mathbf{x}_1} & \cdots & \dfrac{\partial \mathbf{g}_N}{\partial \mathbf{x}_{N-1}} & \dfrac{\partial \mathbf{g}_N}{\partial \mathbf{x}_N} \end{bmatrix}^{-1} \begin{bmatrix} \dfrac{\partial \mathbf{g}_1}{\partial \mathbf{x}_0} \\[2mm] \vdots \\[2mm] \dfrac{\partial \mathbf{g}_{N-1}}{\partial \mathbf{x}_0} \\[2mm] \dfrac{\partial \mathbf{g}_N}{\partial \mathbf{x}_0} \end{bmatrix}. \tag{5.65}
$$

*The properties of discrete points* $\mathbf{x}_k$ $(k = 1, 2, \ldots, N)$ *can be estimated by the eigenvalues of* $DP_{k(k-1)\cdots 1}$ *as*

$$
|DP_{k(k-1)\cdots 1} - \lambda \mathbf{I}_{n \times n}| = 0 \quad (k = 1, 2, \ldots, N). \tag{5.66}
$$

*The eigenvalues of* $DP$ *for such a periodic flow are determined by*

$$
|DP - \lambda \mathbf{I}_{n \times n}| = 0. \tag{5.67}
$$

*Thus, the stability and bifurcation of the periodic flow can be classified by the eigenvalues of* $DP(\mathbf{x}_0^*)$ *with*

$$
([n_1^m, n_1^o] : [n_2^m, n_2^o] : [n_3, \kappa_3] : [n_4, \kappa_4] | n_5 : n_6 : [n_7, l, \kappa_7]). \tag{5.68}
$$

(i)   *If the magnitudes of all eigenvalues of* $DP$ $(|\lambda_i| < 1,\ i = 1, 2, \ldots, n)$ *are less than one, the approximate periodic solution is stable.*

(ii)  *If at least the magnitude of one eigenvalue of* $DP$ $(|\lambda_i| > 1,\ i \in \{1, 2, \ldots, n\})$ *is greater than one, the approximate periodic solution is unstable.*

(iii) *The boundaries between stable and unstable periodic flow with higher-order singularity give bifurcation and stability conditions.*

*Proof* The proof is similar to Theorem 5.1.                                        □

   From the stability and bifurcation analysis, the period-1 flow under the period $T = 2\pi/\Omega$, based on the set of discrete mapping $P_k$ with $\mathbf{g}_k(\mathbf{x}_{k-1}, \mathbf{x}_k, \mathbf{p}) = 0$ $(k = 1, 2, \ldots, N)$, is stable or unstable. If the period-doubling bifurcation occurs, the periodic flow will become a periodic flow under the period $T' = 2T$, and such a periodic flow is called the period-2 flow. Due to the period-doubling, $2N$ nodes of the period-2 flow will be employed to describe the period-2 flow. Thus, consider a mapping structure of the period-2 flow with $2N$ mappings as

$$
P = P_{2N} \circ P_{2N-1} \circ \cdots \circ P_2 \circ P_1 : \mathbf{x}_0 \rightarrow \mathbf{x}_{2N};
$$
$$
\text{with } P_k : \mathbf{x}_{k-1} \rightarrow \mathbf{x}_k (k = 1, 2, \ldots, 2N). \tag{5.69}
$$

For $\mathbf{x}_{2N} = P\mathbf{x}_0$, there is a set of points $\mathbf{x}_k^*$ $(k = 0, 1, \ldots, 2N)$ computed by

$$
\begin{aligned}
&\mathbf{g}_k(\mathbf{x}_{k-1}^*, \mathbf{x}_k^*, \mathbf{p}) = \mathbf{0}, \quad (k = 1, 2, \ldots, 2N); \\
&\mathbf{x}_0^* = \mathbf{x}_{2N}^*.
\end{aligned}
\tag{5.70}
$$

After period-doubling, the period-1 flow becomes period-2 flow. The node points increase to $2N$ points during two periods ($2T$). The period-2 flow can be sketched in Fig. 5.3. The node points are determined through the discrete mapping with mathematical relation in Eq. (5.69). On the other hand,

$$
T' = 2T = \frac{2(2\pi)}{\Omega} = \frac{2\pi}{\omega} \Rightarrow \omega = \frac{\Omega}{2}.
\tag{5.71}
$$

During the period of $T'$, there is a periodic flow, which can be described by node points $\mathbf{x}_k$ $(k = 1, 2, \ldots, N')$. Since the period-1 flow is described by node points $\mathbf{x}_k$ $(k = 1, 2, \ldots, N)$ during the period $T$, due to $T' = 2T$, the period-2 flow can be described by $N' \geq 2N$ nodes. Thus, the corresponding mapping $P_k$ is defined as

$$
P_k : \ \mathbf{x}_{k-1}^{(2)} \rightarrow \mathbf{x}_k^{(2)} \quad (k = 1, 2, \ldots, 2N),
\tag{5.72}
$$

and

$$
\begin{aligned}
&\mathbf{g}_k(\mathbf{x}_{k-1}^{(2)*}, \mathbf{x}_k^{(2)*}, \mathbf{p}) = \mathbf{0}, \quad (k = 1, 2, \ldots, 2N), \\
&\mathbf{x}_0^{(2)*} = \mathbf{x}_{2N}^{(2)*}.
\end{aligned}
\tag{5.73}
$$

**Fig. 5.3** Period-2 flow with $2N$-nodes with short lines. *Solid curve* expected numerical results. The symbols are node points on the periodic flow

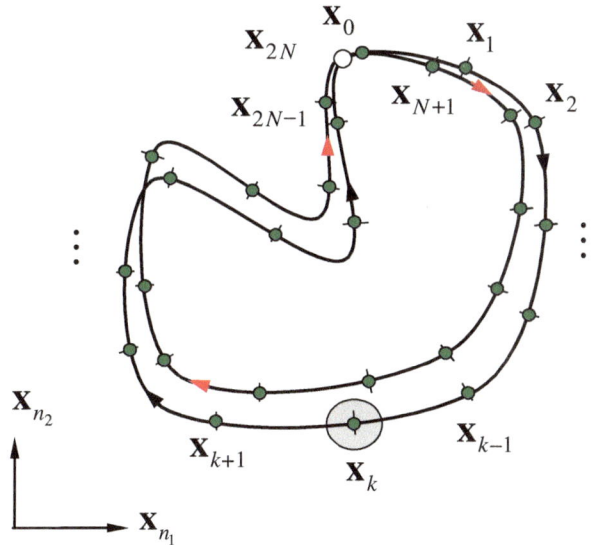

In general, for period $T' = mT$, there is a period-$m$ flow which can be described by $N' \geq mN$. The corresponding mapping $P_k$ is given by

$$P_k : \mathbf{x}_{k-1}^{(m)} \rightarrow \mathbf{x}_k^{(m)} \quad (k = 1, 2, \ldots, mN) \tag{5.74}$$

and

$$\mathbf{g}_k^{(m)}(\mathbf{x}_{k-1}^{(m)*}, \mathbf{x}_k^{(m)*}, \mathbf{p}) = \mathbf{0}, \quad (k = 1, 2, \ldots, mN); \\ \mathbf{x}_0^{(m)*} = \mathbf{x}_{mN}^{(m)*}. \tag{5.75}$$

From the above discussion, the period-$m$ flow in a nonlinear system can be described through $mN$ nodes for period $mT$, as stated in the following theorem.

**Theorem 5.3** *Consider a nonlinear dynamical system in Eq. (5.1). If such a dynamical system has a period-m flow $\mathbf{x}^{(m)}(t)$ with finite norm $||\mathbf{x}^{(m)}||$ and period $mT$ ($T = 2\pi/\Omega$), there is a set of discrete time $t_k$ ($k = 0, 1, \ldots, mN$) with ($N \rightarrow \infty$) during m-periods ($mT$), and the corresponding solution $\mathbf{x}^{(m)}(t_k)$ and vector field $\mathbf{f}(\mathbf{x}^{(m)}(t_k), t_k, \mathbf{p})$ are exact. Suppose a discrete node $\mathbf{x}_k^{(m)}$ is on the approximate solutions of the periodic flow under $||\mathbf{x}^{(m)}(t_k) - \mathbf{x}_k^{(m)}|| \leq \varepsilon_k$ with a small $\varepsilon_k \geq 0$ and*

$$||\mathbf{f}(\mathbf{x}^{(m)}(t_k), t_k, \mathbf{p}) - \mathbf{f}(\mathbf{x}_k^{(m)}, t_k, \mathbf{p})|| \leq \delta_k \tag{5.76}$$

*with a small $\delta_k \geq 0$. During a time interval $t \in [t_{k-1}, t_k]$, there is a mapping $P_k : \mathbf{x}_{k-1}^{(m)} \rightarrow \mathbf{x}_k^{(m)}$ ($k = 1, 2, \ldots, mN$), i.e.,*

$$\mathbf{x}_k^{(m)} = P_k \mathbf{x}_{k-1}^{(m)} \text{ with } \mathbf{g}_k(\mathbf{x}_{k-1}^{(m)}, \mathbf{x}_k^{(m)}, \mathbf{p}) = \mathbf{0}, \quad k = 1, 2, \ldots, mN \tag{5.77}$$

*where $\mathbf{g}_k$ is an implicit vector function. Consider a mapping structure as*

$$P = P_{mN} \circ P_{mN-1} \circ \cdots \circ P_1 : \mathbf{x}_0^{(m)} \rightarrow \mathbf{x}_{mN}^{(m)}; \\ \text{with } P_k : \mathbf{x}_{k-1}^{(m)} \rightarrow \mathbf{x}_k^{(m)} \quad (k = 1, 2, \ldots, mN). \tag{5.78}$$

*For $\mathbf{x}_{mN}^{(m)} = P\mathbf{x}_0^{(m)}$, if there is a set of points $\mathbf{x}_k^{(m)*}$ ($k = 0, 1, \ldots, mN$) computed by*

$$\mathbf{g}_k(\mathbf{x}_{k-1}^{(m)*}, \mathbf{x}_k^{(m)*}, \mathbf{p}) = \mathbf{0}, \quad (k = 1, 2, \ldots, mN) \\ \mathbf{x}_0^{(m)*} = \mathbf{x}_{mN}^{(m)*}, \tag{5.79}$$

*then the points $\mathbf{x}_k^{(m)*}$ ($k = 0, 1, \ldots, mN$) are approximations of points $\mathbf{x}^{(m)}(t_k)$ of the periodic solution. In the neighborhood of $\mathbf{x}_k^{(m)*}$, with $\mathbf{x}_k^{(m)} = \mathbf{x}_k^{(m)*} + \Delta\mathbf{x}_k^{(m)}$, the linearized equation is given by*

$$\Delta \mathbf{x}_k^{(m)} = DP_k \cdot \Delta \mathbf{x}_{k-1}^{(m)} \quad \text{with} \quad \mathbf{g}_k(\mathbf{x}_{k-1}^{(m)*} + \Delta \mathbf{x}_{k-1}^{(m)}, \mathbf{x}_k^{(m)*} + \Delta \mathbf{x}_k^{(m)}, \mathbf{p}) = \mathbf{0}$$
$$(k = 1, 2, \ldots, mN). \tag{5.80}$$

*The resultant Jacobian matrices of the periodic flow are*

$$DP_{k(k-1)\cdots 1} = DP_k \cdot DP_{k-1} \cdots \cdots DP_1, \quad (k = 1, 2, \ldots, mN);$$
$$DP \equiv DP_{mN(mN-1)\cdots 1} = DP_{mN} \cdot DP_{mN-1} \cdots \cdots DP_1 \tag{5.81}$$

*where*

$$DP_k = \left[ \frac{\partial \mathbf{x}_k^{(m)}}{\partial \mathbf{x}_{k-1}^{(m)}} \right]_{(\mathbf{x}_{k-1}^{(m)*}, \mathbf{x}_k^{(m)*})} = - \left[ \frac{\partial \mathbf{g}_k}{\partial \mathbf{x}_k^{(m)}} \right]^{-1} \left[ \frac{\partial \mathbf{g}_k}{\partial \mathbf{x}_{k-1}^{(m)}} \right]\Bigg|_{(\mathbf{x}_{k-1}^{(m)*}, \mathbf{x}_k^{(m)*})}. \tag{5.82}$$

*The eigenvalues of $DP(\mathbf{x}_0^{(m)*})$ and $DP_{k(k-1)\cdots 1}$ for such a periodic flow are determined by*

$$|DP_{k(k-1)\cdots 1} - \bar{\lambda}\mathbf{I}_{n \times n}| = 0, \quad (k = 1, 2, \ldots, mN);$$
$$|DP - \lambda\mathbf{I}_{n \times n}| = 0. \tag{5.83}$$

*Thus, the eigenvalues of $DP_{k(k-1)\ldots 1}$ give the properties of $\mathbf{x}_k$ varying with $\mathbf{x}_0$. The stability and bifurcation of the periodic flow can be classified by the eigenvalues of $DP(\mathbf{x}_0^{(m)*})$ with*

$$([n_1^m, n_1^o] : [n_2^m, n_2^o] : [n_3, \kappa_3] : [n_4, \kappa_4] | n_5 : n_6 : [n_7, l, \kappa_7]). \tag{5.84}$$

(i) *If the magnitudes of all eigenvalues of $DP^{(m)}$ are less than one (i.e., $|\lambda_i| < 1$, $i = 1, 2, \ldots, n$), the approximate period-m solution is stable.*

(ii) *If at least the magnitude of one eigenvalue of $DP^{(m)}$ is greater than one (i.e., $|\lambda_i| > 1$, $i \in \{1, 2, \ldots, n\}$), the approximate period-m solution is unstable.*

(iii) *The boundaries between stable and unstable period-m flow with higher-order singularity give bifurcation and stability conditions.*

*Proof* The discrete mapping for the period-$m$ flow can be developed during $t \in [t_{k-1}, t_k]$ as in Theorem 5.1. The proof is similar to Theorem 5.1. $\qquad\square$

The discrete mapping for a period-$m$ flow with multiple steps can be developed by using many forward and backward nodes. The period-$m$ flow in a nonlinear dynamical system can be determined through the following theorem.

**Theorem 5.4** *Consider a nonlinear dynamical system in Eq. (5.1). If such a dynamical system has a period-m flow $\mathbf{x}^{(m)}(t)$ with finite norm $\|\mathbf{x}^{(m)}\|$ and m-periods $mT(T = 2\pi/\Omega)$, there is a set of discrete time $t_k$ $(k = 0, 1, \ldots, mN)$ with*

$(N \to \infty)$ *during m-period T, and the corresponding solution* $\mathbf{x}^{(m)}(t_k)$ *and vector fields* $\mathbf{f}(\mathbf{x}^{(m)}(t_k), t_k, \mathbf{p})$ *are exact. Suppose a discrete node* $\mathbf{x}_k^{(m)}$ *is on the approximate solution of the periodic flow under* $||\mathbf{x}^{(m)}(t_k) - \mathbf{x}_k^{(m)}|| \le \varepsilon_k$ *with a small* $\varepsilon_k \ge 0$ *and*

$$||\mathbf{f}(\mathbf{x}^{(m)}(t_k), t_k, \mathbf{p}) - \mathbf{f}(\mathbf{x}_k^{(m)}, t_k, \mathbf{p})|| \le \delta_k \qquad (5.85)$$

*with a small* $\delta_k \ge 0$. *During a time interval* $t \in [t_{k-1}, t_k]$, *threw is a mapping* $P_k$ : $\mathbf{x}_{k-1}^{(m)} \to \mathbf{x}_k^{(m)}$ $(k = 1, 2, \ldots, mN)$, *i.e.,*

$$\begin{aligned}
&\mathbf{x}_k^{(m)} = P_k \mathbf{x}_{k-1}^{(m)} \\
&\text{with } \mathbf{g}_k(\mathbf{x}_{s_{kl_1}}^{(m)}, \ldots, \mathbf{x}_{s_{k1}}^{(m)}, \mathbf{x}_{s_{k0}}^{(m)}, \mathbf{x}_{s_{k(-1)}}^{(m)}, \ldots, \mathbf{x}_{s_{k(-l_2)}}^{(m)}, \mathbf{p}) = \mathbf{0}, \\
&s_{kj} = \mathrm{mod}(k - j + mN, mN), j = -l_2, -l_2 + 1, \ldots, l_1 - 1, l_1; \\
&l_1, l_2 \in \{0, 1, 2, \ldots, mN\}, 1 \le l_1 + l_2 \le mN, l_1 \ge 1; (k = 1, 2, \ldots, mN)
\end{aligned} \qquad (5.86)$$

*where* $\mathbf{g}_k$ *is an implicit vector function. Consider a mapping structure as*

$$\begin{aligned}
&P = P_{mN} \circ P_{mN-1} \circ \cdots \circ P_2 \circ P_1 : \ \mathbf{x}_0^{(m)} \to \mathbf{x}_{mN}^{(m)}; \\
&\text{with } P_k : \ \mathbf{x}_{k-1}^{(m)} \to \mathbf{x}_k^{(m)} \ (k = 1, 2, \ldots, mN).
\end{aligned} \qquad (5.87)$$

*For* $\mathbf{x}_{mN}^{(m)} = P\mathbf{x}_0^{(m)}$, *if there is a set of points* $\mathbf{x}_k^{(m)*}$ $(k = 0, 1, \ldots, mN)$ *computed by*

$$\begin{aligned}
&\mathbf{g}_k(\mathbf{x}_{s_{kl_1}}^{(m)*}, \ldots, \mathbf{x}_{s_{k1}}^{(m)*}, \mathbf{x}_{s_{k0}}^{(m)*}, \mathbf{x}_{s_{k(-1)}}^{(m)*}, \ldots, \mathbf{x}_{s_{k(-l_2)}}^{(m)*}, \mathbf{p}) = \mathbf{0}, \\
&(k = 1, 2, \ldots, mN) \\
&\mathbf{x}_0^{(m)*} = \mathbf{x}_{mN}^{(m)*},
\end{aligned} \qquad (5.88)$$

*then the points* $\mathbf{x}_k^{(m)*}(k = 0, 1, \ldots, mN)$ *are approximations of points* $\mathbf{x}^{(m)}(t_k)$ *of the periodic solution. In the neighborhood of* $\mathbf{x}_k^{(m)*}$, *with* $\mathbf{x}_k^{(m)} = \mathbf{x}_k^{(m)*} + \Delta\mathbf{x}_k^{(m)}$, *the linearized equation is given by*

$$\begin{aligned}
&\frac{\partial \mathbf{g}_k}{\partial \mathbf{x}_0^{(m)}} + \cdots + \frac{\partial \mathbf{g}_k}{\partial \mathbf{x}_{k-1}^{(m)}} \frac{\partial \mathbf{x}_{k-1}^{(m)}}{\partial \mathbf{x}_0^{(m)}} + \frac{\partial \mathbf{g}_k}{\partial \mathbf{x}_k^{(m)}} \frac{\partial \mathbf{x}_k^{(m)}}{\partial \mathbf{x}_0^{(m)}} + \frac{\partial \mathbf{g}_k}{\partial \mathbf{x}_{k+1}^{(m)}} \frac{\partial \mathbf{x}_{k+1}^{(m)}}{\partial \mathbf{x}_0^{(m)}} + \cdots + \frac{\partial \mathbf{g}_k}{\partial \mathbf{x}_{mN}^{(m)}} \frac{\partial \mathbf{x}_{mN}^{(m)}}{\partial \mathbf{x}_0^{(m)}} = \mathbf{0}, \\
&\frac{\partial \mathbf{g}_k}{\partial \mathbf{x}_\alpha^{(m)}} = \mathbf{0}(\alpha \ne s_{kj}), \quad j = -l_2, -l_2 + 1, \ldots, l_1 - 1, l_1; \\
&(k = 1, 2, \ldots, mN).
\end{aligned}$$

$$(5.89)$$

*The resultant Jacobian matrices of the periodic flow are*

$$DP_{k(k-1)\ldots 1} = \left[\frac{\partial \mathbf{x}_k^{(m)}}{\partial \mathbf{x}_0^{(m)}}\right]_{(\mathbf{x}_0^{(m)*}, \mathbf{x}_1^{(m)*}, \ldots, \mathbf{x}_{mN}^{(m)*})} \quad (k = 1, 2, \ldots, mN),$$

$$and \quad DP = DP_{(mN)(mN-1)\ldots 1} = \left[\frac{\partial \mathbf{x}_N^{(m)}}{\partial \mathbf{x}_0^{(m)}}\right]_{(\mathbf{x}_0^{(m)*}, \mathbf{x}_1^{(m)*}, \ldots, \mathbf{x}_{mN}^{(m)*})}$$

(5.90)

*where*

$$\begin{bmatrix} \dfrac{\partial \mathbf{x}_1^{(m)}}{\partial \mathbf{x}_0^{(m)}} \\ \vdots \\ \dfrac{\partial \mathbf{x}_{mN-1}^{(m)}}{\partial \mathbf{x}_0^{(m)}} \\ \dfrac{\partial \mathbf{x}_{mN}^{(m)}}{\partial \mathbf{x}_0^{(m)}} \end{bmatrix} = - \begin{bmatrix} \dfrac{\partial \mathbf{g}_1}{\partial \mathbf{x}_1^{(m)}} & \cdots & \dfrac{\partial \mathbf{g}_1}{\partial \mathbf{x}_{mN-1}^{(m)}} & \dfrac{\partial \mathbf{g}_1}{\partial \mathbf{x}_{mN}^{(m)}} \\ \vdots & & \vdots & \vdots \\ \dfrac{\partial \mathbf{g}_{mN-1}}{\partial \mathbf{x}_1^{(m)}} & \cdots & \dfrac{\partial \mathbf{g}_{mN-1}}{\partial \mathbf{x}_{mN-1}^{(m)}} & \dfrac{\partial \mathbf{g}_{mN-1}}{\partial \mathbf{x}_{mN}^{(m)}} \\ \dfrac{\partial \mathbf{g}_{mN}}{\partial \mathbf{x}_1^{(m)}} & \cdots & \dfrac{\partial \mathbf{g}_{mN}}{\partial \mathbf{x}_{mN-1}^{(m)}} & \dfrac{\partial \mathbf{g}_{mN}}{\partial \mathbf{x}_{mN}^{(m)}} \end{bmatrix}^{-1} \begin{bmatrix} \dfrac{\partial \mathbf{g}_1}{\partial \mathbf{x}_0^{(m)}} \\ \vdots \\ \dfrac{\partial \mathbf{g}_{mN-1}}{\partial \mathbf{x}_0^{(m)}} \\ \dfrac{\partial \mathbf{g}_{mN}}{\partial \mathbf{x}_0^{(m)}} \end{bmatrix}.$$

(5.91)

*The properties of discrete points* $\mathbf{x}_k$ $(k = 1, 2, \ldots, mN)$ *can be estimated by the eigenvalues of* $DP_{k(k-1)\cdots 1}$ *as*

$$|DP_{k(k-1)\cdots 1} - \bar{\lambda}\mathbf{I}_{n\times n}| = 0 \quad (k = 1, 2, \ldots, mN).$$

(5.92)

*The eigenvalues of* $DP$ *for such a periodic flow are determined by*

$$|DP - \lambda\mathbf{I}_{n\times n}| = 0.$$

(5.93)

*Thus, the stability and bifurcation of the period-m flow can be classified by the eigenvalues of* $DP(\mathbf{x}_0^{(m)*})$ *with*

$$([n_1^m, n_1^o] : [n_2^m, n_2^o] : [n_3, \kappa_3] : [n_4, \kappa_4]|n_5 : n_6 : [n_7, l, \kappa_7]).$$

(5.94)

(i) *If the magnitudes of all eigenvalues of* $DP(\mathbf{x}_0^{(m)*})$ *are less than one (i.e.,* $|\lambda_i| < 1, i = 1, 2, \ldots, n)$*, the approximate period-m solution is stable.*

(ii) *If at least the magnitude of one eigenvalue of* $DP(\mathbf{x}_0^{(m)*})$ *is greater than one (i.e.,* $|\lambda_i| > 1$*,* $i \in \{1, 2, \ldots, n\})$*, the approximate period-m solution is unstable.*

(iii) *The boundaries between stable and unstable periodic flow with higher-order singularity give bifurcation and stability conditions.*

*Proof* The proof is similar to Theorem 5.1. ☐

## 5.2   Continuous Time-Delay Systems

As in Luo (2014), periodic flows in time-delay nonlinear dynamical systems will be discussed. Two methods are presented herein. The discretized time-delay node is approximated by its neighbored two non-delay nodes through interpolation in discrete maps, which will be discussed first. After then, the discretized delay node between the two non-time-delay nodes will be determined by the integration. The discrete maps of the delay nodes will be obtained. With discrete maps of the non-delay nodes, periodic flows in time-delay systems will be presented.

### 5.2.1   Interpolated Time-Delay Nodes

Consider a time-delay nonlinear dynamical system as

$$\dot{\mathbf{x}} = \mathbf{f}(\mathbf{x}, \mathbf{x}^\tau, t, \mathbf{p}) \in \mathscr{R}^n \tag{5.95}$$

where $\mathbf{f}(\mathbf{x}, \mathbf{x}^\tau, t, \mathbf{p})$ is a $C^r$-continuous nonlinear vector function $(r \geq 1)$ and $\mathbf{x}^\tau = \mathbf{x}(t - \tau)$. If the time-delay nonlinear system has solution points $\mathbf{x}_k \approx \mathbf{x}(t_k)$ and $\mathbf{x}_k^\tau \approx \mathbf{x}(t_k - \tau)$ for $k = 0, 1, 2, \ldots$, as shown in Fig. 5.4. The small circular symbols are the regular solution points, and the large circular symbols are the time-delayed solution points. The delay node $\mathbf{x}_k^\tau \approx \mathbf{x}(t_k - \tau)$ of $\mathbf{x}_k \approx \mathbf{x}(t_k)$ will lie between $\mathbf{x}_{k-l_k}$ and $\mathbf{x}_{k-l_k-1}$ (integer $l_k > 0$). From Eq. (5.95), we have

$$\mathbf{x}(t_k) = \mathbf{x}(t_{k-1}) + \int_{t_{k-1}}^{t_k} \mathbf{f}(\mathbf{x}, \mathbf{x}^\tau, t, \mathbf{p}) \mathrm{d}t. \tag{5.96}$$

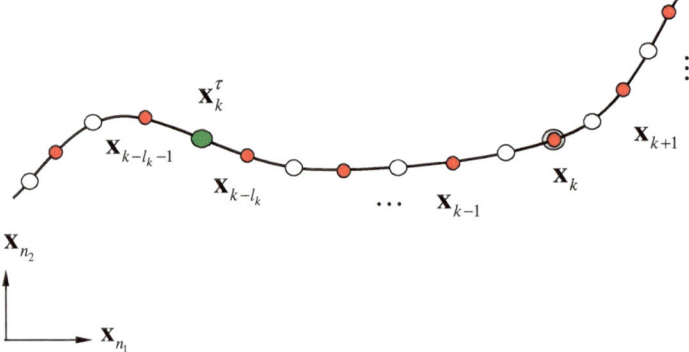

**Fig. 5.4** The discrete points on the solutions of a time-delay dynamical system. The *small circular symbols* are the regular solution points, and the *large circular symbols* are the time-delayed points. The referenced point $\mathbf{x}_k$ and the corresponding time-delay point $\mathbf{x}_k^\tau$ are labeled. The time-delay point $\mathbf{x}_k^\tau$ can be estimated by the two vicinity points $\mathbf{x}_{k-l_k}$ and $\mathbf{x}_{k-l_k-1}$

An interpolation function between $\mathbf{f}(\mathbf{x}_{k-1}, \mathbf{x}_{k-1}^{\tau}, t_{k-1}, \mathbf{p})$ and $\mathbf{f}(\mathbf{x}_k, \mathbf{x}_k^{\tau}, t_k, \mathbf{p})$ is considered to approximate $\mathbf{f}(\mathbf{x}, \mathbf{x}^{\tau}, t, \mathbf{p})$. Equation (5.96) becomes

$$\mathbf{x}_k = \mathbf{x}_{k-1} + \bar{\mathbf{g}}_k(\mathbf{x}_{k-1}, \mathbf{x}_k; \mathbf{x}_{k-1}^{\tau}, \mathbf{x}_k^{\tau}, \mathbf{p}). \tag{5.97}$$

From the above discrete scheme, periodic flows in the time-delay dynamical systems can be investigated herein. If a time-delay nonlinear system has a periodic flow with a period of $T = 2\pi/\Omega$, then such a periodic flow can be expressed by a discrete points through discrete mappings of the time-delay continuous dynamical system as afore-discussed. The method is stated as follows.

**Theorem 5.5** *Consider a time-delay nonlinear dynamical system as*

$$\dot{\mathbf{x}} = \mathbf{f}(\mathbf{x}, \mathbf{x}^{\tau}, t, \mathbf{p}) \in \mathscr{R}^n \tag{5.98}$$

*where $\mathbf{f}(\mathbf{x}, \mathbf{x}^{\tau}, t, \mathbf{p})$ is a $C^r$-continuous nonlinear vector function $(r \geq 1)$ and $\mathbf{x}^{\tau} = \mathbf{x}(t - \tau)$. If such a dynamical system has a periodic flow $\mathbf{x}(t)$ with finite norm $||\mathbf{x}||$ and period $T = 2\pi/\Omega$, there is a set of discrete time $t_k(k = 0, 1, \ldots, N)$ with $(N \to \infty)$ during one period T, and the corresponding solution $\mathbf{x}(t_k)$ with $\mathbf{x}^{\tau}(t_k) = \mathbf{x}(t_k - \tau)$ and vector field $\mathbf{f}(\mathbf{x}(t_k), \mathbf{x}^{\tau}(t_k), t_k, \mathbf{p})$ are exact. Suppose discrete nodes $\mathbf{x}_k$ and $\mathbf{x}_k^{\tau}$ $(k = 0, 1, \ldots, N)$ are on the approximate solution of the periodic flow under $||\mathbf{x}(t_k) - \mathbf{x}_k|| \leq \varepsilon_k$ and $||\mathbf{x}^{\tau}(t_k) - \mathbf{x}_k^{\tau}|| \leq \varepsilon_k^{\tau}$ for small $\varepsilon_k, \varepsilon_k^{\tau} \geq 0$ and*

$$||\mathbf{f}(\mathbf{x}(t_k), \mathbf{x}^{\tau}(t_k), t_k, \mathbf{p}) - \mathbf{f}(\mathbf{x}_k, \mathbf{x}_k^{\tau}, t_k, \mathbf{p})|| \leq \delta_k \tag{5.99}$$

*with a small $\delta_k \geq 0$. During a time interval $t \in [t_k, t_{k+1}]$, there is a mapping $P_k : (\mathbf{x}_{k-1}, \mathbf{x}_{k-1}^{\tau}) \to (\mathbf{x}_k, \mathbf{x}_k^{\tau})$ $(k = 1, 2, \ldots, N)$, i.e.,*

$$
\begin{aligned}
&(\mathbf{x}_k, \mathbf{x}_k^{\tau}) = P_k(\mathbf{x}_{k-1}, \mathbf{x}_{k-1}^{\tau}) \quad with \\
&\mathbf{g}_k(\mathbf{x}_{k-1}, \mathbf{x}_k; \mathbf{x}_{k-1}^{\tau}, \mathbf{x}_k^{\tau}, \mathbf{p}) = \mathbf{0}, \quad (k = 1, 2, \ldots, N) \\
&\mathbf{x}_j^{\tau} = \mathbf{h}_j(\mathbf{x}_{r_j-1}, \mathbf{x}_{r_j}, \theta_{r_j}), j = k, k-1, r_j = j - l_j \\
&(e.g., \ \mathbf{x}_j^{\tau} = \mathbf{x}_{r_j} + \theta_{r_j}(\mathbf{x}_{r_j-1} - \mathbf{x}_{r_j}), \theta_{r_j} = \frac{1}{h_{r_j}}[\tau - \sum_{i=1}^{l_j} h_{r_j+i}]).
\end{aligned}
\tag{5.100}
$$

*where $\mathbf{g}_k$ is an implicit vector function and $\mathbf{h}_j$ is an interpolation vector function. Consider a mapping structure as*

$$
\begin{aligned}
&P = P_N \circ P_{N-1} \circ \cdots \circ P_2 \circ P_1 : \ \mathbf{x}_0 \to \mathbf{x}_N; \\
&with \ P_k : \ (\mathbf{x}_{k-1}, \mathbf{x}_{k-1}^{\tau}) \to (\mathbf{x}_k, \mathbf{x}_k^{\tau}) \quad (k = 1, 2, \ldots, N).
\end{aligned}
\tag{5.101}
$$

*For $(\mathbf{x}_N, \mathbf{x}_N^{\tau}) = P(\mathbf{x}_0, \mathbf{x}_0^{\tau})$, if there is a set of points $(\mathbf{x}_k^*, \mathbf{x}_k^{\tau*})$ $(k = 0, 1, \ldots, N)$ computed by*

$$\left.\begin{aligned}\mathbf{g}_k(\mathbf{x}_{k-1},\mathbf{x}_k;\mathbf{x}_{k-1}^{\tau},\mathbf{x}_k^{\tau},\mathbf{p}) &= \mathbf{0},\\ \mathbf{x}_j^{\tau} = \mathbf{h}_j(\mathbf{x}_{r_j-1},\mathbf{x}_{r_j},\theta_{r_j}), \quad j &= k, k-1\end{aligned}\right\}\quad (k=1,2,\ldots,N)$$

$$\mathbf{x}_{r_j-1} = \mathbf{x}_{\mathrm{mod}(r_j-1+N,N)},\quad \mathbf{x}_{r_j} = \mathbf{x}_{\mathrm{mod}(r_j+N,N)},$$

$$\mathbf{x}_0^* = \mathbf{x}_N^* \quad \text{and} \quad \mathbf{x}_0^{\tau*} = \mathbf{x}_N^{\tau*},$$

$$(5.102)$$

*then the points* $\mathbf{x}_k^*$ *and* $\mathbf{x}_k^{\tau*}$ $(k=0,1,\ldots,N)$ *are approximations of points* $\mathbf{x}(t_k)$ *and* $\mathbf{x}^{\tau}(t_k)$ *of the periodic solution. In the neighborhoods of* $\mathbf{x}_k^*$ *and* $\mathbf{x}_k^{\tau*}$, *with* $\mathbf{x}_k = \mathbf{x}_k^* + \Delta\mathbf{x}_k$ *and* $\mathbf{x}_k^{\tau} = \mathbf{x}_k^{\tau*} + \Delta\mathbf{x}_k^{\tau}$, *the linearized equation is given by*

$$\frac{\partial\mathbf{g}_k}{\partial\mathbf{x}_k}\Delta\mathbf{x}_k + \frac{\partial\mathbf{g}_k}{\partial\mathbf{x}_{k-1}}\Delta\mathbf{x}_{k-1} + \sum_{j=k-1}^{k}\frac{\partial\mathbf{g}_k}{\partial\mathbf{x}_j^{\tau}}\left(\frac{\partial\mathbf{x}_j^{\tau}}{\partial\mathbf{x}_{r_j}}\Delta\mathbf{x}_{r_j} + \frac{\partial\mathbf{x}_j^{\tau}}{\partial\mathbf{x}_{r_j-1}}\Delta\mathbf{x}_{r_j-1}\right) = \mathbf{0}_{1\times n} \quad (5.103)$$

*with* $r_j = j - l_j, j = k-1, k; k = 1, 2, \ldots, N.$

*The resultant Jacobian matrices of the periodic flow are*

$$\mathrm{DP}_{k(k-1)\cdots 1} = \left[\frac{\partial\mathbf{y}_k}{\partial\mathbf{y}_0}\right]_{(\mathbf{y}_0^*,\ldots,\mathbf{y}_k^*)} = \mathbf{A}_k\mathbf{A}_{k-1}\ldots\mathbf{A}_1 \quad (k=1,2,\ldots,N),$$

$$(5.104)$$

$$\text{and } \mathrm{DP} = \mathrm{DP}_{N(N-1)\cdots 1} = \left[\frac{\partial\mathbf{y}_N}{\partial\mathbf{y}_0}\right]_{(\mathbf{y}_0^*,\ldots,\mathbf{y}_N^*)} = \mathbf{A}_N\mathbf{A}_{N-1}\ldots\mathbf{A}_1.$$

*where*

$$\Delta\mathbf{y}_k = \mathbf{A}_k\Delta\mathbf{y}_{k-1}, \quad \mathbf{A}_k \equiv \left[\frac{\partial\mathbf{y}_k}{\partial\mathbf{y}_{k-1}}\right]_{(\mathbf{y}_{k-1}^*,\mathbf{y}_k^*)} \quad (5.105)$$

*and*

$$\mathbf{a}_{kj} = -\left[\frac{\partial\mathbf{g}_k}{\partial\mathbf{x}_k}\right]^{-1}\frac{\partial\mathbf{g}_k}{\partial\mathbf{x}_j}, \quad \mathbf{a}_{kr_j} = -\left[\frac{\partial\mathbf{g}_k}{\partial\mathbf{x}_k}\right]^{-1}\frac{\partial\mathbf{g}_k}{\partial\mathbf{x}_j^{\tau}}\frac{\partial\mathbf{x}_j^{\tau}}{\partial\mathbf{x}_{r_j}},$$

$$\mathbf{a}_{k(r_j-1)} = -\left[\frac{\partial\mathbf{g}_k}{\partial\mathbf{x}_k}\right]^{-1}\frac{\partial\mathbf{g}_k}{\partial\mathbf{x}_j^{\tau}}\frac{\partial\mathbf{x}_j^{\tau}}{\partial\mathbf{x}_{r_j-1}} \quad \text{with } r_j = j - l_j, j = k-1, k;$$

$$\mathbf{y}_k = (\mathbf{x}_k,\mathbf{x}_{k-1},\ldots,\mathbf{x}_{r_{k-1}})^{\mathrm{T}}, \quad \mathbf{y}_{k-1} = (\mathbf{x}_{k-1},\mathbf{x}_{k-2},\ldots,\mathbf{x}_{r_{k-1}-1})^{\mathrm{T}},$$

$$\Delta\mathbf{y}_k = (\Delta\mathbf{x}_k,\Delta\mathbf{x}_{k-1},\ldots,\Delta\mathbf{x}_{r_{k-1}})^{\mathrm{T}}, \quad \Delta\mathbf{y}_{k-1} = (\Delta\mathbf{x}_{k-1},\Delta\mathbf{x}_{k-2},\ldots,\Delta\mathbf{x}_{r_{k-1}-1})^{\mathrm{T}};$$

$$\mathbf{A}_k = \begin{bmatrix}\mathbf{B}_k & (\mathbf{a}_{k(r_{k-1}-1)})_{n\times n} \\ \mathbf{I}_k & \mathbf{0}_k\end{bmatrix}_{n(s+1)\times n(s+1)}, \quad s = 1 + l_{k-1}$$

$$\mathbf{B}_k = [(\mathbf{a}_{k(k-1)})_{n\times n}, \mathbf{0}_{n\times n}, \ldots, (\mathbf{a}_{kr_{k-1}})_{n\times n}],$$

$$\mathbf{I}_k = \mathrm{diag}(\mathbf{I}_{n\times n}, \mathbf{I}_{n\times n}, \ldots, \mathbf{I}_{n\times n})_{ns\times ns}, \quad \mathbf{0}_k = (\underbrace{\mathbf{0}_{n\times n}, \mathbf{0}_{n\times n}\ldots, \mathbf{0}_{n\times n}}_{s})^{\mathrm{T}}.$$

$$(5.106)$$

*The properties of discrete points* $\mathbf{x}_k$ $(k = 1, 2, \ldots, N)$ *can be estimated by the eigenvalues of* $DP_{k(k-1)\ldots 1}$ *as*

$$|DP_{k(k-1)\cdots 1} - \bar{\lambda} \mathbf{I}_{n(s+1) \times n(s+1)}| = 0 \quad (k = 1, 2, \ldots, N). \tag{5.107}$$

*The eigenvalues of* DP *for such a periodic flow are determined by*

$$|DP - \lambda \mathbf{I}_{n(s+1) \times n(s+1)}| = 0. \tag{5.108}$$

*Thus, the stability and bifurcation of the periodic flow can be classified by the eigenvalues of* $DP(\mathbf{y}_0^*)$ *with*

$$([n_1^m, n_1^o] : [n_2^m, n_2^o] : [n_3, \kappa_3] : [n_4, \kappa_4] | n_5 : n_6 : [n_7, l, \kappa_7]). \tag{5.109}$$

(i) *If the magnitudes of all eigenvalues of* DP *are less than one (i.e.,* $|\lambda_i| < 1$, $i = 1, 2, \ldots, n(s+1)$), *the approximate periodic solution is stable.*

(ii) *If at least the magnitude of one eigenvalue of* DP *is greater than one (i.e.,* $|\lambda_i| > 1$, $i \in \{1, 2, \ldots, n(s+1)\}$), *the approximate periodic solution is unstable.*

(iii) *The boundaries between stable and unstable periodic flow with higher-order singularity give bifurcation and stability conditions.*

*Proof* If $\mathbf{f}(\mathbf{x}, \mathbf{x}^\tau, t, \mathbf{p})$ is a $C^r$-continuous nonlinear function vector $(r \geq 1)$, then the velocity $\dot{\mathbf{x}}$ should be $C^r$-continuous $r \geq 1$. If such a time-delay system has a periodic flow $\mathbf{x}(t)$ and $\mathbf{x}^\tau(t)$ with finite norms $||\mathbf{x}||$ and $||\mathbf{x}^\tau||$ with period $T = 2\pi/\Omega$, there is a set of discrete time $t_k$ $(k = 0, 1, \ldots, N)$ with $(N \to \infty)$ during one period $T$. The corresponding solutions $\mathbf{x}(t_k)$ and $\mathbf{x}^\tau(t_k) = \mathbf{x}(t_k - \tau)$ with vector fields $\mathbf{f}(\mathbf{x}(t_k), \mathbf{x}^\tau(t_k), t_k, \mathbf{p})$ are exact. Consider a time interval $t \in [t_{k-1}, t_k]$,

$$\mathbf{x}(t) = \mathbf{x}(t_{k-1}) + \int_{t_{k-1}}^{t} \mathbf{f}(\mathbf{x}, \mathbf{x}^\tau, t, \mathbf{p}) dt. \tag{5.110}$$

For the time interval $[t_{k-1}, t_k]$ divided into $s$-nodes $t_{k(i)} = t_{k-1} + c_i h_k$ with $c_i \in [0, 1]$ and $\mathbf{f}(\mathbf{x}(t_{k(i)}), \mathbf{x}^\tau(t_{k(i)}), t_{k(i)}, \mathbf{p})$ $(i = 1, \ldots, s)$ with $\mathbf{x}^\tau(t_{k(i)}) = \mathbf{x}(t_{k(i)} - \tau)$, there is an approximate function $\mathbf{P}(t, \mathbf{C})$ with unknown $\mathbf{C} = (C_1, \ldots, C_s)^T$ and $C_i$ $(i = 1, \ldots, s)$, and the following condition is satisfied, i.e.,

$$\mathbf{f}(\mathbf{x}(t_{k(i)}), \mathbf{x}^\tau(t_{k(i)}), t_{k(i)}, \mathbf{p}) = \mathbf{P}(t_{k(i)}, t_{k(i)} - \tau, \mathbf{C})$$
$$|\frac{\partial \mathbf{P}}{\partial \mathbf{C}}| \neq 0, \quad i = 1, 2, \ldots, s. \tag{5.111}$$

The unknowns $\mathbf{C}(t_k) = (C_1, \ldots, C_s)^T$ with $\mathbf{t}_k = (t_{k(1)}, \ldots, t_{k(s)})^T = t_k(1, 1, \ldots, 1)^T + h_k(c_1, \ldots, c_s)^T$ are determined. For a small $\delta > 0$, if there is a relation

$$|\mathbf{P}(t, t - \tau, \mathbf{C}(\mathbf{t}_k)) - \mathbf{f}(\mathbf{x}, \mathbf{x}^\tau, t, \mathbf{p})| \le \delta \tag{5.112}$$

for $t \in [t_{k-1}, t_k]$, Eq. (5.110) can be approximated as

$$\mathbf{x}(t) = \mathbf{x}(t_{k-1}) + \int_{t_{k-1}}^{t} [\mathbf{P}(t, t - \tau, \mathbf{C}(\mathbf{t}_k)) + O(\delta)]dt;$$

$$\bar{\mathbf{x}}(t) = \bar{\mathbf{x}}(t_{k-1}) + \int_{t_{k-1}}^{t} \mathbf{P}(t, t - \tau, \mathbf{C}(\mathbf{t}_k))dt \tag{5.113}$$

and

$$\bar{\mathbf{x}}(t_k) = \bar{\mathbf{x}}(t_{k-1}) + \int_{t_{k-1}}^{t_k} \mathbf{P}(t, t - \tau, \mathbf{C}(\mathbf{t}_k))dt. \tag{5.114}$$

Let $\bar{\mathbf{x}}(t_{k-1}) = \mathbf{x}_{k-1}$, $\bar{\mathbf{x}}(t_k) = \mathbf{x}_k$ and $\bar{\mathbf{x}}^\tau(t_k) = \mathbf{x}_k^\tau$. For any small $\{\varepsilon_{k-1}, \varepsilon_{k-1}^\tau\} > 0$ and $\varepsilon_k > 0$, under $||\mathbf{x}(t_{k-1}) - \mathbf{x}_{k-1}|| \le \varepsilon_{k-1}, ||\mathbf{x}^\tau(t_{k-1}) - \mathbf{x}_{k-1}^\tau|| \le \varepsilon_{k-1}^\tau$ and $||\mathbf{x}(t_k) - \mathbf{x}_k|| \le \varepsilon_k$, Eq. (5.114) gives

$$\mathbf{x}_k = \mathbf{x}_{k-1} + \bar{\mathbf{g}}_k(\mathbf{x}_{k-1}, \mathbf{x}_k; \mathbf{x}_{k-1}^\tau, \mathbf{x}_k^\tau, \mathbf{p}),$$

$$\bar{\mathbf{g}}_k(\mathbf{x}_{k-1}, \mathbf{x}_k; \mathbf{x}_{k-1}^\tau, \mathbf{x}_k^\tau, \mathbf{p}) = \int_{t_{k-1}}^{t_k} \mathbf{P}(t, t - \tau, \mathbf{C}(\mathbf{t}_k))dt;$$

$$\mathbf{x}_j^\tau = \mathbf{h}_j(\mathbf{x}_{r_j-1}, \mathbf{x}_{r_j}, \theta_{r_j}) \quad \text{with } \theta_{r_j} = \frac{1}{h_{r_j}}[\tau - \sum_{i=1}^{l_j} h_{r_j+i}] \tag{5.115}$$

$$\text{for } r_j = j - l_j, j = k - 1, k.$$

Thus, a discrete mapping relation is obtained by

$$\mathbf{g}_k(\mathbf{x}_{k-1}, \mathbf{x}_{k-1}^\tau, \mathbf{x}_k, \mathbf{x}_k^\tau, \mathbf{p}) \equiv \mathbf{x}_k - \mathbf{x}_{k-1} - \bar{\mathbf{g}}_k(\mathbf{x}_{k-1}, \mathbf{x}_{k-1}^\tau, \mathbf{x}_k, \mathbf{x}_k^\tau, \mathbf{p}) = \mathbf{0}. \tag{5.116}$$

From the discrete mapping, two points $\mathbf{x}(t_{k-1})$ and $\mathbf{x}(t_k)$ for the time interval $t \in [t_{k-1}, t_k]$ $(k = 0, 1, \ldots, N)$ can be approximated by $\mathbf{x}_{k-1}$ and $\mathbf{x}_k$, respectively. If $\mathbf{f}(\mathbf{x}, \mathbf{x}^\tau, t, \mathbf{p})$ is a $C^r$-continuous nonlinear vector function, we have $||\mathbf{f}||_\mathbf{x} \le L$ and $||\mathbf{f}||_{\mathbf{x}^\tau} \le L^\tau$ ($L$ and $L^\tau$ constant). Thus for $j = k - 1, k$

$$||\mathbf{f}(\mathbf{x}(t_j), \mathbf{x}^\tau(t_j), t_j, \mathbf{p}) - \mathbf{f}(\mathbf{x}_j, \mathbf{x}_j^\tau, t_j, \mathbf{p})||$$
$$\le L||\mathbf{x}(t_j) - \mathbf{x}_j|| + L^\tau||\mathbf{x}^\tau(t_j) - \mathbf{x}_j^\tau|| \le L\varepsilon_j + L^\tau\varepsilon_j^\tau = \delta_j. \tag{5.117}$$

Once a mapping $P_k : (\mathbf{x}_{k-1}, \mathbf{x}_{k-1}^\tau) \to (\mathbf{x}_k, \mathbf{x}_k^\tau)$ with $\mathbf{g}_k(\mathbf{x}_{k-1}, \mathbf{x}_{k-1}^\tau, \mathbf{x}_k, \mathbf{x}_k^\tau, \mathbf{p}) = \mathbf{0}$ $(k = 1, 2, \ldots, N)$ exists, the periodic flow can be formed by $P : \mathbf{x}_0 \to \mathbf{x}_N$ with $P = P_N \circ \cdots \circ P_2 \circ P_1$, i.e., for $P_k : (\mathbf{x}_{k-1}, \mathbf{x}_{k-1}^\tau) \to (\mathbf{x}_k, \mathbf{x}_k^\tau)$, we have

$$\left.\begin{array}{l} \mathbf{g}_k(\mathbf{x}_{k-1}, \mathbf{x}_{k-1}^{\tau}, \mathbf{x}_k, \mathbf{x}_k^{\tau}, \mathbf{p}) = \mathbf{0}, \\ \mathbf{x}_j^{\tau} = \mathbf{h}_j(\mathbf{x}_{r_j-1}, \mathbf{x}_{r_j}, \theta_{r_j}), \\ j = k, k - 1; r_j = j - l_j \end{array}\right\} \quad (k = 1, 2, \ldots, N). \quad (5.118)$$

With the periodicity condition,

$$\begin{aligned} \mathbf{x}_{r_j} &= \mathbf{x}_{\mathrm{mod}(r_j+N,N)}, \quad j = k, k - 1 \\ \mathbf{x}_0 &= \mathbf{x}_N \quad \text{and} \quad \mathbf{x}_0^{\tau} = \mathbf{x}_N^{\tau}. \end{aligned} \quad (5.119)$$

Solving Eqs. (5.118) and (5.119) gives $\mathbf{x}_k^*$ and $\mathbf{x}_k^{\tau*}$ ($k = 0, 1, 2, \ldots, N$) to get the period-1 flow. For the stability of such a periodic flow, consider $\mathbf{x}_k = \mathbf{x}_k^* + \Delta\mathbf{x}_k$ and $\mathbf{x}_k^{\tau} = \mathbf{x}_k^{\tau*} + \Delta\mathbf{x}_k^{\tau}$ ($k = 1, 2, \ldots, N$) for $\mathbf{x}_k \in U(\mathbf{x}_k^*)$ and $\mathbf{x}_k^{\tau} \in U(\mathbf{x}_k^{\tau*})$. Equation (5.118) becomes

$$\begin{aligned} &\mathbf{g}_k(\mathbf{x}_{k-1}^* + \Delta\mathbf{x}_{k-1}, \mathbf{x}_k^* + \Delta\mathbf{x}_k; \mathbf{x}_{k-1}^{\tau*} + \Delta\mathbf{x}_{k-1}^{\tau}, \mathbf{x}_k^{\tau*} + \Delta\mathbf{x}_k^{\tau}, \mathbf{p}) = \mathbf{0}, \\ &\Delta\mathbf{x}_{k-1} = \Delta\mathbf{x}_{k-1}, \Delta\mathbf{x}_{k-2} = \Delta\mathbf{x}_{k-2}, \ldots, \Delta\mathbf{x}_{r_{k-1}} = \Delta\mathbf{x}_{r_{k-1}} \\ &(k = 1, 2, \ldots, N). \end{aligned} \quad (5.120)$$

Thus, differentiation of $\mathbf{g}_k(\mathbf{x}_{k-1}, \mathbf{x}_{k-1}^{\tau}, \mathbf{x}_k, \mathbf{x}_k^{\tau}, \mathbf{p}) = \mathbf{0}$ gives

$$\begin{aligned} &\sum_{j=k-1}^{k} \left[\frac{\partial\mathbf{g}_k}{\partial\mathbf{x}_j}\right]_{(\mathbf{x}_0^*, \mathbf{x}_0^{\tau*}, \ldots, \mathbf{x}_k^*, \mathbf{x}_k^{\tau*})} \Delta\mathbf{x}_j + \left[\frac{\partial\mathbf{g}_k}{\partial\mathbf{x}_j^{\tau}}\right]_{(\mathbf{x}_0^*, \mathbf{x}_0^{\tau*}, \ldots, \mathbf{x}_k^{\tau*}, \mathbf{x}_k^*)} \Delta\mathbf{x}_j^{\tau} = \mathbf{0}_{n \times 1} \\ &\Delta\mathbf{x}_{k-1} = \Delta\mathbf{x}_{k-1}, \quad \Delta\mathbf{x}_{k-2} = \Delta\mathbf{x}_{k-2}, \ldots, \quad \Delta\mathbf{x}_{r_{k-1}} = \Delta\mathbf{x}_{r_{k-1}} \\ &(k = 1, 2, \ldots, N) \end{aligned} \quad (5.121)$$

with

$$\Delta\mathbf{x}_j^{\tau} = \frac{\partial\mathbf{x}_j^{\tau}}{\partial\mathbf{x}_{r_j}} \Delta\mathbf{x}_{r_j} + \frac{\partial\mathbf{x}_j^{\tau}}{\partial\mathbf{x}_{r_j-1}} \Delta\mathbf{x}_{r_j-1} \quad (5.122)$$

$$\text{with } r_j = j - l_j, j = k - 1, k.$$

Let

$$\begin{aligned} &\mathbf{a}_{kj} = -\left[\frac{\partial\mathbf{g}_k}{\partial\mathbf{x}_k}\right]^{-1} \frac{\partial\mathbf{g}_k}{\partial\mathbf{x}_j}, \quad \mathbf{a}_{kr_j} = -\left[\frac{\partial\mathbf{g}_k}{\partial\mathbf{x}_k}\right]^{-1} \frac{\partial\mathbf{g}_k}{\partial\mathbf{x}_j^{\tau}} \frac{\partial\mathbf{x}_j^{\tau}}{\partial\mathbf{x}_{r_j}}, \\ &\mathbf{a}_{k(r_j-1)} = -\left[\frac{\partial\mathbf{g}_k}{\partial\mathbf{x}_k}\right]^{-1} \frac{\partial\mathbf{g}_k}{\partial\mathbf{x}_j^{\tau}} \frac{\partial\mathbf{x}_j^{\tau}}{\partial\mathbf{x}_{r_j-1}} \quad \text{with } r_j = j - l_j, j = k - 1, k; \\ &\mathbf{y}_k = (\mathbf{x}_k, \mathbf{x}_{k-1}, \ldots, \mathbf{x}_{k-l_{k-1}})^{\mathrm{T}}, \quad \mathbf{y}_{k-1} = (\mathbf{x}_{k-1}, \mathbf{x}_{k-2}, \ldots, \mathbf{x}_{k-1-l_{k-1}})^{\mathrm{T}}, \\ &\Delta\mathbf{y}_k = (\Delta\mathbf{x}_k, \Delta\mathbf{x}_{k-1}, \ldots, \Delta\mathbf{x}_{k-l_{k-1}})^{\mathrm{T}}, \quad \Delta\mathbf{y}_{k-1} = (\Delta\mathbf{x}_{k-1}, \Delta\mathbf{x}_{k-2}, \ldots, \Delta\mathbf{x}_{k-1-l_{k-1}})^{\mathrm{T}}. \end{aligned}$$

$$(5.123)$$

Thus,

$$\Delta \mathbf{y}_k = \mathbf{A}_k \Delta \mathbf{y}_{k-1} \equiv \left[ \frac{\partial \mathbf{y}_k}{\partial \mathbf{y}_{k-1}} \right]_{(\mathbf{y}_{k-1}^*, \mathbf{y}_k^*)} \Delta \mathbf{y}_{k-1}. \tag{5.124}$$

Finally, we have

$$\mathbf{A}_k = \begin{bmatrix} \mathbf{B}_k & (\mathbf{a}_{k(r_{k-1}-1)})_{n \times n} \\ \mathbf{I}_k & \mathbf{0}_k \end{bmatrix}_{n(s+1) \times n(s+1)}, \quad s = 1 + l_{k-1}$$

$$\mathbf{B}_k = [(\mathbf{a}_{k(k-1)})_{n \times n}, \mathbf{0}_{n \times n}, \dots, (\mathbf{a}_{kr_{k-1}})_{n \times n}], \tag{5.125}$$

$$\mathbf{I}_k = \mathrm{diag}(\mathbf{I}_{n \times n}, \mathbf{I}_{n \times n}, \dots, \mathbf{I}_{n \times n})_{ns \times ns}, \quad \mathbf{0}_k = \underbrace{(\mathbf{0}_{n \times n}, \mathbf{0}_{n \times n}, \dots, \mathbf{0}_{n \times n})}_{s}^{\mathrm{T}}.$$

From the forgoing equation, we have $[\partial \mathbf{y}_k / \partial \mathbf{y}_0]$. Thus, the linearized equation based on the initial point $\mathbf{y}_0$ in Eq. (5.120) gives

$$\Delta \mathbf{y}_k = \mathrm{DP}_{k(k-1)\cdots 1} \Delta \mathbf{y}_0 = \left[ \frac{\partial \mathbf{y}_k}{\partial \mathbf{y}_0} \right]_{(\mathbf{y}_0^*, \mathbf{y}_1^*, \dots, \mathbf{y}_{k-1}^*, \mathbf{y}_k^*)} \Delta \mathbf{y}_0 \quad (k = 1, 2, \dots, N)$$

$$\Delta \mathbf{y}_N = \mathrm{DP} \Delta \mathbf{y}_0 = \mathrm{DP}_{N(N-1)\cdots 1} \Delta \mathbf{y}_0 = \left[ \frac{\partial \mathbf{y}_k}{\partial \mathbf{y}_0} \right]_{(\mathbf{y}_0^*, \mathbf{y}_1^*, \dots, \mathbf{y}_{N-1}^*, \mathbf{y}_N^*)} \Delta \mathbf{y}_0 \tag{5.126}$$

Setting $\Delta \mathbf{y}_k = \bar{\lambda} \Delta \mathbf{y}_0$ and $\Delta \mathbf{y}_N = \lambda \Delta \mathbf{y}_0$, the forgoing equation becomes

$$(\mathrm{DP}_{k(k-1)\cdots 1} - \bar{\lambda} \mathbf{I}_{n(s+1) \times n(s+1)}) \Delta \mathbf{y}_0 = \mathbf{0},$$

$$(\mathrm{DP} - \lambda \mathbf{I}_{n(s+1) \times n(s+1)}) \Delta \mathbf{y}_0 = \mathbf{0}, \tag{5.127}$$

For any non-trivial solution ($\|\Delta \mathbf{y}_0\| \neq 0$), we have

$$|\mathrm{DP}_{k(k-1)\cdots 1} - \bar{\lambda} \mathbf{I}_{(s+1) \times n(s+1)}| = 0,$$

$$|\mathrm{DP} - \lambda \mathbf{I}_{(s+1) \times n(s+1)}| = 0. \tag{5.128}$$

Thus, the eigenvalues of $\mathrm{DP}_{k(k-1)\dots 1}$ give the changes of $\Delta \mathbf{y}_k$ with $\Delta \mathbf{y}_0$. In addition, the eigenvalues of DP are computed for the periodic solution due to $\mathbf{y}_N^* = \mathbf{y}_0^*$. From the stability and bifurcation theory of dynamical systems at fixed points in discrete nonlinear systems, the stability and bifurcation of the periodic solution of the time-delay nonlinear system can be classified as stated in the theorem. This theorem is proved.                                                                                       □

For a time-delay system, a periodic solution is represented by $N$ discrete points $\mathbf{x}_k$ ($k = 0, 1, 2, \dots, N$) and the corresponding time-delay points $\mathbf{x}_k^\tau$ ($k = 0, 1, 2, \dots, N$), as shown in Fig. 5.5. The small, filled circular symbols are for discrete nodes, and the large, hollow circular symbols are for time-delay nodes. The time-delay nodes are obtained by interpolation. The periodicity requires $\mathbf{x}_N = \mathbf{x}_0$ and $\mathbf{x}_N^\tau = \mathbf{x}_0^\tau$. To

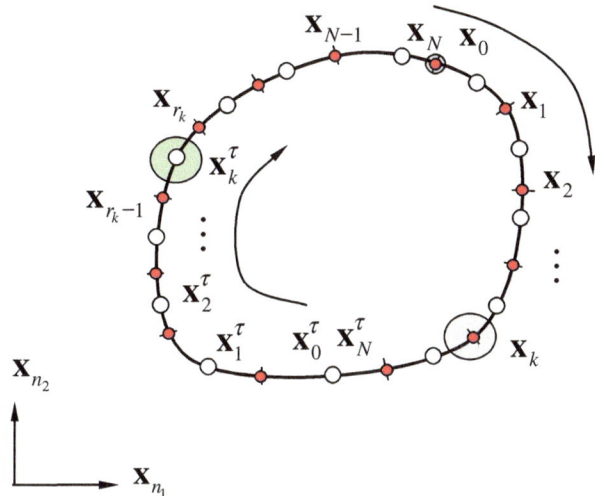

**Fig. 5.5** Period-1 flow with $N$-nodes for a time-delay system. The small, *filled circular symbols* are for non-time-delay nodes, and the large, *hollow circular symbols* are for time-delay nodes. The referenced point $\mathbf{x}_k$ and the corresponding time-delay point $\mathbf{x}_k^\tau$ are labeled. The time-delay point $\mathbf{x}_k^\tau$ can be estimated by the two vicinity points $\mathbf{x}_{r_k}$ and $\mathbf{x}_{r_k-1}$ where $r_k = \text{mod}(k - l_k + N, N)$

reduce computation, the time-delay points $\mathbf{x}_k^\tau$ $(k = 0, 1, 2, \ldots, N)$ are interpolated by $\mathbf{x}_{\text{mod}(k+N-l_k-1,N)}$ and $\mathbf{x}_{\text{mod}(k+N-l_k,N)}$. For $k = 0$, $\mathbf{x}_0^\tau$ is interpolated by $\mathbf{x}_{N-l_0-1}$ and $\mathbf{x}_{N-l_0}$ for periodic flow. For $k = N$, $\mathbf{x}_N^\tau$ is interpolated by $\mathbf{x}_{N-1}$ and $\mathbf{x}_N = \mathbf{x}_0$ for periodic flow. In fact, $\mathbf{x}_k^\tau$ can be interpolated by multiple nodes around two points of $\mathbf{x}_{\text{mod}(k+N-l_k,N)}$ and $\mathbf{x}_{\text{mod}(k+N-l_k-1,N)}$. For instance, $s_1 + s_2 + 1$ nodes, $\mathbf{x}_{\text{mod}(k+N-l_k-r,N)}$ $(r = -s_2, -s_2 + 1, \ldots, s_1 - 1, s_1)$, are used for interpolation of the time-delay $\mathbf{x}_k^\tau$. At least, two points $\mathbf{x}_{\text{mod}(k+N-l_k,N)}$ and $\mathbf{x}_{\text{mod}(k+N-l_k-1,N)}$ should be used for interpolation with a better approximation.

From the forgoing theorem, a set of nonlinear, time-delay, discrete mappings $P_k$ with $\mathbf{g}_k(\mathbf{x}_{k-1}, \mathbf{x}_{k-1}^\tau, \mathbf{x}_k, \mathbf{x}_k^\tau, \mathbf{p}) = \mathbf{0}$ $(k = 1, 2, \ldots, N)$ and interpolated time-delay nodes are developed for a periodic flow. Such a mapping can be used for numerical simulations. For given $\mathbf{x}_{k-1}$, $\mathbf{x}_{k-1}^\tau$, and $\mathbf{x}_k^\tau$, one can compute $\mathbf{x}_k$ through the algebraic equation $\mathbf{g}_k(\mathbf{x}_{k-1}, \mathbf{x}_{k-1}^\tau, \mathbf{x}_k, \mathbf{x}_k^\tau, \mathbf{p}) = \mathbf{0}$ plus the interpolated time-delay nodes.

In addition to a one-step time-delay mapping of $P_k$, one can develop a multi-step (or $r$-steps) time-delay mapping of $P_k$ with

$$\mathbf{g}_k(\mathbf{x}_{k-r}, \ldots, \mathbf{x}_{k-1}, \mathbf{x}_k; \mathbf{x}_{k-r}^\tau, \ldots, \mathbf{x}_{k-1}^\tau, \mathbf{x}_k^\tau, \mathbf{p}) = \mathbf{0},$$

$$\mathbf{x}_{k-j}^\tau = \mathbf{h}_{k-j}(\mathbf{x}_{r_{kj}-1}, \mathbf{x}_{r_{kj}}, \theta_{r_{kj}}), \quad \theta_{r_{kj}} = \frac{1}{h_{r_{kj}}}[\tau - \sum_{i=1}^{l_j} h_{r_{kj}+i}],$$

$$r_{kj} = k - j - l_j, \quad j = 0, 1, 2, \ldots, r; \tag{5.129}$$

$$r \in \{1, 2, \ldots, k\} \quad \text{and} \quad k = 1, 2, \ldots, N.$$

(i)  If $r = 1$, we have $j = 0, 1$. So the one-step time-delay mapping is recovered from the multi-step time-delay mapping.

(ii)  If $r = 2$, we have $j = 0, 1, 2$. So the two-step time-delay mapping is obtained from the multi-step time-delay mapping as

$$\mathbf{g}_k(\mathbf{x}_{k-2}, \mathbf{x}_{k-1}, \mathbf{x}_k; \mathbf{x}_{k-1}^\tau, \mathbf{x}_{k-2}^\tau, \mathbf{x}_k^\tau, \mathbf{p}) = \mathbf{0}$$

$$\mathbf{x}_{k-j}^\tau = \mathbf{h}_{k-j}(\mathbf{x}_{r_{kj}-1}, \mathbf{x}_{r_{kj}}, \theta_{r_{kj}}), \quad \theta_{r_{kj}} = \frac{1}{h_{r_{kj}}}[\tau - \sum_{i=1}^{l_j} h_{r_{kj}+i}], \tag{5.130}$$

$$r_{kj} = k - j - l_j, \quad j = 0, 1, 2; \quad k = 1, 2, \ldots, N$$

which can be expanded as

$$\mathbf{g}_1(\mathbf{x}_0, \mathbf{x}_1; \mathbf{x}_0^\tau, \mathbf{x}_1^\tau, \mathbf{p}) = \mathbf{0},$$

$$\vdots \tag{5.131}$$

$$\mathbf{g}_k(\mathbf{x}_{k-2}, \mathbf{x}_{k-1}, \mathbf{x}_k; \mathbf{x}_{k-2}^\tau, \mathbf{x}_{k-1}^\tau, \mathbf{x}_k^\tau, \mathbf{p}) = \mathbf{0}, \quad (k = 1, 2, \ldots, N).$$

(iii)  If $r = k$, the $k$-steps time-delay mapping is obtained, i.e.,

$$\mathbf{g}_k(\mathbf{x}_0, \mathbf{x}_1, \ldots, \mathbf{x}_k; \mathbf{x}_0^\tau, \mathbf{x}_1^\tau, \ldots, \mathbf{x}_k^\tau, \mathbf{p}) = \mathbf{0}$$

$$\mathbf{x}_{k-j}^\tau = \mathbf{h}_{k-j}(\mathbf{x}_{r_{kj}-1}, \mathbf{x}_{r_{kj}}, \theta_{r_{kj}}), \quad \theta_{r_{kj}} = \frac{1}{h_{r_{kj}}}[\tau - \sum_{i=1}^{l_j} h_{r_{kj}+i}], \tag{5.132}$$

$$r_{kj} = k - j - l_j, \quad j = 0, 1, \ldots k - 1, k; k = 1, 2, \ldots, N.$$

and the forgoing equations can be expanded as

$$\mathbf{g}_1(\mathbf{x}_0, \mathbf{x}_1; \mathbf{x}_0^\tau, \mathbf{x}_1^\tau, \mathbf{p}) = \mathbf{0},$$

$$\vdots \tag{5.133}$$

$$\mathbf{g}_k(\mathbf{x}_0, \mathbf{x}_1, \ldots, \mathbf{x}_k, \mathbf{x}_0^\tau, \mathbf{x}_1^\tau, \ldots, \mathbf{x}_k^\tau, \mathbf{p}) = \mathbf{0}$$
$$(k = 1, 2, \ldots, N).$$

From the multi-step (or $r$-steps) mapping of $P_k$ without $k - j \geq 0$, with the periodicity condition ($\mathbf{x}_0 = \mathbf{x}_N$ and $\mathbf{x}_0^\tau = \mathbf{x}_N^\tau$), the periodic flow can be obtained via

$$\mathbf{g}_k(\mathbf{x}_{s_{kr}}, \ldots, \mathbf{x}_{s_{k1}}, \mathbf{x}_{s_{k0}}; \mathbf{x}_{s_{kr}}^\tau, \ldots, \mathbf{x}_{s_{k1}}^\tau, \mathbf{x}_{s_{k0}}^\tau, \mathbf{p}) = \mathbf{0};$$

$$\mathbf{x}_{s_{kj}}^\tau = \mathbf{h}_{s_{kj}}(\mathbf{x}_{r_{kj}-1}, \mathbf{x}_{r_{kj}}, \theta_{r_{kj}}), \quad \theta_{r_{kj}} = \frac{1}{h_{r_{kj}}}[\tau - \sum_{i=1}^{l_{s_{kj}}} h_{r_{kj}+i}],$$

$$r_{kj} = k - j - l_{s_{kj}}, s_{kj} = k - j, \quad j = 0, 1, 2, \ldots, r,$$
$$r \in \{1, 2, \ldots, N\} \quad \text{and} \quad k = 1, 2, \ldots, N; \tag{5.134}$$
$$\mathbf{x}_{r_{kj}-1} = \mathbf{x}_{\text{mod}(r_{kj}-1+N,N)}, \quad \mathbf{x}_{r_{kj}} = \mathbf{x}_{\text{mod}(r_{kj}+N,N)}, (\mathbf{x}_0, \mathbf{x}_0^{\tau}) = (\mathbf{x}_N, \mathbf{x}_N^{\tau}).$$

Suppose node points $\mathbf{x}_k^*$ and $\mathbf{x}_k^{\tau*}$ ($k = 0, 1, 2, \ldots, N$) of periodic flows are obtained, the corresponding stability and bifurcation can be analyzed in the neighborhoods of $\mathbf{x}_k^*$ and $\mathbf{x}_k^{\tau*}$ with $\mathbf{x}_k = \mathbf{x}_k^* + \Delta\mathbf{x}_k$ and $\mathbf{x}_k^{\tau} = \mathbf{x}_k^{\tau*} + \Delta\mathbf{x}_k^{\tau}$ plus interpolated time-delay node $\mathbf{x}_{s_{kj}}^{\tau} = \mathbf{h}_{s_{kj}}(\mathbf{x}_{r_{kj}-1}, \mathbf{x}_{r_{kj}}, \theta_{r_{kj}})$. That is,

$$\sum_{j=0}^{r} \frac{\partial\mathbf{g}_k}{\partial\mathbf{x}_{s_{kj}}} \Delta\mathbf{x}_{s_{kj}} + \frac{\partial\mathbf{g}_k}{\partial\mathbf{x}_{s_{kj}}^{\tau}} \frac{\partial\mathbf{x}_{s_{kj}}^{\tau}}{\partial\mathbf{x}_{r_{kj}-1}} \Delta\mathbf{x}_{r_{kj}-1} + \frac{\partial\mathbf{g}_k}{\partial\mathbf{x}_{s_{kj}}^{\tau}} \frac{\partial\mathbf{x}_{s_{kj}}^{\tau}}{\partial\mathbf{x}_{r_{kj}}} \Delta\mathbf{x}_{r_{kj}} = \mathbf{0}_{1\times n} \tag{5.135}$$
$$(k = 1, 2, \ldots, N; r \in \{1, 2, \ldots, k\}).$$

Let

$$\mathbf{a}_{ks_{kj}} = -\left[\frac{\partial\mathbf{g}_k}{\partial\mathbf{x}_k}\right]^{-1} \frac{\partial\mathbf{g}_k}{\partial\mathbf{x}_{s_{kj}}}, \quad \mathbf{a}_{kr_{kj}} = -\left[\frac{\partial\mathbf{g}_k}{\partial\mathbf{x}_k}\right]^{-1} \frac{\partial\mathbf{g}_k}{\partial\mathbf{x}_{s_{kj}}^{\tau}} \frac{\partial\mathbf{x}_{s_{kj}}^{\tau}}{\partial\mathbf{x}_{r_{kj}}},$$

$$\mathbf{a}_{k(r_{kj}-1)} = -\left[\frac{\partial\mathbf{g}_k}{\partial\mathbf{x}_k}\right]^{-1} \frac{\partial\mathbf{g}_k}{\partial\mathbf{x}_{s_{kj}}^{\tau}} \frac{\partial\mathbf{x}_{s_{kj}}^{\tau}}{\partial\mathbf{x}_{r_{kj}-1}}$$

with $r_{kj} = k - j - l_{s_{kj}}, s_{kj} = k - j, \quad j = 0, 1, 2, \ldots, r;$

$$\mathbf{y}_k = (\mathbf{x}_k, \mathbf{x}_{k-1}, \ldots, \mathbf{x}_{r_{kr}})^{\text{T}}, \quad \mathbf{y}_{k-1} = (\mathbf{x}_{k-1}, \mathbf{x}_{k-2}, \ldots, \mathbf{x}_{r_{kr}-1})^{\text{T}},$$
$$\Delta\mathbf{y}_k = (\Delta\mathbf{x}_k, \Delta\mathbf{x}_{k-1}, \ldots, \Delta\mathbf{x}_{r_{kr}})^{\text{T}}, \quad \Delta\mathbf{y}_{k-1} = (\Delta\mathbf{x}_{k-1}, \Delta\mathbf{x}_{k-2}, \ldots, \Delta\mathbf{x}_{r_{kr}-1})^{\text{T}}. \tag{5.136}$$

Thus,

$$\mathbf{A}_k = \begin{bmatrix} \mathbf{B}_k & (\mathbf{a}_{k(r_{kr}-1)})_{n\times n} \\ \mathbf{I}_k & \mathbf{0}_k \end{bmatrix}_{n(s+1)\times n(s+1)}, \quad s = r + l_{s_{kr}}$$
$$\mathbf{B}_k = ((\mathbf{a}_{k(k-1)})_{n\times n}, (\mathbf{a}_{k(k-2)})_{n\times n}, \ldots, (\mathbf{a}_{k(k-r)})_{n\times n}, \ldots, (\mathbf{a}_{kr_{kr}})_{n\times n}) \tag{5.137}$$
$$\mathbf{I}_k = \text{diag}(\mathbf{I}_{n\times n}, \mathbf{I}_{n\times n}, \ldots, \mathbf{I}_{n\times n})_{ns\times ns}, \quad \mathbf{0}_k = (\underbrace{\mathbf{0}_{n\times n}, \mathbf{0}_{n\times n}, \ldots, \mathbf{0}_{n\times n}}_{ns})^{\text{T}}.$$

Finally, we have

$$\Delta\mathbf{y}_k = \mathbf{A}_k\Delta\mathbf{y}_{k-1}. \tag{5.138}$$

From the mapping structure, we have

$$\Delta \mathbf{y}_N = \mathrm{DP} \cdot \Delta \mathbf{y}_0 \quad \text{and} \quad \mathrm{DP} = \left[\frac{\partial \mathbf{y}_N}{\partial \mathbf{y}_0}\right] = \mathbf{A}_N \mathbf{A}_{N-1} \ldots \mathbf{A}_1. \qquad (5.139)$$

Letting $\Delta \mathbf{y}_N = \lambda \Delta \mathbf{y}_0$, we have

$$(\mathrm{DP} - \lambda \mathbf{I}_{n(s+1) \times n(s+1)}) \Delta \mathbf{y}_0 = \mathbf{0}. \qquad (5.140)$$

The eigenvalue of DP is given by $|\mathrm{DP} - \lambda \mathbf{I}_{n(s+1) \times n(s+1)}| = 0$. In addition, we have

$$\Delta \mathbf{y}_k = \mathrm{DP}_{k(k-1)\cdots 1} \cdot \Delta \mathbf{y}_0 \quad \text{and} \quad \mathrm{DP}_{k(k-1)\cdots 1} = \left[\frac{\partial \mathbf{y}_k}{\partial \mathbf{y}_0}\right] = \mathbf{A}_k \mathbf{A}_{k-1} \ldots \mathbf{A}_1 \qquad (5.141)$$

$$(k = 1, 2, \ldots, N).$$

Letting $\Delta \mathbf{y}_k = \bar{\lambda} \Delta \mathbf{y}_0$, we have

$$(\mathrm{DP}_{k(k-1)\cdots 1} - \bar{\lambda} \mathbf{I}_{n(s+1) \times n(s+1)}) \Delta \mathbf{y}_0 = \mathbf{0}. \qquad (5.142)$$

The eigenvalues of $\mathrm{DP}_{k(k-1)\cdots 1}$ are given by $|\mathrm{DP}_{k(k-1)\cdots 1} - \bar{\lambda} \mathbf{I}_{n(s+1) \times n(s+1)}| = 0$. Such eigenvalues tell effects of variation of $\mathbf{y}_0$ on node points $\mathbf{y}_k$ in the corresponding neighborhood. The neighborhood of $\mathbf{x}_k^*$ (i.e., $U_k(\mathbf{x}_k^*)$) is presented in Fig. 5.6 as large circle. Since the time-delay points are interpolated by regular nodes, the variation of time-delay points can be determined by neighborhoods of such regular node points. In such a neighborhood, the eigenvalues can be used to

**Fig. 5.6** Neighborhoods of $N$-nodes for a period-1 flow of a time-delay system. *Solid curve* gives numerical results. The local *shaded* area is a small neighborhood at the $k$th node of the solution. The *red symbols* are for discrete node points of the periodic flow. The *hollow symbols* are for time-delay nodes of the periodic flow

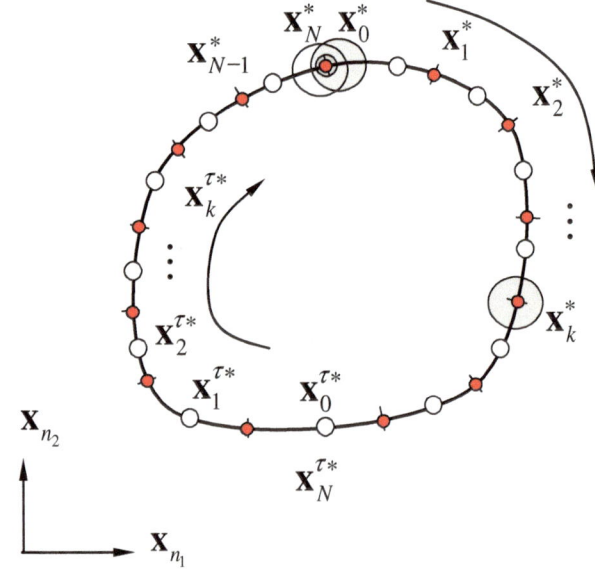

measure the effects $\Delta \mathbf{y}_k$ of $\mathbf{y}_k^*$ varying with $\Delta \mathbf{y}_0$ of $\mathbf{y}_0^*$. The eigenvalues of DP are given by $|\text{DP} - \lambda \mathbf{I}_{n(s+1) \times n(s+1)}| = 0$, which implies the stability and bifurcation of the period-1 flow.

(i) If $r = 1$, Eqs. (5.135) becomes

$$\sum_{j=0}^{1} \frac{\partial \mathbf{g}_k}{\partial \mathbf{x}_{S_{kj}}} \Delta \mathbf{x}_{S_{kj}} + \frac{\partial \mathbf{g}_k}{\partial \mathbf{x}_{S_{kj}}^\tau} \frac{\partial \mathbf{x}_{S_{kj}}^\tau}{\partial \mathbf{x}_{r_{kj}-1}} \Delta \mathbf{x}_{r_{kj}-1} + \frac{\partial \mathbf{g}_k}{\partial \mathbf{x}_{S_{kj}}^\tau} \frac{\partial \mathbf{x}_{S_{kj}}^\tau}{\partial \mathbf{x}_{r_{kj}}} \Delta \mathbf{x}_{r_{kj}} = \mathbf{0}_{1 \times n} \quad (5.143)$$

with $r_{kj} = k - j - l_{S_{kj}}$, $s_{kj} = k - j$, $j = 0, 1$; $(k = 1, 2, \ldots, N)$.

Setting

$$\mathbf{a}_{k S_{kj}} = -\left[\frac{\partial \mathbf{g}_k}{\partial \mathbf{x}_k}\right]^{-1} \frac{\partial \mathbf{g}_k}{\partial \mathbf{x}_{S_{kj}}}, \quad \mathbf{a}_{k r_{kj}} = -\left[\frac{\partial \mathbf{g}_k}{\partial \mathbf{x}_k}\right]^{-1} \frac{\partial \mathbf{g}_k}{\partial \mathbf{x}_{S_{kj}}^\tau} \frac{\partial \mathbf{x}_{S_{kj}}^\tau}{\partial \mathbf{x}_{r_{kj}}},$$

$$\mathbf{a}_{k(r_{kj}-1)} = -\left[\frac{\partial \mathbf{g}_k}{\partial \mathbf{x}_k}\right]^{-1} \frac{\partial \mathbf{g}_k}{\partial \mathbf{x}_{S_{kj}}^\tau} \frac{\partial \mathbf{x}_{S_{kj}}^\tau}{\partial \mathbf{x}_{r_{kj}-1}} \quad \text{with } r_{kj} = k - j - l_{S_{kj}}, s_{kj} = k - j, j = 0, 1;$$

$$\mathbf{y}_k = (\mathbf{x}_k, \mathbf{x}_{k-1}, \ldots, \mathbf{x}_{r_{k1}})^\text{T}, \quad \mathbf{y}_{k-1} = (\mathbf{x}_{k-1}, \mathbf{x}_{k-2}, \ldots, \mathbf{x}_{r_{k1}-1})^\text{T},$$

$$\Delta \mathbf{y}_k = (\Delta \mathbf{x}_k, \Delta \mathbf{x}_{k-1}, \ldots, \Delta \mathbf{x}_{r_{k1}})^\text{T}, \quad \Delta \mathbf{y}_{k-1} = (\Delta \mathbf{x}_{k-1}, \Delta \mathbf{x}_{k-2}, \ldots, \Delta \mathbf{x}_{r_{k1}-1})^\text{T}.$$

$$(5.144)$$

Thus,

$$\mathbf{A}_k = \begin{bmatrix} \mathbf{B}_k & (\mathbf{a}_{k(r_{k1}-1)})_{n \times n} \\ \mathbf{I}_k & \mathbf{0}_k \end{bmatrix}_{n(s+1) \times n(s+1)}, \quad s = 1 + l_{k-1}$$

$$\mathbf{B}_k = [(\mathbf{a}_{k(k-1)})_{n \times n}, \mathbf{0}_{n \times n}, \ldots, (\mathbf{a}_{k r_{k0}})_{n \times n}, (\mathbf{a}_{k r_{k1}})_{n \times n}]$$

$$\mathbf{I}_k = \text{diag}(\mathbf{I}_{n \times n}, \mathbf{I}_{n \times n}, \ldots, \mathbf{I}_{n \times n})_{ns \times ns}, \quad \mathbf{0}_k = (\underbrace{\mathbf{0}_{n \times n}, \mathbf{0}_{n \times n}, \ldots, \mathbf{0}_{n \times n}}_{s})^\text{T}.$$

$$(5.145)$$

Finally, we have

$$\Delta \mathbf{y}_k = \mathbf{A}_k \Delta \mathbf{y}_{k-1}. \quad (5.146)$$

So we have

$$\text{DP} = \left[\frac{\partial \mathbf{y}_N}{\partial \mathbf{y}_0}\right] = \mathbf{A}_N \mathbf{A}_{N-1} \ldots \mathbf{A}_1. \quad (5.147)$$

(ii) For $r = k$, Eq. (5.135) with periodicity condition $(\mathbf{x}_0 = \mathbf{x}_N)$ gives node points $\mathbf{x}_k^*$ $(k = 0, 1, 2, \ldots, N)$. The corresponding stability and bifurcation can be analyzed in the neighborhoods of $\mathbf{x}_k^*$ and $\mathbf{x}_k^{\tau*}$ with $\mathbf{x}_k = \mathbf{x}_k^* + \Delta \mathbf{x}_k$ and $\mathbf{x}_k^\tau = \mathbf{x}_k^{\tau*} + \Delta \mathbf{x}_k^\tau$. Equation (5.136) becomes

$$\sum_{j=0}^{k} \frac{\partial \mathbf{g}_k}{\partial \mathbf{x}_{s_{kj}}} \Delta \mathbf{x}_{s_{kj}} + \frac{\partial \mathbf{g}_k}{\partial \mathbf{x}_{s_{kj}}^{\tau}} \frac{\partial \mathbf{x}_{s_{kj}}^{\tau}}{\partial \mathbf{x}_{r_{kj}-1}} \Delta \mathbf{x}_{r_{kj}-1} + \frac{\partial \mathbf{g}_k}{\partial \mathbf{x}_{s_{kj}}^{\tau}} \frac{\partial \mathbf{x}_{s_{kj}}^{\tau}}{\partial \mathbf{x}_{r_{k-j}}} \Delta \mathbf{x}_{r_{k-j}} = \mathbf{0}_{1 \times n} \tag{5.148}$$

$$j = 0, 1, \ldots, k; \quad k = 1, 2, \ldots, N.$$

Thus,

$$\mathbf{a}_{k s_{kj}} = -\left[\frac{\partial \mathbf{g}_k}{\partial \mathbf{x}_k}\right]^{-1} \frac{\partial \mathbf{g}_k}{\partial \mathbf{x}_{s_{kj}}}, \quad \mathbf{a}_{k r_{kj}} = -\left[\frac{\partial \mathbf{g}_k}{\partial \mathbf{x}_k}\right]^{-1} \frac{\partial \mathbf{g}_k}{\partial \mathbf{x}_{s_{kj}}^{\tau}} \frac{\partial \mathbf{x}_{s_{kj}}^{\tau}}{\partial \mathbf{x}_{r_{kj}}},$$

$$\mathbf{a}_{k(r_{kj}-1)} = -\left[\frac{\partial \mathbf{g}_k}{\partial \mathbf{x}_k}\right]^{-1} \frac{\partial \mathbf{g}_k}{\partial \mathbf{x}_{s_{kj}}^{\tau}} \frac{\partial \mathbf{x}_{s_{kj}}^{\tau}}{\partial \mathbf{x}_{r_{kj}-1}}$$

with $r_{kj} = k - j - l_{s_{kj}}, s_{kj} = k - j, \quad j = 0, 1, 2, \ldots, k;$

$$\mathbf{y}_k = (\mathbf{x}_k, \mathbf{x}_{k-1}, \ldots, \mathbf{x}_{k-l_{k-r}})^{\mathrm{T}}, \quad \mathbf{y}_{k-1} = (\mathbf{x}_{k-1}, \mathbf{x}_{k-2}, \ldots, \mathbf{x}_{k-1-l_{k-r}})^{\mathrm{T}},$$

$$\Delta \mathbf{y}_k = (\Delta \mathbf{x}_k, \Delta \mathbf{x}_{k-1}, \ldots, \Delta \mathbf{x}_{k-l_{k-r}})^{\mathrm{T}}, \quad \Delta \mathbf{y}_{k-1} = (\Delta \mathbf{x}_{k-1}, \Delta \mathbf{x}_{k-2}, \ldots, \Delta \mathbf{x}_{k-1-l_{k-r}})^{\mathrm{T}}.$$

$$\tag{5.149}$$

Finally, we have

$$\Delta \mathbf{y}_k = \mathbf{A}_k \Delta \mathbf{y}_{k-1}, \quad \mathbf{A}_k = \left[\frac{\partial \mathbf{y}_k}{\partial \mathbf{y}_{k-1}}\right]_{(\mathbf{y}_{k-1}^*, \mathbf{y}_k^*)}. \tag{5.150}$$

Using $\partial \mathbf{y}_k / \partial \mathbf{y}_0$, the eigenvalues are determined by

$$|DP_{k(k-1)\ldots 1} - \lambda \mathbf{I}_{n(s+1) \times n(s+1)}| = 0 \quad \text{with } DP_{k(k-1)\ldots 1} = \left[\frac{\partial \mathbf{y}_k}{\partial \mathbf{y}_0}\right] = \mathbf{A}_k \mathbf{A}_{k-1} \ldots \mathbf{A}_1. \tag{5.151}$$

which is used to measure the properties of node points on the period-1 flow for the time-delay system.

The multi-step mappings are developed from the afore-determined nodes of periodic motion. During time interval $[t_0, t_0 + T]$, the periodic flow can be determined by

$$\mathbf{x}(t) = \mathbf{x}(t_l) + \int_{t_l}^{t} \mathbf{f}(\mathbf{x}, \mathbf{x}^{\tau}, t, \mathbf{p}) dt, \quad l \in \{0, 1, 2, \ldots, k-1\}. \tag{5.152}$$

For such a periodic flow, all $N$-nodes during the time interval $t \in [t_0, t_0 + T]$ are selected, and nodes $\mathbf{x}(t_k)(k = 0, 1, \ldots, N)$. Under $\|\mathbf{x}(t_k) - \mathbf{x}_k\| \le \varepsilon_k$ with $\varepsilon_k \ge 0$,

$$\|\mathbf{f}(\mathbf{x}(t_k), \mathbf{x}^{\tau}(t_k), t_k, \mathbf{p}) - \mathbf{f}(\mathbf{x}_k, \mathbf{x}_k^{\tau}, t_k, \mathbf{p})\| \le \delta_k. \tag{5.153}$$

If $\mathbf{x}_0, \ldots, \mathbf{x}_N$ and $\mathbf{x}_0^\tau, \ldots, \mathbf{x}_N^\tau$ are given, $\mathbf{f}(\mathbf{x}_k, \mathbf{x}_k^\tau, t_k, \mathbf{p})$ $(k = 0, 1, \ldots, N)$ can be determined. An interpolation polynomial $\mathbf{P}(t, \mathbf{x}_0, \ldots, \mathbf{x}_N, \mathbf{x}_0^\tau, \ldots, \mathbf{x}_N^\tau, t_0, \ldots, t_N, \mathbf{p})$ is used for an approximation of $\mathbf{f}(\mathbf{x}, \mathbf{x}^\tau, t, \mathbf{p})$. That is,

$$\mathbf{f}(\mathbf{x}, \mathbf{x}^\tau, t, \mathbf{p}) \approx \mathbf{P}(t, \mathbf{x}_0, \ldots, \mathbf{x}_N, \mathbf{x}_0^\tau, \ldots, \mathbf{x}_N^\tau, t_0, \ldots, t_N, \mathbf{p}) \tag{5.154}$$

and $\mathbf{x}(t_k) \approx \mathbf{x}_k$ $(k = 0, 1, \ldots, N)$ can be computed by

$$\mathbf{x}_k = \mathbf{x}_{k-1} + \int_{t_{k-1}}^{t_k} \mathbf{P}(t, \mathbf{x}_0, \ldots, \mathbf{x}_N, \mathbf{x}_0^\tau, \ldots, \mathbf{x}_N^\tau, t_0, \ldots, t_N, \mathbf{p}) \mathrm{d}t. \tag{5.155}$$

Therefore, we have

$$\mathbf{x}_k = \mathbf{x}_{k-1} + \bar{\mathbf{g}}_k(\mathbf{x}_0, \ldots, \mathbf{x}_N, \mathbf{x}_0^\tau, \ldots, \mathbf{x}_N^\tau, \mathbf{p}). \tag{5.156}$$

The mapping $P_k$ $(k \in \{1, 2, \ldots, N\})$ is

$$\begin{aligned} &\mathbf{g}_k(\mathbf{x}_k, \ldots, \mathbf{x}_{k-N}, \mathbf{x}_k^\tau, \ldots, \mathbf{x}_{k-N}^\tau, \mathbf{p}) = \mathbf{0}, \\ &\mathbf{x}_j^\tau = \mathbf{h}_j(\mathbf{x}_{r_j-1}, \mathbf{x}_{r_j}), r_j = j - l_j, \quad j = 0, 1, 2, \ldots, N. \end{aligned} \tag{5.157}$$

The periodic motions are determined by the mapping $P_k$ $(k = 1, 2, \ldots, N)$ and periodicity conditions

$$\left. \begin{aligned} &\mathbf{g}_k(\mathbf{x}_k^*, \ldots, \mathbf{x}_{k-N}^*, \mathbf{x}_k^{\tau*}, \ldots, \mathbf{x}_{k-N}^{\tau*}, \mathbf{p}) = \mathbf{0}; \\ &\mathbf{x}_{k-j}^{*\tau} = \mathbf{h}_{k-j}(\mathbf{x}_{r_j-1}^*, \mathbf{x}_{r_j-1}^*), r_j = k - j - l_j \end{aligned} \right\} $$
$$(j = 0, 1, 2, \ldots, N), (k = 1, 2, \ldots, N) \tag{5.158}$$
$$\mathbf{x}_{r_j-1}^* = \mathbf{x}_{\mathrm{mod}(r_j-1+N,N)}^*, \quad \mathbf{x}_{r_j}^* = \mathbf{x}_{\mathrm{mod}(r_j+N,N)}^*;$$
$$\mathbf{x}_{k-j}^* = \mathbf{x}_{\mathrm{mod}(k-j+N,N)}^*, \quad \mathbf{x}_0^* = \mathbf{x}_N^*, \mathbf{x}_0^{\tau*} = \mathbf{x}_N^{\tau*}.$$

From the forgoing equation, node points $\mathbf{x}_k^*$ and $\mathbf{x}_k^{\tau*}$ $(k = 0, 1, 2, \ldots, N)$ can be determined. The corresponding dynamical characteristics in the neighborhood of $\mathbf{x}_k^*$ with $\mathbf{x}_k = \mathbf{x}_k^* + \Delta\mathbf{x}_k$ are discussed by variation of $\mathbf{x}_0$ in the neighborhood of $\mathbf{x}_0^*$ with $\mathbf{x}_0 = \mathbf{x}_0^* + \Delta\mathbf{x}_0$. The derivative of Eq. (5.158) with respect to $\mathbf{x}_0$ gives

$$\sum_{j=k-N}^{k} \frac{\partial \mathbf{g}_k}{\partial \mathbf{x}_j} \Delta\mathbf{x}_j + \frac{\partial \mathbf{g}_k}{\partial \mathbf{x}_j^\tau} \frac{\partial \mathbf{x}_j^\tau}{\partial \mathbf{x}_{r_j-1}} \Delta\mathbf{x}_{r_j-1} + \frac{\partial \mathbf{g}_k}{\partial \mathbf{x}_j^\tau} \frac{\partial \mathbf{x}_j^\tau}{\partial \mathbf{x}_{r_j}} \Delta\mathbf{x}_{r_j} = \mathbf{0}_{1 \times n} \tag{5.159}$$
$$(k = 1, 2, \ldots, N).$$

Herein, the following vectors and matrices are defined as

$$\mathbf{a}_{ks_{kj}} = -\left[\frac{\partial \mathbf{g}_k}{\partial \mathbf{x}_k}\right]^{-1}\frac{\partial \mathbf{g}_k}{\partial \mathbf{x}_{s_{kj}}}, \quad \mathbf{a}_{kr_{kj}} = -\left[\frac{\partial \mathbf{g}_k}{\partial \mathbf{x}_k}\right]^{-1}\frac{\partial \mathbf{g}_k}{\partial \mathbf{x}_{s_{kj}}^\tau}\frac{\partial \mathbf{x}_{s_{kj}}^\tau}{\partial \mathbf{x}_{r_{kj}}},$$

$$\mathbf{a}_{k(r_{kj}-1)} = -\left[\frac{\partial \mathbf{g}_k}{\partial \mathbf{x}_k}\right]^{-1}\frac{\partial \mathbf{g}_k}{\partial \mathbf{x}_{s_{kj}}^\tau}\frac{\partial \mathbf{x}_{s_{kj}}^\tau}{\partial \mathbf{x}_{r_{kj}-1}}$$

$$\text{with } r_{kj} = k - j - l_{s_{kj}}, \quad s_{kj} = k - j, \quad j = 0, 1, 2, \ldots, N;$$

$$\mathbf{y}_k = (\mathbf{x}_k, \mathbf{x}_{k-1}, \ldots, \mathbf{x}_{k-l_{k-N}})^\mathrm{T}, \quad \mathbf{y}_{k-1} = (\mathbf{x}_{k-1}, \mathbf{x}_{k-2}, \ldots, \mathbf{x}_{k-1-l_{k-N}})^\mathrm{T},$$

$$\Delta\mathbf{y}_k = (\Delta\mathbf{x}_k, \Delta\mathbf{x}_{k-1}, \ldots, \Delta\mathbf{x}_{k-l_{k-N}})^\mathrm{T}, \quad \Delta\mathbf{y}_{k-1} = (\Delta\mathbf{x}_{k-1}, \Delta\mathbf{x}_{k-2}, \ldots, \Delta\mathbf{x}_{k-1-l_{k-N}})^\mathrm{T}.$$

$$(5.160)$$

From the above discussion, the discrete mapping can be developed through many forward and backward nodes. The periodic flow in the nonlinear time-delay system can be determined through the following theorem.

**Theorem 5.6** *Consider a nonlinear time-delay system in Eq. (5.98). If such a system has a periodic flow* $\mathbf{x}(t)$ *with finite norm* $\|\mathbf{x}\|$ *and one period* $T = 2\pi/\Omega$, *there is a set of discrete time* $t_k$ ($k = 0, 1, \ldots, N$) *with* ($N \to \infty$) *during one period* $T$, *and the corresponding solution* $\mathbf{x}(t_k)$ *and vector field* $\mathbf{f}(\mathbf{x}(t_k), \mathbf{x}^\tau(t_k), t_k, \mathbf{p})$ *are exact. Suppose discrete nodes* $\mathbf{x}_k$ *and* $\mathbf{x}_k^\tau$ *are on the approximate solution of the periodic flow under* $\|\mathbf{x}(t_k) - \mathbf{x}_k\| \le \varepsilon_k$ *and* $\|\mathbf{x}^\tau(t_k) - \mathbf{x}_k^\tau\| \le \varepsilon_k^\tau$ *with small* $\varepsilon_k, \varepsilon_k^\tau \ge 0$ *and*

$$\|\mathbf{f}(\mathbf{x}(t_k), \mathbf{x}^\tau(t_k), t_k, \mathbf{p}) - \mathbf{f}(\mathbf{x}_k, \mathbf{x}_k^\tau, t_k, \mathbf{p})\| \le \delta_k \quad (5.161)$$

*with a small* $\delta_k \ge 0$. *During a time interval* $t \in [t_{k-1}, t_k]$, *there is a mapping* $P_k :$ $(\mathbf{x}_{k-1}, \mathbf{x}_{k-1}^\tau) \to (\mathbf{x}_k, \mathbf{x}_k^\tau)$ ($k = 1, 2, \ldots, N$), *i.e.*,

$$P_k : (\mathbf{x}_{k-1}, \mathbf{x}_{k-1}^\tau) \to (\mathbf{x}_k, \mathbf{x}_k^\tau) \quad \text{with}$$

$$\mathbf{g}_k(\mathbf{x}_{s_{kr_1}}, \ldots, \mathbf{x}_{s_{k0}}, \ldots, \mathbf{x}_{s_{k(-r_2)}}, \mathbf{x}_{s_{kr_1}}^\tau, \ldots, \mathbf{x}_{s_{k0}}^\tau, \ldots, \mathbf{x}_{s_{k(-r_2)}}^\tau, \mathbf{p}) = \mathbf{0},$$

$$\mathbf{x}_{s_{kj}}^\tau = \mathbf{h}_{s_{kj}}(\mathbf{x}_{r_{kj}-1}, \mathbf{x}_{r_{kj}}, \theta_{r_{kj}}), \quad \theta_{r_{kj}} = \frac{1}{h_{r_{kj}}}[\tau - \sum_{i=1}^{l_{s_{kj}}} h_{r_{kj}+i}]; \quad (5.162)$$

$$r_{kj} = k - j - l_{s_{kj}}, s_{kj} = k - j; j = -r_2, -r_2 + 1, \ldots - 1, 0, 1, \ldots, r_1 - 1, r_1;$$

$$r_1, r_2 \in \{0, 1, 2, \ldots, N\}; 1 \le r_1 + r_2 \le N, r_1 \ge 1; (k = 1, 2, \ldots, N).$$

*where* $\mathbf{g}_k$ *is an implicit vector function and* $\mathbf{h}_j$ *is an interpolation vector function. Consider a mapping structure as*

$$P = P_N \circ P_{N-1} \circ \cdots \circ P_2 \circ P_1 : (\mathbf{x}_0, \mathbf{x}_0^\tau) \to (\mathbf{x}_N, \mathbf{x}_N^\tau);$$

$$\text{with } P_k : (\mathbf{x}_{k-1}, \mathbf{x}_{k-1}^\tau) \to (\mathbf{x}_k, \mathbf{x}_k^\tau)(k = 1, 2, \ldots, N).$$

$$(5.163)$$

For $(\mathbf{x}_N, \mathbf{x}_N^\tau) = P(\mathbf{x}_0, \mathbf{x}_0^\tau)$, if there is a set of points $\mathbf{x}_k^*$ and $\mathbf{x}_k^{\tau*}(k = 0, 1, \ldots, N)$ computed by

$$\mathbf{g}_k(\mathbf{x}_{s_{kr_1}}^*, \ldots, \mathbf{x}_{s_{k0}}^*, \ldots, \mathbf{x}_{s_{k(-r_2)2}}^*, \mathbf{x}_{s_{kr_1}}^{\tau*}, \ldots, \mathbf{x}_{s_{k0}}^{\tau*}, \ldots, \mathbf{x}_{s_{k(-r_2)}}^{\tau*}, \mathbf{p}) = 0,$$

$$\mathbf{x}_{s_{kj}}^{\tau*} = \mathbf{h}_{s_{kj}}(\mathbf{x}_{r_{kj}-1}^*, \mathbf{x}_{r_{kj}}^*, \theta_{r_{kj}}), \quad \theta_{r_{kj}} = \frac{1}{h_{r_{kj}}}[\tau - \sum_{i=1}^{l_{s_{kj}}} h_{r_{kj}+i,}];$$

$$\mathbf{x}_{r_{kj}-1}^* = \mathbf{x}_{\mathrm{mod}(r_{kj}-1+N,N)}^*, \quad \mathbf{x}_{r_{kj}}^* = \mathbf{x}_{\mathrm{mod}(r_{kj}+N,N)}^*;$$

$$\mathbf{x}_{s_{kj}}^* = \mathbf{x}_{\mathrm{mod}(s_{kj}+N,N)}^*; \quad (\mathbf{x}_0^*, \mathbf{x}_0^{\tau*}) = (\mathbf{x}_N^*, \mathbf{x}_N^{\tau*}); \tag{5.164}$$

then the points $\mathbf{x}_k^*$ and $\mathbf{x}_k^{\tau*}$ $(k = 0, 1, \ldots, N)$ are approximations of points $\mathbf{x}(t_k)$ and $\mathbf{x}^\tau(t_k)$ of the periodic solution. In the neighborhoods of $\mathbf{x}_k^*$ and $\mathbf{x}_k^{\tau*}$, with $\mathbf{x}_k = \mathbf{x}_k^* + \Delta\mathbf{x}_k$ and $\mathbf{x}_k^\tau = \mathbf{x}_k^{\tau*} + \Delta\mathbf{x}_k^\tau$, the linearized equation is given by

$$\sum_{j=r_1}^{-r_2} \frac{\partial \mathbf{g}_k}{\partial \mathbf{x}_{s_{kj}}} \Delta\mathbf{x}_{s_{kj}} + \frac{\partial \mathbf{g}_k}{\partial \mathbf{x}_{s_{kj}}^\tau} \frac{\partial \mathbf{x}_{k_{kj}}^\tau}{\partial \mathbf{x}_{r_{kj}-1}} \Delta\mathbf{x}_{r_{kj}-1} + \frac{\partial \mathbf{g}_k}{\partial \mathbf{x}_{s_{kj}}^\tau} \frac{\partial \mathbf{x}_{s_{kj}}^\tau}{\partial \mathbf{x}_{r_{kj}}} \Delta\mathbf{x}_{r_{kj}} = \mathbf{0}$$

$$\text{with} \quad \frac{\partial \mathbf{g}_k}{\partial \mathbf{x}_\alpha} = \mathbf{0} \quad \text{and} \quad \frac{\partial \mathbf{g}_k}{\partial \mathbf{x}_\alpha^\tau} = \mathbf{0}(\alpha \neq s_{kj}, j = -r_2, -r_2 + 1, \ldots, r_1 - 1, r_1)$$

$$(k = 1, 2, \ldots, N).$$

$$\tag{5.165}$$

The resultant Jacobian matrices of the periodic flow are

$$\mathrm{DP}_{k(k-1)\ldots1} = \left[\frac{\partial \mathbf{y}_k}{\partial \mathbf{y}_0}\right]_{(\mathbf{y}_0^*, \mathbf{y}_1^*, \ldots, \mathbf{y}_k^*)} = \mathbf{A}_k \mathbf{A}_{k-1} \ldots \mathbf{A}_1 \quad (k = 1, 2, \ldots, N),$$

$$\text{and} \quad \mathrm{DP} = \mathrm{DP}_{N(N-1)\ldots1} = \left[\frac{\partial \mathbf{y}_N}{\partial \mathbf{y}_0}\right]_{(\mathbf{y}_0^*, \mathbf{y}_1^*, \ldots, \mathbf{y}_N^*)} = \mathbf{A}_N \mathbf{A}_{N-1} \ldots \mathbf{A}_1 \tag{5.166}$$

where

$$\Delta\mathbf{y}_k = \mathbf{A}_k \Delta\mathbf{y}_{k-1}, \quad \mathbf{A}_k = \left[\frac{\partial \mathbf{y}_k}{\partial \mathbf{y}_{k-1}}\right]_{(\mathbf{y}_{k-1}^*, \mathbf{y}_k^*)} \tag{5.167}$$

and

$$\mathbf{a}_{ks_{kj}} = -\left[\frac{\partial \mathbf{g}_k}{\partial \mathbf{x}_{k+r_2}}\right]^{-1} \frac{\partial \mathbf{g}_k}{\partial \mathbf{x}_{s_{kj}}}, \quad \mathbf{a}_{kr_{kj}} = -\left[\frac{\partial \mathbf{g}_k}{\partial \mathbf{x}_{k+r_2}}\right]^{-1} \frac{\partial \mathbf{g}_k}{\partial \mathbf{x}_{s_{kj}}^\tau} \frac{\partial \mathbf{x}_{s_{kj}}^\tau}{\partial \mathbf{x}_{r_{kj}}},$$

$$\mathbf{a}_{k(r_{kj}-1)} = -\left[\frac{\partial \mathbf{g}_k}{\partial \mathbf{x}_{k+r_2}}\right]^{-1} \frac{\partial \mathbf{g}_k}{\partial \mathbf{x}_{s_{kj}}^\tau} \frac{\partial \mathbf{x}_{s_{kj}}^\tau}{\partial \mathbf{x}_{r_{kj}-1}} \quad \text{with} \quad r_{kj} = k - j - l_{s_{kj}}, s_{kj} = k - j;$$

$$j = -r_2, -r_2 + 1, \ldots - 1, 0, 1, \ldots, r_1 - 1, r_1; r_1, r_2 \in \{0, 1, 2, \ldots, N\};$$
$$1 \le r_1 + r_2 \le N, r_1 \ge 1; (k = 1, 2, \ldots, N).$$
$$\mathbf{y}_k = (\mathbf{x}_{k+r_2}, \mathbf{x}_{k+r_2-1}, \ldots, \mathbf{x}_{r_{kr_1}})^{\mathrm{T}}, \quad \mathbf{y}_{k-1} = (\mathbf{x}_{k+r_2-1}, \mathbf{x}_{k+r_2-1}, \ldots, \mathbf{x}_{r_{k(r_1-1)}})^{\mathrm{T}},$$
$$\Delta \mathbf{y}_k = (\Delta \mathbf{x}_{k+r_2}, \Delta \mathbf{x}_{k+r_2-1}, \ldots, \Delta \mathbf{x}_{r_{kr_1}})^{\mathrm{T}}, \quad \Delta \mathbf{y}_{k-1} = (\Delta \mathbf{x}_{k+r_2-1}, \Delta \mathbf{x}_{k+r_2-1}, \ldots, \Delta \mathbf{x}_{r_{k(r_1-1)}})^{\mathrm{T}}$$

$$(5.168)$$

*and*

$$\mathbf{A}_k = \begin{bmatrix} \mathbf{B}_k & (\mathbf{a}_{k(r_{kr_1}-1)})_{n \times n} \\ \mathbf{I}_k & \mathbf{0}_k \end{bmatrix}_{n(s+1) \times n(s+1)}, \quad s = (r_1 + r_2 + l_{s_{kr_1}})$$

$$\mathbf{B}_k = ((\mathbf{a}_{k(k+r_2-1)})_{n \times n}, (\mathbf{a}_{k(k+r_2-1)})_{n \times n}, \ldots, (\mathbf{a}_{k(k-r)})_{n \times n}, \ldots, (\mathbf{a}_{kr_{kr_1}})_{n \times n}) \quad (5.169)$$

$$\mathbf{I}_k = \mathrm{diag}(\mathbf{I}_{n \times n}, \mathbf{I}_{n \times n}, \ldots, \mathbf{I}_{n \times n})_{(ns \times ns)}, \quad \mathbf{0}_k = (\underbrace{\mathbf{0}_{n \times n}, \mathbf{0}_{n \times n}, \ldots, \mathbf{0}_{n \times n}}_{s})^{\mathrm{T}}.$$

*The properties of discrete points* $\mathbf{x}_k$ $(k = 1, 2, \ldots, N)$ *can be estimated by the eigenvalues of* $\mathrm{DP}_{k(k-1)\ldots 1}$ *as*

$$|\mathrm{DP}_{k(k-1)\ldots 1} - \bar{\lambda} \mathbf{I}_{n(s+1) \times n(s+1)}| = 0 \quad (k = 1, 2, \ldots, N). \tag{5.170}$$

*The eigenvalues of* $\mathrm{DP}$ *for such periodic flow are determined by*

$$|\mathrm{DP} - \lambda \mathbf{I}_{n(s+1) \times n(s+1)}| = 0. \tag{5.171}$$

*Thus, the stability and bifurcation of the periodic flow can be classified by the eigenvalues of* $\mathrm{DP}(\mathbf{y}_0^*)$ *with*

$$([n_1^m, n_1^o] : [n_2^m, n_2^o] : [n_3, \kappa_3] : [n_4, \kappa_4] | n_5 : n_6 : [n_7, l, \kappa_7]). \tag{5.172}$$

(i)   *If the magnitudes of all eigenvalues of* $\mathrm{DP}$ *are less than one* $(|\lambda_i| < 1,$ $i = 1, 2, \ldots, n(s+1))$*, the approximate periodic solution is stable.*
(ii)  *If at least the magnitude of one eigenvalue of* $\mathrm{DP}$ *is greater than one* $(|\lambda_i| > 1, i \in \{1, 2, \ldots, n(s+1)\})$*, the approximate periodic solution is unstable.*
(iii) *The boundaries between stable and unstable periodic flow with higher-order singularity give bifurcation and stability conditions.*

*Proof* The proof is similar to Theorem 5.5.                                                   □

From the forgoing theorem, the stability and bifurcation analysis for the period-1 flow of the time-delay system can be completed from discrete mappings $P_k$ with $\mathbf{g}_k(\mathbf{x}_{k-1}, \mathbf{x}_k, \mathbf{p}) = \mathbf{0}$ and $\mathbf{x}_j^{\tau} = \mathbf{h}_k(\mathbf{x}_{r_j-1}, \mathbf{x}_{r_j}, \theta_j)$ $(j = k - 1, k; k = 1, 2, \ldots, N)$ under the period $T = 2\pi/\Omega$. If the period-doubling bifurcation occurs, the periodic flow will become a period-2 flow under the period $T' = 2T$. $2N$ regular nodes of the period-2 flow will be employed. $2N$ time-delay nodes of the period-2 flow will be converted into the regular nodes through interpolation. Thus, consider a mapping structure of the period-2 flow of the time-delay system with $2N$ mappings.

$$P^{(2)} = P \circ P = P_{2N} \circ P_{2N-1} \circ \cdots \circ P_2 \circ P_1 : (\mathbf{x}_0, \mathbf{x}_0^\tau) \rightarrow (\mathbf{x}_{2N}, \mathbf{x}_{2N}^\tau);$$
$$\text{with } P_k : (\mathbf{x}_{k-1}, \mathbf{x}_{k-1}^\tau) \rightarrow (\mathbf{x}_k, \mathbf{x}_k^\tau)(k = 1, 2, \ldots, 2N). \tag{5.173}$$

For $(\mathbf{x}_{2N}, \mathbf{x}_{2N}^\tau) = P^{(2)}(\mathbf{x}_0, \mathbf{x}_0^\tau)$, there is a set of points $(\mathbf{x}_k^*, \mathbf{x}_k^{*\tau})$ computed by

$$\left.\begin{array}{l} \mathbf{g}_k(\mathbf{x}_{k-1}^*, \mathbf{x}_k^*; \mathbf{x}_{k-1}^{\tau*}, \mathbf{x}_k^{\tau*}, \mathbf{p}) = \mathbf{0}, \\ \mathbf{x}_r^{\tau*} = \mathbf{h}_r(\mathbf{x}_{s_r-1}^*, \mathbf{x}_{s_r}^*, \theta_{s_r}), \\ s_r = r - l_r; r = k, k-1 \end{array}\right\} \quad (k = 1, 2, \ldots, 2N) \tag{5.174}$$
$$\mathbf{x}_{s_r-1}^* = \mathbf{x}_{\mathrm{mod}(s_r-1+2N,2N)}^*, \mathbf{x}_{s_r}^* = \mathbf{x}_{\mathrm{mod}(s_r+2N,2N)}^*$$
$$(\mathbf{x}_0^*, \mathbf{x}_0^{*\tau}) = (\mathbf{x}_{2N}^*, \mathbf{x}_{2N}^{\tau*}).$$

After period-doubling, the period-1 flow becomes period-2 flow. The node points increase to $2N$ points during two periods $(2T)$. The period-2 flow is sketched in Fig. 5.7. The node points are determined in Eq. (5.174). On the other hand,

$$T' = 2T = \frac{2(2\pi)}{\Omega} = \frac{2\pi}{\omega} \Rightarrow \omega = \frac{\Omega}{2}. \tag{5.175}$$

During a period $T'$, there is a periodic flow described by node points $\mathbf{x}_k$ $(k = 0, 1, \ldots, N')$. Due to $T' = 2T$, the period-2 flow can be described by $N' \geq 2N$

**Fig. 5.7** Period-2 flow with $2N$-nodes with short lines. *Solid curve* is for a numerical result. The *filled symbols* are for discrete node points on the periodic flow, and the *hollow symbols* are for time-delay nodes on the periodic flow

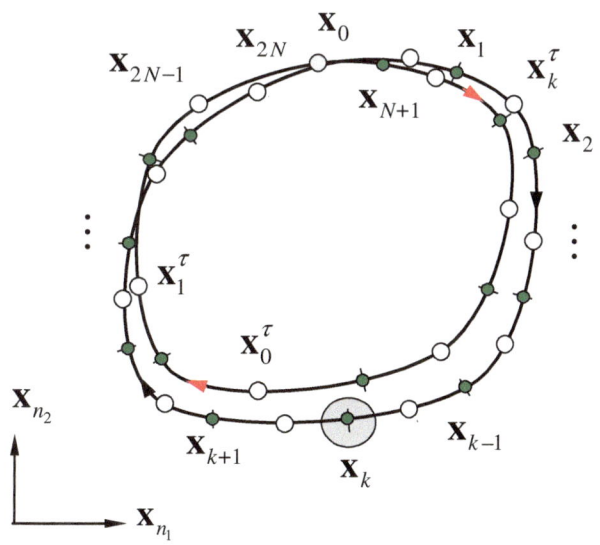

nodes. The time-delay nodes can be interpolated by the regular points. Thus, the corresponding mapping $P_k$ is defined as

$$P_k : (\mathbf{x}_{k-1}^{(2)}, \mathbf{x}_{k-1}^{\tau(2)}) \rightarrow (\mathbf{x}_k^{(2)}, \mathbf{x}_k^{\tau(2)})  \quad (k = 1, 2, \ldots, 2N) \qquad (5.176)$$

and

$$\left.\begin{aligned}
&\mathbf{g}_k(\mathbf{x}_{k-1}^{(2)*}, \mathbf{x}_k^{(2)*}; \mathbf{x}_{k-1}^{\tau(2)*}, \mathbf{x}_k^{\tau(2)*}, \mathbf{p}) = \mathbf{0}, \\
&\mathbf{x}_r^{\tau(2)*} = \mathbf{h}_r(\mathbf{x}_{s_r-1}^{(2)*}, \mathbf{x}_{s_r}^{(2)*}, \theta_{s_r}) \\
&s_r = r - l_r, r = k, k - 1
\end{aligned}\right\} \quad (k = 1, 2, \ldots, 2N) \qquad (5.177)$$

$$\mathbf{x}_{s_r-1}^{(2)*} = \mathbf{x}_{\mathrm{mod}(s_r-1+2N,2N)}^{(2)*}, \mathbf{x}_{s_r}^{(2)*} = \mathbf{x}_{\mathrm{mod}(s_r+2N,2N)}^{(2)*}, \mathbf{x}_0^{(2)*} = \mathbf{x}_{2N}^{(2)*}.$$

In general, for period $T' = mT$, there is a period-$m$ flow which can be described with $N' \geq mN$. The corresponding mapping $P_k$ is

$$P_k : (\mathbf{x}_{k-1}^{(m)}, \mathbf{x}_{k-1}^{\tau(m)}) \rightarrow (\mathbf{x}_k^{(m)}, \mathbf{x}_k^{\tau(m)})  \quad (k = 1, 2, \ldots, mN) \qquad (5.178)$$

and

$$\left.\begin{aligned}
&\mathbf{g}_k(\mathbf{x}_{k-1}^{(m)*}, \mathbf{x}_k^{(m)*}; \mathbf{x}_{k-1}^{\tau(m)*}, \mathbf{x}_k^{\tau(m)*}\mathbf{p}) = \mathbf{0}, \\
&\mathbf{x}_r^{\tau(m)*} = \mathbf{h}_r(\mathbf{x}_{s_r-1}^{(m)*}, \mathbf{x}_{s_r}^{(m)*}, \theta_{s_r}), \\
&r = k, k - 1; s_r = r - l_r
\end{aligned}\right\} \quad (k = 1, 2, \ldots, mN) \qquad (5.179)$$

$$\mathbf{x}_{s_r-1}^{(m)*} = \mathbf{x}_{\mathrm{mod}(s_r-1+mN,mN)}^{(m)*}, \quad \mathbf{x}_{s_r}^{(m)*} = \mathbf{x}_{\mathrm{mod}(s_r+mN,mN)}^{(m)*}; \mathbf{x}_0^{(m)*} = \mathbf{x}_{mN}^{(m)*}.$$

From the above discussion, the period-$m$ flow in a time-delay, nonlinear dynamical system can be described through $mN$ nodes for period $mT$.

**Theorem 5.7** *Consider a time-delay nonlinear dynamical system in* Eq. (5.98). *If such a time-delay dynamical system has a period-m flow* $\mathbf{x}^{(m)}(t)$ *with finite norm* $||\mathbf{x}^{(m)}||$ *and period* $mT$ *$(T = 2\pi/\Omega)$, there is a set of discrete time* $t_k$ *$(k = 0, 1, \ldots, mN)$ with $(N \rightarrow \infty)$ during m-periods $(mT)$, and the corresponding solution* $\mathbf{x}^{(m)}(t_k)$ *and vector field* $\mathbf{f}(\mathbf{x}^{(m)}(t_k), t_k, \mathbf{p})$ *are exact. Suppose discrete nodes* $\mathbf{x}_k^{(m)}$ *and* $\mathbf{x}_k^{\tau(m)}$ *are on the approximate solution of the period-m flow under* $||\mathbf{x}^{(m)}(t_k) - \mathbf{x}_k^{(m)}|| \leq \varepsilon_k$ *and* $||\mathbf{x}^{\tau(m)}(t_k) - \mathbf{x}_k^{\tau(m)}|| \leq \varepsilon_k^\tau$ *with small* $\varepsilon_k, \varepsilon_k^\tau \geq 0$ *and*

$$||\mathbf{f}(\mathbf{x}^{(m)}(t_k), \mathbf{x}^{\tau(m)}(t_k), t_k, \mathbf{p}) - \mathbf{f}(\mathbf{x}_k^{(m)}, \mathbf{x}_k^{\tau(m)}t_k, \mathbf{p})|| \leq \delta_k \qquad (5.180)$$

*with a small* $\delta_k \geq 0$. *During a time interval* $t \in [t_{k-1}, t_k]$, *there is a mapping* $P_k :$ $(\mathbf{x}_{k-1}^{(m)}, \mathbf{x}_{k-1}^{\tau(m)}) \rightarrow (\mathbf{x}_k^{(m)}, \mathbf{x}_k^{\tau(m)})$ $(k = 1, 2, \ldots, mN)$, *i.e.,*

$$(\mathbf{x}_k^{(m)}, \mathbf{x}_k^{\tau(m)}) = P_k(\mathbf{x}_{k-1}^{(m)}, \mathbf{x}_{k-1}^{\tau(m)}) \quad \text{with } \mathbf{g}_k(\mathbf{x}_{k-1}^{(m)}, \mathbf{x}_k^{(m)}; \mathbf{x}_{k-1}^{\tau(m)}, \mathbf{x}_k^{\tau(m)}, \mathbf{p}) = \mathbf{0},$$

$$\mathbf{x}_j^{\tau(m)} = \mathbf{h}_j(\mathbf{x}_{r_j-1}^{(m)}, \mathbf{x}_{r_j}^{(m)}, \theta_{r_j}), \quad j = k, k-1; r_j = j - l_j, k = 1, 2, \ldots, mN;$$

$$(\text{e.g.,} \quad \mathbf{x}_r^{\tau(m)} = \mathbf{x}_{s_r}^{(m)} + \theta_r(\mathbf{x}_{r_r-1}^{(m)} - \mathbf{x}_{r_r}^{(m)}), \quad \theta_r = \frac{1}{h_{r_j}}[\tau - \sum_{i=1}^{l_{r_j}} h_{r_j+i}]).$$

(5.181)

*where $\mathbf{g}_k$ is an implicit vector function and $\mathbf{h}_j$ is an interpolation vector function. Consider a mapping structure as*

$$P = P_{mN} \circ P_{mN-1} \circ \cdots \circ P_2 \circ P_1 : \mathbf{x}_0^{(m)} \to \mathbf{x}_{mN}^{(m)};$$

$$\text{with } P_k : (\mathbf{x}_{k-1}^{(m)}, \mathbf{x}_{k-1}^{\tau(m)}) \to (\mathbf{x}_k^{(m)}, \mathbf{x}_k^{\tau(m)}) \ (k = 1, 2, \ldots, mN).$$

(5.182)

*For $\mathbf{x}_{mN}^{(m)} = P(\mathbf{x}_0^{(m)}, \mathbf{x}_0^{\tau(m)})$, if there is a set of points $(\mathbf{x}_k^{(m)*}, \mathbf{x}_k^{\tau(m)*}) \ (k = 0, 1, \ldots, mN)$ computed by*

$$\left.\begin{array}{l} \mathbf{g}_k(\mathbf{x}_{k-1}^{(m)*}, \mathbf{x}_k^{(m)*}; \mathbf{x}_{k-1}^{\tau(m)*}, \mathbf{x}_k^{\tau(m)*}, \mathbf{p}) = \mathbf{0}, \\[2mm] \mathbf{x}_j^{\tau(m)*} = \mathbf{h}_j(\mathbf{x}_{r_j-1}^{(m)*}, \mathbf{x}_{r_j}^{(m)*}, \theta_{r_j}), j = k, k-1 \end{array}\right\} \quad (k = 1, 2, \ldots, mN)$$

$$\mathbf{x}_0^{(m)*} = \mathbf{x}_{mN}^{(m)*} \quad \text{and} \quad \mathbf{x}_0^{\tau(m)*} = \mathbf{x}_{mN}^{\tau(m)*},$$

(5.183)

*then the points $\mathbf{x}_k^{(m)*}$ and $\mathbf{x}_k^{\tau(m)*}$ $(k = 0, 1, \ldots, mN)$ are approximations of points $\mathbf{x}^{(m)}(t_k)$ and $\mathbf{x}^{\tau(m)}(t_k)$ of the periodic solution. In the neighborhoods of $\mathbf{x}_k^{(m)*}$ and $\mathbf{x}_k^{\tau(m)*}$, with $\mathbf{x}_k^{(m)} = \mathbf{x}_k^{(m)*} + \Delta\mathbf{x}_k^{(m)}$ and $\mathbf{x}_k^{\tau(m)} = \mathbf{x}_k^{\tau(m)*} + \Delta\mathbf{x}_k^{\tau(m)}$, the linearized equation is given by*

$$\sum_{j=k-1}^{k} \frac{\partial \mathbf{g}_k}{\partial \mathbf{x}_j^{(m)}} \Delta\mathbf{x}_j^{(m)} + \frac{\partial \mathbf{g}_k}{\partial \mathbf{x}_j^{\tau(m)}} \left(\frac{\partial \mathbf{x}_j^{\tau(m)}}{\partial \mathbf{x}_{r_j}^{\tau(m)}} \Delta\mathbf{x}_{r_j}^{\tau(m)} + \frac{\partial \mathbf{x}_j^{\tau(m)}}{\partial \mathbf{x}_{r_j-1}^{\tau(m)}} \Delta\mathbf{x}_{r_j-1}^{\tau(m)}\right) = \mathbf{0}_{1 \times n}$$

(5.184)

*with $r_j = j - l_j, j = k-1, k; (k = 1, 2, \ldots, mN)$.*

*The resultant Jacobian matrices of the period-m flow are*

$$DP_{k(k-1)\ldots 1} = \left[\frac{\partial \mathbf{y}_k^{(m)}}{\partial \mathbf{y}_0^{(m)}}\right]_{(\mathbf{x}_0^{(m)*}, \ldots, \mathbf{x}_k^{(m)*})} = \mathbf{A}_k \mathbf{A}_{k-1} \ldots \mathbf{A}_1 \quad (k = 1, 2, \ldots, mN),$$

$$\text{and } DP = DP_{mN(mN-1)\ldots 1} = \left[\frac{\partial \mathbf{y}_{mN}^{(m)}}{\partial \mathbf{y}_0^{(m)}}\right]_{(\mathbf{y}_0^{(m)*}, \ldots, \mathbf{y}_{1mN}^{(m)*})} = \mathbf{A}_{mN} \mathbf{A}_{mN-1} \ldots \mathbf{A}_1$$

(5.185)

*where*

$$\Delta \mathbf{y}_k^{(m)} = \mathbf{A}_k^{(m)} \Delta \mathbf{y}_{k-1}^{(m)}, \quad \mathbf{A}_k^{(m)} \equiv \left[ \frac{\partial \mathbf{y}_k^{(m)}}{\partial \mathbf{y}_{k-1}^{(m)}} \right]_{(\mathbf{y}_{k-1}^{(m)*}, \mathbf{y}_k^{(m)*})}, \tag{5.186}$$

*and*

$$\mathbf{a}_{kj}^{(m)} = - \left[ \frac{\partial \mathbf{g}_k}{\partial \mathbf{x}_k^{(m)}} \right]^{-1} \frac{\partial \mathbf{g}_k}{\partial \mathbf{x}_j^{(m)}}, \quad \mathbf{a}_{kr_j}^{(m)} = - \left[ \frac{\partial \mathbf{g}_k}{\partial \mathbf{x}_k^{(m)}} \right]^{-1} \frac{\partial \mathbf{g}_k}{\partial \mathbf{x}_j^{(m)\tau}} \frac{\partial \mathbf{x}_j^{(m)\tau}}{\partial \mathbf{x}_{r_j}^{(m)\tau}},$$

$$\mathbf{a}_{k(r_j-1)}^{(m)} = - \left[ \frac{\partial \mathbf{g}_k}{\partial \mathbf{x}_k^{(m)}} \right]^{-1} \frac{\partial \mathbf{g}_k}{\partial \mathbf{x}_j^{(m)\tau}} \frac{\partial \mathbf{x}_j^{(m)\tau}}{\partial \mathbf{x}_{r_j-1}^{(m)\tau}} \quad \text{with } r_j = j - l_j, \quad j = k - 1, k;$$

$$\mathbf{y}_k^{(m)} = (\mathbf{x}_k^{(m)}, \mathbf{x}_{k-1}^{(m)}, \ldots, \mathbf{x}_{r_{k-1}}^{(m)})^{\mathrm{T}}, \quad \mathbf{y}_{k-1}^{(m)} = (\mathbf{x}_{k-1}^{(m)}, \mathbf{x}_{k-2}^{(m)}, \ldots, \mathbf{x}_{r_{k-1}-1}^{(m)})^{\mathrm{T}},$$

$$\Delta \mathbf{y}_k^{(m)} = (\Delta \mathbf{x}_k^{(m)}, \Delta \mathbf{x}_{k-1}^{(m)}, \ldots, \Delta \mathbf{x}_{r_{k-1}}^{(m)})^{\mathrm{T}}, \quad \Delta \mathbf{y}_{k-1}^{(m)} = (\Delta \mathbf{x}_{k-1}^{(m)}, \Delta \mathbf{x}_{k-2}^{(m)}, \ldots, \Delta \mathbf{x}_{r_{k-1}-1}^{(m)})^{\mathrm{T}},$$

$$\mathbf{A}_k^{(m)} = \begin{bmatrix} \mathbf{B}_k^{(m)} & (\mathbf{a}_{k(r_{k-1}-1)}^{(m)})_{n \times n} \\ \mathbf{I}_k^{(m)} & \mathbf{0}_k^{(m)} \end{bmatrix}_{n(s+1) \times n(s+1)}, \quad s = 1 + l_{k-1};$$

$$\mathbf{B}_k^{(m)} = [(\mathbf{a}_{k(k-1)}^{(m)})_{n \times n}, \mathbf{0}_{n \times n}, \ldots, (\mathbf{a}_{k(r_k-1)}^{(m)})_{n \times n}],$$

$$\mathbf{I}_k^{(m)} = \mathrm{diag}(\mathbf{I}_{n \times n}, \mathbf{I}_{n \times n}, \ldots, \mathbf{I}_{n \times n})_{ns \times ns}, \quad \mathbf{0}_k^{(m)} = (\underbrace{\mathbf{0}_{n \times n}, \mathbf{0}_{n \times n} \ldots, \mathbf{0}_{n \times n}}_{s})^{\mathrm{T}}. \tag{5.187}$$

*The properties of discrete points* $\mathbf{x}_k^{(m)}$ *($k = 1, 2, \ldots, mN$) can be estimated by the eigenvalues of* $\mathrm{DP}_{k(k-1)\ldots 1}$ *as*

$$|\mathrm{DP}_{k(k-1)\ldots 1} - \bar{\lambda} \mathbf{I}_{n(s+1) \times n(s+1)}| = 0 \quad (k = 1, 2, \ldots, mN), \tag{5.188}$$

*The eigenvalues of* $\mathrm{DP}$ *for such a periodic flow are determined by*

$$|\mathrm{DP} - \lambda \mathbf{I}_{n(s+1) \times n(s+1)}| = 0, \tag{5.189}$$

*Thus, the stability and bifurcation of the period-m flow can be classified by the eigenvalues of* $\mathrm{DP}(\mathbf{y}_0^*)$ *with*

$$([n_1^m, n_1^o] : [n_2^m, n_2^o] : [n_3, \kappa_3] : [n_4, \kappa_4] | n_5 : n_6 : [n_7, l, \kappa_7]). \tag{5.190}$$

(i) *If the magnitudes of all eigenvalues of* $\mathrm{DP}$ *are less than one (i.e.,* $|\lambda_i| < 1, i = 1, 2, \ldots, n(s+1)$*), the approximate periodic solution is stable.*

(ii) *If at least the magnitude of one eigenvalue of* $\mathrm{DP}$ *is greater than one (i.e.,* $|\lambda_i| > 1, i \in \{1, 2, \ldots, n(s+1)\}$*), the approximate periodic solution is unstable.*

(iii)  *The boundaries between stable and unstable periodic flow with higher-order singularity give bifurcation and stability conditions.*

*Proof* The proof is similar to Theorem 5.5.                                        □

The period-$m$ flow in a time-delay dynamical system can be determined by the discrete mapping for a period-$m$ flow with multiple steps as follows.

**Theorem 5.8** *Consider a time-delay nonlinear dynamical system in Eq. (5.98). If such a system has a period-m flow $\mathbf{x}^{(m)}(t)$ with finite norm $||\mathbf{x}^{(m)}||$ and m-periods $mT(T = 2\pi/\Omega)$, there is a set of discrete time $t_k$ $(k = 0, 1, \ldots, mN)$ with $(N \to \infty)$ during m-periods $(mT)$, and the corresponding solutions $\mathbf{x}^{(m)}(t_k)$ and $\mathbf{x}^{\tau(m)}(t_k)$ with vector field $\mathbf{f}(\mathbf{x}^{(m)}(t_k), \mathbf{x}^{\tau(m)}(t_k), t_k, \mathbf{p})$ are exact. Suppose discrete nodes $\mathbf{x}_k^{(m)}$ and $\mathbf{x}_k^{\tau(m)}$ $(k = 0, 1, 2, \ldots, mN)$ are on the approximate solution of the periodic flow under $||\mathbf{x}^{(m)}(t_k) - \mathbf{x}_k^{(m)}|| \leq \varepsilon_k$ and $||\mathbf{x}^{\tau(m)}(t_k) - \mathbf{x}_k^{\tau(m)}|| \leq \varepsilon_k^\tau$ with small $\varepsilon_k, \varepsilon_k^\tau \geq 0$ and*

$$||\mathbf{f}(\mathbf{x}^{(m)}(t_k), \mathbf{x}^{\tau(m)}(t_k), t_k, \mathbf{p}) - \mathbf{f}(\mathbf{x}_k^{(m)}, \mathbf{x}_k^{\tau(m)} t_k, \mathbf{p})|| \leq \delta_k \qquad (5.191)$$

*with a small $\delta_k \geq 0$. During a time interval $t \in [t_{k-1}, t_k]$, there is a mapping $P_k : (\mathbf{x}_{k-1}^{(m)}, \mathbf{x}_{k-1}^{\tau(m)}) \to (\mathbf{x}_k^{(m)}, \mathbf{x}_k^{\tau(m)})$ $(k = 1, 2, \ldots, mN)$ as*

$$(\mathbf{x}_k^{(m)}, \mathbf{x}_k^{\tau(m)}) = P_k(\mathbf{x}_{k-1}^{(m)}, \mathbf{x}_{k-1}^{\tau(m)}) \ \ with$$

$$\mathbf{g}_k(\mathbf{x}_{s_{kr_1}}^{(m)}, \ldots, \mathbf{x}_{s_{k0}}^{(m)}, \ldots, \mathbf{x}_{s_{k(-r_2)}}^{(m)}, \mathbf{x}_{s_{kr_1}}^{\tau(m)}, \ldots, \mathbf{x}_{s_{k0}}^{\tau(m)}, \ldots, \mathbf{x}_{s_{k(-r_2)}}^{\tau(m)}, \mathbf{p}) = \mathbf{0},$$

$$\mathbf{x}_{s_{kj}}^{\tau(m)} = \mathbf{h}_{s_{kj}}(\mathbf{x}_{r_{kj}-1}^{(m)}, \mathbf{x}_{r_{kj}}^{(m)}, \theta_{r_{kj}}), \theta_{r_{kj}} = \frac{1}{h_{r_{kj}}}[\tau - \sum_{i=1}^{l_{s_{kj}}} h_{r_{kj}+i}],$$

$$r_{kj} = k - j - l_{s_{kj}}, s_{kj} = k - j; j = -r_2, -r_2 + 1, \ldots - 1, 0, 1, \ldots, r_1 - 1, r_1;$$

$$r_1, r_2 \in \{0, 1, 2, \ldots, mN\}, 1 \leq r_1 + r_2 \leq mN, r_1 \geq 1; (k = 1, 2, \ldots, mN).$$

$$(5.192)$$

*where $\mathbf{g}_k$ is an implicit vector function and $\mathbf{h}_j$ is an interpolation vector function. Consider a mapping structure as*

$$P = P_{mN} \circ P_{mN-1} \circ \cdots \circ P_2 \circ P_1 : (\mathbf{x}_0^{(m)}, \mathbf{x}_0^{\tau(m)}) \to (\mathbf{x}_{mN}^{(m)}, \mathbf{x}_{mN}^{\tau(m)});$$

$$with \ P_k : (\mathbf{x}_{k-1}^{(m)}, \mathbf{x}_{k-1}^{\tau(m)}) \to (\mathbf{x}_k^{(m)}, \mathbf{x}_k^{\tau(m)})(k = 1, 2, \ldots, mN). \qquad (5.193)$$

*For $(\mathbf{x}_{mN}^{(m)}, \mathbf{x}_{mN}^{\tau(m)}) = P(\mathbf{x}_0^{(m)}, \mathbf{x}_0^{\tau(m)})$, if there is a set of points $\mathbf{x}_k^*$ $(k = 0, 1, \ldots, N)$ computed by*

$$\mathbf{g}_k(\mathbf{x}_{s_{kr_1}}^{(m)*}, \ldots, \mathbf{x}_{s_{k0}}^{(m)*}, \ldots, \mathbf{x}_{s_{k(-r_2)2}}^{(m)*}, \mathbf{x}_{s_{kr_1}}^{\tau(m)*}, \ldots, \mathbf{x}_{s_{k0}}^{\tau(m)*}, \ldots, \mathbf{x}_{s_{k(-r_2)}}^{\tau(m)*}, \mathbf{p}) = \mathbf{0},$$

$$\mathbf{x}_{s_{kj}}^{\tau(m)*} = \mathbf{h}_{s_{kj}}(\mathbf{x}_{r_{kj}-1}^{(m)*}, \mathbf{x}_{r_{kj}}^{(m)*}, \theta_{r_{kj}}), \quad \theta_{r_{kj}} = \frac{1}{h_{r_{kj}}}[\tau - \sum_{i=1}^{l_{s_{kj}}} h_{r_{kj}+i}];$$

$$\mathbf{x}_{r_{kj}-1}^{(m)*} = \mathbf{x}_{\mathrm{mod}(r_{kj}-1+mN,mN)}^{(m)*}, \quad \mathbf{x}_{r_{kj}}^{(m)*} = \mathbf{x}_{\mathrm{mod}(r_{kj}+mN,mN)}^{(m)*}; \tag{5.194}$$

$$\mathbf{x}_{s_{kj}}^{(m)*} = \mathbf{x}_{\mathrm{mod}(s_{kj}+mN,mN)}^{(m)*}, \quad (\mathbf{x}_0^{*(m)}, \mathbf{x}_0^{\tau(m)*}) = (\mathbf{x}_{mN}^{(m)*}, \mathbf{x}_{mN}^{\tau(m)*})$$

*then the points $\mathbf{x}_k^{(m)*}$ and $\mathbf{x}_k^{\tau(m)*}$ $(k = 0, 1, \ldots, mN)$ are approximations of points $\mathbf{x}^{(m)}(t_k)$ and $\mathbf{x}^{\tau(m)}(t_k)$ of the periodic solution. In the neighborhood of $\mathbf{x}_k^{*(m)}$ and $\mathbf{x}_k^{\tau(m)*}$, with $\mathbf{x}_k^{(m)} = \mathbf{x}_k^{(m)*} + \Delta\mathbf{x}_k^{(m)}$ and $\mathbf{x}_k^{\tau(m)} = \mathbf{x}_k^{\tau(m)*} + \Delta\mathbf{x}_k^{\tau(m)}$, the linearized equation is given by*

$$\sum_{j=r_1}^{-r_2} \frac{\partial\mathbf{g}_k}{\partial\mathbf{x}_{s_{kj}}^{(m)}}\Delta\mathbf{x}_{s_{kj}}^{(m)} + \frac{\partial\mathbf{g}_k}{\partial\mathbf{x}_{s_{kj}}^{\tau(m)}}\frac{\partial\mathbf{x}_{s_{kj}}^{\tau(m)}}{\partial\mathbf{x}_{r_{kj}-1}^{(m)}}\mathbf{x}_{r_{kj}-1}^{(m)} + \frac{\partial\mathbf{g}_k}{\partial\mathbf{x}_{s_{kj}}^{\tau(m)}}\frac{\partial\mathbf{x}_{s_{kj}}^{\tau(m)}}{\partial\mathbf{x}_{r_{kj}}^{(m)}}\Delta\mathbf{x}_{r_{kj}}^{(m)} = \mathbf{0}$$

*with $\dfrac{\partial\mathbf{g}_k}{\partial\mathbf{x}_\alpha^{(m)}} = \mathbf{0}$ and $\dfrac{\partial\mathbf{g}_k}{\partial\mathbf{x}_\alpha^{\tau(m)}} = \mathbf{0}(\alpha \neq s_{kj}), \quad j = -r_2, -r_2 + 1, \ldots, r_1 - 1, r_1;$*

*$(k = 1, 2, \ldots, mN)$.*

$$(5.195)$$

*The resultant Jacobian matrices of the periodic flow are*

$$\mathrm{DP}_{k(k-1)\ldots1} = \left[\frac{\partial\mathbf{y}_k^{(m)}}{\partial\mathbf{y}_0^{(m)}}\right]_{(\mathbf{y}_0^{(m)*}, \mathbf{y}_1^{(m)*}, \ldots, \mathbf{y}_k^{(m)*})} = \mathbf{A}_k^{(m)}\mathbf{A}_{k-1}^{(m)}\ldots\mathbf{A}_1^{(m)}$$

*$(k = 1, 2, \ldots, mN)$,* $$(5.196)$$

*and* $\quad \mathrm{DP} = \mathrm{DP}_{mN(mN-1)\ldots1} = \left[\dfrac{\partial\mathbf{y}_{mN}^{(m)}}{\partial\mathbf{y}_0^{(m)}}\right]_{(\mathbf{y}_0^{(m)*}, \mathbf{y}_1^{(m)*}, \ldots, \mathbf{y}_{mN}^{(m)*})} = \mathbf{A}_N^{(m)}\mathbf{A}_{N-1}^{(m)}\ldots\mathbf{A}_1^{(m)}$

*where*

$$\Delta\mathbf{y}_k^{(m)} = \mathbf{A}_k^{(m)}\Delta\mathbf{y}_{k-1}^{(m)}, \quad \mathbf{A}_k^{(m)} = \left[\frac{\partial\mathbf{y}_k^{(m)}}{\partial\mathbf{y}_{k-1}^{(m)}}\right]_{(\mathbf{y}_{k-1}^{(m)*}, \mathbf{y}_k^{(m)*})} \tag{5.197}$$

*and*

$$\mathbf{a}_{ks_{kj}}^{(m)} = -\left[\frac{\partial\mathbf{g}_k}{\partial\mathbf{x}_{k+r_2}^{(m)}}\right]^{-1}\frac{\partial\mathbf{g}_k}{\partial\mathbf{x}_{ks_{kj}}^{(m)}}, \quad \mathbf{a}_{kr_{kj}}^{(m)} = -\left[\frac{\partial\mathbf{g}_k}{\partial\mathbf{x}_{k+r_2}^{(m)}}\right]^{-1}\frac{\partial\mathbf{g}_k}{\partial\mathbf{x}_{ks_{kj}}^{(m)\tau}}\frac{\partial\mathbf{x}_{ks_{kj}}^{(m)\tau}}{\partial\mathbf{x}_{r_{kj}}^{(m)}}, \tag{5.198}$$

$$\mathbf{a}_{k(r_{kj}-1)}^{(m)} = -\left[\frac{\partial \mathbf{g}_k}{\partial \mathbf{x}_{k+r_2}^{(m)}}\right]^{-1} \frac{\partial \mathbf{g}_k}{\partial \mathbf{x}_{ks_{kj}}^{(m)\tau}} \frac{\partial \mathbf{x}_{ks_{kj}}^{(m)\tau}}{\partial \mathbf{x}_{r_{kj}-1}^{(m)}} \quad \text{with } r_{kj} = k - j - l_{s_{kj}}, s_{kj} = k - j;$$

$$j = -r_2, -r_2 + 1, \ldots -1, 0, 1, \ldots, r_1 - 1, r_1; r_1, r_2 \in \{0, 1, 2, \ldots, N\};$$

$$1 \le r_1 + r_2 \le N, r_1 \ge 1; (k = 1, 2, \ldots, N);$$

$$\mathbf{y}_k^{(m)} = (\mathbf{x}_{k+r_2}^{(m)}, \mathbf{x}_{k+r_2-1}^{(m)}, \ldots, \mathbf{x}_{r_{kr_1}}^{(m)})^{\mathrm{T}}, \quad \mathbf{y}_{k-1}^{(m)} = (\mathbf{x}_{k+r_2-1}^{(m)}, \mathbf{x}_{k+r_2-2}^{(m)}, \ldots, \mathbf{x}_{r_{kr_1}-1}^{(m)})^{\mathrm{T}},$$

$$\Delta\mathbf{y}_k^{(m)} = (\Delta\mathbf{x}_{k+r_2}^{(m)}, \Delta\mathbf{x}_{k+r_2-1}^{(m)}, \ldots, \Delta\mathbf{x}_{r_{kr_1}}^{(m)})^{\mathrm{T}},$$

$$\Delta\mathbf{y}_{k-1}^{(m)} = (\Delta\mathbf{x}_{k+r_2-1}^{(m)}, \Delta\mathbf{x}_{k+r_2-2}^{(m)}, \ldots, \Delta\mathbf{x}_{r_{kr_1}-1}^{(m)})^{\mathrm{T}},$$

$$\text{(5.198)}$$

*and*

$$\mathbf{A}_k^{(m)} = \begin{bmatrix} \mathbf{B}_k^{(m)} & (\mathbf{a}_{k(r_{kr_1}-1)}^{(m)})_{n \times n} \\ \mathbf{I}_k^{(m)} & \mathbf{0}_k^{(m)} \end{bmatrix}_{n(s+1) \times n(s+1)}, \quad s = r_1 + r_2 + l_{s_{kr_1}};$$

$$\mathbf{B}_k^{(m)} = [(\mathbf{a}_{k(k+r_2-1)}^{(m)})_{n \times n}, \mathbf{0}_{n \times n}, \ldots, (\mathbf{a}_{kr_{kr_1}}^{(m)})_{n \times n}],$$

$$\mathbf{I}_k^{(m)} = \mathrm{diag}(\mathbf{I}_{n \times n}, \mathbf{I}_{n \times n}, \ldots, \mathbf{I}_{n \times n})_{ns \times ns}, \quad \mathbf{0}_k^{(m)} = (\underbrace{\mathbf{0}_{n \times n}, \mathbf{0}_{n \times n} \ldots, \mathbf{0}_{n \times n}}_{s})^{\mathrm{T}}. \quad \text{(5.199)}$$

*The properties of discrete points $\mathbf{x}_k$ $(k = 1, 2, \ldots, mN)$ can be estimated by the eigenvalues of $DP_{k(k-1)\ldots1}$ as*

$$|DP_{k(k-1)\ldots1} - \bar{\lambda}\mathbf{I}_{n(s+1) \times n(s+1)}| = 0 \quad (k = 1, 2, \ldots, mN). \quad \text{(5.200)}$$

*The eigenvalues of $DP$ for such periodic flow are determined by*

$$|DP - \lambda\mathbf{I}_{n(s+1) \times n(s+1)}| = 0. \quad \text{(5.201)}$$

*Thus, the stability and bifurcation of the periodic flow can be classified by the eigenvalues of $DP(\mathbf{y}_0^*)$ with*

$$([n_1^m, n_1^o] : [n_2^m, n_2^o] : [n_3, \kappa_3] : [n_4, \kappa_4]|n_5 : n_6 : [n_7, l, \kappa_7]). \quad \text{(5.202)}$$

(i) *If the magnitudes of all eigenvalues of $DP$ are less than one $(|\lambda_i| < 1, i = 1, 2, \ldots, n(s+1))$, the approximate periodic solution is stable.*

(ii) *If at least the magnitude of one eigenvalue of $DP$ is greater than one $(|\lambda_i| > 1, i \in \{1, 2, \ldots, n(s+1)\})$, the approximate periodic solution is unstable.*

(iii) *The boundaries between stable and unstable periodic flow with higher-order singularity give bifurcation and stability conditions.*

*Proof* The proof is similar to Theorem 5.5. $\qquad\square$

## 5.2.2  Integrated Time-Delay Nodes

If a time-delay nonlinear system has approximate solution points $\mathbf{x}_k \approx \mathbf{x}(t_k)$ and $\mathbf{x}_k^\tau \approx \mathbf{x}(t_k - \tau)$ for $k = 0, 1, 2, \ldots$, as shown in Fig. 5.8, the small circular symbols are the regular solution points, and the large circular symbols are time-delay points. Between $\mathbf{x}_k$ and $\mathbf{x}_{k+1}$, there is a time-delay point $\mathbf{x}_{k+s_k}^\tau \approx \mathbf{x}(t_{k+s_k} - \tau)$ where $(t_{k+s_k} - \tau) \in [t_k, t_{k+1}]$ with an integer $s_k$. From Eq. (5.95), we have

$$\mathbf{x}(t_k) = \mathbf{x}(t_{k-1}) + \int_{t_{k-1}}^{t_k} \mathbf{f}(\mathbf{x}, \mathbf{x}^\tau, t, \mathbf{p}) dt,$$

$$\mathbf{x}(t_{k-1+s_{k-1}} - \tau) = \mathbf{x}(t_{k-1}) + \int_{t_{k-1}}^{t_{k-1+s_{k-1}} - \tau} \mathbf{f}(\mathbf{x}, \mathbf{x}^\tau, t, \mathbf{p}) dt. \tag{5.203}$$

Consider an interpolation function between $\mathbf{f}(\mathbf{x}_k, \mathbf{x}_k^\tau, t_k, \mathbf{p})$ and $\mathbf{f}(\mathbf{x}_{k+1}, \mathbf{x}_{k+1}^\tau, t_{k+1}, \mathbf{p})$ to approximate $\mathbf{f}(\mathbf{x}, \mathbf{x}^\tau, t, \mathbf{p})$. Equation (5.203) becomes

$$\begin{aligned}
\mathbf{x}_k &\approx \mathbf{x}_{k-1} + \bar{\mathbf{g}}_k(\mathbf{x}_{k-1}, \mathbf{x}_k; \mathbf{x}_{k-1}^\tau, \mathbf{x}_k^\tau, \mathbf{p}), \\
\mathbf{x}_{k-1+s_{k-1}}^\tau &\approx \mathbf{x}_{k-1} + \bar{\mathbf{h}}_k(\mathbf{x}_{k-1}, \mathbf{x}_k; \mathbf{x}_{k-1}^\tau, \mathbf{x}_k^\tau, \mathbf{p}).
\end{aligned} \tag{5.204}$$

From the above discrete scheme for non-delay nodes and delay nodes, periodic flows in time-delay dynamical systems can be discussed. If a time-delay system has a periodic flow with a period of $T = 2\pi/\Omega$, then such a periodic flow can be described by discrete points. The method is stated as follows.

**Theorem 5.9** *Consider a time-delay nonlinear dynamical system as*

$$\dot{\mathbf{x}} = \mathbf{f}(\mathbf{x}, \mathbf{x}^\tau, t, \mathbf{p}) \in \mathscr{R}^n \tag{5.205}$$

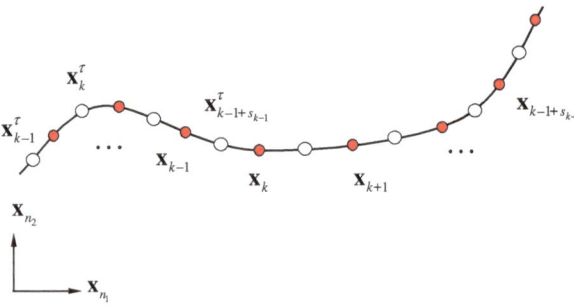

**Fig. 5.8** The discrete points on the solutions of a time-delay dynamical system. The *small circular symbols* are the regular solution points, and the *large circular symbols* are the time-delayed points

*where* $\mathbf{f}(\mathbf{x}, \mathbf{x}^{\tau}, t, \mathbf{p})$ *is a $C^r$-continuous nonlinear vector function $(r \geq 1)$ and* $\mathbf{x}^{\tau} = \mathbf{x}(t - \tau)$. *If such a time-delay dynamical system has a periodic flow $\mathbf{x}(t)$ with finite norm $||\mathbf{x}||$ and period $T = 2\pi/\Omega$, there is a set of discrete time $t_k(k = (0, 1, \ldots, N)$ with $(N \rightarrow \infty)$ during one period $T$, and the corresponding solutions $\mathbf{x}(t_k)$ and $\mathbf{x}^{\tau}(t_k) = \mathbf{x}(t_k - \tau)$ with vector field $\mathbf{f}(\mathbf{x}(t_k), \mathbf{x}^{\tau}(t_k), t_k, \mathbf{p})$ are exact. Suppose discrete nodes $\mathbf{x}_k$ and $\mathbf{x}_k^{\tau}$ are on the approximate solutions of the periodic flow under $||\mathbf{x}(t_k) - \mathbf{x}_k|| \leq \varepsilon_k$ and $||\mathbf{x}^{\tau}(t_k) - \mathbf{x}_k^{\tau}|| \leq \varepsilon_k^{\tau}$ with small $\varepsilon_k, \varepsilon_k^{\tau} \geq 0$ and*

$$||\mathbf{f}(\mathbf{x}(t_k), \mathbf{x}^{\tau}(t_k), t_k, \mathbf{p}) - \mathbf{f}(\mathbf{x}_k, \mathbf{x}_k^{\tau}, t_k, \mathbf{p})|| \leq \delta_k \qquad (5.206)$$

*with a small $\delta_k \geq 0$. During a time interval $t \in [t_{k-1}, t_k]$, there is a mapping $P_k : (\mathbf{x}_{k-1}, \mathbf{x}_{k-1}^{\tau}) \rightarrow (\mathbf{x}_k, \mathbf{x}_k^{\tau})$ $(k = 1, 2, \ldots, N)$ as*

$$(\mathbf{x}_k, \mathbf{x}_k^{\tau}) = P_k(\mathbf{x}_{k-1}, \mathbf{x}_{k-1}^{\tau}) \quad \text{with}$$

$$\left. \begin{array}{l} \mathbf{g}_k(\mathbf{x}_{k-1}, \mathbf{x}_k; \mathbf{x}_{k-1}^{\tau}, \mathbf{x}_k^{\tau}, \mathbf{p}) = \mathbf{0}, \\ \mathbf{h}_k(\mathbf{x}_{k-1}, \mathbf{x}_k; \mathbf{x}_{k-1}^{\tau}, \mathbf{x}_k^{\tau}, \mathbf{x}_{r_k}^{\tau}, \mathbf{p}) = \mathbf{0} \end{array} \right\} \quad k = 1, 2, \ldots, N$$

$$r_k = \text{mod}(k - 1 + s_{k-1}, N), \quad \text{and} \quad \mathbf{x}_{r_k}^{\tau} \approx \mathbf{x}(t_{k-1+s_{k-1}} - \tau), t_{k+s_{k-1}-1} \in [t_{k-1}, t_k] \tag{5.207}$$

*where $\mathbf{g}_k$ and $\mathbf{h}_k$ are implicit vector functions for regular and time-delay nodes, respectively. Consider a mapping structure as*

$$\begin{aligned} &P = P_N \circ P_{N-1} \circ \cdots \circ P_2 \circ P_1 : (\mathbf{x}_0, \mathbf{x}_0^{\tau}) \rightarrow (\mathbf{x}_N, \mathbf{x}_N^{\tau}); \\ &\text{with } P_k : (\mathbf{x}_{k-1}, \mathbf{x}_{k-1}^{\tau}) \rightarrow (\mathbf{x}_k, \mathbf{x}_k^{\tau}) \quad (k = 1, 2, \ldots, N). \end{aligned} \tag{5.208}$$

*For $(\mathbf{x}_N, \mathbf{x}_N^{\tau})^T = P(\mathbf{x}_0, \mathbf{x}_0^{\tau})$, if there is a set of points $(\mathbf{x}_k^*, \mathbf{x}_k^{\tau*})$ $(k = 0, 1, \ldots, N)$ computed by*

$$\left. \begin{array}{l} \mathbf{g}_k(\mathbf{x}_{k-1}, \mathbf{x}_k; \mathbf{x}_{k-1}^{\tau}, \mathbf{x}_k^{\tau}, \mathbf{p}) = \mathbf{0}, \\ \mathbf{h}_k(\mathbf{x}_{k-1}, \mathbf{x}_k; \mathbf{x}_{k-1}^{\tau}, \mathbf{x}_k^{\tau}, \mathbf{x}_{r_k}^{\tau}, \mathbf{p}) = \mathbf{0} \end{array} \right\} \quad (k = 1, 2, \ldots, N)$$

$$\text{and} \quad r_k = \text{mod}(k - 1 + s_{k-1}, N), \mathbf{x}_0^* = \mathbf{x}_N^* \quad \text{and} \quad \mathbf{x}_0^{\tau*} = \mathbf{x}_N^{\tau*}, \tag{5.209}$$

*then the points $\mathbf{x}_k^*$ and $\mathbf{x}_k^{\tau*}$ $(k = 0, 1, \ldots, N)$ are approximations of points $\mathbf{x}(t_k)$ and $\mathbf{x}^{\tau}(t_k)$ of the periodic solution. In the neighborhoods of $\mathbf{x}_k^*$ and $\mathbf{x}_k^{\tau*}$, with $\mathbf{x}_k = \mathbf{x}_k^* + \Delta\mathbf{x}_k$ and $\mathbf{x}_k^{\tau} = \mathbf{x}_k^{\tau*} + \Delta\mathbf{x}_k^{\tau}$, the linearized equation is given by*

$$\frac{\partial\mathbf{g}_k}{\partial\mathbf{x}_{k-1}}\frac{\partial\mathbf{x}_{k-1}}{\partial\mathbf{x}_0} + \frac{\partial\mathbf{g}_k}{\partial\mathbf{x}_k}\frac{\partial\mathbf{x}_k}{\partial\mathbf{x}_0} + \frac{\partial\mathbf{g}_k}{\partial\mathbf{x}_{k-1}^{\tau}}\frac{\partial\mathbf{x}_{k-1}^{\tau}}{\partial\mathbf{x}_0} + \frac{\partial\mathbf{g}_k}{\partial\mathbf{x}_k^{\tau}}\frac{\partial\mathbf{x}_k^{\tau}}{\partial\mathbf{x}_0} = \mathbf{0},$$

$$\frac{\partial\mathbf{g}_k}{\partial\mathbf{x}_{k-1}}\frac{\partial\mathbf{x}_{k-1}}{\partial\mathbf{x}_0^{\tau}} + \frac{\partial\mathbf{g}_k}{\partial\mathbf{x}_k}\frac{\partial\mathbf{x}_k}{\partial\mathbf{x}_0^{\tau}} + \frac{\partial\mathbf{g}_k}{\partial\mathbf{x}_{k-1}^{\tau}}\frac{\partial\mathbf{x}_{k-1}^{\tau}}{\partial\mathbf{x}_0^{\tau}} + \frac{\partial\mathbf{g}_k}{\partial\mathbf{x}_k^{\tau}}\frac{\partial\mathbf{x}_k^{\tau}}{\partial\mathbf{x}_0^{\tau}} = \mathbf{0};$$

$$\frac{\partial \mathbf{h}_k}{\partial \mathbf{x}_{k-1}} \frac{\partial \mathbf{x}_{k-1}}{\partial \mathbf{x}_0} + \frac{\partial \mathbf{h}_k}{\partial \mathbf{x}_k} \frac{\partial \mathbf{x}_k}{\partial \mathbf{x}_0} + \frac{\partial \mathbf{h}_k}{\partial \mathbf{x}_{k-1}^\tau} \frac{\partial \mathbf{x}_{k-1}^\tau}{\partial \mathbf{x}_0} + \frac{\partial \mathbf{h}_k}{\partial \mathbf{x}_k^\tau} \frac{\partial \mathbf{x}_k^\tau}{\partial \mathbf{x}_0} + \frac{\partial \mathbf{h}_k}{\partial \mathbf{x}_{r_k}^\tau} \frac{\partial \mathbf{x}_{r_k}^\tau}{\partial \mathbf{x}_0} = \mathbf{0},$$

$$\frac{\partial \mathbf{h}_k}{\partial \mathbf{x}_{k-1}} \frac{\partial \mathbf{x}_{k-1}}{\partial \mathbf{x}_0^\tau} + \frac{\partial \mathbf{h}_k}{\partial \mathbf{x}_k} \frac{\partial \mathbf{x}_k}{\partial \mathbf{x}_0^\tau} + \frac{\partial \mathbf{h}_k}{\partial \mathbf{x}_{k-1}^\tau} \frac{\partial \mathbf{x}_{k-1}^\tau}{\partial \mathbf{x}_0^\tau} + \frac{\partial \mathbf{h}_k}{\partial \mathbf{x}_k^\tau} \frac{\partial \mathbf{x}_k^\tau}{\partial \mathbf{x}_0^\tau} + \frac{\partial \mathbf{h}_k}{\partial \mathbf{x}_{r_k}^\tau} \frac{\partial \mathbf{x}_{r_k}^\tau}{\partial \mathbf{x}_0^\tau} = \mathbf{0} \qquad (5.210)$$

with $r_k = \mathrm{mod}(k - 1 + s_{k-1}, N)$   and   $(k = 1, 2, \ldots, N)$.

*The resultant Jacobian matrices of the periodic flow are*

$$\mathrm{DP}_{k(k-1)\ldots 1} = \begin{bmatrix} \frac{\partial \mathbf{x}_k}{\partial \mathbf{x}_0} & \frac{\partial \mathbf{x}_k}{\partial \mathbf{x}_0^\tau} \\ \frac{\partial \mathbf{x}_k}{\partial \mathbf{x}_0} & \frac{\partial \mathbf{x}_k}{\partial \mathbf{x}_0^\tau} \end{bmatrix}_{(\mathbf{x}_0^*, \mathbf{x}_0^{\tau*}, \ldots, \mathbf{x}_N^*, \mathbf{x}_N^{\tau*})}, \qquad \mathrm{DP} = \begin{bmatrix} \frac{\partial \mathbf{x}_N}{\partial \mathbf{x}_0} & \frac{\partial \mathbf{x}_N}{\partial \mathbf{x}_0^\tau} \\ \frac{\partial \mathbf{x}_N}{\partial \mathbf{x}_0} & \frac{\partial \mathbf{x}_N}{\partial \mathbf{x}_0^\tau} \end{bmatrix}_{(\mathbf{x}_0^*, \mathbf{x}_0^{\tau*}, \ldots, \mathbf{x}_N^*, \mathbf{x}_N^{\tau*})}$$

$$(k = 1, 2, \ldots, N)$$

$$(5.211)$$

*where*

$$\mathbf{y} = \mathbf{A}^{-1} \mathbf{b} \ and \ \mathbf{y}^\tau = \mathbf{A}^{-1} \mathbf{b}^\tau \qquad (5.212)$$

*and*

$$\mathbf{A} = (\mathbf{A}_{kl})_{2nN \times 2nN},$$

$$\mathbf{y} = (\mathbf{y}_1, \mathbf{y}_2, \ldots, \mathbf{y}_N)^\mathrm{T}, \quad \mathbf{y}^\tau = (\mathbf{y}_1^\tau, \mathbf{y}_2^\tau, \ldots, \mathbf{y}_N^\tau)^\mathrm{T},$$

$$\mathbf{b} = (\mathbf{b}_1, \mathbf{b}_2, \ldots, \mathbf{b}_N)^\mathrm{T}, \quad \mathbf{b}^\tau = (\mathbf{b}_1^\tau, \mathbf{b}_2^\tau, \ldots, \mathbf{b}_N^\tau)^\mathrm{T};$$

$$\mathbf{A}_{kl} = \sum_{l_k} \mathbf{A}_{kl_k} \delta_l^{l_k} \quad for \ l_k = k - 1, k, r_k; l_k > 0;$$

$$\mathbf{b}_k = -\sum_{l_k} \left[ \frac{\partial \mathbf{g}_k}{\partial \mathbf{x}_0}, \frac{\partial \mathbf{h}_k}{\partial \mathbf{x}_0} \right]^\mathrm{T} \delta_0^{l_k}, \quad \mathbf{b}_k^\tau = -\sum_{l_k} \left[ \frac{\partial \mathbf{g}_k}{\partial \mathbf{x}_0^\tau}, \frac{\partial \mathbf{h}_k}{\partial \mathbf{x}_0^\tau} \right]^\mathrm{T} \delta_0^{l_k};$$

$$\mathbf{A}_{kj} = \begin{bmatrix} \mathbf{a}_{kj} & \mathbf{a}_{kj}^\tau \\ \mathbf{b}_{kj} & \mathbf{b}_{kj}^\tau \end{bmatrix}, \quad \mathbf{A}_{kr_k} = \begin{bmatrix} \mathbf{0}_{n \times n} & \mathbf{0}_{n \times n} \\ \mathbf{0}_{n \times n} & \mathbf{b}_{kr_k}^\tau \end{bmatrix}, \quad (j = k - 1, k);$$

$$\mathbf{a}_{kj} = \left[ \frac{\partial \mathbf{g}_k}{\partial \mathbf{x}_j} \right], \quad \mathbf{a}_{kj}^\tau = \left[ \frac{\partial \mathbf{g}_k}{\partial \mathbf{x}_j^\tau} \right], \quad \mathbf{b}_{kj} = \left[ \frac{\partial \mathbf{h}_k}{\partial \mathbf{x}_j} \right], \quad \mathbf{b}_{kj}^\tau = \left[ \frac{\partial \mathbf{h}_k}{\partial \mathbf{x}_j^\tau} \right], \quad \mathbf{b}_{kr_k}^\tau = \left[ \frac{\partial \mathbf{g}_k}{\partial \mathbf{x}_{r_k}^\tau} \right];$$

$$\mathbf{y}_k = \left[ \frac{\partial \mathbf{x}_k}{\partial \mathbf{x}_0}, \frac{\partial \mathbf{x}_k^\tau}{\partial \mathbf{x}_0} \right]^\mathrm{T}, \quad \mathbf{y}_k^\tau = \left[ \frac{\partial \mathbf{x}_k}{\partial \mathbf{x}_0^\tau}, \frac{\partial \mathbf{x}_k^\tau}{\partial \mathbf{x}_0^\tau} \right]^\mathrm{T},$$

$$(k = 1, 2, \ldots, N).$$

$$(5.213)$$

*The properties of discrete points* $\mathbf{x}_k$ $(k = 1, 2, \ldots, N)$ *can be estimated by the eigenvalues of* $DP_{k(k-1)\ldots1}$ *as*

$$|DP_{k(k-1)\ldots1} - \lambda \mathbf{I}_{2n \times 2n}| = 0 \quad (k = 1, 2, \ldots, N). \tag{5.214}$$

*The eigenvalues of* $DP$ *for such periodic flow are determined by*

$$|DP - \lambda \mathbf{I}_{2n \times 2n}| = 0. \tag{5.215}$$

*Thus, the stability and bifurcation of the periodic flow can be classified by the eigenvalues of* $DP(\mathbf{x}_0^*)$ *with*

$$([n_1^m, n_1^o] : [n_2^m, n_2^o] : [n_3, \kappa_3] : [n_4, \kappa_4] | n_5 : n_6 : [n_7, l, \kappa_7]). \tag{5.216}$$

  (i) *If the magnitudes of all eigenvalues of* $DP$ *are less than one (i.e.,* $|\lambda_i| < 1$, $i = 1, 2, \ldots, 2n$), *the approximate periodic solution is stable.*
 (ii) *If at least the magnitude of one eigenvalue of* $DP$ *is greater than one (i.e.,* $|\lambda_i| > 1$, $i \in \{1, 2, \ldots, 2n\}$), *the approximate periodic solution is unstable.*
(iii) *The boundaries between stable and unstable periodic flow with higher-order singularity give bifurcation and stability conditions.*

*Proof* If $\mathbf{f}(\mathbf{x}, \mathbf{x}^\tau, t, \mathbf{p})$ is a $C^r$-continuous nonlinear function vector $(r \geq 1)$, then the velocity $\dot{\mathbf{x}}$ should be $C^r$-continuous $(r \geq 1)$. If a time-delay system has a periodic flow $\mathbf{x}(t)$ with finite norms $||\mathbf{x}||$ with period $T = 2\pi/\Omega$, there is a set of discrete time $t_k$ $(k = 0, 1, \ldots, N)$ with $(N \to \infty)$ during one period $T$. The solutions $\mathbf{x}(t_k)$ and $\mathbf{x}^\tau(t_k) = \mathbf{x}(t_k - \tau)$ with vector fields $\mathbf{f}(\mathbf{x}(t_k), \mathbf{x}^\tau(t_k), t_k, \mathbf{p})$ are exact. Consider a time interval $t \in [t_{k-1}, t_k]$,

$$\mathbf{x}(t) = \mathbf{x}(t_{k-1}) + \int_{t_{k-1}}^{t} \mathbf{f}(\mathbf{x}, \mathbf{x}^\tau, t, \mathbf{p}) dt. \tag{5.217}$$

For the time interval divided into $s$-nodes $t_{k(i)} = t_{k-1} + c_i h_k$ with $c_i \in [0, 1]$ and $\mathbf{f}(\mathbf{x}(t_{k(i)}), \mathbf{x}^\tau(t_{k(i)}), t_{k(i)}, \mathbf{p})$ $(i = 1, \ldots, s)$ with $\mathbf{x}^\tau(t_{k(i)}) = \mathbf{x}(t_{k(i)} - \tau)$, there is an approximate function $\mathbf{P}(t, \mathbf{C})$ with unknown $\mathbf{C} = (C_1, \ldots, C_s)^T$ and $C_i$ $(i = 1, \ldots, s)$, and the following condition is satisfied, i.e.,

$$\mathbf{f}(\mathbf{x}(t_{k(i)}), \mathbf{x}^\tau(t_{k(i)}), t_{k(i)}, \mathbf{p}) = \mathbf{P}(t_{k(i)}, t_{k(i)} - \tau, \mathbf{C}),$$
$$i = 1, 2, \ldots, s; |\frac{\partial \mathbf{P}}{\partial \mathbf{C}}| \neq 0. \tag{5.218}$$

The unknowns $\mathbf{C}(t_k) = (C_1, \ldots, C_s)^T$ with $\mathbf{t}_k = (t_{k(1)}, \ldots, t_{k(s)})^T = t_k(1, 1, \ldots, 1)^T + h_k(c_1, \ldots, c_s)^T$ are determined. For a small $\delta > 0$, if there is a relation

$$|\mathbf{P}(t, t - \tau, \mathbf{C}(\mathbf{t}_k)) - \mathbf{f}(\mathbf{x}, \mathbf{x}^\tau, t, \mathbf{p})| \leq \delta \tag{5.219}$$

for $t \in [t_{k-1}, t_k]$, Eq. (5.217) can be approximated as

$$\mathbf{x}(t) = \mathbf{x}(t_{k-1}) + \int_{t_{k-1}}^{t} [\mathbf{P}(t, t - \tau, \mathbf{C}(\mathbf{t}_k)) + O(\delta)] dt;$$

$$\bar{\mathbf{x}}(t) = \bar{\mathbf{x}}(t_{k-1}) + \int_{t_{k-1}}^{t} \mathbf{P}(t, t - \tau, \mathbf{C}(\mathbf{t}_k)) dt \tag{5.220}$$

and

$$\bar{\mathbf{x}}(t_k) = \bar{\mathbf{x}}(t_{k-1}) + \int_{t_{k-1}}^{t_k} \mathbf{P}(t, t - \tau, \mathbf{C}(\mathbf{t}_k)) dt,$$

$$\bar{\mathbf{x}}(t_{k-1+s_{k-1}} - \tau) = \bar{\mathbf{x}}(t_{k-1}) + \int_{t_{k-1}}^{t_{k-1+s_{k-1}} - \tau} \mathbf{P}(t, t - \tau, \mathbf{C}(\mathbf{t}_k)) dt. \tag{5.221}$$

Let $\bar{\mathbf{x}}(t_k) = \mathbf{x}_k$, $\bar{\mathbf{x}}(t_{k-1}) = \mathbf{x}_{k-1}, \bar{\mathbf{x}}^\tau(t_{k-1}) = \mathbf{x}_{k-1}^\tau$ and $\bar{\mathbf{x}}^\tau(t_k) = \mathbf{x}_k^\tau$. For any small $\{\varepsilon_{k-1}, \varepsilon_{k-1}^\tau\} > 0$ and $\{\varepsilon_k, \varepsilon_k^\tau\} > 0$, under $||\mathbf{x}(t_{k-1}) - \mathbf{x}_{k-1}|| \leq \varepsilon_{k-1}$, $||\mathbf{x}^\tau(t_{k-1}) - \mathbf{x}_{k-1}^\tau|| \leq \varepsilon_{k-1}^\tau, ||\mathbf{x}(t_k) - \mathbf{x}_k|| \leq \varepsilon_k$, and $||\mathbf{x}^\tau(t_k) - \mathbf{x}_k^\tau|| \leq \varepsilon_k^\tau$, Eq. (5.221) gives

$$\mathbf{x}_k = \mathbf{x}_{k-1} + \bar{\mathbf{g}}_k(\mathbf{x}_{k-1}, \mathbf{x}_k; \mathbf{x}_{k-1}^\tau, \mathbf{x}_k^\tau, \mathbf{p}),$$

$$\mathbf{x}_{r_k}^\tau = \mathbf{x}_{k-1} + \bar{\mathbf{h}}_k(\mathbf{x}_{k-1}, \mathbf{x}_k; \mathbf{x}_{k-1}^\tau, \mathbf{x}_k^\tau, \mathbf{p}),$$

$$\bar{\mathbf{g}}_k(\mathbf{x}_k, \mathbf{x}_{k+1}; \mathbf{x}_k^\tau, \mathbf{x}_{k+1}^\tau, \mathbf{p}) = \int_{t_{k-1}}^{t_k} \mathbf{P}(t, t - \tau, \mathbf{C}(\mathbf{t}_k)) dt;$$

$$\bar{\mathbf{h}}_k(\mathbf{x}_k, \mathbf{x}_{k+1}; \mathbf{x}_k^\tau, \mathbf{x}_{k+1}^\tau, \mathbf{p}) = \int_{t_{k-1}}^{t_{r_k} - \tau} \mathbf{P}(t, t - \tau, \mathbf{C}(\mathbf{t}_k)) dt; \tag{5.222}$$

$$r_k = \mathrm{mod}(k - 1 + s_{k-1}, N).$$

Thus, a discrete mapping relation is obtained by

$$\begin{aligned} &\mathbf{g}_k(\mathbf{x}_{k-1}, \mathbf{x}_k; \mathbf{x}_{k-1}^\tau, \mathbf{x}_k^\tau, \mathbf{p}) \\ &\equiv \mathbf{x}_k - \mathbf{x}_{k-1} - \bar{\mathbf{g}}_k(\mathbf{x}_{k-1}, \mathbf{x}_k; \mathbf{x}_{k-1}^\tau, \mathbf{x}_k^\tau, \mathbf{p}) = \mathbf{0}, \\ &\mathbf{h}_k(\mathbf{x}_{k-1}, \mathbf{x}_k; \mathbf{x}_{k-1}^\tau, \mathbf{x}_k^\tau, \mathbf{x}_{r_k}^\tau, \mathbf{p}) \\ &\equiv \mathbf{x}_{r_k}^\tau - \mathbf{x}_{k-1} - \bar{\mathbf{h}}_k(\mathbf{x}_{k-1}, \mathbf{x}_k; \mathbf{x}_{k-1}^\tau, \mathbf{x}_k^\tau, \mathbf{x}_{r_k}^\tau, \mathbf{p}) = \mathbf{0}. \end{aligned} \tag{5.223}$$

From the discrete mapping, two points $\mathbf{x}(t_{k-1})$ and $\mathbf{x}(t_k)$ for the time interval $t \in [t_{k-1}, t_k]$ $k = 1, 2, \ldots, N$ can be approximated by $\mathbf{x}_{k-1}$ and $\mathbf{x}_k$, respectively. If $\mathbf{f}(\mathbf{x}, \mathbf{x}^\tau, t, \mathbf{p})$ *is a* $C^r$-*continuous nonlinear vector function, we have* $\|\mathbf{f}\|_{\mathbf{x}} \leq L$ and $\|\mathbf{f}\|_{\mathbf{x}^\tau} \leq L^\tau$ ($L$ and $L^\tau$ constant). Thus,

$$
\begin{aligned}
&\|\mathbf{f}(\mathbf{x}(t_{k-1}), \mathbf{x}^\tau(t_{k-1}), t_{k-1}, \mathbf{p}) - \mathbf{f}(\mathbf{x}_{k-1}, \mathbf{x}_{k-1}^\tau, t_{k-1}, \mathbf{p})\| \\
&\leq L \|\mathbf{x}(t_{k-1}) - \mathbf{x}_{k-1}\| + L^\tau \|\mathbf{x}^\tau(t_{k-1}) - \mathbf{x}_{k-1}^\tau\| \\
&\leq L\varepsilon_{k-1} + L^\tau \varepsilon_{k-1}^\tau = \delta_{k-1},
\end{aligned}
\tag{5.224}
$$

$$
\begin{aligned}
&\|\mathbf{f}(\mathbf{x}(t_k), \mathbf{x}^\tau(t_k), t_k, \mathbf{p}) - \mathbf{f}(\mathbf{x}_k, \mathbf{x}_k^\tau, t_k, \mathbf{p})\| \\
&\leq L \|\mathbf{x}(t_k) - \mathbf{x}_k\| + L^\tau \|\mathbf{x}^\tau(t_k) - \mathbf{x}_k^\tau\| \\
&\leq L\varepsilon_k + L^\tau \varepsilon_k^\tau = \delta_k.
\end{aligned}
\tag{5.225}
$$

Once the mapping $P_k : (\mathbf{x}_{k-1}, \mathbf{x}_{k-1}^\tau) \rightarrow (\mathbf{x}_k, \mathbf{x}_k^\tau)$ exists with

$$
\left.
\begin{aligned}
&\mathbf{g}_k(\mathbf{x}_{k-1}, \mathbf{x}_k; \mathbf{x}_{k-1}^\tau, \mathbf{x}_k^\tau, \mathbf{p}) = \mathbf{0}, \\
&\mathbf{h}_k(\mathbf{x}_{k-1}, \mathbf{x}_k; \mathbf{x}_{k-1}^\tau, \mathbf{x}_k^\tau, \mathbf{x}_{r_k}^\tau, \mathbf{p}) = \mathbf{0} \\
&\text{with } r_k = \text{mod}(k - 1 + s_{k-1}, N)
\end{aligned}
\right\} \quad \text{for } k = 1, 2, \ldots, N;
\tag{5.226}
$$

the periodic flow is formed by $P : (\mathbf{x}_0, \mathbf{x}_0^\tau) \rightarrow (\mathbf{x}_N, \mathbf{x}_N^\tau)$ with $P = P_N \circ P_{N-1} \circ \ldots \circ P_2 \circ P_1$, i.e.,

$$
\begin{aligned}
&\mathbf{g}_k(\mathbf{x}_{k-1}, \mathbf{x}_k; \mathbf{x}_{k-1}^\tau, \mathbf{x}_k^\tau, \mathbf{p}) = \mathbf{0}, \\
&\mathbf{h}_k(\mathbf{x}_{k-1}, \mathbf{x}_k; \mathbf{x}_{k-1}^\tau, \mathbf{x}_k^\tau, \mathbf{x}_{r_k}^\tau, \mathbf{p}) = \mathbf{0}; \\
&(k = 1, 2, \ldots, N).
\end{aligned}
\tag{5.227}
$$

With the periodicity condition, we have

$$
\mathbf{x}_0 = \mathbf{x}_N, \quad \mathbf{x}_0^\tau = \mathbf{x}_N^\tau.
\tag{5.228}
$$

Solving Eqs. (5.227) and (5.227) gives $\mathbf{x}_k^*$ and $\mathbf{x}_k^{\tau*}$ ($k = 1, 2, \ldots, N$) to get the period-1 flow. For the stability of such a periodic flow, consider $\mathbf{x}_k = \mathbf{x}_k^* + \Delta\mathbf{x}_k$ and $\mathbf{x}_k^\tau = \mathbf{x}_k^{\tau*} + \Delta\mathbf{x}_k^\tau$ ($k = 1, 2, \ldots, N$) for $\mathbf{x}_k \in U(\mathbf{x}_k^*)$ and $\mathbf{x}_k^\tau \in U(\mathbf{x}_k^{\tau*})$. Equation (5.226) becomes

$$
\begin{aligned}
&\mathbf{g}_k(\mathbf{x}_{k-1}^* + \Delta\mathbf{x}_{k-1}, \mathbf{x}_k^* + \Delta\mathbf{x}_k; \mathbf{x}_{k-1}^{\tau*} + \Delta\mathbf{x}_{k-1}^\tau, \mathbf{x}_k^{\tau*} + \Delta\mathbf{x}_k^\tau, \mathbf{p}) = \mathbf{0}, \\
&\mathbf{h}_k(\mathbf{x}_{k-1}^* + \Delta\mathbf{x}_{k-1}, \mathbf{x}_k^* + \Delta\mathbf{x}_k; \mathbf{x}_{k-1}^{\tau*} + \Delta\mathbf{x}_{k-1}^\tau, \mathbf{x}_k^{\tau*} + \Delta\mathbf{x}_k^\tau, \mathbf{x}_{r_k}^{\tau*} + \Delta\mathbf{x}_{r_k}^\tau, \mathbf{p}) = \mathbf{0}; \\
&(k = 1, 2, \ldots, N).
\end{aligned}
$$

$$
\tag{5.229}
$$

Thus, derivatives of $\mathbf{g}_k(\mathbf{x}_{k-1}, \mathbf{x}_{k-1}^{\tau}, \mathbf{x}_k, \mathbf{x}_k^{\tau}, \mathbf{p}) = \mathbf{0}$ with respect to $\mathbf{x}_0$ gives

$$\mathbf{y} = \mathbf{A}^{-1}\mathbf{b} \quad \text{and} \quad \mathbf{y}^{\tau} = \mathbf{A}^{-1}\mathbf{b}^{\tau} \tag{5.230}$$

where

$$\begin{aligned}
\mathbf{A} &= (\mathbf{A}_{kl})_{2nN \times 2nN}, \\
\mathbf{y} &= (\mathbf{y}_1, \mathbf{y}_2, \ldots, \mathbf{y}_N)^{\mathrm{T}}, \quad \mathbf{y}^{\tau} = (\mathbf{y}_1^{\tau}, \mathbf{y}_2^{\tau}, \ldots, \mathbf{y}_N^{\tau})^{\mathrm{T}}, \\
\mathbf{b} &= (\mathbf{b}_1, \mathbf{b}_2, \ldots, \mathbf{b}_N)^{\mathrm{T}}, \quad \mathbf{b}^{\tau} = (\mathbf{b}_1^{\tau}, \mathbf{b}_2^{\tau}, \ldots, \mathbf{b}_N^{\tau})^{\mathrm{T}};
\end{aligned} \tag{5.231}$$

and

$$\begin{aligned}
\mathbf{A}_{kl} &= \sum_{l_k} \mathbf{A}_{kl_k} \delta_l^{l_k} \quad \text{for } l_k = k-1, k, r_k; l_k > 0 \\
\mathbf{y}_k &= \left[ \frac{\partial \mathbf{x}_k}{\partial \mathbf{x}_0}, \frac{\partial \mathbf{x}_k^{\tau}}{\partial \mathbf{x}_0} \right]^{\mathrm{T}}, \quad \mathbf{y}_k^{\tau} = \left[ \frac{\partial \mathbf{x}_k}{\partial \mathbf{x}_0^{\tau}}, \frac{\partial \mathbf{x}_k^{\tau}}{\partial \mathbf{x}_0^{\tau}} \right]^{\mathrm{T}}; \\
\mathbf{A}_{kj} &= \begin{bmatrix} \mathbf{a}_{kj} & \mathbf{a}_{kj}^{\tau} \\ \mathbf{b}_{kj} & \mathbf{b}_{kj}^{\tau} \end{bmatrix}, \quad \mathbf{A}_{kr_k} = \begin{bmatrix} \mathbf{0}_{n \times n} & \mathbf{0}_{n \times n} \\ \mathbf{0}_{n \times n} & \mathbf{b}_{kr_k}^{\tau} \end{bmatrix} (j = k-1, k), \\
\mathbf{b}_k &= -\sum_{l_k} \left[ \frac{\partial \mathbf{g}_k}{\partial \mathbf{x}_0}, \frac{\partial \mathbf{h}_k}{\partial \mathbf{x}_0} \right]^{\mathrm{T}} \delta_0^{l_k}, \quad \mathbf{b}_k^{\tau} = -\sum_{l_k} \left[ \frac{\partial \mathbf{g}_k}{\partial \mathbf{x}_0^{\tau}}, \frac{\partial \mathbf{h}_k}{\partial \mathbf{x}_0^{\tau}} \right]^{\mathrm{T}} \delta_0^{l_k}
\end{aligned} \tag{5.232}$$

with

$$\mathbf{a}_{kj} = \left[ \frac{\partial \mathbf{g}_k}{\partial \mathbf{x}_j} \right], \mathbf{a}_{kj}^{\tau} = \left[ \frac{\partial \mathbf{g}_k}{\partial \mathbf{x}_j^{\tau}} \right], \quad \mathbf{b}_{kj} = \left[ \frac{\partial \mathbf{h}_k}{\partial \mathbf{x}_j} \right], \quad \mathbf{b}_{kj}^{\tau} = \left[ \frac{\partial \mathbf{h}_k}{\partial \mathbf{x}_j^{\tau}} \right], \quad \mathbf{b}_{kr_k}^{\tau} = \left[ \frac{\partial \mathbf{h}_k}{\partial \mathbf{x}_{r_k}^{\tau}} \right]$$

$$(k = 1, 2, \ldots, N; j = k-1, k).$$

$$\tag{5.233}$$

From the forgoing equation, we have $\mathbf{y}_k$ and $\mathbf{y}_k^{\tau}$. Thus, the linearized equation based on the initial point $\mathbf{x}_0$ and $\mathbf{x}_0^{\tau}$ in Eq. (5.229) gives

$$\begin{aligned}
\begin{bmatrix} \Delta \mathbf{x}_k \\ \Delta \mathbf{x}_k^{\tau} \end{bmatrix} &= \mathrm{DP}_{k(k-1)\cdots 1} \begin{bmatrix} \Delta \mathbf{x}_0 \\ \Delta \mathbf{x}_0^{\tau} \end{bmatrix} \quad (k = 1, 2, \cdots, N), \\
\begin{bmatrix} \Delta \mathbf{x}_N \\ \Delta \mathbf{x}_N^{\tau} \end{bmatrix} &= \mathrm{DP} \begin{bmatrix} \Delta \mathbf{x}_0 \\ \Delta \mathbf{x}_0^{\tau} \end{bmatrix} = \mathrm{DP}_{N(N-1)\cdots 1} \begin{bmatrix} \Delta \mathbf{x}_0 \\ \Delta \mathbf{x}_0^{\tau} \end{bmatrix};
\end{aligned} \tag{5.234}$$

where

$$\mathrm{DP}_{k(k-1)\cdots 1} = \begin{bmatrix} \frac{\partial \mathbf{x}_k}{\partial \mathbf{x}_0} & \frac{\partial \mathbf{x}_k}{\partial \mathbf{x}_0^{\tau}} \\ \frac{\partial \mathbf{x}_k}{\partial \mathbf{x}_0} & \frac{\partial \mathbf{x}_k}{\partial \mathbf{x}_0^{\tau}} \end{bmatrix}_{(\mathbf{x}_0^*, \mathbf{x}_0^{\tau*}, \ldots, \mathbf{x}_N^*, \mathbf{x}_N^{\tau*})} \quad (k = 1, 2, \ldots, N),$$

$$\text{DP} = \begin{bmatrix} \dfrac{\partial \mathbf{x}_N}{\partial \mathbf{x}_0} & \dfrac{\partial \mathbf{x}_N}{\partial \mathbf{x}_0^\tau} \\[3mm] \dfrac{\partial \mathbf{x}_N}{\partial \mathbf{x}_0} & \dfrac{\partial \mathbf{x}_N}{\partial \mathbf{x}_0^\tau} \end{bmatrix}_{(\mathbf{x}_0^*, \mathbf{x}_0^{\tau*}, \ldots, \mathbf{x}_N^*, \mathbf{x}_N^{\tau*})}. \tag{5.235}$$

Setting $(\Delta \mathbf{x}_k, \Delta \mathbf{x}_k^\tau)^{\mathrm{T}} = \lambda^{(k)}(\Delta \mathbf{x}_0, \Delta \mathbf{x}_0^\tau)^{\mathrm{T}}$ and $(\Delta \mathbf{x}_N, \Delta \mathbf{x}_N^\tau)^{\mathrm{T}} = \lambda(\Delta \mathbf{x}_0, \Delta \mathbf{x}_0^\tau)^{\mathrm{T}}$, the forgoing equation becomes

$$(\text{DP}_{k(k-1)\cdots 1} - \lambda^{(k)} \mathbf{I}_{2n \times 2n}) \begin{bmatrix} \Delta \mathbf{x}_0 \\ \Delta \mathbf{x}_0^\tau \end{bmatrix} = \mathbf{0},$$

$$(\text{DP} - \lambda \mathbf{I}_{2n \times 2n}) \begin{bmatrix} \Delta \mathbf{x}_0 \\ \Delta \mathbf{x}_0^\tau \end{bmatrix} = \mathbf{0}. \tag{5.236}$$

For any non-trivial solution $||\Delta \mathbf{x}_0|| + ||\Delta \mathbf{x}_0^\tau|| \neq 0$, we have

$$|\text{DP}_{k(k-1)\cdots 1} - \lambda^{(k)} \mathbf{I}_{2n \times 2n}| = 0 \quad \text{and} \quad |\text{DP} - \lambda \mathbf{I}_{2n \times 2n}| = 0. \tag{5.237}$$

Thus, the eigenvalues of $DP_{k(k-1)\cdots 1}$ give changes of $(\Delta \mathbf{x}_k, \Delta \mathbf{x}_k^\tau)$ with $(\Delta \mathbf{x}_0, \Delta \mathbf{x}_0^\tau)$. In addition, the eigenvalues of DP are computed for the periodic solution due to $\mathbf{x}_N^* = \mathbf{x}_0^*$ and $\mathbf{x}_N^{\tau*} = \mathbf{x}_0^{\tau*}$. From the stability and bifurcation theory of dynamical systems at fixed points in discrete nonlinear systems with time delay, the stability and bifurcation of the periodic solution can be classified as stated in the theorem. This theorem is proved.                                                                         □

For a time-delay system, a periodic solution is represented by $N$ discrete points $(\mathbf{x}_k, k = 0, 1, 2, \ldots, N)$ and the corresponding time-delay points $(\mathbf{x}_k^\tau, k = 0, 1, 2, \ldots, N)$, as shown in Fig. 5.9. The time-delay nodes are obtained by the integration. Thus, we have two sets of discrete mappings. The small, filled circular symbols are for discrete nodes, and the large, hollow circular symbols are for time-delay nodes. The periodicity requires $\mathbf{x}_N = \mathbf{x}_0$ and $\mathbf{x}_N^\tau = \mathbf{x}_0^\tau$.

From the forgoing theorem, a set of nonlinear, time-delay, and discrete mappings $P_k$ with $\mathbf{g}_k(\mathbf{x}_{k-1}, \mathbf{x}_k, \mathbf{x}_{k-1}^\tau, \mathbf{x}_k^\tau, \mathbf{p}) = \mathbf{0}$ and $\mathbf{h}_k(\mathbf{x}_{k-1}, \mathbf{x}_k, \mathbf{x}_{k-1}^\tau, \mathbf{x}_k^\tau, \mathbf{x}_{r_k}^\tau, \mathbf{p}) = \mathbf{0}$ ($k = 1, 2, \cdots, N$) are developed for a periodic flow. In addition to a one-step time-delay mapping of $P_k$, one can develop a multi-step (or $r$-steps) time-delay mapping of $P_k$ with

$$\mathbf{g}_k(\mathbf{x}_{k-r}, \ldots, \mathbf{x}_{k-1} \mathbf{x}_k; \mathbf{x}_{k-r}^\tau, \ldots, \mathbf{x}_{k-1}^\tau, \mathbf{x}_k^\tau, \mathbf{p}) = \mathbf{0},$$
$$\mathbf{h}_k(\mathbf{x}_{k-r}, \ldots, \mathbf{x}_{k-1}, \mathbf{x}_k; \mathbf{x}_{k-r}^\tau, \ldots, \mathbf{x}_{k-1}^\tau, \mathbf{x}_k^\tau, \mathbf{x}_{r_k}^\tau, \mathbf{p}) = \mathbf{0}; \tag{5.238}$$
$$k = 1, 2, \ldots, N; r_k = \mathrm{mod}(k - 1 + s_{k-1}, N) \quad \text{and} \quad r \in \{1, 2, \ldots, k\}.$$

(i)  If $r = 1$, the one-step time-delay mapping is recovered from the multi-step time-delay mapping.

(ii)  If $r = 2$, the two-step time-delay mapping is obtained from the multi-step time-delay mapping as

$$\mathbf{g}_k(\mathbf{x}_{k-2}, \mathbf{x}_{k-1}, \mathbf{x}_k; \mathbf{x}_{k-2}^\tau, \mathbf{x}_{k-1}^\tau, \mathbf{x}_k^\tau, \mathbf{p}) = \mathbf{0},$$

$$\mathbf{h}_k(\mathbf{x}_{k-2}, \mathbf{x}_{k-1}, \mathbf{x}_k; \mathbf{x}_{k-2}^\tau, \mathbf{x}_{k-1}^\tau, \mathbf{x}_k^\tau, \mathbf{x}_{r_k}^\tau, \mathbf{p}) = \mathbf{0}, \qquad (5.239)$$

$$(k = 1, 2, \ldots, N, r_k = \mathrm{mod}(k - 1 + s_{k-1}, N))$$

which can be expanded as

$$\mathbf{g}_1(\mathbf{x}_0, \mathbf{x}_1; \mathbf{x}_0^\tau, \mathbf{x}_1^\tau, \mathbf{p}) = \mathbf{0},$$

$$\mathbf{h}_1(\mathbf{x}_0, \mathbf{x}_1; \mathbf{x}_0^\tau, \mathbf{x}_1^\tau, \mathbf{x}_{r_1}^\tau, \mathbf{p}) = \mathbf{0};$$

$$\vdots \qquad\qquad\qquad (5.240)$$

$$\mathbf{g}_k(\mathbf{x}_{k-2}, \mathbf{x}_{k-1}, \mathbf{x}_k; \mathbf{x}_{k-2}^\tau, \mathbf{x}_{k-1}^\tau, \mathbf{x}_k^\tau, \mathbf{p}) = \mathbf{0},$$

$$\mathbf{h}_k(\mathbf{x}_{k-2}, \mathbf{x}_{k-1}, \mathbf{x}_k; \mathbf{x}_{k-2}^\tau, \mathbf{x}_{k-1}^\tau, \mathbf{x}_k^\tau, \mathbf{x}_{r_k}^\tau, \mathbf{p}) = \mathbf{0};$$

$$(k = 1, 2, \ldots, N, r_k = \mathrm{mod}(k - 1 + s_{k-1}, N)).$$

(iii)  If $r = k$, the $k$-steps time-delay mapping is obtained. That is,

$$\mathbf{g}_k(\mathbf{x}_0, \mathbf{x}_1, \ldots, \mathbf{x}_k; \mathbf{x}_0^\tau, \mathbf{x}_1^\tau, \ldots, \mathbf{x}_k^\tau, \mathbf{p}) = \mathbf{0},$$

$$\mathbf{h}_k(\mathbf{x}_0, \mathbf{x}_1, \ldots, \mathbf{x}_k; \mathbf{x}_0^\tau, \mathbf{x}_1^\tau, \ldots, \mathbf{x}_k^\tau, \mathbf{x}_{r_k}^\tau, \mathbf{p}) = \mathbf{0} \qquad (5.241)$$

$$(k = 1, 2, \ldots, N, r_k = \mathrm{mod}(k - 1 + s_{k-1}, N))$$

and the forgoing equations can be expanded as

$$\mathbf{g}_1(\mathbf{x}_0, \mathbf{x}_1; \mathbf{x}_0^\tau, \mathbf{x}_1^\tau, \mathbf{p}) = \mathbf{0},$$

$$\mathbf{h}_1(\mathbf{x}_0, \mathbf{x}_1; \mathbf{x}_0^\tau, \mathbf{x}_1^\tau, \mathbf{x}_{r_1}^\tau, \mathbf{p}) = \mathbf{0};$$

$$\vdots \qquad\qquad\qquad (5.242)$$

$$\mathbf{g}_k(\mathbf{x}_0, \mathbf{x}_1, \ldots, \mathbf{x}_k; \mathbf{x}_0^\tau, \mathbf{x}_1^\tau, \ldots, \mathbf{x}_k^\tau, \mathbf{p}) = \mathbf{0},$$

$$\mathbf{h}_k(\mathbf{x}_0, \mathbf{x}_1, \ldots, \mathbf{x}_k; \mathbf{x}_0^\tau, \mathbf{x}_1^\tau, \ldots, \mathbf{x}_k^\tau, \mathbf{x}_{r_k}^\tau, \mathbf{p}) = \mathbf{0};$$

$$(k = 1, 2, \ldots, N, r_k = \mathrm{mod}(k - 1 + s_{k-1}, N)).$$

From the multi-step (or $r$-steps) mapping of $P_k$ without $k - r \geq 0$, with the periodicity condition ($\mathbf{x}_0 = \mathbf{x}_N$ and $\mathbf{x}_0^\tau = \mathbf{x}_N^\tau$), the periodic flow can be obtained via

$$\left. \begin{array}{l} \mathbf{g}_k(\mathbf{x}_{k-r}, \cdots, \mathbf{x}_{k-1}, \mathbf{x}_k; \mathbf{x}_{k-r}^\tau, \cdots, \mathbf{x}_{k-1}^\tau, \mathbf{x}_k^\tau, \mathbf{p}) = \mathbf{0}, \\[4pt] \mathbf{h}_k(\mathbf{x}_{k-r}, \cdots, \mathbf{x}_{k-1}, \mathbf{x}_k; \mathbf{x}_{k-r}^\tau, \cdots, \mathbf{x}_{k-1}^\tau, \mathbf{x}_k^\tau, \mathbf{x}_{r_k}^\tau, \mathbf{p}) = \mathbf{0} \end{array} \right\}$$

$$(k = 1, 2, \cdots, N; \ r_k = \mathrm{mod}(k - 1 + s_{k-1}, N) \text{ and } r \in \{1, 2, \cdots, k\}), \qquad (5.243)$$

$$\mathbf{x}_0 = \mathbf{x}_N \quad \text{and} \quad \mathbf{x}_0^\tau = \mathbf{x}_N^\tau.$$

Suppose node points $\mathbf{x}_k^*$ ($k = 0, 1, \ldots, N$) of periodic flows are obtained, the corresponding stability and bifurcation can be analyzed in the neighborhood of $\mathbf{x}_k^*$ with $\mathbf{x}_k = \mathbf{x}_k^* + \Delta\mathbf{x}_k$ and $\mathbf{x}_k^\tau = \mathbf{x}_k^{\tau*} + \Delta\mathbf{x}_k^\tau$, that is,

$$\sum_{j=0}^{r} \frac{\partial\mathbf{g}_k}{\partial\mathbf{x}_{k-j}} \frac{\partial\mathbf{x}_{k-j}}{\partial\mathbf{x}_0} + \frac{\partial\mathbf{g}_k}{\partial\mathbf{x}_{k-j}^\tau} \frac{\partial\mathbf{x}_{k-j}^\tau}{\partial\mathbf{x}_0} = \mathbf{0}_{n\times n},$$

$$\sum_{j=0}^{r} \frac{\partial\mathbf{h}_k}{\partial\mathbf{x}_{k-j}} \frac{\partial\mathbf{x}_{k-j}}{\partial\mathbf{x}_0} + \frac{\partial\mathbf{h}_k}{\partial\mathbf{x}_{k-j}^\tau} \frac{\partial\mathbf{x}_{k-j}^\tau}{\partial\mathbf{x}_0} + \frac{\partial\mathbf{h}_k}{\partial\mathbf{x}_{r_k}^\tau} \frac{\partial\mathbf{x}_{r_k}^\tau}{\partial\mathbf{x}_0} = \mathbf{0}_{n\times n};$$

$$\sum_{j=0}^{r} \frac{\partial\mathbf{g}_k}{\partial\mathbf{x}_{k-j}} \frac{\partial\mathbf{x}_{k-j}}{\partial\mathbf{x}_0^\tau} + \frac{\partial\mathbf{g}_k}{\partial\mathbf{x}_{k-j}^\tau} \frac{\partial\mathbf{x}_{k-j}^\tau}{\partial\mathbf{x}_0^\tau} = \mathbf{0}_{n\times n}, \qquad (5.244)$$

$$\sum_{j=0}^{r} \frac{\partial\mathbf{h}_k}{\partial\mathbf{x}_{k-j}} \frac{\partial\mathbf{x}_{k-j}}{\partial\mathbf{x}_0^\tau} + \frac{\partial\mathbf{h}_k}{\partial\mathbf{x}_{k-j}^\tau} \frac{\partial\mathbf{x}_{k-j}^\tau}{\partial\mathbf{x}_0^\tau} + \frac{\partial\mathbf{h}_k}{\partial\mathbf{x}_{r_k}^\tau} \frac{\partial\mathbf{x}_{r_k}^\tau}{\partial\mathbf{x}_0^\tau} = \mathbf{0}_{n\times n};$$

with $r_k = \mathrm{mod}(k - 1 + s_{k-1}, N)$, $(k = 1, 2, \ldots, N; r \in \{1, 2, \ldots, k\})$.

Let

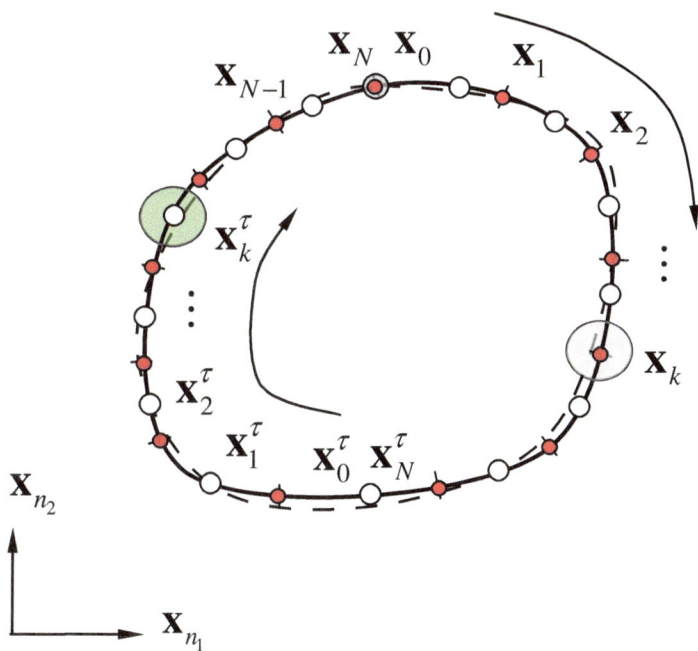

**Fig. 5.9** Period-1 flow with $N$-nodes for a time-delay system. The *small, filled circular symbols* are for non-time-delay discrete nodes, and the *large, hollow circular symbols* are for time-delay discrete nodes. The *dashed curve* is the expected exact solution for such a time-delay system

$$\mathbf{a}_{kj} = \left[\frac{\partial \mathbf{g}_k}{\partial \mathbf{x}_j}\right], \quad \mathbf{a}_{kj}^{\tau} = \left[\frac{\partial \mathbf{g}_k}{\partial \mathbf{x}_j^{\tau}}\right],$$

$$\mathbf{b}_{kj} = \left[\frac{\partial \mathbf{h}_k}{\partial \mathbf{x}_j}\right], \quad \mathbf{b}_{kj}^{\tau} = \left[\frac{\partial \mathbf{h}_k}{\partial \mathbf{x}_j^{\tau}}\right], \quad \mathbf{b}_{kr_k}^{\tau} = \left[\frac{\partial \mathbf{h}_k}{\partial \mathbf{x}_{r_k}^{\tau}}\right], \tag{5.245}$$

$$(k = 1, 2, \ldots, N), j = k - r, \ldots, k - 1, k;$$

$$\mathbf{A} = (\mathbf{A}_{kl})_{2nN \times 2nN},$$

$$\mathbf{y} = (\mathbf{y}_1, \mathbf{y}_2, \ldots, \mathbf{y}_N)^{\mathrm{T}}, \quad \mathbf{y}^{\tau} = (\mathbf{y}_1^{\tau}, \mathbf{y}_2^{\tau}, \ldots, \mathbf{y}_N^{\tau})^{\mathrm{T}},$$

$$\mathbf{b} = (\mathbf{b}_1, \mathbf{b}_2, \ldots, \mathbf{b}_N)^{\mathrm{T}}, \quad \mathbf{b}^{\tau} = (\mathbf{b}_1^{\tau}, \mathbf{b}_2^{\tau}, \ldots, \mathbf{b}_N^{\tau})^{\mathrm{T}};$$

$$\mathbf{A}_{kl} = \sum_{l_k} \mathbf{A}_{kl_k} \delta_l^{l_k} \quad \text{for } l_k = k - r, \ldots, k - 1, k, r_k; l_k > 0;$$

$$\mathbf{b}_k = -\sum_{l_k} \left[\frac{\partial \mathbf{g}_k}{\partial \mathbf{x}_0}, \frac{\partial \mathbf{h}_k}{\partial \mathbf{x}_0}\right]^{\mathrm{T}} \delta_0^{l_k}, \quad \mathbf{b}_k^{\tau} = -\sum_{l_k} \left[\frac{\partial \mathbf{g}_k}{\partial \mathbf{x}_0^{\tau}}, \frac{\partial \mathbf{h}_k}{\partial \mathbf{x}_0^{\tau}}\right]^{\mathrm{T}} \delta_0^{l_k}; \tag{5.246}$$

$$\mathbf{y}_j = \left[\frac{\partial \mathbf{x}_j}{\partial \mathbf{x}_0}, \frac{\partial \mathbf{x}_j^{\tau}}{\partial \mathbf{x}_0}\right]^{\mathrm{T}}, \quad \mathbf{y}_j^{\tau} = \left[\frac{\partial \mathbf{x}_j}{\partial \mathbf{x}_0^{\tau}}, \frac{\partial \mathbf{x}_j^{\tau}}{\partial \mathbf{x}_0^{\tau}}\right]^{\mathrm{T}};$$

$$\mathbf{A}_{kj} = \begin{bmatrix} \mathbf{a}_{kj} & \mathbf{a}_{kj}^{\tau} \\ \mathbf{b}_{kj} & \mathbf{b}_{kj}^{\tau} \end{bmatrix}, \quad \mathbf{A}_{kr_k} = \begin{bmatrix} \mathbf{0}_{n \times n} & \mathbf{0}_{n \times n} \\ \mathbf{0}_{n \times n} & \mathbf{b}_{kr_k}^{\tau} \end{bmatrix};$$

$$(j = k - r, \ldots, k - 1, k).$$

Finally, Eq. (5.244) becomes

$$\mathbf{y} = \mathbf{A}^{-1}\mathbf{b} \quad \text{and} \quad \mathbf{y}^{\tau} = \mathbf{A}^{-1}\mathbf{b}^{\tau}. \tag{5.247}$$

From the mapping structure, we have

$$\begin{bmatrix} \Delta \mathbf{x}_N \\ \Delta \mathbf{x}_N^{\tau} \end{bmatrix} = \mathrm{DP} \begin{bmatrix} \Delta \mathbf{x}_0 \\ \Delta \mathbf{x}_0^{\tau} \end{bmatrix} = \mathrm{DP}_{N(N-1)\ldots 1} \begin{bmatrix} \Delta \mathbf{x}_0 \\ \Delta \mathbf{x}_0^{\tau} \end{bmatrix},$$

$$\text{with } \mathrm{DP} = \begin{bmatrix} \frac{\partial \mathbf{x}_N}{\partial \mathbf{x}_0} & \frac{\partial \mathbf{x}_N}{\partial \mathbf{x}_0^{\tau}} \\ \frac{\partial \mathbf{x}_N}{\partial \mathbf{x}_0} & \frac{\partial \mathbf{x}_N}{\partial \mathbf{x}_0^{\tau}} \end{bmatrix}_{(\mathbf{x}_0^*, \ldots, \mathbf{x}_N^*; \mathbf{x}_0^{\tau*}, \ldots, \mathbf{x}_N^{\tau*})}. \tag{5.248}$$

Letting $(\Delta \mathbf{x}_N, \Delta \mathbf{x}_N^{\tau})^{\mathrm{T}} = \lambda (\Delta \mathbf{x}_0, \Delta \mathbf{x}_0^{\tau})^{\mathrm{T}}$, we have

$$(\mathrm{DP} - \lambda \mathbf{I}_{2n \times 2n}) \begin{bmatrix} \Delta \mathbf{x}_0 \\ \Delta \mathbf{x}_0^{\tau} \end{bmatrix} = \mathbf{0}. \tag{5.249}$$

The eigenvalue of DP is given by $|DP - \lambda \mathbf{I}_{2n \times 2n}| = 0$. In addition, we have

$$
\begin{bmatrix} \Delta \mathbf{x}_k \\ \Delta \mathbf{x}_k^\tau \end{bmatrix} = DP_{k(k-1)\cdots 1} \begin{bmatrix} \Delta \mathbf{x}_0 \\ \Delta \mathbf{x}_0^\tau \end{bmatrix} \quad (k = 1, 2, \ldots, N),
$$

$$
\text{with } DP_{k(k-1)\cdots 1} = \begin{bmatrix} \frac{\partial \mathbf{x}_k}{\partial \mathbf{x}_0} & \frac{\partial \mathbf{x}_k}{\partial \mathbf{x}_0^\tau} \\ \frac{\partial \mathbf{x}_k}{\partial \mathbf{x}_0} & \frac{\partial \mathbf{x}_k}{\partial \mathbf{x}_0^\tau} \end{bmatrix}_{(\mathbf{x}_0^*, \ldots, \mathbf{x}_N^*; \mathbf{x}_0^{\tau*}, \ldots, \mathbf{x}_N^{\tau*})}. \tag{5.250}
$$

Letting $(\Delta \mathbf{x}_k, \Delta \mathbf{x}_k^\tau)^{\mathrm{T}} = \bar{\lambda}(\Delta \mathbf{x}_0, \Delta \mathbf{x}_0^\tau)^{\mathrm{T}}$, we have

$$
(DP_{k(k-1)\cdots 1} - \bar{\lambda} \mathbf{I}_{2n \times 2n}) \begin{bmatrix} \Delta \mathbf{x}_0 \\ \Delta \mathbf{x}_0^\tau \end{bmatrix} = \mathbf{0}. \tag{5.251}
$$

The eigenvalues of $DP_{k(k-1)\cdots 1}$ are given by $|DP_{k(k-1)\cdots 1} - \bar{\lambda} \mathbf{I}_{2n \times 2n}| = 0$. Such eigenvalues still tell effects of variation of $(\mathbf{x}_0, \mathbf{x}_0^\tau)$ on node points $(\mathbf{x}_k, \mathbf{x}_k^\tau)$ in their vicinity. The neighborhoods of $\mathbf{x}_k^*$ and $\mathbf{x}_k^{\tau*}$ (i.e., $U(\mathbf{x}_k^*)$ and $U(\mathbf{x}_k^{\tau*})$) are presented in Fig. 5.10 as large circles. In the neighborhoods, the eigenvalues can be used to measure the effects $\Delta \mathbf{x}_k$ and $\Delta \mathbf{x}_k^\tau$ of $\mathbf{x}_k^*$ and $\mathbf{x}_k^{\tau*}$ varying with $\Delta \mathbf{x}_0$ and $\Delta \mathbf{x}_0^\tau$ at $\mathbf{x}_0^*$ and $\mathbf{x}_0^{\tau*}$.

(i) If $r = 1$, Eqs. (5.244) becomes

$$
\sum_{j=0}^{1} \frac{\partial \mathbf{g}_k}{\partial \mathbf{x}_{k-j}} \frac{\partial \mathbf{x}_{k-j}}{\partial \mathbf{x}_0} + \frac{\partial \mathbf{g}_k}{\partial \mathbf{x}_{k-j}^\tau} \frac{\partial \mathbf{x}_{k-j}^\tau}{\partial \mathbf{x}_0} = \mathbf{0}_{n \times n},
$$

$$
\sum_{j=0}^{1} \frac{\partial \mathbf{h}_k}{\partial \mathbf{x}_{k-j}} \frac{\partial \mathbf{x}_{k-j}}{\partial \mathbf{x}_0} + \frac{\partial \mathbf{h}_k}{\partial \mathbf{x}_{k-j}^\tau} \frac{\partial \mathbf{x}_{k-j}^\tau}{\partial \mathbf{x}_0} + \frac{\partial \mathbf{h}_k}{\partial \mathbf{x}_{r_k}^\tau} \frac{\partial \mathbf{x}_{r_k}^\tau}{\partial \mathbf{x}_0} = \mathbf{0}_{n \times n};
$$

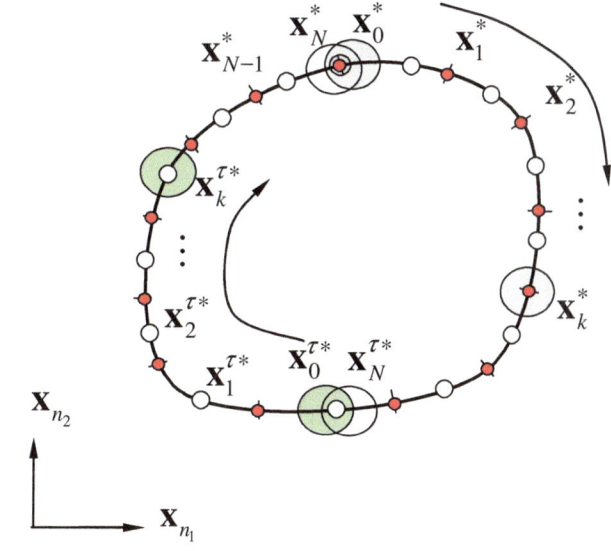

Fig. 5.10 Neighborhoods of $N$-nodes for a period-1 flow of a time-delay system. *Solid curve* is for a numerical result. The local *shaded* area is a small neighborhood at the $k$th node of the solution. The *red symbols* are node points on the periodic flow, and the *hollow symbols* are for time-delay nodes on the periodic flow

$$\sum_{j=0}^{1} \frac{\partial \mathbf{g}_k}{\partial \mathbf{x}_{k-j}} \frac{\partial \mathbf{x}_{k-j}}{\partial \mathbf{x}_0^\tau} + \frac{\partial \mathbf{g}_k}{\partial \mathbf{x}_{k-j}^\tau} \frac{\partial \mathbf{x}_{k-j}^\tau}{\partial \mathbf{x}_0^\tau} = \mathbf{0}_{n\times n},$$

$$\sum_{j=0}^{1} \frac{\partial \mathbf{h}_k}{\partial \mathbf{x}_{k-j}} \frac{\partial \mathbf{x}_{k-j}}{\partial \mathbf{x}_0^\tau} + \frac{\partial \mathbf{h}_k}{\partial \mathbf{x}_{k-j}^\tau} \frac{\partial \mathbf{x}_{k-j}^\tau}{\partial \mathbf{x}_0^\tau} + \frac{\partial \mathbf{h}_k}{\partial \mathbf{x}_{r_k}^\tau} \frac{\partial \mathbf{x}_{r_k}^\tau}{\partial \mathbf{x}_0^\tau} = \mathbf{0}_{n\times n}; \tag{5.252}$$

with $r_k = \mathrm{mod}(k - 1 + s_{k-1}, N)$, $(k = 1, 2, \ldots, N)$.

Let

$$\mathbf{a}_{kj} = \left[\frac{\partial \mathbf{g}_k}{\partial \mathbf{x}_j}\right], \quad \mathbf{a}_{kj}^\tau = \left[\frac{\partial \mathbf{g}_k}{\partial \mathbf{x}_j^\tau}\right],$$

$$\mathbf{b}_{kj} = \left[\frac{\partial \mathbf{h}_k}{\partial \mathbf{x}_j}\right], \quad \mathbf{b}_{kj}^\tau = \left[\frac{\partial \mathbf{h}_k}{\partial \mathbf{x}_j^\tau}\right], \quad \mathbf{b}_{kr_k}^\tau = \left[\frac{\partial \mathbf{h}_k}{\partial \mathbf{x}_{r_k}^\tau}\right] \tag{5.253}$$

$$(k = 1, 2, \ldots, N), j = k - 1, k.$$

Thus,

$$\mathbf{A} = (\mathbf{A}_{kl})_{2nN \times 2nN},$$

$$\mathbf{y} = (\mathbf{y}_1, \mathbf{y}_2, \ldots, \mathbf{y}_N)^{\mathrm{T}}, \quad \mathbf{y}^\tau = (\mathbf{y}_1^\tau, \mathbf{y}_2^\tau, \ldots, \mathbf{y}_N^\tau)^{\mathrm{T}},$$

$$\mathbf{b} = (\mathbf{b}_1, \mathbf{b}_2, \ldots, \mathbf{b}_N)^{\mathrm{T}}, \quad \mathbf{b}^\tau = (\mathbf{b}_1^\tau, \mathbf{b}_2^\tau, \ldots, \mathbf{b}_N^\tau)^{\mathrm{T}};$$

$$\mathbf{A}_{kl} = \sum_{l_k} \mathbf{A}_{kl_k} \delta_l^{l_k} \quad \text{for } l_k = k - 1, k, r_k; l_k > 0;$$

$$\mathbf{b}_k = -\sum_{l_k} \left[\frac{\partial \mathbf{g}_k}{\partial \mathbf{x}_0}, \frac{\partial \mathbf{h}_k}{\partial \mathbf{x}_0}\right]^{\mathrm{T}} \delta_0^{l_k}, \quad \mathbf{b}_k^\tau = -\sum_{l_k} \left[\frac{\partial \mathbf{g}_k}{\partial \mathbf{x}_0^\tau}, \frac{\partial \mathbf{h}_k}{\partial \mathbf{x}_0^\tau}\right]^{\mathrm{T}} \delta_0^{l_k}; \tag{5.254}$$

$$\mathbf{y}_j = \left[\frac{\partial \mathbf{x}_j}{\partial \mathbf{x}_0}, \frac{\partial \mathbf{x}_j^\tau}{\partial \mathbf{x}_0}\right]^{\mathrm{T}}, \quad \mathbf{y}_j^\tau = \left[\frac{\partial \mathbf{x}_j}{\partial \mathbf{x}_0^\tau}, \frac{\partial \mathbf{x}_j^\tau}{\partial \mathbf{x}_0^\tau}\right]^{\mathrm{T}};$$

$$\mathbf{A}_{kj} = \begin{bmatrix} \mathbf{a}_{kj} & \mathbf{a}_{kj}^\tau \\ \mathbf{b}_{kj} & \mathbf{b}_{kj}^\tau \end{bmatrix}, \quad \mathbf{A}_{kr_k} = \begin{bmatrix} \mathbf{0}_{n\times n} & \mathbf{0}_{n\times n} \\ \mathbf{0}_{n\times n} & \mathbf{b}_{kr_k}^\tau \end{bmatrix};$$

$$(j = k - 1, k).$$

Finally, Eq. (5.252) becomes

$$\mathbf{y} = \mathbf{A}^{-1}\mathbf{b} \quad \text{and} \quad \mathbf{y}^\tau = \mathbf{A}^{-1}\mathbf{b}^\tau. \tag{5.255}$$

So we have

$$\mathrm{DP} = \begin{bmatrix} \frac{\partial \mathbf{x}_N}{\partial \mathbf{x}_0} & \frac{\partial \mathbf{x}_N}{\partial \mathbf{x}_0^\tau} \\ \frac{\partial \mathbf{x}_N}{\partial \mathbf{x}_0} & \frac{\partial \mathbf{x}_N}{\partial \mathbf{x}_0^\tau} \end{bmatrix}_{(\mathbf{x}_0^*, \ldots, \mathbf{x}_N^*, \mathbf{x}_0^{\tau*}, \ldots, \mathbf{x}_N^{\tau*})}. \tag{5.256}$$

(ii)  For $r = k$, Eq. (5.244) with periodicity condition $(\mathbf{x}_0 = \mathbf{x}_N)$ gives node points $\mathbf{x}_k^*$ $(k = 0, 1, 2, \ldots, N)$. The corresponding stability and bifurcation can be analyzed in the neighborhoods of $\mathbf{x}_k^*$ and $\mathbf{x}_k^{\tau*}$ with $\mathbf{x}_k = \mathbf{x}_k^* + \Delta\mathbf{x}_k$ and $\mathbf{x}_k^{\tau} = \mathbf{x}_k^{\tau*} + \Delta\mathbf{x}_k^{\tau}$ for the periodic motion. Equation (5.244) becomes

$$\sum_{j=0}^{k} \frac{\partial \mathbf{g}_k}{\partial \mathbf{x}_{k-j}} \frac{\partial \mathbf{x}_{k-j}}{\partial \mathbf{x}_0} + \frac{\partial \mathbf{g}_k}{\partial \mathbf{x}_{k-j}^{\tau}} \frac{\partial \mathbf{x}_{k-j}^{\tau}}{\partial \mathbf{x}_0} = \mathbf{0}_{n \times n},$$

$$\sum_{j=0}^{k} \frac{\partial \mathbf{h}_k}{\partial \mathbf{x}_{k-j}} \frac{\partial \mathbf{x}_{k-j}}{\partial \mathbf{x}_0} + \frac{\partial \mathbf{h}_k}{\partial \mathbf{x}_{k-j}^{\tau}} \frac{\partial \mathbf{x}_{k-j}^{\tau}}{\partial \mathbf{x}_0} + \frac{\partial \mathbf{h}_k}{\partial \mathbf{x}_{r_k}^{\tau}} \frac{\partial \mathbf{x}_{r_k}^{\tau}}{\partial \mathbf{x}_0} = \mathbf{0}_{n \times n};$$

$$\sum_{j=0}^{k} \frac{\partial \mathbf{g}_k}{\partial \mathbf{x}_{k-j}} \frac{\partial \mathbf{x}_{k-j}}{\partial \mathbf{x}_0^{\tau}} + \frac{\partial \mathbf{g}_k}{\partial \mathbf{x}_{k-j}^{\tau}} \frac{\partial \mathbf{x}_{k-j}^{\tau}}{\partial \mathbf{x}_0^{\tau}} = \mathbf{0}_{n \times n}, \qquad (5.257)$$

$$\sum_{j=0}^{k} \frac{\partial \mathbf{h}_k}{\partial \mathbf{x}_{k-j}} \frac{\partial \mathbf{x}_{k-j}}{\partial \mathbf{x}_0^{\tau}} + \frac{\partial \mathbf{h}_k}{\partial \mathbf{x}_{k-j}^{\tau}} \frac{\partial \mathbf{x}_{k-j}^{\tau}}{\partial \mathbf{x}_0^{\tau}} + \frac{\partial \mathbf{h}_k}{\partial \mathbf{x}_{r_k}^{\tau}} \frac{\partial \mathbf{x}_{r_k}^{\tau}}{\partial \mathbf{x}_0^{\tau}} = \mathbf{0}_{n \times n};$$

with $r_k = \mathrm{mod}(k - 1 + s_{k-1}, N), (k = 1, 2, \ldots, N)$.

Let

$$\mathbf{a}_{kj} = \left[ \frac{\partial \mathbf{g}_k}{\partial \mathbf{x}_j} \right], \quad \mathbf{a}_{kj}^{\tau} = \left[ \frac{\partial \mathbf{g}_k}{\partial \mathbf{x}_j^{\tau}} \right],$$

$$\mathbf{b}_{kj} = \left[ \frac{\partial \mathbf{h}_k}{\partial \mathbf{x}_j} \right], \quad \mathbf{b}_{kj}^{\tau} = \left[ \frac{\partial \mathbf{h}_k}{\partial \mathbf{x}_j^{\tau}} \right], \quad \mathbf{b}_{kr_k}^{\tau} = \left[ \frac{\partial \mathbf{h}_k}{\partial \mathbf{x}_{r_k}^{\tau}} \right] \qquad (5.258)$$

$$(k = 1, 2, \ldots, N), j = 0, 1, \ldots, k - 1, k.$$

Finally, we have

$$\mathbf{A} = (\mathbf{A}_{kl})_{2nN \times 2nN},$$

$$\mathbf{y} = (\mathbf{y}_1, \mathbf{y}_2, \ldots, \mathbf{y}_N)^{\mathrm{T}}, \quad \mathbf{y}^{\tau} = (\mathbf{y}_1^{\tau}, \mathbf{y}_2^{\tau}, \ldots, \mathbf{y}_N^{\tau})^{\mathrm{T}},$$

$$\mathbf{b} = (\mathbf{b}_1, \mathbf{b}_2, \ldots, \mathbf{b}_N)^{\mathrm{T}}, \quad \mathbf{b}^{\tau} = (\mathbf{b}_1^{\tau}, \mathbf{b}_2^{\tau}, \ldots, \mathbf{b}_N^{\tau})^{\mathrm{T}};$$

$$\mathbf{A}_{kl} = \sum_{l_k} \mathbf{A}_{kl_k} \delta_l^{l_k} \quad \text{for } l_k = 0, 1, \ldots, k - 1, k, r_k; l_k > 0;$$

$$\mathbf{b}_k = -\sum_{l_k} \left[ \frac{\partial \mathbf{g}_k}{\partial \mathbf{x}_0}, \frac{\partial \mathbf{h}_k}{\partial \mathbf{x}_0} \right]^{\mathrm{T}} \delta_0^{l_k}, \quad \mathbf{b}_k^{\tau} = -\sum_{l_k} \left[ \frac{\partial \mathbf{g}_k}{\partial \mathbf{x}_0^{\tau}}, \frac{\partial \mathbf{h}_k}{\partial \mathbf{x}_0^{\tau}} \right]^{\mathrm{T}} \delta_0^{l_k};$$

$$\mathbf{y}_j = \left[\frac{\partial \mathbf{x}_j}{\partial \mathbf{x}_0}, \frac{\partial \mathbf{x}_j^\tau}{\partial \mathbf{x}_0}\right]^{\mathrm{T}}, \quad \mathbf{y}_j^\tau = \left[\frac{\partial \mathbf{x}_j}{\partial \mathbf{x}_0^\tau}, \frac{\partial \mathbf{x}_j^\tau}{\partial \mathbf{x}_0^\tau}\right]^{\mathrm{T}};$$

$$\mathbf{A}_{kj} = \begin{bmatrix} \mathbf{a}_{kj} & \mathbf{a}_{kj}^\tau \\ \mathbf{b}_{kj} & \mathbf{b}_{kj}^\tau \end{bmatrix}, \quad \mathbf{A}_{kr_k} = \begin{bmatrix} \mathbf{0}_{n\times n} & \mathbf{0}_{n\times n} \\ \mathbf{0}_{n\times n} & \mathbf{b}_{kr_k}^\tau \end{bmatrix}; \tag{5.259}$$

$$(j = 0, 1, \ldots, k-1, k).$$

Thus, the eigenvalues are determined by

$$|DP_{k(k-1)\cdots 1} - \lambda \mathbf{I}_{2n\times 2n}| = 0$$

$$\text{with } DP_{k(k-1)\cdots 1} = \begin{bmatrix} \frac{\partial \mathbf{x}_k}{\partial \mathbf{x}_0} & \frac{\partial \mathbf{x}_k}{\partial \mathbf{x}_0^\tau} \\ \frac{\partial \mathbf{x}_k}{\partial \mathbf{x}_0} & \frac{\partial \mathbf{x}_k}{\partial \mathbf{x}_0^\tau} \end{bmatrix}_{(\mathbf{x}_0^*,\ldots,\mathbf{x}_N^*,\mathbf{x}_0^{\tau*},\ldots,\mathbf{x}_N^{\tau*})}. \tag{5.260}$$

for the properties of node points on the periodic flow of the time-delay system.

The multi-step mappings are developed from the previous determined nodes of periodic motion. During time interval $t \in [t_0, t_0 + T]$, the periodic flow is

$$\mathbf{x}(t) = \mathbf{x}(t_l) + \int_{t_l}^{t} \mathbf{f}(\mathbf{x}, \mathbf{x}^\tau, t, \mathbf{p}) dt, \quad l \in \{0, 1, \ldots, k-1\}. \tag{5.261}$$

For such a periodic flow, $N$-nodes during the time interval $t \in [t_0, t_0 + T]$ are selected, and the corresponding points $\mathbf{x}(t_k)$ $(k = 1, 2, \ldots, N)$. Under $\|\mathbf{x}(t_k) - \mathbf{x}_k\| \leq \varepsilon_k$ with $\varepsilon_k \geq 0$,

$$\|\mathbf{f}(\mathbf{x}(t_k), \mathbf{x}^\tau(t_k), t_k, \mathbf{p}) - \mathbf{f}(\mathbf{x}_k, \mathbf{x}_k^\tau, t_k, \mathbf{p})\| \leq \delta_k. \tag{5.262}$$

Suppose that $\mathbf{x}_0, \ldots, \mathbf{x}_N$ and $\mathbf{x}_0^\tau, \ldots, \mathbf{x}_N^\tau$ are given, $\mathbf{f}(\mathbf{x}_k, \mathbf{x}_k^\tau, t_k, \mathbf{p})$ $(k = 0, 1, \ldots, N)$ can be determined. An interpolation polynomial $\mathbf{P}(t, \mathbf{x}_0, \ldots, \mathbf{x}_N, \mathbf{x}_0^\tau, \ldots, \mathbf{x}_N^\tau, t_0, \ldots, t_N, \mathbf{p})$ is determined, which can be used to approximate $\mathbf{f}(\mathbf{x}, \mathbf{x}^\tau, t, \mathbf{p})$. That is,

$$\mathbf{f}(\mathbf{x}, \mathbf{x}^\tau, t, \mathbf{p}) \approx \mathbf{P}(t, \mathbf{x}_0, \ldots, \mathbf{x}_N, \mathbf{x}_0^\tau, \ldots, \mathbf{x}_N^\tau, t_0, \ldots, t_N, \mathbf{p}) \tag{5.263}$$

and $\mathbf{x}(t_k) \approx \mathbf{x}_k$ can be computed by

$$\mathbf{x}_k = \mathbf{x}_{k-1} + \int_{t_{k-1}}^{t_k} \mathbf{P}(t, \mathbf{x}_0, \ldots, \mathbf{x}_N; \mathbf{x}_0^\tau, \ldots, \mathbf{x}_N^\tau, t_0, \ldots, t_N, \mathbf{p}) dt,$$

$$\mathbf{x}_{k-1+s_{k-1}}^\tau = \mathbf{x}_{k-1} + \int_{t_{k-1}}^{t_{k-1+s_{k-1}}-\tau} \mathbf{P}(t, \mathbf{x}_0, \ldots, \mathbf{x}_N; \mathbf{x}_0^\tau, \ldots, \mathbf{x}_N^\tau, t_0, \ldots, t_N, \mathbf{p}) dt. \tag{5.264}$$

Therefore, we have

$$
\begin{aligned}
\mathbf{x}_k &= \mathbf{x}_{k-1} + \bar{\mathbf{g}}_k(\mathbf{x}_0, \ldots, \mathbf{x}_N; \mathbf{x}_0^\tau, \ldots, \mathbf{x}_N^\tau, \mathbf{p}), \\
\mathbf{x}_{k-1+s_{k-1}}^\tau &= \mathbf{x}_{k-1} + \bar{\mathbf{h}}_k(\mathbf{x}_0, \ldots, \mathbf{x}_N; \mathbf{x}_0^\tau, \ldots, \mathbf{x}_N^\tau, \mathbf{p}).
\end{aligned}
\tag{5.265}
$$

The mapping $P_k$ $(k \in \{1, 2, \ldots, N\})$ is

$$
\begin{aligned}
\mathbf{g}_k(\mathbf{x}_0, \ldots, \mathbf{x}_N; \mathbf{x}_0^\tau, \ldots, \mathbf{x}_N^\tau, \mathbf{p}) &= \mathbf{0}, \\
\mathbf{h}_k(\mathbf{x}_0, \ldots, \mathbf{x}_N; \mathbf{x}_0^\tau, \ldots, \mathbf{x}_N^\tau, \mathbf{x}_{r_k}^\tau, \mathbf{p}) &= \mathbf{0}; \\
r_k &= \mathrm{mod}(k - 1 + s_{k-1}, N).
\end{aligned}
\tag{5.266}
$$

The periodic flow is determined by the mappings and periodicity conditions, i.e.,

$$
\left.
\begin{aligned}
\mathbf{g}_k(\mathbf{x}_0, \ldots, \mathbf{x}_N; \mathbf{x}_0^\tau, \ldots, \mathbf{x}_N^\tau, \mathbf{p}) &= \mathbf{0}, \\
\mathbf{h}_k(\mathbf{x}_0, \ldots, \mathbf{x}_N; \mathbf{x}_0^\tau, \ldots, \mathbf{x}_N^\tau, \mathbf{x}_{r_k}^\tau, \mathbf{p}) &= \mathbf{0}
\end{aligned}
\right\} \quad \text{for } k = 1, 2, \ldots, N
$$

$$
\mathbf{x}_0 = \mathbf{x}_N \quad \text{and} \quad \mathbf{x}_0^\tau = \mathbf{x}_N^\tau.
\tag{5.267}
$$

From the forgoing equation, node points $\mathbf{x}_k^*$ and $\mathbf{x}_k^{\tau*}$ $(k = 0, 1, 2, \ldots, N)$ can be determined. The corresponding stability and bifurcation is discussed in the neighborhood of $\mathbf{x}_k^*$ and $\mathbf{x}_k^{\tau*}$ with $\mathbf{x}_k = \mathbf{x}_k^* + \Delta\mathbf{x}_k$ and $\mathbf{x}_k^\tau = \mathbf{x}_k^{\tau*} + \Delta\mathbf{x}_k^\tau$. The derivative of Eq. (5.267) with respect to $\mathbf{x}_0$ gives

$$
\begin{aligned}
\sum_{j=1}^N \frac{\partial \mathbf{g}_k}{\partial \mathbf{x}_j} \frac{\partial \mathbf{x}_j}{\partial \mathbf{x}_0} + \frac{\partial \mathbf{g}_k}{\partial \mathbf{x}_j^\tau} \frac{\partial \mathbf{x}_j^\tau}{\partial \mathbf{x}_0} &= \mathbf{0}_{n \times n}, \\
\sum_{j=1}^N \frac{\partial \mathbf{h}_k}{\partial \mathbf{x}_{k-j}} \frac{\partial \mathbf{x}_j}{\partial \mathbf{x}_0} + \frac{\partial \mathbf{h}_k}{\partial \mathbf{x}_j^\tau} \frac{\partial \mathbf{x}_j^\tau}{\partial \mathbf{x}_0} + \frac{\partial \mathbf{h}_k}{\partial \mathbf{x}_{r_k}^\tau} \frac{\partial \mathbf{x}_{r_k}^\tau}{\partial \mathbf{x}_0} &= \mathbf{0}_{n \times n}; \\
\sum_{j=1}^N \frac{\partial \mathbf{g}_k}{\partial \mathbf{x}_j} \frac{\partial \mathbf{x}_j}{\partial \mathbf{x}_0^\tau} + \frac{\partial \mathbf{g}_k}{\partial \mathbf{x}_j^\tau} \frac{\partial \mathbf{x}_j^\tau}{\partial \mathbf{x}_0^\tau} &= \mathbf{0}_{n \times n}, \\
\sum_{j=1}^N \frac{\partial \mathbf{h}_k}{\partial \mathbf{x}_j} \frac{\partial \mathbf{x}_j}{\partial \mathbf{x}_0^\tau} + \frac{\partial \mathbf{h}_k}{\partial \mathbf{x}_j^\tau} \frac{\partial \mathbf{x}_j^\tau}{\partial \mathbf{x}_0^\tau} + \frac{\partial \mathbf{h}_k}{\partial \mathbf{x}_{r_k}^\tau} \frac{\partial \mathbf{x}_{r_k}^\tau}{\partial \mathbf{x}_0^\tau} &= \mathbf{0}_{n \times n};
\end{aligned}
\tag{5.268}
$$

with $r_k = \mathrm{mod}(k - 1 + s_{k-1}, N), k = 1, 2, \ldots, N$.

In other words,

$$
\begin{aligned}
\mathbf{A} &= (\mathbf{A}_{kl})_{2nN \times 2nN}, \\
\mathbf{y} &= (\mathbf{y}_1, \mathbf{y}_2, \ldots, \mathbf{y}_N)^\mathrm{T}, \quad \mathbf{y}^\tau = (\mathbf{y}_1^\tau, \mathbf{y}_2^\tau, \ldots, \mathbf{y}_N^\tau)^\mathrm{T}, \\
\mathbf{b} &= (\mathbf{b}_1, \mathbf{b}_2, \ldots, \mathbf{b}_N)^\mathrm{T}, \quad \mathbf{b}^\tau = (\mathbf{b}_1^\tau, \mathbf{b}_2^\tau, \ldots, \mathbf{b}_N^\tau)^\mathrm{T}; \\
\mathbf{A}_{kl} &= \sum_{l_k} \mathbf{A}_{kl_k} \delta_l^{l_k} \quad \text{for } l_k = 1, 2, \ldots, N, r_k;
\end{aligned}
$$

$$\mathbf{y}_k = \left[\frac{\partial \mathbf{x}_k}{\partial \mathbf{x}_0}, \frac{\partial \mathbf{x}_k^\tau}{\partial \mathbf{x}_0}\right]^{\mathrm{T}}, \quad \mathbf{y}_k^\tau = \left[\frac{\partial \mathbf{x}_k}{\partial \mathbf{x}_0^\tau}, \frac{\partial \mathbf{x}_k^\tau}{\partial \mathbf{x}_0^\tau}\right]^{\mathrm{T}};$$

$$\mathbf{b}_k = -\sum_{l_k}\left[\frac{\partial \mathbf{g}_k}{\partial \mathbf{x}_0}, \frac{\partial \mathbf{h}_k}{\partial \mathbf{x}_0}\right]^{\mathrm{T}}\delta_0^{l_k}, \quad \mathbf{b}_k^\tau = -\sum_{l_k}\left[\frac{\partial \mathbf{g}_k}{\partial \mathbf{x}_0^\tau}, \frac{\partial \mathbf{h}_k}{\partial \mathbf{x}_0^\tau}\right]^{\mathrm{T}}\delta_0^{l_k};$$

$$\mathbf{A}_{kj} = \begin{bmatrix} \mathbf{a}_{kj} & \mathbf{a}_{kj}^\tau \\ \mathbf{b}_{kj} & \mathbf{b}_{kj}^\tau \end{bmatrix}, \quad \mathbf{A}_{kr_k} = \begin{bmatrix} \mathbf{0}_{n\times n} & \mathbf{0}_{n\times n} \\ \mathbf{0}_{n\times n} & \mathbf{b}_{kr_k}^\tau \end{bmatrix};$$

$$(j = 1, 2, \ldots, N).$$

(5.269)

From the above discussion, the discrete mapping can be developed through many forward and backward nodes. The periodic flow in the time-delay nonlinear dynamical system can be determined through the following theorem.

**Theorem 5.10** *Consider a time-delay nonlinear dynamical system in* Eq. (5.205). *If such a system has a periodic flow* $\mathbf{x}(t)$ *with finite norm* $\|\mathbf{x}\|$ *and period* $T = 2\pi/\Omega$, *there is a set of discrete time* $t_k(k = 0, 1, \ldots, N)$ *with* $(N \to \infty)$ *during one period* $T$, *and the corresponding solutions* $\mathbf{x}(t_k)$ *and* $\mathbf{x}^\tau(t_k) = \mathbf{x}(t_k - \tau)$ *with vector field* $\mathbf{f}(\mathbf{x}(t_k), \mathbf{x}^\tau(t_k), t_k, \mathbf{p})$ *are exact. Suppose discrete nodes* $\mathbf{x}_k$ *and* $\mathbf{x}_k^\tau$ *are on the approximate solution of the periodic flow under* $\|\mathbf{x}(t_k) - \mathbf{x}_k\| \le \varepsilon_k$ *and* $\|\mathbf{x}^\tau(t_k) - \mathbf{x}_k^\tau\| \le \varepsilon_k^\tau$ *with small* $\varepsilon_k, \varepsilon_k^\tau \ge 0$ *and*

$$\|\mathbf{f}(\mathbf{x}(t_k), \mathbf{x}^\tau(t_k), t_k, \mathbf{p}) - \mathbf{f}(\mathbf{x}_k, \mathbf{x}_k^\tau, t_k, \mathbf{p})\| \le \delta_k \tag{5.270}$$

*with a small* $\delta_k \ge 0$. *During a time interval* $t \in [t_{k-1}, t_k]$, *there is a mapping* $P_k:$ $(\mathbf{x}_{k-1}, \mathbf{x}_{k-1}^\tau) \to (\mathbf{x}_k, \mathbf{x}_k^\tau)$ $(k = 1, 2, \ldots, N)$, *i.e.,*

$$(\mathbf{x}_k, \mathbf{x}_k^\tau) = P_k(\mathbf{x}_{k-1}, \mathbf{x}_{k-1}^\tau) \text{ with}$$
$$\mathbf{g}_k(\mathbf{x}_{s_{kl_1}}, \ldots, \mathbf{x}_{s_{k0}}, \ldots, \mathbf{x}_{s_{k(-l_2)}}; \mathbf{x}_{s_{kl_1}}^\tau, \ldots, \mathbf{x}_{s_{k0}}^\tau, \ldots, \mathbf{x}_{s_{k(-l_2)}}^\tau, \mathbf{p}) = \mathbf{0},$$
$$\mathbf{h}_k(\mathbf{x}_{s_{kl_1}}, \ldots, \mathbf{x}_{s_{k0}}, \ldots, \mathbf{x}_{s_{k(-l_2)}}; \mathbf{x}_{s_{kl_1}}^\tau, \ldots, \mathbf{x}_{s_{k0}}^\tau, \ldots, \mathbf{x}_{s_{k(-l_2)}}^\tau, \mathbf{x}_{r_k}^\tau, \mathbf{p}) = \mathbf{0},$$
$$s_{kj} = \mathrm{mod}(k - j + N, N), j = -l_2, -l_2 + 1, \ldots, l_1 - 1, l_1;$$
$$r_k = \mathrm{mod}(k - 1 + s_{k-1}, N); l_1, l_2 \in \{0, 1, 2, \ldots, N\};$$
$$1 \le l_1 + l_2 \le N, l_1 \ge 1; (k = 1, 2, \ldots, N)$$

(5.271)

*where* $\mathbf{g}_k$ *and* $\mathbf{h}_k$ *are implicit vector functions for regular and time-delay nodes, respectively. Consider a mapping structure as*

$$P = P_N \circ P_{N-1} \circ \cdots \circ P_2 \circ P_1: \ (\mathbf{x}_0, \mathbf{x}_0^\tau) \to (\mathbf{x}_N, \mathbf{x}_N^\tau);$$
$$\text{with } P_k: \ (\mathbf{x}_{k-1}, \mathbf{x}_{k-1}^\tau) \to (\mathbf{x}_k, \mathbf{x}_k^\tau)(k = 1, 2, \ldots, N).$$

(5.272)

*For* $(\mathbf{x}_N, \mathbf{x}_N^\tau) = P(\mathbf{x}_0, \mathbf{x}_0^\tau)$, *if there is a set of points* $\mathbf{x}_k^*$ $(k = 0, 1, \ldots, N)$ *given by*

$$\left.\begin{array}{l} \mathbf{g}_k\big(\mathbf{x}^*_{s_{kl_1}},\ldots,\mathbf{x}^*_{s_{k0}},\ldots,\mathbf{x}^*_{s_{k(-l_2)}};\mathbf{x}^{\tau*}_{s_{kl_1}},\ldots,\mathbf{x}^{\tau*}_{s_{k0}},\ldots,\mathbf{x}^{\tau*}_{s_{k(-l_2)}},\mathbf{p}\big)=\mathbf{0}, \\[2mm] \mathbf{h}_k\big(\mathbf{x}^*_{s_{kl_1}},\ldots,\mathbf{x}^*_{s_{k0}},\ldots,\mathbf{x}^*_{s_{k(-l_2)}};\mathbf{x}^{\tau*}_{s_{kl_1}},\ldots,\mathbf{x}^{\tau*}_{s_{k0}},\ldots,\mathbf{x}^{\tau*}_{s_{k(-l_2)}},\mathbf{x}^{\tau*}_{r_k},\mathbf{p}\big)=\mathbf{0}, \end{array}\right\} \tag{5.273}$$

$$s_{kj}=\mathrm{mod}(k-j+N,N),k=1,2,\ldots,N;$$

$$\mathbf{x}^*_0=\mathbf{x}^*_N\quad and\quad \mathbf{x}^{\tau*}_0=\mathbf{x}^{\tau*}_N,$$

*then the points* $\mathbf{x}^*_k$ *and* $\mathbf{x}^{\tau*}_k$ $(k=0,1,\ldots,N)$ *are approximations of points* $\mathbf{x}(t_k)$ *and* $\mathbf{x}^\tau(t_k)$ *of the periodic solution. In the neighborhoods of* $\mathbf{x}^*_k$ *and* $\mathbf{x}^{\tau*}_k$, *with* $\mathbf{x}_k=\mathbf{x}^*_k+\Delta\mathbf{x}_k$ *and* $\mathbf{x}^\tau_k=\mathbf{x}^{\tau*}_k+\Delta\mathbf{x}^\tau_k$, *the linearized equation is given by*

$$\frac{\partial\mathbf{g}_k}{\partial\mathbf{x}_0}+\sum_{j=1}^{N}\frac{\partial\mathbf{g}_k}{\partial\mathbf{x}_j}\frac{\partial\mathbf{x}_j}{\partial\mathbf{x}_0}+\sum_{j=1}^{N}\frac{\partial\mathbf{g}_k}{\partial\mathbf{x}^\tau_j}\frac{\partial\mathbf{x}^\tau_j}{\partial\mathbf{x}_0}=\mathbf{0},$$

$$\frac{\partial\mathbf{h}_k}{\partial\mathbf{x}_0}+\sum_{j=1}^{N}\frac{\partial\mathbf{h}_k}{\partial\mathbf{x}_j}\frac{\partial\mathbf{x}_j}{\partial\mathbf{x}_0}+\sum_{j=1}^{N}\frac{\partial\mathbf{h}_k}{\partial\mathbf{x}^\tau_j}\frac{\partial\mathbf{x}^\tau_j}{\partial\mathbf{x}_0}+\frac{\partial\mathbf{h}_k}{\partial\mathbf{x}^\tau_{r_k}}\frac{\partial\mathbf{x}^\tau_{r_k}}{\partial\mathbf{x}_0}=\mathbf{0};$$

$$\frac{\partial\mathbf{g}_k}{\partial\mathbf{x}^\tau_0}+\sum_{j=1}^{N}\frac{\partial\mathbf{g}_k}{\partial\mathbf{x}_j}\frac{\partial\mathbf{x}_j}{\partial\mathbf{x}^\tau_0}+\sum_{j=1}^{N}\frac{\partial\mathbf{g}_k}{\partial\mathbf{x}^\tau_j}\frac{\partial\mathbf{x}^\tau_j}{\partial\mathbf{x}^\tau_0}=\mathbf{0}, \tag{5.274}$$

$$\frac{\partial\mathbf{h}_k}{\partial\mathbf{x}^\tau_0}+\sum_{j=1}^{N}\frac{\partial\mathbf{h}_k}{\partial\mathbf{x}_j}\frac{\partial\mathbf{x}_j}{\partial\mathbf{x}^\tau_0}+\sum_{j=1}^{N}\frac{\partial\mathbf{h}_k}{\partial\mathbf{x}^\tau_j}\frac{\partial\mathbf{x}^\tau_j}{\partial\mathbf{x}^\tau_0}+\frac{\partial\mathbf{h}_k}{\partial\mathbf{x}^\tau_{r_k}}\frac{\partial\mathbf{x}^\tau_{r_k}}{\partial\mathbf{x}^\tau_0}=\mathbf{0};$$

*with*

$$\frac{\partial\mathbf{g}_k}{\partial\mathbf{x}_\alpha}=\mathbf{0}\quad and\quad \frac{\partial\mathbf{g}_k}{\partial\mathbf{x}^\tau_\alpha}=\mathbf{0}(\alpha\neq s_{kj}),$$

$$\frac{\partial\mathbf{h}_k}{\partial\mathbf{x}_\alpha}=\mathbf{0}\quad and\quad \frac{\partial\mathbf{h}_k}{\partial\mathbf{x}^\tau_\alpha}=\mathbf{0}(\alpha\neq s_{kj}), \tag{5.275}$$

$$k=1,2,\ldots,N;j=-l_2,-l_2+1,\ldots,l_1-1,l_1.$$

*The resultant Jacobian matrices of the periodic flow are*

$$\mathrm{DP}_{k(k-1)\ldots1}=\begin{bmatrix}\dfrac{\partial\mathbf{x}_k}{\partial\mathbf{x}_0}&\dfrac{\partial\mathbf{x}_k}{\partial\mathbf{x}^\tau_0}\\[2mm]\dfrac{\partial\mathbf{x}_k}{\partial\mathbf{x}_0}&\dfrac{\partial\mathbf{x}_k}{\partial\mathbf{x}^\tau_0}\end{bmatrix}_{(\mathbf{x}^*_0,\mathbf{x}^{\tau*}_0,\ldots,\mathbf{x}^*_N,\mathbf{x}^{\tau*}_N)},\quad \mathrm{DP}=\begin{bmatrix}\dfrac{\partial\mathbf{x}_N}{\partial\mathbf{x}_0}&\dfrac{\partial\mathbf{x}_N}{\partial\mathbf{x}^\tau_0}\\[2mm]\dfrac{\partial\mathbf{x}_N}{\partial\mathbf{x}_0}&\dfrac{\partial\mathbf{x}_N}{\partial\mathbf{x}^\tau_0}\end{bmatrix}_{(\mathbf{x}^*_0,\mathbf{x}^{\tau*}_0,\ldots,\mathbf{x}^*_N,\mathbf{x}^{\tau*}_N)}$$

$$(k=1,2,\ldots,N). \tag{5.276}$$

*where*

$$\mathbf{y}=\mathbf{A}^{-1}\mathbf{b}\quad and\quad \mathbf{y}^\tau=\mathbf{A}^{-1}\mathbf{b}^\tau \tag{5.277}$$

*and*

$$\mathbf{A} = (\mathbf{A}_{kl})_{2nN \times 2nN},$$

$$\mathbf{A}_{kl} = \sum_{l_k} \mathbf{A}_{kl_k} \delta_l^{l_k} \quad \text{for } l_k = s_{kl_1}, \dots, s_{k0}, \dots, s_{k(-l_2)}, r_k; l_k \neq 0;$$

$$\mathbf{A}_{kj} = \begin{bmatrix} \mathbf{a}_{kj} & \mathbf{a}_{kj}^\tau \\ \mathbf{b}_{kj} & \mathbf{b}_{kj}^\tau \end{bmatrix}, \quad \mathbf{A}_{kr_k} = \begin{bmatrix} \mathbf{0}_{n \times n} & \mathbf{0}_{n \times n} \\ \mathbf{0}_{n \times n} & \mathbf{b}_{kr_k}^\tau \end{bmatrix},$$

$$\mathbf{a}_{kj} = \left[\frac{\partial \mathbf{g}_k}{\partial \mathbf{x}_j}\right], \quad \mathbf{a}_{kj}^\tau = \left[\frac{\partial \mathbf{g}_k}{\partial \mathbf{x}_j^\tau}\right]; \qquad\qquad\qquad\qquad (5.278)$$

$$\mathbf{b}_{kj} = \left[\frac{\partial \mathbf{h}_k}{\partial \mathbf{x}_j}\right], \quad \mathbf{b}_{kj}^\tau = \left[\frac{\partial \mathbf{h}_k}{\partial \mathbf{x}_j^\tau}\right], \quad \mathbf{b}_{kr_k}^\tau = \left[\frac{\partial \mathbf{h}_k}{\partial \mathbf{x}_{r_k}^\tau}\right]$$

$$(j = s_{kl_1}, \dots, s_{k0}, \dots, s_{k(-l_2)});$$

*and*

$$\mathbf{y} = (\mathbf{y}_1, \mathbf{y}_2, \dots, \mathbf{y}_N)^\mathrm{T}, \quad \mathbf{y}^\tau = (\mathbf{y}_1^\tau, \mathbf{y}_2^\tau, \dots, \mathbf{y}_N^\tau)^\mathrm{T},$$

$$\mathbf{b} = (\mathbf{b}_1, \mathbf{b}_2, \dots, \mathbf{b}_N)^\mathrm{T}, \quad \mathbf{b}^\tau = (\mathbf{b}_1^\tau, \mathbf{b}_2^\tau, \dots, \mathbf{b}_N^\tau)^\mathrm{T};$$

$$\mathbf{b}_k = -\sum_{l_k} \left[\frac{\partial \mathbf{g}_k}{\partial \mathbf{x}_0}, \frac{\partial \mathbf{h}_k}{\partial \mathbf{x}_0}\right]^\mathrm{T} \delta_0^{l_k}, \quad \mathbf{b}_k^\tau = -\sum_{l_k} \left[\frac{\partial \mathbf{g}_k}{\partial \mathbf{x}_0^\tau}, \frac{\partial \mathbf{h}_k}{\partial \mathbf{x}_0^\tau}\right]^\mathrm{T} \delta_0^{l_k};$$

$$\mathbf{y}_k = \left[\frac{\partial \mathbf{x}_k}{\partial \mathbf{x}_0}, \frac{\partial \mathbf{x}_k^\tau}{\partial \mathbf{x}_0}\right]^\mathrm{T}, \quad \mathbf{y}_k^\tau = \left[\frac{\partial \mathbf{x}_k}{\partial \mathbf{x}_0^\tau}, \frac{\partial \mathbf{x}_k^\tau}{\partial \mathbf{x}_0^\tau}\right]^\mathrm{T}, \qquad\qquad (5.279)$$

$$(k = 1, 2, \dots, N).$$

*The properties of discrete points $\mathbf{x}_k$ and $\mathbf{x}_k^\tau$ ($k = 1, 2, \dots, N$) can be estimated by the eigenvalues of $\mathrm{DP}_{k(k-1)\cdots 1}$ as*

$$|\mathrm{DP}_{k(k-1)\cdots 1} - \bar{\lambda}\mathbf{I}_{2n \times 2n}| = 0 \quad (k = 1, 2, \dots, N). \qquad (5.280)$$

*The eigenvalues of DP for such a periodic flow in the time-delay system are determined by*

$$|\mathrm{DP} - \lambda\mathbf{I}_{2n \times 2n}| = 0. \qquad\qquad\qquad (5.281)$$

*Thus, the stability and bifurcation of the periodic flow can be classified by the eigenvalues of $\mathrm{DP}(\mathbf{x}_0^*, \mathbf{x}_0^{\tau*})$ with*

$$([n_1^m, n_1^o] : [n_2^m, n_2^o] : [n_3, \kappa_3] : [n_4, \kappa_4]|n_5 : n_6 : [n_7, l, \kappa_7]). \qquad (5.282)$$

(i)  *If the magnitudes of all eigenvalues of DP are less than one (i.e., $|\lambda_i| < 1$, $i = 1, 2, \dots, 2n$), the approximate periodic solution is stable.*

(ii) *If at least the magnitude of one eigenvalue of DP is greater than one (i.e., $|\lambda_i| > 1$, $i \in \{1, 2, \ldots, 2n\}$), the approximate periodic solution is unstable.*

(iii) *The boundaries between stable and unstable periodic flow with higher-order singularity give bifurcation and stability conditions.*

*Proof* The proof is similar to Theorem 5.9. □

As discussed in the previous section, once the period-doubling bifurcation of the period-1 flow occurs, the period-1 flow will become a new periodic flow under the period $T' = 2T$. Thus, consider a mapping structure of the period-2 flow with $2N$ mappings

$$P^{(2)} = P \circ P = P_{2N} \circ P_{2N-1} \circ \cdots \circ P_2 \circ P_1 : \ (\mathbf{x}_0, \mathbf{x}_0^{\tau}) \to (\mathbf{x}_{2N}, \mathbf{x}_{2N}^{\tau});$$
$$\text{with } P_k : \ (\mathbf{x}_{k-1}, \mathbf{x}_{k-1}^{\tau}) \to (\mathbf{x}_k, \mathbf{x}_k^{\tau})(k = 1, 2, \ldots, 2N). \tag{5.283}$$

For $(\mathbf{x}_{2N}, \mathbf{x}_{2N}^{\tau}) = P^{(2)}(\mathbf{x}_0, \mathbf{x}_0^{\tau})$, points $(\mathbf{x}_k^*, \mathbf{x}_k^{\tau*})$ $(k = 0, 1, \ldots, 2N)$ are computed by

$$\left.\begin{array}{l} \mathbf{g}_k(\mathbf{x}_{k-1}^*, \mathbf{x}_k^*; \mathbf{x}_{k-1}^{\tau*}, \mathbf{x}_k^{\tau*}, \mathbf{p}) = \mathbf{0}, \\ \mathbf{h}_k(\mathbf{x}_{k-1}^*, \mathbf{x}_k^*; \mathbf{x}_{k-1}^{\tau*}, \mathbf{x}_k^{\tau*}, \mathbf{x}_{r_k}^{\tau*}, \mathbf{p}) = \mathbf{0} \end{array}\right\} \ (k = 1, 2, \ldots, 2N)$$
$$r_k = \text{mod}(k - 1 + s_{k-1}, 2N),$$
$$\mathbf{x}_0^* = \mathbf{x}_{2N}^*, \mathbf{x}_0^{\tau*} = \mathbf{x}_{2N}^{\tau*}. \tag{5.284}$$

After period-doubling, the period-1 flow becomes a period-2 flow. The node points increase to $2N$ points during two periods $(2T)$. The node points are determined through the discrete mapping in Eq. (5.283). On the other hand,

$$T' = 2T = \frac{2(2\pi)}{\Omega} = \frac{2\pi}{\omega} \Rightarrow \omega = \frac{\Omega}{2}. \tag{5.285}$$

Similarly, during the period of $T'$, a periodic flow can be described by node points $\mathbf{x}_k$ and $\mathbf{x}_k^{\tau}$ $(k = 1, 2, \ldots, N')$. Due to $T' = 2T$, the period-2 flow can be described by $N' \geq 2N$ nodes. Thus, the corresponding mapping $P_k$ is defined as

$$P_k : \ (\mathbf{x}_{k-1}^{(2)}, \mathbf{x}_{k-1}^{\tau(2)}) \to (\mathbf{x}_k^{(2)}, \mathbf{x}_k^{\tau(2)}) \ (k = 1, 2, \ldots, 2N) \tag{5.286}$$

and

$$\left.\begin{array}{l} \mathbf{g}_k(\mathbf{x}_{k-1}^{(2)*}, \mathbf{x}_k^{(2)*}; \mathbf{x}_{k-1}^{\tau(2)*}, \mathbf{x}_k^{\tau(2)*}, \mathbf{p}) = \mathbf{0}, \\ \mathbf{h}_k(\mathbf{x}_{k-1}^{(2)*}, \mathbf{x}_k^{(2)*}; \mathbf{x}_{k-1}^{\tau(2)*}, \mathbf{x}_k^{\tau(2)*}, \mathbf{x}_{r_k}^{\tau(2)*}, \mathbf{p}) = \mathbf{0} \end{array}\right\} \ (k = 1, 2, \ldots, 2N)$$
$$\mathbf{x}_0^{(2)*} = \mathbf{x}_{2N}^{(2)*}, \mathbf{x}_0^{\tau(2)*} = \mathbf{x}_{2N}^{\tau(2)*}. \tag{5.287}$$

In general, for period $T' = mT$, there is a period-$m$ flow which can be described by $N' \geq mN$. The corresponding mapping $P_k$ is

$$P_k : \ (\mathbf{x}_{k-1}^{(m)}, \mathbf{x}_{k-1}^{\tau(m)}) \longrightarrow (\mathbf{x}_k^{(m)}, \mathbf{x}_k^{\tau(m)}) \quad (k = 1, 2, \ldots, mN) \tag{5.288}$$

and

$$\left.
\begin{aligned}
\mathbf{g}_k(\mathbf{x}_{k-1}^{(m)*}, \mathbf{x}_k^{(m)*}; \mathbf{x}_{k-1}^{\tau(m)*}, \mathbf{x}_k^{\tau(m)*}, \mathbf{p}) &= \mathbf{0}, \\
\mathbf{h}_k(\mathbf{x}_{k-1}^{(m)*}, \mathbf{x}_k^{(m)*}; \mathbf{x}_{k-1}^{\tau(m)*}, \mathbf{x}_k^{\tau(m)*}, \mathbf{x}_{r_k}^{\tau(m)*}, \mathbf{p}) &= \mathbf{0}
\end{aligned}
\right\} \quad (k = 1, 2, \ldots, mN)$$
$$r_k = \mathrm{mod}(k - 1 + s_{k-1}, mN)$$
$$\mathbf{x}_0^{(m)*} = \mathbf{x}_{mN}^{(m)*}, \quad \mathbf{x}_0^{\tau(m)*} = \mathbf{x}_{mN}^{\tau(m)*}. \tag{5.289}$$

From the above discussion, the period-$m$ flow in a time-delay nonlinear system can be described through $mN$ regular nodes and $mN$ time-delay nodes for period $mT$. The method is stated as follows.

**Theorem 5.11** *Consider a time-delay nonlinear system in Eq. (5.205). If such a system has a period-m flow $\mathbf{x}^{(m)}(t)$ with finite norm $||\mathbf{x}^{(m)}||$ and period mT $(T = 2\pi/\Omega)$, there is a set of discrete time $t_k$ $(k = 0, 1, \ldots, mN)$ with $(N \to \infty)$ during m-period mT, and the corresponding solutions $\mathbf{x}^{(m)}(t_k)$ and $\mathbf{x}^{\tau(m)}(t_k)$ with vector field $\mathbf{f}(\mathbf{x}^{(m)}(t_k), \mathbf{x}^{\tau(m)}(t_k), t_k, \mathbf{p})$ are exact. Suppose discrete nodes $\mathbf{x}_k^{(m)}$ and $\mathbf{x}_k^{\tau(m)}$ are on the approximate solution of the periodic flow under $||\mathbf{x}^{(m)}(t_k) - \mathbf{x}_k^{(m)}|| \leq \varepsilon_k$ and $||\mathbf{x}^{\tau(m)}(t_k) - \mathbf{x}_k^{\tau(m)}|| \leq \varepsilon_k^\tau$ with small $\varepsilon_k, \varepsilon_k^\tau \geq 0$ and*

$$||\mathbf{f}(\mathbf{x}^{(m)}(t_k), \mathbf{x}^{\tau(m)}(t_k), t_k, \mathbf{p}) - \mathbf{f}(\mathbf{x}_k^{(m)}, \mathbf{x}_k^{\tau(m)} t_k, \mathbf{p})|| \leq \delta_k \tag{5.290}$$

*with a small $\delta_k \geq 0$. During a time interval $t \in [t_{k-1}, t_k]$, there is a mapping $P_k :$ $(\mathbf{x}_{k-1}^{(m)}, \mathbf{x}_{k-1}^{\tau(m)}) \to (\mathbf{x}_k^{(m)}, \mathbf{x}_k^{\tau(m)})$ $(k = 1, 2, \ldots, mN)$, i.e.,*

$$\begin{aligned}
(\mathbf{x}_k^{(m)}, \mathbf{x}_k^{\tau(m)}) &= P_k(\mathbf{x}_{k-1}^{(m)}, \mathbf{x}_{k-1}^{\tau(m)}) \quad with \\
\mathbf{g}_k(\mathbf{x}_{k-1}^{(m)}, \mathbf{x}_k^{(m)}; \mathbf{x}_{k-1}^{\tau(m)}, \mathbf{x}_k^{\tau(m)}, \mathbf{p}) &= \mathbf{0}, \\
\mathbf{h}_k(\mathbf{x}_{k-1}^{(m)}, \mathbf{x}_k^{(m)}; \mathbf{x}_{k-1}^{\tau(m)}, \mathbf{x}_k^{\tau(m)}, \mathbf{x}_{r_k}^{\tau(m)}, \mathbf{p}) &= \mathbf{0}; \\
r_k = \mathrm{mod}(k + s_k, mN), \quad (k &= 1, 2, \ldots, mN)
\end{aligned} \tag{5.291}$$

*where $\mathbf{g}_k$ and $\mathbf{h}_k$ are implicit vector functions for regular and time-delay nodes, respectively. Consider a mapping structure as*

$$\begin{aligned}
P = P_N \circ P_{N-1} \circ \cdots \circ P_2 \circ P_1 : \ (\mathbf{x}_0^{(m)}, \mathbf{x}_0^{\tau(m)}) &\to (\mathbf{x}_{mN}^{(m)}, \mathbf{x}_{mN}^{\tau(m)}); \\
with \ P_k : \ (\mathbf{x}_{k-1}^{(m)}, \mathbf{x}_{k-1}^{\tau(m)}) &\to (\mathbf{x}_k^{(m)}, \mathbf{x}_k^{\tau(m)})(k = 1, 2, \ldots, mN).
\end{aligned} \tag{5.292}$$

*For* $(\mathbf{x}_{mN}^{(m)}, \mathbf{x}_{mN}^{\tau(m)}) = P(\mathbf{x}_0^{(m)}, \mathbf{x}_0^{\tau(m)})$, *if there is a set of points* $(\mathbf{x}_k^{(m)}, \mathbf{x}_k^{\tau(m)})(k = 0, 1, \ldots, mN)$ *computed by*

$$\mathbf{g}_k(\mathbf{x}_{k-1}^{(m)*}, \mathbf{x}_k^{(m)*}; \mathbf{x}_{k-1}^{\tau(m)*}, \mathbf{x}_k^{\tau(m)*}, \mathbf{p}) = \mathbf{0},$$

$$\mathbf{h}_k(\mathbf{x}_{k-1}^{(m)*}, \mathbf{x}_k^{(m)*}; \mathbf{x}_{k-1}^{\tau(m)*}, \mathbf{x}_k^{\tau(m)*}, \mathbf{x}_{r_k}^{\tau(m)*}, \mathbf{p}) = \mathbf{0}, (k = 1, 2, \ldots, mN); \qquad (5.293)$$

$$\mathbf{x}_0^{(m)*} = \mathbf{x}_{mN}^{(m)*}, \quad \mathbf{x}_0^{\tau(m)*} = \mathbf{x}_{mN}^{\tau(m)*},$$

*then the points* $\mathbf{x}_k^{(m)*}$ *and* $\mathbf{x}_k^{\tau(m)*}$ $(k = 0, 1, \ldots, mN)$ *are approximations of points* $\mathbf{x}^{(m)}(t_k)$ *and* $\mathbf{x}^{\tau(m)}(t_k)$ *of the periodic solution. In the neighborhoods of* $\mathbf{x}_k^{(m)*}$ *and* $\mathbf{x}_k^{\tau(m)*}$, *with* $\mathbf{x}_k^{(m)} = \mathbf{x}_k^{(m)*} + \Delta\mathbf{x}_k^{(m)}$ *and* $\mathbf{x}_k^{\tau(m)} = \mathbf{x}_k^{\tau(m)*} + \Delta\mathbf{x}_k^{\tau(m)}$, *the linearized equation is given by*

$$\frac{\partial \mathbf{g}_k}{\partial \mathbf{x}_{k-1}^{(m)}}\frac{\partial \mathbf{x}_{k-1}^{(m)}}{\partial \mathbf{x}_0^{(m)}} + \frac{\partial \mathbf{g}_k}{\partial \mathbf{x}_k^{(m)}}\frac{\partial \mathbf{x}_k^{(m)}}{\partial \mathbf{x}_0^{(m)}} + \frac{\partial \mathbf{g}_k}{\partial \mathbf{x}_{k-1}^{\tau(m)}}\frac{\partial \mathbf{x}_{k-1}^{\tau(m)}}{\partial \mathbf{x}_0^{(m)}} + \frac{\partial \mathbf{g}_k}{\partial \mathbf{x}_k^{\tau(m)}}\frac{\partial \mathbf{x}_k^{\tau(m)}}{\partial \mathbf{x}_0^{(m)}} = \mathbf{0},$$

$$\frac{\partial \mathbf{g}_k}{\partial \mathbf{x}_{k-1}^{(m)}}\frac{\partial \mathbf{x}_{k-1}^{(m)}}{\partial \mathbf{x}_0^{\tau(m)}} + \frac{\partial \mathbf{g}_k}{\partial \mathbf{x}_k^{(m)}}\frac{\partial \mathbf{x}_k^{(m)}}{\partial \mathbf{x}_0^{\tau(m)}} + \frac{\partial \mathbf{g}_k}{\partial \mathbf{x}_{k-1}^{\tau(m)}}\frac{\partial \mathbf{x}_{k-1}^{\tau(m)}}{\partial \mathbf{x}_0^{\tau(m)}} + \frac{\partial \mathbf{g}_k}{\partial \mathbf{x}_k^{\tau(m)}}\frac{\partial \mathbf{x}_k^{\tau(m)}}{\partial \mathbf{x}_0^{\tau(m)}} = \mathbf{0};$$

$$\frac{\partial \mathbf{h}_k}{\partial \mathbf{x}_{k-1}^{(m)}}\frac{\partial \mathbf{x}_{k-1}^{(m)}}{\partial \mathbf{x}_0^{(m)}} + \frac{\partial \mathbf{h}_k}{\partial \mathbf{x}_k^{(m)}}\frac{\partial \mathbf{x}_k^{(m)}}{\partial \mathbf{x}_0^{(m)}} + \frac{\partial \mathbf{h}_k}{\partial \mathbf{x}_{k-1}^{\tau(m)}}\frac{\partial \mathbf{x}_{k-1}^{\tau(m)}}{\partial \mathbf{x}_0^{(m)}}$$

$$+ \frac{\partial \mathbf{h}_k}{\partial \mathbf{x}_k^{\tau(m)}}\frac{\partial \mathbf{x}_k^{\tau(m)}}{\partial \mathbf{x}_0^{(m)}} + \frac{\partial \mathbf{h}_k}{\partial \mathbf{x}_{r_k}^{\tau(m)}}\frac{\partial \mathbf{x}_{r_k}^{\tau(m)}}{\partial \mathbf{x}_0^{(m)}} = \mathbf{0},$$

$$\frac{\partial \mathbf{h}_k}{\partial \mathbf{x}_{k-1}^{(m)}}\frac{\partial \mathbf{x}_{k-1}^{(m)}}{\partial \mathbf{x}_0^{\tau(m)}} + \frac{\partial \mathbf{h}_k}{\partial \mathbf{x}_k^{(m)}}\frac{\partial \mathbf{x}_k^{(m)}}{\partial \mathbf{x}_0^{\tau(m)}} + \frac{\partial \mathbf{h}_k}{\partial \mathbf{x}_{k-1}^{\tau(m)}}\frac{\partial \mathbf{x}_{k-1}^{\tau(m)}}{\partial \mathbf{x}_0^{\tau(m)}} \qquad (5.294)$$

$$+ \frac{\partial \mathbf{h}_k}{\partial \mathbf{x}_k^{\tau(m)}}\frac{\partial \mathbf{x}_k^{\tau(m)}}{\partial \mathbf{x}_0^{\tau(m)}} + \frac{\partial \mathbf{h}_k}{\partial \mathbf{x}_{r_k}^{\tau(m)}}\frac{\partial \mathbf{x}_{r_k}^{\tau(m)}}{\partial \mathbf{x}_0^{\tau(m)}} = \mathbf{0}$$

*with* $r_k = \mathrm{mod}(k - 1 + s_{k-1}, mN)$ *and* $(k = 1, 2, \ldots, mN)$.

*The resultant Jacobian matrices of the periodic flow are*

$$\mathrm{DP}_{k(k-1)\ldots 1} = \begin{bmatrix} \dfrac{\partial \mathbf{x}_k^{(m)}}{\partial \mathbf{x}_0^{(m)}} & \dfrac{\partial \mathbf{x}_k^{(m)}}{\partial \mathbf{x}_0^{\tau(m)}} \\[2ex] \dfrac{\partial \mathbf{x}_k^{(m)}}{\partial \mathbf{x}_0^{(m)}} & \dfrac{\partial \mathbf{x}_k^{(m)}}{\partial \mathbf{x}_0^{\tau(m)}} \end{bmatrix}_{(\mathbf{x}_0^{(m)*}, \mathbf{x}_0^{\tau(m)*}\ldots\mathbf{x}_{mN}^{(m)*}, \mathbf{x}_{mN}^{\tau(m)*})} \qquad (k = 1, 2, \ldots, mN),$$

$$\mathrm{DP} = \begin{bmatrix} \dfrac{\partial \mathbf{x}_{mN}^{(m)}}{\partial \mathbf{x}_0^{(m)}} & \dfrac{\partial \mathbf{x}_{mN}^{(m)}}{\partial \mathbf{x}_0^{\tau(m)}} \\[2ex] \dfrac{\partial \mathbf{x}_{mN}^{(m)}}{\partial \mathbf{x}_0^{(m)}} & \dfrac{\partial \mathbf{x}_{mN}^{(m)}}{\partial \mathbf{x}_0^{\tau(m)}} \end{bmatrix}_{(\mathbf{x}_0^{(m)*}, \mathbf{x}_0^{\tau(m)*}\ldots\mathbf{x}_{mN}^{(m)*}, \mathbf{x}_{mN}^{\tau(m)*})} \qquad (5.295)$$

*where*

$$\mathbf{y}^{(m)} = (\mathbf{A}^{(m)})^{-1}\mathbf{b}^{(m)} \quad and \quad \mathbf{y}^{\tau(m)} = (\mathbf{A}^{(m)})^{-1}\mathbf{b}^{\tau(m)} \tag{5.296}$$

*and*

$$\mathbf{A}^{(m)} = (\mathbf{A}_{kl}^{(m)})_{2nmN \times 2nmN},$$

$$\mathbf{A}_{kl}^{(m)} = \sum_{l_k} \mathbf{A}_{kl_k}^{(m)} \delta_l^{l_k} \quad for\ l_k = k-1, k, r_k; l_k > 0;$$

$$\mathbf{A}_{kj}^{(m)} = \begin{bmatrix} \mathbf{a}_{kj}^{(m)} & \mathbf{a}_{kj}^{\tau(m)} \\ \mathbf{b}_{kj}^{(m)} & \mathbf{b}_{kj}^{\tau(m)} \end{bmatrix}, \quad \mathbf{A}_{kr_k}^{(m)} = \begin{bmatrix} \mathbf{0}_{n \times n} & \mathbf{0}_{n \times n} \\ \mathbf{0}_{n \times n} & \mathbf{b}_{kr_k}^{\tau(m)} \end{bmatrix};$$

$$\mathbf{a}_{kj}^{(m)} = \left[\frac{\partial \mathbf{g}_k}{\partial \mathbf{x}_j^{(m)}}\right], \quad \mathbf{a}_{kj}^{\tau(m)} = \left[\frac{\partial \mathbf{g}_k}{\partial \mathbf{x}_j^{\tau(m)}}\right], \tag{5.297}$$

$$\mathbf{b}_{kj}^{(m)} = \left[\frac{\partial \mathbf{h}_k}{\partial \mathbf{x}_j^{\tau(m)}}\right], \quad \mathbf{b}_{kj}^{\tau(m)} = \left[\frac{\partial \mathbf{h}_k}{\partial \mathbf{x}_j^{\tau(m)}}\right], \quad \mathbf{b}_{kr_k}^{\tau(m)} = \left[\frac{\partial \mathbf{h}_k}{\partial \mathbf{x}_{r_k}^{\tau(m)}}\right];$$

$$(j = k-1, k) \quad and \quad (k = 1, 2, \ldots, mN)$$

*and*

$$\mathbf{y}^{(m)} = (\mathbf{y}_1^{(m)}, \mathbf{y}_2^{(m)}, \ldots, \mathbf{y}_{mN}^{(m)})^{\mathrm{T}}, \quad \mathbf{y}^{\tau(m)} = (\mathbf{y}_1^{\tau(m)}, \mathbf{y}_2^{\tau(m)}, \ldots, \mathbf{y}_{mN}^{\tau(m)})^{\mathrm{T}},$$

$$\mathbf{b}^{(m)} = (\mathbf{b}_1^{(m)}, \mathbf{b}_2^{(m)}, \ldots, \mathbf{b}_{mN}^{(m)})^{\mathrm{T}}, \quad \mathbf{b}^{\tau(m)} = (\mathbf{b}_1^{\tau(m)}, \mathbf{b}_2^{\tau(m)}, \ldots, \mathbf{b}_{mN}^{\tau(m)})^{\mathrm{T}};$$

$$\mathbf{b}_k^{(m)} = -\sum_{l_k} \left[\frac{\partial \mathbf{g}_k}{\partial \mathbf{x}_0^{(m)}}, \frac{\partial \mathbf{h}_k}{\partial \mathbf{x}_0^{(m)}}\right]^{\mathrm{T}} \delta_0^{l_k}, \quad \mathbf{b}_k^{\tau(m)} = -\sum_{l_k} \left[\frac{\partial \mathbf{g}_k}{\partial \mathbf{x}_0^{\tau(m)}}, \frac{\partial \mathbf{h}_k}{\partial \mathbf{x}_0^{\tau(m)}}\right]^{\mathrm{T}} \delta_0^{l_k} \tag{5.298}$$

$$\mathbf{y}_k^{(m)} = \left[\frac{\partial \mathbf{x}_k^{(m)}}{\partial \mathbf{x}_0^{(m)}}, \frac{\partial \mathbf{x}_k^{\tau(m)}}{\partial \mathbf{x}_0^{(m)}}\right]^{\mathrm{T}}, \quad \mathbf{y}_k^{\tau(m)} = \left[\frac{\partial \mathbf{x}_k^{(m)}}{\partial \mathbf{x}_0^{\tau(m)}}, \frac{\partial \mathbf{x}_k^{\tau(m)}}{\partial \mathbf{x}_0^{\tau(m)}}\right]^{\mathrm{T}}$$

$$(k = 1, 2, \ldots, mN)$$

*The properties of discrete points* $(\mathbf{x}_k^{(m)}, \mathbf{x}_k^{\tau(m)})$ $(k = 1, 2, \ldots, mN)$ *can be estimated by the eigenvalues of* $\mathrm{DP}_{k(k-1)\cdots 1}$ *as*

$$|\mathrm{DP}_{k(k-1)\cdots 1}^{(m)} - \bar{\lambda}\mathbf{I}_{2n \times 2n}| = 0. \tag{5.299}$$

*The eigenvalues of DP for such a periodic flow in the time-delay system are determined by*

$$|\mathrm{DP}^{(m)} - \lambda\mathbf{I}_{2n \times 2n}| = 0. \tag{5.300}$$

*Thus, the stability and bifurcation of the periodic flow can be classified by the eigenvalues of* $\mathrm{DP}^{(m)}(\mathbf{x}_0^{(m)*}, \mathbf{x}_0^{\tau(m)*})$ *with*

$$([n_1^m, n_1^o] : [n_2^m, n_2^o] : [n_3, \kappa_3] : [n_4, \kappa_4] | n_5 : n_6 : [n_7, l, \kappa_7]). \tag{5.301}$$

(i) *If the magnitudes of all eigenvalues of* $\mathrm{DP}^{(m)}$ *are less than one (i.e.,* $|\lambda_i| < 1, i = 1, 2, \ldots, 2n$), *the approximate period-m solution is stable.*

(ii) *If at least the magnitude of one eigenvalue of* $\mathrm{DP}^{(m)}$ *is greater than one (i.e.,* $|\lambda_i| > 1, i \in \{1, 2, \ldots, 2n\}$), *the approximate period-m solution is unstable.*

(iii) *The boundaries between stable and unstable period-m flow with higher-order singularity give bifurcation and stability conditions.*

*Proof* The discrete mapping for the period-$m$ flow for the time-delay nonlinear system can be developed during $t \in [t_{k-1}, t_k]$ as in Theorem 5.9. The proof is similar to Theorem 5.9. □

The discrete mapping for a period-$m$ flow with multiple steps can be developed by using many forward and backward nodes. The period-$m$ flow in the time-delay nonlinear system can be obtained by the following theorem.

**Theorem 5.12** *Consider a time-delay nonlinear dynamical system in Eq. (5.205). If such a system has a period-m flow* $\mathbf{x}^{(m)}(t)$ *with finite norm* $||\mathbf{x}^{(m)}||$ *and period mT* $(T = 2\pi/\Omega)$, *there is a set of discrete time* $t_k$ *(k = 0, 1, ..., mN) with* $(N \to \infty)$ *during m-period mT, and the corresponding solutions* $\mathbf{x}^{(m)}(t_k)$ *and* $\mathbf{x}^{\tau(m)}(t_k)$ *with vector fields* $\mathbf{f}(\mathbf{x}^{(m)}(t_k), \mathbf{x}^{\tau(m)}(t_k), t_k, \mathbf{p})$ *are exact. Suppose discrete nodes* $\mathbf{x}_k^{(m)}$ *and* $\mathbf{x}_k^{\tau(m)}$ *are on the approximate solution of the periodic flow under* $||\mathbf{x}^{(m)}(t_k) - \mathbf{x}_k^{(m)}|| \leq \varepsilon_k$ *and* $||\mathbf{x}^{\tau(m)}(t_k) - \mathbf{x}_k^{\tau(m)}|| \leq \varepsilon_k^{\tau}$ *with small* $\varepsilon_k, \varepsilon_k^{\tau} \geq 0$ *and*

$$||\mathbf{f}(\mathbf{x}^{(m)}(t_k), \mathbf{x}^{\tau(m)}(t_k), t_k, \mathbf{p}) - \mathbf{f}(\mathbf{x}_k^{(m)}, \mathbf{x}_k^{\tau(m)} t_k, \mathbf{p})|| \leq \delta_k \tag{5.302}$$

*with a small* $\delta_k \geq 0$. *During a time interval* $t \in [t_{k-1}, t_k]$, *there is a mapping* $P_k : (\mathbf{x}_{k-1}^{(m)}, \mathbf{x}_{k-1}^{\tau(m)}) \to (\mathbf{x}_k^{(m)}, \mathbf{x}_k^{\tau(m)})$ *(k = 1, 2, ..., mN), i.e.,*

$$
\begin{aligned}
&(\mathbf{x}_k^{(m)}, \mathbf{x}_k^{\tau(m)}) = P_k(\mathbf{x}_{k-1}^{(m)}, \mathbf{x}_{k-1}^{\tau(m)}) \text{ with} \\
&\mathbf{g}_k(\mathbf{x}_{s_{kl_1}}^{(m)}, \ldots, \mathbf{x}_{s_{k0}}^{(m)}, \ldots, \mathbf{x}_{s_{k(-l_2)}}^{(m)}; \mathbf{x}_{s_{kl_1}}^{\tau(m)}, \ldots, \mathbf{x}_{s_{k0}}^{\tau(m)}, \ldots, \mathbf{x}_{s_{k(-l_2)}}^{\tau(m)}, \mathbf{p}) = \mathbf{0}, \\
&\mathbf{h}_k(\mathbf{x}_{s_{kl_1}}^{(m)}, \ldots, \mathbf{x}_{s_{k0}}^{(m)}, \ldots, \mathbf{x}_{s_{k(-l_2)}}^{(m)}; \mathbf{x}_{s_{kl_1}}^{\tau(m)}, \ldots, \mathbf{x}_{s_{k0}}^{\tau(m)}, \ldots, \mathbf{x}_{s_{k(-l_2)}}^{\tau(m)}, \mathbf{x}_{r_k}^{\tau(m)}, \mathbf{p}) = \mathbf{0}, \\
&s_{kj} = \mathrm{mod}(k - j + mN, mN), j = -l_2, -l_2 + 1, \ldots, l_1 - 1, l_1; \\
&r_k = \mathrm{mod}(k - 1 + s_{k-1}, mN); l_1, l_2 \in \{0, 1, 2, \ldots, mN\}; \\
&1 \leq l_1 + l_2 \leq mN, l_1 \geq 1; (k = 1, 2, \ldots, mN)
\end{aligned}
\tag{5.303}
$$

*where* $\mathbf{g}_k$ *and* $\mathbf{h}_k$ *are implicit vector functions for regular and time-delay nodes, respectively. Consider a mapping structure as*

$$P = P_{mN} \circ P_{mN-1} \circ \cdots \circ P_2 \circ P_1 : (\mathbf{x}_0^{(m)}, \mathbf{x}_0^{\tau(m)}) \rightarrow (\mathbf{x}_{mN}^{(m)}, \mathbf{x}_{mN}^{\tau(m)});$$

$$\text{with } P_k : (\mathbf{x}_{k-1}^{(m)}, \mathbf{x}_{k-1}^{\tau(m)}) \rightarrow (\mathbf{x}_k^{(m)}, \mathbf{x}_k^{\tau(m)}) \ (k = 1, 2, \ldots, mN).$$

(5.304)

*For* $(\mathbf{x}_k^{(m)}, \mathbf{x}_k^{\tau(m)}) = P_k(\mathbf{x}_{k-1}^{(m)}, \mathbf{x}_{k-1}^{\tau(m)})$, *if there is a set of points* $(\mathbf{x}_k^{(m)*}, \mathbf{x}_k^{\tau(m)*})$ $(k = 0, 1, \ldots, mN)$ *computed by*

$$\left.\begin{array}{l} \mathbf{g}_k(\mathbf{x}_{s_{kl_1}}^{(m)*}, \ldots, \mathbf{x}_{s_{k0}}^{(m)*}, \ldots, \mathbf{x}_{s_{k(-l_2)}}^{(m)*}; \mathbf{x}_{s_{kl_1}}^{\tau(m)*}, \ldots, \mathbf{x}_{s_{k0}}^{\tau(m)*}, \ldots, \mathbf{x}_{s_{k(-l_2)}}^{\tau(m)*}, \mathbf{p}) = \mathbf{0}, \\[2mm] \mathbf{h}_k(\mathbf{x}_{s_{kl_1}}^{(m)*}, \ldots, \mathbf{x}_{s_{k0}}^{(m)*}, \ldots, \mathbf{x}_{s_{k(-l_2)}}^{(m)*}; \mathbf{x}_{s_{kl_1}}^{\tau(m)*}, \ldots, \mathbf{x}_{s_{k0}}^{\tau(m)*}, \ldots, \mathbf{x}_{s_{k(-l_2)}}^{\tau(m)*}, \mathbf{x}_{r_k}^{\tau(m)*}, \mathbf{p}) = \mathbf{0} \end{array}\right\}$$

$(k = 1, 2, \ldots, mN)$

$$\mathbf{x}_0^{(m)*} = \mathbf{x}_{mN}^{(m)*} \quad and \quad \mathbf{x}_0^{\tau(m)*} = \mathbf{x}_{mN}^{\tau(m)*},$$

(5.305)

*then the points* $\mathbf{x}_k^{(m)*}$ *and* $\mathbf{x}_k^{\tau(m)*}$ $(k = 0, 1, \ldots, mN)$ *are approximations of points* $\mathbf{x}^{(m)}(t_k)$ *and* $\mathbf{x}^{\tau(m)}(t_k)$ *of the periodic solution. In the neighborhoods of* $\mathbf{x}_k^{(m)*}$ *and* $\mathbf{x}_k^{\tau(m)*}$, *with* $\mathbf{x}_k^{(m)} = \mathbf{x}_k^{(m)*} + \Delta\mathbf{x}_k^{(m)}$ *and* $\mathbf{x}_k^{\tau(m)} = \mathbf{x}_k^{\tau(m)*} + \Delta\mathbf{x}_k^{\tau(m)}$, *the linearized equation is given by*

$$\frac{\partial\mathbf{g}_k}{\partial\mathbf{x}_0^{(m)}} + \sum_{j=1}^{N} \frac{\partial\mathbf{g}_k}{\partial\mathbf{x}_j^{(m)}} \frac{\partial\mathbf{x}_j^{(m)}}{\partial\mathbf{x}_0^{(m)}} + \sum_{j=1}^{N} \frac{\partial\mathbf{g}_k}{\partial\mathbf{x}_j^{\tau(m)}} \frac{\partial\mathbf{x}_j^{\tau(m)}}{\partial\mathbf{x}_0^{(m)}} = \mathbf{0},$$

$$\frac{\partial\mathbf{h}_k}{\partial\mathbf{x}_0^{(m)}} + \sum_{j=1}^{N} \frac{\partial\mathbf{h}_k}{\partial\mathbf{x}_j^{(m)}} \frac{\partial\mathbf{x}_j^{(m)}}{\partial\mathbf{x}_0^{(m)}} + \sum_{j=1}^{N} \frac{\partial\mathbf{h}_k}{\partial\mathbf{x}_j^{\tau(m)}} \frac{\partial\mathbf{x}_j^{\tau(m)}}{\partial\mathbf{x}_0^{(m)}} + \frac{\partial\mathbf{h}_k}{\partial\mathbf{x}_{r_k}^{\tau(m)}} \frac{\partial\mathbf{x}_{r_k}^{\tau(m)}}{\partial\mathbf{x}_0^{(m)}} = \mathbf{0};$$

$$\frac{\partial\mathbf{g}_k}{\partial\mathbf{x}_0^{\tau(m)}} + \sum_{j=1}^{N} \frac{\partial\mathbf{g}_k}{\partial\mathbf{x}_j^{(m)}} \frac{\partial\mathbf{x}_j^{(m)}}{\partial\mathbf{x}_0^{\tau(m)}} + \sum_{j=1}^{N} \frac{\partial\mathbf{g}_k}{\partial\mathbf{x}_j^{\tau(m)}} \frac{\partial\mathbf{x}_j^{\tau(m)}}{\partial\mathbf{x}_0^{\tau(m)}} = \mathbf{0},$$

$$\frac{\partial\mathbf{h}_k}{\partial\mathbf{x}_0^{\tau(m)}} + \sum_{j=1}^{N} \frac{\partial\mathbf{h}_k}{\partial\mathbf{x}_j^{(m)}} \frac{\partial\mathbf{x}_j^{(m)}}{\partial\mathbf{x}_0^{\tau(m)}} + \sum_{j=1}^{N} \frac{\partial\mathbf{h}_k}{\partial\mathbf{x}_j^{\tau(m)}} \frac{\partial\mathbf{x}_j^{\tau(m)}}{\partial\mathbf{x}_0^{\tau(m)}} + \frac{\partial\mathbf{h}_k}{\partial\mathbf{x}_{r_k}^{\tau(m)}} \frac{\partial\mathbf{x}_{r_k}^{\tau(m)}}{\partial\mathbf{x}_0^{\tau(m)}} = \mathbf{0};$$

(5.306)

*with*

$$\frac{\partial\mathbf{g}_k}{\partial\mathbf{x}_\alpha^{(m)}} = \mathbf{0} \quad and \quad \frac{\partial\mathbf{g}_k}{\partial\mathbf{x}_\alpha^{\tau(m)}} = \mathbf{0}(\alpha \neq s_{kj}),$$

$$\frac{\partial\mathbf{h}_k}{\partial\mathbf{x}_\alpha^{(m)}} = \mathbf{0} \quad and \quad \frac{\partial\mathbf{h}_k}{\partial\mathbf{x}_\alpha^{\tau(m)}} = \mathbf{0}(\alpha \neq s_{kj}),$$

$$s_{kj} = \mod(k - j + mN, mN), j = -l_2, -l_2 + 1, \ldots, l_1 - 1, l_1;$$

$$r_k = \mod(k - 1 + s_{k-1}, mN).$$

(5.307)

*The resultant Jacobian matrices of the periodic flow are*

$$
\mathrm{DP}_{k(k-1)\cdots1}^{(m)} = \begin{bmatrix} \dfrac{\partial \mathbf{x}_k^{(m)}}{\partial \mathbf{x}_0^{(m)}} & \dfrac{\partial \mathbf{x}_k^{(m)}}{\partial \mathbf{x}_0^{\tau(m)}} \\[2mm] \dfrac{\partial \mathbf{x}_k^{(m)}}{\partial \mathbf{x}_0^{(m)}} & \dfrac{\partial \mathbf{x}_k^{(m)}}{\partial \mathbf{x}_0^{\tau(m)}} \end{bmatrix}_{(\mathbf{x}_0^{(m)*},\mathbf{x}_0^{\tau(m)*},\dots,\mathbf{x}_{mN}^{(m)*},\mathbf{x}_{mN}^{\tau(m)*})} \qquad (k = 1, 2, \dots, mN),
$$

$$
\mathrm{DP}^{(m)} = \begin{bmatrix} \dfrac{\partial \mathbf{x}_{mN}^{(m)}}{\partial \mathbf{x}_0^{(m)}} & \dfrac{\partial \mathbf{x}_{mN}^{(m)}}{\partial \mathbf{x}_0^{\tau(m)}} \\[2mm] \dfrac{\partial \mathbf{x}_{mN}^{(m)}}{\partial \mathbf{x}_0^{(m)}} & \dfrac{\partial \mathbf{x}_{mN}^{(m)}}{\partial \mathbf{x}_0^{\tau(m)}} \end{bmatrix}_{(\mathbf{x}_0^{(m)*},\mathbf{x}_0^{\tau(m)*},\dots,\mathbf{x}_{mN}^{(m)*},\mathbf{x}_{mN}^{\tau(m)*})}
$$

$$(5.308)$$

*where*

$$
\mathbf{y}^{(m)} = (\mathbf{A}^{(m)})^{-1}\mathbf{b}^{(m)} \quad and \quad \mathbf{y}^{\tau(m)} = (\mathbf{A}^{(m)})^{-1}\mathbf{b}^{\tau(m)} \qquad (5.309)
$$

*and*

$$
\mathbf{A}^{(m)} = (\mathbf{A}_{kl}^{(m)})_{2mnN\times2mnN},
$$
$$
\mathbf{A}_{kl}^{(m)} = \sum_{l_k}\mathbf{A}_{kl_k}^{(m)}\delta_l^{l_k} \quad \text{for } l_k = s_{kr}, r_k; l_k \neq 0;
$$

$$
\mathbf{A}_{kj}^{(m)} = \begin{bmatrix} \mathbf{a}_{kj}^{(m)} & \mathbf{a}_{kj}^{\tau(m)} \\[1mm] \mathbf{b}_{kj}^{(m)} & \mathbf{b}_{kj}^{\tau(m)} \end{bmatrix}, \quad \mathbf{A}_{kr_k} = \begin{bmatrix} \mathbf{0}_{n\times n} & \mathbf{0}_{n\times n} \\[1mm] \mathbf{0}_{n\times n} & \mathbf{b}_{kr_k}^{\tau(m)} \end{bmatrix},
$$

$$
\mathbf{a}_{kj}^{(m)} = \left[\dfrac{\partial \mathbf{g}_k}{\partial \mathbf{x}_j^{(m)}}\right], \quad \mathbf{a}_{kj}^{\tau(m)} = \left[\dfrac{\partial \mathbf{g}_k}{\partial \mathbf{x}_j^{\tau(m)}}\right],
$$

$$
\mathbf{b}_{kj}^{(m)} = \left[\dfrac{\partial \mathbf{h}_k}{\partial \mathbf{x}_j^{(m)}}\right], \quad \mathbf{b}_{kj}^{\tau(m)} = \left[\dfrac{\partial \mathbf{h}_k}{\partial \mathbf{x}_j^{\tau(m)}}\right], \quad \mathbf{b}_{kr_k}^{\tau(m)} = \left[\dfrac{\partial \mathbf{h}_k}{\partial \mathbf{x}_{r_k}^{\tau(m)}}\right]
$$

$$
(j = s_{kr}), r = -l_2, -l_2 + 1, \dots, l_1 - 1, l_1;
$$

$$(5.310)$$

*and*

$$
\mathbf{y}^{(m)} = (\mathbf{y}_1^{(m)}, \mathbf{y}_2^{(m)}, \dots, \mathbf{y}_{mN}^{(m)})^{\mathrm{T}}, \quad \mathbf{y}^{\tau(m)} = (\mathbf{y}_1^{\tau(m)}, \mathbf{y}_2^{\tau(m)}, \dots, \mathbf{y}_{mN}^{\tau(m)})^{\mathrm{T}},
$$
$$
\mathbf{b}^{(m)} = (\mathbf{b}_1^{(m)}, \mathbf{b}_2^{(m)}, \dots, \mathbf{b}_{mN}^{(m)})^{\mathrm{T}}, \quad \mathbf{b}^{\tau(m)} = (\mathbf{b}_1^{\tau(m)}, \mathbf{b}_2^{\tau(m)}, \dots, \mathbf{b}_{mN}^{\tau(m)})^{\mathrm{T}};
$$

$$(5.311)$$

$$\mathbf{b}_k^{(m)} = -\sum_{l_k} \left[ \frac{\partial \mathbf{g}_k}{\partial \mathbf{x}_0^{(m)}}, \frac{\partial \mathbf{h}_k}{\partial \mathbf{x}_0^{(m)}} \right]^{\mathrm{T}} \delta_0^{l_k}, \quad \mathbf{b}_k^{\tau(m)} = -\sum_{l_k} \left[ \frac{\partial \mathbf{g}_k}{\partial \mathbf{x}_0^{\tau(m)}}, \frac{\partial \mathbf{h}_k}{\partial \mathbf{x}_0^{\tau(m)}} \right]^{\mathrm{T}} \delta_0^{l_k}$$

$$\mathbf{y}_k^{(m)} = \left[ \frac{\partial \mathbf{x}_k^{(m)}}{\partial \mathbf{x}_0^{(m)}}, \frac{\partial \mathbf{x}_k^{\tau(m)}}{\partial \mathbf{x}_0^{(m)}} \right]^{\mathrm{T}}, \quad \mathbf{y}_k^{\tau(m)} = \left[ \frac{\partial \mathbf{x}_k^{(m)}}{\partial \mathbf{x}_0^{\tau(m)}}, \frac{\partial \mathbf{x}_k^{\tau(m)}}{\partial \mathbf{x}_0^{\tau(m)}} \right]^{\mathrm{T}},$$

$$(k = 1, 2, \ldots, mN).$$

(5.311)

*The properties of discrete points* $\mathbf{x}_k^{(m)}$ *and* $\mathbf{x}_k^{\tau(m)}$ $(k = 1, 2, \ldots, mN)$ *can be estimated by the eigenvalues of* $\mathrm{DP}_{k(k-1)\cdots1}$ *as*

$$|\mathrm{DP}_{k(k-1)\cdots1} - \bar{\lambda} \mathbf{I}_{2n\times2n}| = 0. \tag{5.312}$$

*The eigenvalues of* $\mathrm{DP}$ *for such a periodic flow in the time-delay system are determined by*

$$|\mathrm{DP} - \lambda \mathbf{I}_{2n\times2n}| = 0. \tag{5.313}$$

*Thus, the stability and bifurcation of the periodic flow in the time-delay system can be classified by the eigenvalues of* $\mathrm{DP}(\mathbf{x}_0^{(m)*}, \mathbf{x}_0^{\tau(m)*})$ *with*

$$([n_1^m, n_1^o] : [n_2^m, n_2^o] : [n_3, \kappa_3] : [n_4, \kappa_4] | n_5 : n_6 : [n_7, l, \kappa_7]). \tag{5.314}$$

(i)   *If the magnitudes of all eigenvalues of* $\mathrm{DP}^{(m)}$ *are less than one (i.e.,* $|\lambda_i| < 1, i = 1, 2, \ldots, 2n$), *the approximate periodic solution is stable.*

(ii)  *If at least the magnitude of one eigenvalue of* $\mathrm{DP}^{(m)}$ *is greater than one (i.e.,* $|\lambda_i| > 1, i \in \{1, 2, \ldots, 2n\}$), *the approximate periodic solution is unstable.*

(iii) *The boundaries between stable and unstable periodic flow with higher-order singularity give bifurcation and stability conditions.*

**Proof** The discrete mapping for the period-$m$ flow for the time-delay nonlinear system can be developed during $t \in [t_{k-1}, t_k]$ as in Theorem 5.9. The proof is similar to Theorem 5.9.                                                                           □

## 5.3   Discrete Fourier Series

Consider a nonlinear dynamical system with/without time delay. If such a dynamical system has a period-m flow $\mathbf{x}^{(m)}(t)$ with finite norm $||\mathbf{x}^{(m)}||$ and period $mT$ $(T = 2\pi/\Omega)$, then

$$\mathbf{x}^{(m)}(t + mT) = \mathbf{x}^{(m)}(t). \tag{5.315}$$

From the Fourier series theory of periodic function, a definition is introduced.

**Definition 5.1** *Consider a nonlinear dynamical system with/without time delay, and such a dynamical system has a flow* $\mathbf{x}(t)$ *on the time interval* $t \in (0, T)$. *Assume there are node points* $t_j$ ($j = 0, 1, 2, \ldots, N$) *with* $t_0 = 0$ *and* $t_N = T$. *If* $\mathbf{x}(t_j)$ *is finite* ($j = 0, 1, 2, \ldots, N$) *and* $\mathbf{x}(t)$ *is continuous for* $t \in (t_{i-1}, t_i)$ ($i = 1, 2, \ldots, N$), *such a flow* $\mathbf{x}(t)$ *is called to be piecewise continuous on the time interval* $t \in (0, T)$.

**Definition 5.2** *Consider a nonlinear dynamical system with/without time delay, and such a dynamical system has a period-m flow* $\mathbf{x}^{(m)}(t)$ *with finite norm* $\|\mathbf{x}^{(m)}\|$ *and period* $mT(T = 2\pi/\Omega)$. *If* $\mathbf{x}^{(m)}(t)$ *is a piecewise continuous flow on* $t \in (0, mT)$, *there is the Fourier series* $\mathbf{S}^{(m)}(t) \in \mathscr{R}^n$ *for the period-m flow* $\mathbf{x}^{(m)}(t) \in \mathscr{R}^n$ *as*

$$\mathbf{S}^{(m)}(t) = \mathbf{a}_0^{(m)} + \sum_{j=1}^{\infty} \mathbf{b}_{j/m} \cos(\frac{j}{m}\Omega t) + \mathbf{c}_{j/m} \sin(\frac{j}{m}\Omega t). \tag{5.316}$$

*If* $\mathbf{S}^{(m)}(t) = \mathbf{x}^{(m)}(t)$, *the coefficients* $\mathbf{a}_0^{(m)}, \mathbf{b}_{j/m}, \mathbf{c}_{j/m}$ *in Eq.* (5.316) *are by the Euler's formulas*

$$\mathbf{a}_0^{(m)} = \frac{1}{mT} \int_0^{mT} \mathbf{x}^{(m)}(t) dt,$$

$$\mathbf{b}_{j/m} = \frac{2}{mT} \int_0^{mT} \mathbf{x}^{(m)}(t) \cos(\frac{j}{m}\Omega t) dt \quad (j = 1, 2, \ldots), \tag{5.317}$$

$$\mathbf{c}_{j/m} = \frac{2}{mT} \int_0^{mT} \mathbf{x}^{(m)}(t) \cos(\frac{j}{m}\Omega t) dt \quad (j = 1, 2, \ldots)$$

and

$$\mathbf{a}_0^{(m)} = (a_{01}^{(m)}, a_{02}^{(m)}, \ldots, a_{0n}^{(m)})^{\mathrm{T}} \in \mathscr{R}^n;$$
$$\mathbf{b}_{j/m} = (b_{j/m1}, b_{j/m2} \ldots, b_{j/mn})^{\mathrm{T}} \in \mathscr{R}^n, \tag{5.318}$$
$$\mathbf{c}_{j/m} = (c_{j/m1}, c_{j/m2} \ldots, c_{j/mn})^{\mathrm{T}} \in \mathscr{R}^n.$$

**Theorem 5.13** *Consider a nonlinear dynamical system with/without time delay, and such a dynamical system has a period-m flow* $\mathbf{x}^{(m)}(t)$ *with finite norm* $\|\mathbf{x}^{(m)}\|$ *and period* $mT$ ($T = 2\pi/\Omega$). *If* $D^{(l+1)}\mathbf{x}^{(m)}(t)$ ($l \geq 0$) *is a piecewise continuous flow on* $t \in (0, mT)$ *and has a left-hand derivative and right-hand derivative* $D^{(l+1)}\mathbf{x}^{(m)}(t)$ *with* $\|D^{(l+1)}\mathbf{x}^{(m)}(t)\| < K$ *at each point in such time interval, then the Fourier series* $\mathbf{S}^{(m)}(t) \in \mathscr{R}^n$ *for the period-m flow* $\mathbf{x}^{(m)}(t) \in \mathscr{R}^n$ *is convergent with order l, and* $\mathbf{x}^{(m)}(t)$ *is continuous with the lth-order differentiation. Thus,* $\mathbf{S}^{(m)}(t) = \mathbf{x}^{(m)}(t)$, *i.e.,*

$$\mathbf{x}^{(m)}(t) = \mathbf{a}_0^{(m)} + \sum_{j=1}^{\infty} \mathbf{b}_{j/m} \cos(\frac{j}{m}\Omega t) + \mathbf{c}_{j/m} \sin(\frac{j}{m}\Omega t). \qquad (5.319)$$

If $\mathbf{x}^{(m)}(t)$ is discontinuous at $t = t_i$, then the following equation exists

$$\mathbf{x}^{(m)}(t_i) = \frac{1}{2}[\mathbf{x}^{(m)}(t_i^-) + \mathbf{x}^{(m)}(t_i^+)]. \qquad (5.320)$$

where $\mathbf{x}^{(m)}(t_i^-)$ and $\mathbf{x}^{(m)}(t_i^+)$ are the left-hand and right-hand limits, respectively. Thus, the Fourier series of $\mathbf{x}^{(m)}(t)$ can be expressed as in Eq. (5.319).

*Proof* The proof can be found from Kreyszig (1988). Since the basis of the Fourier series is continuous with infinite derivatives, if the period-m flow of a dynamical system can be expressed by the Fourier series, then the period-m flow should be continuous. Suppose a period-m flow is continuous, expressed by

$$\mathbf{x}^{(m)}(t) = \mathbf{a}_0^{(m)} + \sum_{j=1}^{\infty} \mathbf{b}_{j/m} \cos(\frac{j}{m}\Omega t) + \mathbf{c}_{j/m} \sin(\frac{j}{m}\Omega t).$$

(i)  The forgoing equation is averaged in the time interval $t \in (0, mT)$; thus,

$$\int_0^{mT} \mathbf{x}^{(m)}(t)dt = \int_0^{mT} [\mathbf{a}_0^{(m)} + \sum_{j=1}^{\infty} \mathbf{b}_{j/m} \cos(\frac{j}{m}\Omega t) + \mathbf{c}_{j/m} \sin(\frac{j}{m}\Omega t)]dt$$

$$= \mathbf{a}_0^{(m)}mT + \sum_{j=1}^{\infty} \mathbf{b}_{j/m} \int_0^{mT} \cos(\frac{j}{m}\Omega t)dt + \mathbf{c}_{j/m} \int_0^{mT} \sin(\frac{j}{m}\Omega t)dt$$

$$= \mathbf{a}_0^{(m)}mT.$$

Therefore, we have

$$\mathbf{a}_0^{(m)} = \frac{1}{mT} \int_0^{mT} \mathbf{x}^{(m)}(t)dt.$$

(ii)  Multiplication of $\cos(l\Omega t/m)$ to the Fourier series expression gives

$$\mathbf{x}^{(m)}(t)\cos(\frac{l}{m}\Omega t) = \mathbf{a}_0^{(m)}\cos(\frac{l}{m}\Omega t)$$

$$+ \sum_{j=1}^{\infty} \mathbf{b}_{j/m}\frac{1}{2}[\cos(\frac{j-l}{m}\Omega t) + \cos(\frac{j+l}{m}\Omega t)]$$

$$+ \mathbf{c}_{j/m}\frac{1}{2}[\sin(\frac{j-l}{m}\Omega t) + \sin(\frac{j+l}{m}\Omega t)].$$

The integration of the forgoing equation gives

$$\int_0^{mT} \mathbf{x}^{(m)}(t)\cos(\frac{l}{m}\Omega t)dt = \mathbf{a}_0^{(m)}\int_0^{mT}\cos(\frac{l}{m}\Omega t)dt$$

$$+\sum_{j=1}^{\infty}\{\mathbf{b}_{j/m}\int_0^{mT}\frac{1}{2}[\cos(\frac{j-l}{m}\Omega t)+\cos(\frac{j+l}{m}\Omega t)]dt$$

$$+\mathbf{c}_{j/m}\int_0^{mT}\frac{1}{2}[\sin(\frac{j-l}{m}\Omega t)+\sin(\frac{j+l}{m}\Omega t)]dt\}$$

$$=\frac{mT}{2}\mathbf{b}_{j/m}.$$

If $j\neq l$, all integrals in the right-hand side are zero. For $j=l$, only the integral for the term of $\cos[(j-l)\Omega t/m]$ is not zero, and other integrals are zero. They are based on the orthogonality of the basis of sine and cosine in the Fourier series expansion. So, we have

$$\mathbf{b}_{j/m}=\frac{2}{mT}\int_0^{mT}\mathbf{x}^{(m)}(t)\cos(\frac{j}{m}\Omega t)dt.$$

(iii)  Multiplication of $\sin(l\Omega t/m)$ to the Fourier expression gives

$$\mathbf{x}^{(m)}(t)\sin(\frac{l}{m}\Omega t) = \mathbf{a}_0^{(m)}\sin(\frac{l}{m}\Omega t)$$

$$+\sum_{j=1}^{\infty}\mathbf{b}_{j/m}\frac{1}{2}[\sin(\frac{j+l}{m}\Omega t)-\sin(\frac{j-l}{m}\Omega t)]$$

$$+\mathbf{c}_{j/m}\frac{1}{2}[\cos(\frac{j-l}{m}\Omega t)-\cos(\frac{j+l}{m}\Omega t)].$$

The integration of the forgoing equation gives

$$\int_0^{mT}\mathbf{x}^{(m)}(t)\sin(\frac{l}{m}\Omega t)dt = \mathbf{a}_0^{(m)}\int_0^{mT}\sin(\frac{l}{m}\Omega t)dt$$

$$+\sum_{j=1}^{\infty}\{\mathbf{b}_{j/m}\int_0^{mT}\frac{1}{2}[\sin(\frac{j+l}{m}\Omega t)-\sin(\frac{j-l}{m}\Omega t)]dt$$

$$+ \mathbf{c}_{j/m} \int_0^{mT} \frac{1}{2} [\cos(\frac{j-l}{m}\Omega t) - \cos(\frac{j+l}{m}\Omega t)] dt\}$$

$$= \frac{mT}{2} \mathbf{b}_{j/m}.$$

which is also based on the orthogonality of the basis of sine and cosine in the Fourier series expansion. Thus, we have

$$\mathbf{c}_{j/m} = \frac{2}{mT} \int_0^{mT} \mathbf{x}^{(m)}(t) \sin(\frac{j}{m}\Omega t) dt.$$

In Eq. (5.320), the piecewise flow is enforced to be continuous, which can be expanded by the Fourier series. This theorem is proved. □

*Remarks* (i) The piecewise continuous periodic flow in a dynamical system cannot be expressed to the Fourier series expansion. Such piecewise continuous periodic flow should be investigated through the discontinuous dynamical systems theory (e.g., Filippov 1988; Luo 2009, 2011). (ii) If a periodic flow possesses the $k^{th}$ derivatives that are continuous, then the Fourier series expansion of the periodic flow is convergent with $1/j^k$. The detailed discussion of the Fourier series theory for periodic functions can be referred to Churchill (1941).

**Definition 5.4** *Consider a nonlinear dynamical system with/without time delay, and such a dynamical system has a period-m flow* $\mathbf{x}^{(m)}(t)$ *with finite norm* $\|\mathbf{x}^{(m)}\|$ *and period mT* $(T = 2\pi/\Omega)$. *If* $\mathbf{x}^{(m)}(t)$ *is a continuous flow on* $t \in (0, mT)$, *there is the finite Fourier series,* $\mathbf{T}_M^{(m)}(t) \in \mathcal{R}^n$ *for the period-m flow* $\mathbf{x}^{(m)}(t) \in \mathcal{R}^n$ *as*

$$\mathbf{T}_M^{(m)}(t) = \mathbf{a}_0^{(m)} + \sum_{j=1}^M \mathbf{b}_{j/m} \cos(\frac{j}{m}\Omega t) + \mathbf{c}_{j/m} \sin(\frac{j}{m}\Omega t) \qquad (5.321)$$

*which is called a trigonometric polynomial of order M.*

From discrete mapping structures, the node points of periodic flows are computed. Consider the node points of period-m flows as $\mathbf{x}_k^{(m)} = (x_{1k}^{(m)}, x_{2k}^{(m)}, \ldots, x_{nk}^{(m)})^{\mathrm{T}}$ for $k = 0, 1, 2, \ldots, mN$ in a nonlinear dynamical system. The approximate expression for period-m flow is determined by the Fourier series as

$$\mathbf{x}^{(m)}(t) \approx \mathbf{a}_0^{(m)} + \sum_{j=1}^M \mathbf{b}_{j/m} \cos(\frac{j}{m}\Omega t) + \mathbf{c}_{j/m} \sin(\frac{j}{m}\Omega t). \qquad (5.322)$$

There are $(2M + 1)$ unknown vector coefficients of $\mathbf{a}_0^{(m)}, \mathbf{b}_{j/m}, \mathbf{c}_{j/m}$. To determine such unknowns, at least we have the given nodes $\mathbf{x}_k^{(m)}$ $(k = 0, 1, 2, \ldots, mN)$ with $mN + 1 \geq 2M + 1$. In other words, we have $M \leq mN/2$. The node points $\mathbf{x}_k^{(m)}$ on the period-m flow can be expressed by the finite Fourier series, for $t_k \in [0, mT]$

$$
\mathbf{x}^{(m)}(t_k) \equiv \mathbf{x}_k^{(m)} = \mathbf{a}_0^{(m)} + \sum_{j=1}^{mN/2} \mathbf{b}_{j/m} \cos(\frac{j}{m}\Omega t_k) + \mathbf{c}_{j/m} \sin(\frac{j}{m}\Omega t_k)
$$

$$
= \mathbf{a}_0^{(m)} + \sum_{j=1}^{mN/2} \mathbf{b}_{j/m} \cos(\frac{j}{m}\frac{2k\pi}{N}) + \mathbf{c}_{j/m} \sin(\frac{j}{m}\frac{2k\pi}{N}) \qquad (5.323)
$$

$$
(k = 0, 1, \ldots, mN).
$$

**Theorem 5.14** *Consider a nonlinear dynamical system with/without time delay, and such a dynamical system has a period-m flow $\mathbf{x}^{(m)}(t)$ with finite norm $\|\mathbf{x}^{(m)}\|$ and period mT $(T = 2\pi/\Omega)$. If the node points of period-m flows in a nonlinear dynamical system are $\mathbf{x}_k^{(m)} = (x_{1k}^{(m)}, x_{2k}^{(m)}, \ldots, x_{nk}^{(m)})^{\mathrm{T}}$ for $k = 0, 1, 2, \ldots, mN$ with*

$$
t_k = k\Delta t = \frac{2k\pi}{\Omega N} \quad \text{with} \ \Delta t = \frac{T}{N} = \frac{2\pi}{\Omega N}, \qquad (5.324)
$$

*then, there is a trigonometric polynomial $\mathbf{T}_M^{(m)}(t)$, and $\mathbf{x}^{(m)}(t)$ can be approximated by $\mathbf{T}_M^{(m)}(t)$ under the minimization of $\sum_{k=0}^{mN} [(\mathbf{x}^{(m)}(t_k) - \mathbf{T}_{mN/2}^{(m)}(t_k)]^2$ (i.e., $\mathbf{x}^{(m)}(t) \approx \mathbf{T}_{mN/2}^{(m)}(t))$. That is,*

$$
\mathbf{x}^{(m)}(t) \approx \mathbf{a}_0^{(m)} + \sum_{j=1}^{mN/2} \mathbf{b}_{j/m} \cos(\frac{j}{m}\Omega t) + \mathbf{c}_{j/m} \sin(\frac{j}{m}\Omega t) \qquad (5.325)
$$

*where*

$$
\mathbf{a}_0^{(m)} = \frac{1}{N} \sum_{k=0}^{mN} \mathbf{x}_k^{(m)},
$$

$$
\mathbf{b}_{j/m} = \frac{2}{mN} \sum_{k=0}^{mN} \mathbf{x}_k^{(m)} \cos(k\frac{2j\pi}{mN}), \qquad (5.326)
$$

$$
\mathbf{c}_{j/m} = \frac{2}{mN} \sum_{k=0}^{mN} \mathbf{x}_k^{(m)} \sin(k\frac{2j\pi}{mN})
$$

$$
(j = 1, 2, \ldots, mN/2).
$$

*Proof* Let

$$F = \sum_{k=0}^{mN} \left[ \left( \mathbf{x}^{(m)}(t_k) - \mathbf{T}_{mN/2}^{(m)}(t_k) \right) \right]^2,$$

where

$$\mathbf{T}_{mN/2}^{(m)}(t) = \mathbf{a}_0^{(m)} + \sum_{j=1}^{mN/2} \mathbf{b}_{j/m} \cos(\frac{j}{m}\Omega t) + \mathbf{c}_{j/m} \sin(\frac{j}{m}\Omega t).$$

Taking the derivative of function $F$ gives the following three cases.

(i)  For the constant term $\mathbf{a}_0^{(m)}$, $\partial F/\partial \mathbf{a}_0^{(m)} = \mathbf{0}$ gives

$$\sum_{k=0}^{mN} \left[ \mathbf{x}^{(m)}(t_k) - \mathbf{T}_M^{(m)}(t_k) \right] = \mathbf{0}.$$

Further,

$$\sum_{k=0}^{mN} \mathbf{x}^{(m)}(t_k) - \sum_{k=0}^{mN} \mathbf{a}_0^{(m)} - \sum_{j=1}^{mN/2} \mathbf{b}_{j/m} \sum_{k=0}^{mN} \cos(\frac{j}{m}\Omega t_k)$$

$$- \sum_{j=1}^{mN/2} \mathbf{c}_{j/m} \sum_{k=0}^{mN} \sin(\frac{j}{m}\Omega t_k) = \mathbf{0},$$

where

$$\sum_{k=0}^{mN} \cos(\frac{j}{m}\Omega t_k) \frac{N}{T} \Delta t = \frac{N}{T} \sum_{k=0}^{mN} \cos(\frac{j}{m}\Omega t_k) \Delta t$$

$$\approx \frac{N}{T} \int_0^{mT} \cos(\frac{j}{m}\Omega t) dt = 0,$$

$$\sum_{k=0}^{mN} \sin(\frac{j}{m}\Omega t_k) \frac{N}{T} \Delta t = \frac{N}{T} \sum_{k=0}^{mN} \sin(\frac{j}{m}\Omega t_k) \Delta t$$

$$\approx \frac{N}{T} \int_0^{mT} \sin(\frac{j}{m}\Omega t) dt = 0.$$

Thus,

$$\mathbf{a}_0^{(m)} = \frac{1}{mN} \sum_{k=0}^{mN} \mathbf{x}^{(m)}(t_k).$$

(ii)   For the cosine term $\mathbf{b}_{j/m}, \partial F / \partial \mathbf{b}_{j/m} = \mathbf{0}$ gives

$$\sum_{k=0}^{mN} [\mathbf{x}^{(m)}(t_k) - \mathbf{T}_{mN/2}^{(m)}(t_k)] \cos(\frac{j}{m} \Omega t_k) = \mathbf{0}.$$

Thus,

$$\sum_{k=0}^{mN} \mathbf{x}^{(m)}(t_k) \cos(\frac{j}{m} \Omega t_k) - \mathbf{a}_0^{(m)} \sum_{k=0}^{mN} \cos(\frac{j}{m} \Omega t_k)$$

$$- \sum_{j_1=1}^{mN/2} \mathbf{b}_{j_1/m} \sum_{k=0}^{mN} \frac{1}{2} [\cos(\frac{j_1-j}{m} \Omega t_k) + \cos(\frac{j_1+j}{m} \Omega t_k)]$$

$$- \sum_{j_1=1}^{mN/2} \mathbf{c}_{j_1/m} \sum_{k=0}^{mN} \frac{1}{2} [\sin(\frac{j_1-j}{m} \Omega t_k) + \sin(\frac{j_1+j}{m} \Omega t_k)] = \mathbf{0}.$$

If $j_1 \neq j$, from the previous discussion, we have

$$\sum_{k=0}^{mN} \cos(\frac{j_1-j}{m} \Omega t_k) \approx 0 \quad \text{and} \quad \sum_{k=0}^{mN} \cos(\frac{j_1+j}{m} \Omega t_k) \approx 0,$$

$$\sum_{k=0}^{mN} \sin(\frac{j_1-j}{m} \Omega t_k) \approx 0 \quad \text{and} \quad \sum_{k=0}^{mN} \sin(\frac{j_1+j}{m} \Omega t_k) \approx 0.$$

However, if $j_1 = j$, we have

$$\sum_{k=0}^{mN} \cos(\frac{j_1-j}{m} \Omega t_k) = mN \quad \text{and} \quad \sum_{k=0}^{mN} \cos(\frac{j_1+j}{m} \Omega t_k) \approx 0,$$

$$\sum_{k=0}^{mN} \sin(\frac{j_1-j}{m} \Omega t_k) = 0 \quad \text{and} \quad \sum_{k=0}^{mN} \sin(\frac{j_1+j}{m} \Omega t_k) \approx 0.$$

Thus,

$$\sum_{k=0}^{mN} \mathbf{x}^{(m)}(t_k) \cos(\frac{j}{m} \Omega t_k) - \frac{1}{2} \mathbf{b}_{j/m} mN \approx \mathbf{0}.$$

That is,

$$\mathbf{b}_{j/m} \approx \frac{2}{mN} \sum_{k=0}^{mN} \mathbf{x}^{(m)}(t_k) \cos(\frac{j}{m}\Omega t_k).$$

(iii)  For the sine term $\mathbf{c}_{j/m}$, $\partial F / \partial \mathbf{c}_{j/m} = \mathbf{0}$ gives

$$\sum_{k=0}^{mN} \left[ \mathbf{x}^{(m)}(t_k) - \mathbf{T}^{(m)}_{mN/2}(t_k) \right] \sin(\frac{j}{m}\Omega t_k) = \mathbf{0}.$$

Thus,

$$\sum_{k=0}^{mN} \mathbf{x}^{(m)}(t_k) \cos(\frac{j}{m}\Omega t_k) - \mathbf{a}_0^{(m)} \sum_{k=0}^{mN} \cos(\frac{j}{m}\Omega t_k)$$

$$- \sum_{j_1=1}^{mN/2} \mathbf{b}_{j_1/m} \sum_{k=0}^{mN} \frac{1}{2} \left[ \sin(\frac{j_1+j}{m}\Omega t_k) - \sin(\frac{j_1-j}{m}\Omega t_k) \right]$$

$$- \sum_{j_1=1}^{mN/2} \mathbf{c}_{j_1/m} \sum_{k=0}^{mN} \frac{1}{2} \left[ \cos(\frac{j_1-j}{m}\Omega t_k) - \cos(\frac{j_1+j}{m}\Omega t_k) \right] = \mathbf{0}.$$

Similarly, from the previous discussion, we have

$$\sum_{k=0}^{mN} \mathbf{x}^{(m)}(t_k) \sin(\frac{j}{m}\Omega t_k) - \frac{1}{2}\mathbf{c}_{j/m}mN \approx \mathbf{0}.$$

That is,

$$\mathbf{c}_{j/m} \approx \frac{2}{mN} \sum_{k=0}^{mN} \mathbf{x}^{(m)}(t_k) \sin(\frac{j}{m}\Omega t_k).$$

This theorem is proved.                                                  □

In the above theorem, the coefficients for discrete Fourier series can be computed by direct use of Euler's formulas through the discrete nodes. For a period-m flow $\mathbf{x}^{(m)}(t)$ with finite norm $||\mathbf{x}^{(m)}||$ and period $mT$ $(T = 2\pi/\Omega)$, consider the nodes of period-m flows in a nonlinear system are $\mathbf{x}_k^{(m)} = (x_{1k}^{(m)}, x_{2k}^{(m)}, \ldots, x_{nk}^{(m)})^{\mathrm{T}}$ for $k = 0, 1, 2, \ldots, mN$. The integration in the coefficients of the Fourier series is by the interpolation of the discrete nodes. Let $h = \Delta t = T/N$ where $T = 2\pi/\Omega$ and

$\mathbf{x}^{(m)}(t_0) = \mathbf{x}^{(m)}(t_{mN})$. For simplicity, let $t_0 = 0$. Application of the trapezoidal rules to the Euler's formulas of the Fourier series produces the discrete Euler's formulas.

(i)   The constant term $\mathbf{a}_0^{(m)}$ is discussed as follows.

$$
\begin{aligned}
\mathbf{a}_0^{(m)} &= \frac{1}{mT} \int_0^{mT} \mathbf{x}^{(m)}(t)\mathrm{d}t \\
&= \frac{1}{mT} [\frac{1}{2}\mathbf{x}^{(m)}(t_0) + \mathbf{x}^{(m)}(t_1) + \cdots + \mathbf{x}^{(m)}(t_{mN-1}) + \frac{1}{2}\mathbf{x}^{(m)}(t_{mN})]h \\
&\quad - \frac{h^3}{12mT} \sum_{k=1}^{mN} \frac{\mathrm{d}^2\mathbf{x}^{(m)}(t)}{\mathrm{d}t^2}\Big|_{t=t_k^c}
\end{aligned}
$$

$$(5.327)$$

where $t_k^c \in [t_{k-1}, t_k]$ for $k = 1, 2, \ldots, mN$. Letting $\max_k ||\mathrm{d}^2\mathbf{x}^{(m)}(t)/\mathrm{d}t^2|_{t=t_k^c}|| = L$,

$$
||\mathbf{a}_0^{(m)} - \frac{1}{mN} \sum_{k=0}^{mN} \mathbf{x}^{(m)}(t_k)|| \le \frac{h^2}{12} L. \tag{5.328}
$$

Thus,

$$
\mathbf{a}_0^{(m)} \approx \frac{1}{mN} \sum_{k=0}^{mN} \mathbf{x}^{(m)}(t_k) \approx \frac{1}{mN} \sum_{k=0}^{mN} \mathbf{x}_k^{(m)}. \tag{5.329}
$$

(ii)   The cosine terms coefficients $\mathbf{b}_{j/m}$ $(j = 1, 2, \ldots, mN/2)$ are discussed.

$$
\begin{aligned}
\mathbf{b}_{j/m} &= \frac{2}{mT} \int_0^{mT} \mathbf{x}^{(m)}(t) \cos(\frac{j}{m}\Omega t)\mathrm{d}t \\
&= \frac{2}{mT} [\frac{1}{2}\mathbf{x}^{(m)}(t_0) \cos(\frac{j}{m}\Omega t_0) + \mathbf{x}^{(m)}(t_1) \cos(\frac{j}{m}\Omega t_1) + \cdots \\
&\quad + \mathbf{x}^{(m)}(t_{mN-1}) \cos(\frac{j}{m}\Omega t_{mN-1}) + \frac{1}{2}\mathbf{x}^{(m)}(t_{mN-1}) \cos(\frac{j}{m}\Omega t_{mN})]h \\
&\quad - \frac{h^3}{6mT} \sum_{k=1}^{mN} \frac{\mathrm{d}^2}{\mathrm{d}t^2}[\mathbf{x}^{(m)}(t) \cos(\frac{j}{m}\Omega t)]|_{t=t_k^c}.
\end{aligned}
$$

$$(5.330)$$

From the forgoing equation, we have

$$
||\mathbf{b}_{j/m} - \frac{2}{mN} \sum_{k=0}^{mN} \mathbf{x}^{(m)}(t_k) \cos(\frac{j}{m}\Omega t_k)|| \le \frac{h^2}{6} L_1. \tag{5.331}
$$

where $\max\limits_{k}||d^2[\mathbf{x}^{(m)}(t)\cos(j\Omega t/m)]/dt^2|_{t=t_k^c}|| = L_1$. Thus, the cosine coefficients in discrete Fourier series is

$$\mathbf{b}_{j/m} \approx \frac{2}{mN}\sum_{k=0}^{mN}\mathbf{x}^{(m)}(t_k)\cos(\frac{j}{m}\Omega t_k) \approx \frac{2}{mN}\sum_{k=0}^{mN}\mathbf{x}_k^{(m)}\cos(\frac{j}{m}\Omega t_k). \qquad (5.332)$$

(iii)   The sine terms coefficients $\mathbf{c}_{j/m}$ $(j=1,2,\ldots,mN/2)$ can be discussed similarly. That is,

$$\begin{aligned}
\mathbf{c}_{j/m} &= \frac{2}{mT}\int_{0}^{mT}\mathbf{x}^{(m)}(t)\sin(\frac{j}{m}\Omega t)dt \\
&= \frac{2}{mT}[\frac{1}{2}\mathbf{x}^{(m)}(t_0)\sin(\frac{j}{m}\Omega t_0) + \mathbf{x}^{(m)}(t_1)\sin(\frac{j}{m}\Omega t_1) + \cdots \\
&\quad + \mathbf{x}^{(m)}(t_{mN-1})\sin(\frac{j}{m}\Omega t_{mN-1}) + \frac{1}{2}\mathbf{x}^{(m)}(t_{mN})\sin(\frac{j}{m}\Omega t_{mN})]h \\
&\quad - \frac{h^3}{6mT}\sum_{k=1}^{mN}\frac{d^2}{dt^2}[\mathbf{x}^{(m)}(t)\sin(\frac{j}{m}\Omega t)]|_{t=t_k^c}.
\end{aligned} \qquad (5.333)$$

From the forgoing equation, we have

$$||\mathbf{b}_{j/m} - \frac{2}{mN}\sum_{k=0}^{mN}\mathbf{x}^{(m)}(t_k)\sin(\frac{j}{m}\Omega t_k)|| \leq \frac{h^2}{6}L_2. \qquad (5.334)$$

where $\max\limits_{k}||d^2[\mathbf{x}^{(m)}(t)\sin(j\Omega t/m)]/dt^2|_{t=t_k^c}|| = L_2$. Thus, the cosine coefficients in discrete Fourier series are

$$\mathbf{b}_{j/m} \approx \frac{2}{mN}\sum_{k=0}^{mN}\mathbf{x}^{(m)}(t_k)\cos(\frac{j}{m}\Omega t_k) \approx \frac{2}{mN}\sum_{k=0}^{mN}\mathbf{x}_k^{(m)}\cos(\frac{j}{m}\Omega t_k). \qquad (5.335)$$

In fact, other interpolation can be used to obtain the Euler's formulas, which is not presented.

The harmonic amplitudes and harmonic phases for period-$m$ motion are

$$A_{j/ms} = \sqrt{b_{j/ms}^2 + c_{j/ms}^2}, \quad \varphi_{j/ms} = \arctan\frac{c_{j/ms}}{b_{j/ms}}, \quad (s=1,2,\ldots,n). \qquad (5.336)$$

Thus, the approximation of period-$m$ motion in Eq. (5.325) is given by

$$\mathbf{x}^{(m)}(t) \approx \mathbf{a}_0^{(m)} + \sum_{j=1}^{mN/2}\mathbf{b}_{j/m}\cos(\frac{j}{m}\Omega t) + \mathbf{c}_{j/m}\sin(\frac{j}{m}\Omega t). \qquad (5.337)$$

The forgoing equation can be expressed as

$$x_s^{(m)}(t) = a_{0s}^{(m)} + \sum_{j=1}^{mN/2} A_{j/ms} \cos(\frac{j}{m}\Omega t - \varphi_{j/ms})$$ (5.338)

$$(s = 1, 2, \ldots, n).$$

# References

Churchill, R. N. (1941). *Fourier Series and Boundary Value Problems*. New York: McGraw-Hill.

Filippov, A. F. (1988). *Differential Equations with Discontinuous Righthand Sides*. Dordrecht: Kluwer Academic.

Kreyszig, E. (1988). *Advanced Engineering Mathematics*. New York: John Wiley & Sons.

Luo, A. C. J. (2009). *Discontinuous Dynamical Systems on Time-varying Domains*. Beijing/Heidelberg: Higher Education Press/Springer.

Luo, A. C. J. (2011). *Discontinuous Dynamical Systems*. Beijing/Heidelberg: Higher Education Press/Springer.

Luo, A. C. J. (2012a). *Discrete and Switching Dynamical Systems*. Beijing/Glen Carbon: Higher Education Press/L&H Scientific.

Luo, A. C. J. (2012b). *Regularity and Complexity in Dynamical Systems*. New York: Springer.

Luo, A. C. J. (2014). Periodic flows to chaos based on implicit mappings of nonlinear dynamical systems, *International Journal of Bifurcation and Chaos*, in press.

# Chapter 6
# Periodic Motions to Chaos in Duffing Oscillator

This chapter will present periodic motions in the Duffing oscillator investigated through the mapping structures of discrete implicit maps. The discrete implicit maps will be obtained from the differential equation of the Duffing oscillator. From mapping structures, bifurcation trees of periodic motions in such a nonlinear oscillator will be predicted analytically through nonlinear algebraic equations of implicit maps, and the corresponding stability and bifurcation analysis of periodic motion in the bifurcation trees will be carried out. The bifurcation trees of periodic motions will also be presented through the harmonic amplitudes of the discrete Fourier series. Finally, from the analytical prediction, numerical simulation results of periodic motions will be presented for the verification of the analytical prediction. The harmonic amplitude spectrums will also be presented, and the corresponding analytical expressions of periodic motions can be obtained approximately.

## 6.1 Period-1 Motions

As period-1 motion in Luo and Guo (2015), consider the Duffing oscillator as

$$\ddot{x} + \delta\dot{x} - \alpha x + \beta x^3 = Q_0 \cos \Omega t. \tag{6.1}$$

The state equation of the above equation in state space is

$$\dot{x} = y \quad \text{and} \quad \dot{y} = Q_0 \cos \Omega t - \delta\dot{x} + \alpha x - \beta x^3. \tag{6.2}$$

The differential equation in Eq. (6.1) can be discretized by a midpoint scheme for the time interval $t \in [t_{k-1}, t_k]$ to form a map $P_k$ $(k = 0, 1, 2, \ldots)$ as

$$P_k: (x_{k-1}, y_{k-1}) \to (x_k, y_k) \Rightarrow (x_k, y_k) = P_k(x_{k-1}, y_{k-1}) \tag{6.3}$$

© Higher Education Press, Beijing and Springer-Verlag Berlin Heidelberg 2015
A.C.J. Luo, *Discretization and Implicit Mapping Dynamics*,
Nonlinear Physical Science, DOI 10.1007/978-3-662-47275-0_6

with the implicit relation as

$$x_k = x_{k-1} + \frac{1}{2}h(y_{k-1} + y_k),$$

$$y_k = y_{k-1} + h[Q_0 \cos \Omega(t_{k-1} + \frac{1}{2}h) - \frac{1}{2}\delta(y_{k-1} + y_k) \qquad (6.4)$$

$$+ \frac{1}{2}\alpha(x_{k-1} + x_k) - \frac{1}{8}\beta(x_{k-1} + x_k)^3].$$

For the midpoint scheme, the local error is $O(h^3)$. To predict the periodic solution in such a Duffing oscillator analytically, consider a mapping structure as

$$P = \underbrace{P_N \circ P_{N-1} \circ \ldots \circ P_2 \circ P_1}_{N-\text{actions}} : (x_0, y_0) \rightarrow (x_N, y_N) \qquad (6.5)$$

with

$$\begin{aligned}
&P_1 : (x_0, y_0) \rightarrow (x_1, y_1) \Rightarrow (x_1, y_1) = P_1(x_0, y_0),\\
&P_2 : (x_1, y_1) \rightarrow (x_2, y_2) \Rightarrow (x_2, y_2) = P_2(x_1, y_1),\\
&\quad\vdots \qquad\qquad\qquad\qquad\qquad\qquad\qquad\qquad\qquad (6.6)\\
&P_{N-1} : (x_{N-2}, y_{N-2}) \rightarrow (x_{N-1}, y_{N-1}) \Rightarrow (x_{N-1}, y_{N-1}) = P_{N-1}(x_{N-2}, y_{N-2}),\\
&P_N : (x_{N-1}, y_{N-1}) \rightarrow (x_N, y_N) \Rightarrow (x_N, y_N) = P_N(x_{N-1}, y_{N-1}).
\end{aligned}$$

For $t_k = t_0 + kh$ with given $t_0$ and $h$, from Eq. (6.4), the corresponding algebraic equations are

$$\left.\begin{aligned}
&x_1 = x_0 + \frac{1}{2}h(y_0 + y_1),\\
&y_1 = y_0 + h[Q_0 \cos \Omega(t_0 + \frac{1}{2}h) - \frac{1}{2}\delta(y_0 + y_1)\\
&\qquad + \frac{1}{2}\alpha(x_0 + x_1) - \frac{1}{8}\beta(x_0 + x_1)^3]
\end{aligned}\right\} \quad \text{for } P_1;$$

$$\vdots$$

$$\left.\begin{aligned}
&x_k = x_{k-1} + \frac{1}{2}h(y_{k-1} + y_k),\\
&y_k = y_{k-1} + h[Q_0 \cos \Omega(t_{k-1} + \frac{1}{2}h) - \frac{1}{2}\delta(y_{k-1} + y_k)\\
&\qquad + \frac{1}{2}\alpha(x_{k-1} + x_k) - \frac{1}{8}\beta(x_{k-1} + x_k)^3]
\end{aligned}\right\} \quad \text{for } P_k; \qquad (6.7)$$

$$\vdots$$

$$\left.\begin{aligned}
&x_N = x_{N-1} + \frac{1}{2}h(y_{N-1} + y_N),\\
&y_N = y_{N-1} + h[Q_0 \cos \Omega(t_{N-1} + \frac{1}{2}h) - \frac{1}{2}\delta(y_{N-1} + y_N)\\
&\qquad + \frac{1}{2}\alpha(x_{N-1} + x_N) - \frac{1}{8}\beta(x_{N-1} + x_N)^3]
\end{aligned}\right\} \quad \text{for } P_N.$$

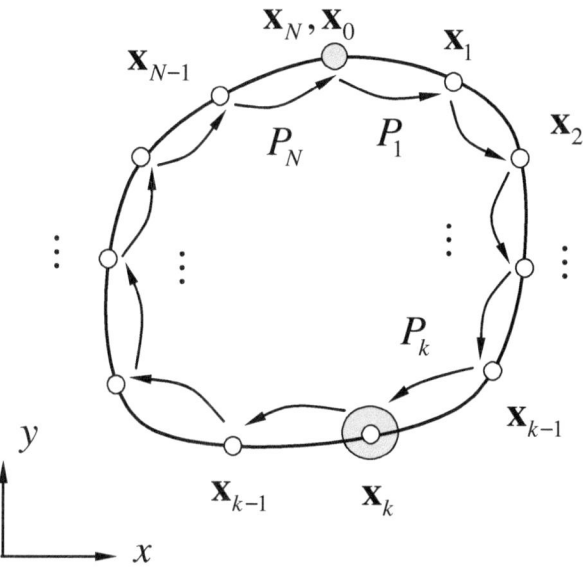

**Fig. 6.1** Period-1 motion with $N$-nodes of the Duffing oscillator. The mapping structures are depicted through single mappings with the *arrowed curves*. The *circular symbols* represent the node points of the period-1 motion

With periodicity conditions, we have

$$(x_N, y_N) = (x_0, y_0). \tag{6.8}$$

From Eqs. (6.7) and (6.8), node values of period-1 motion for the discretized Duffing oscillator can be determined by $2(N + 1)$ equations. Such a periodic solution can be sketched in Fig. 6.1. The node points are depicted by the circular symbols, labeled by $\mathbf{x}_k = (x_k, y_k)^{\mathrm{T}}$ $(k = 0, 1, 2, \ldots, N)$, and the initial and final points are equal for periodicity. The mappings are depicted through the curves with arrows. Once the period-1 motion $\mathbf{x}_k^*$ is obtained, the stability of period-1 motion can be discussed by the corresponding Jacobian matrix. Consider a small perturbation in the neighborhood of $\mathbf{x}_k^*$, $\mathbf{x}_k = \mathbf{x}_k^* + \Delta\mathbf{x}_k$, $(k = 0, 1, 2, \ldots, N)$. For the mapping structure in Eq. (6.5), we have

$$\Delta\mathbf{x}_N = DP\Delta\mathbf{x}_0 = \underbrace{DP_N \cdot DP_{N-1} \cdot \ldots \cdot DP_2 \cdot DP_1}_{N\text{-muplication}} \Delta\mathbf{x}_0. \tag{6.9}$$

with

$$\Delta\mathbf{x}_1 = DP_1\Delta\mathbf{x}_0 \equiv \left[\frac{\partial\mathbf{x}_1}{\partial\mathbf{x}_0}\right]_{(\mathbf{x}_0^*,\mathbf{x}_1^*)} \Delta\mathbf{x}_0$$

$$\Delta \mathbf{x}_2 = DP_2 \Delta \mathbf{x}_1 \equiv \left[ \frac{\partial \mathbf{x}_2}{\partial \mathbf{x}_1} \right]_{(\mathbf{x}_1^*, \mathbf{x}_2^*)} \Delta \mathbf{x}_1,$$

$$\vdots$$

$$\Delta \mathbf{x}_{N-1} = DP_{N-1} \Delta \mathbf{x}_{N-2} \equiv \left[ \frac{\partial \mathbf{x}_{N-1}}{\partial \mathbf{x}_{N-2}} \right]_{(\mathbf{x}_{N-2}^*, \mathbf{x}_{N-1}^*)} \Delta \mathbf{x}_{N-2}, \qquad (6.10)$$

$$\Delta \mathbf{x}_N = DP_N \Delta \mathbf{x}_{N-1} \equiv \left[ \frac{\partial \mathbf{x}_N}{\partial \mathbf{x}_{N-1}} \right]_{(\mathbf{x}_{N-1}^*, \mathbf{x}_N^*)} \Delta \mathbf{x}_{N-1};$$

where

$$DP_k = \left[ \frac{\partial \mathbf{x}_k}{\partial \mathbf{x}_{k-1}} \right]_{(\mathbf{x}_k^*, \mathbf{x}_{k-1}^*)} = \begin{bmatrix} \frac{\partial x_k}{\partial x_{k-1}} & \frac{\partial x_k}{\partial y_{k-1}} \\ \frac{\partial y_k}{\partial x_{k-1}} & \frac{\partial y_k}{\partial y_{k-1}} \end{bmatrix}_{(\mathbf{x}_k^*, \mathbf{x}_{k-1}^*)} \quad \text{for } k = 1, 2, \dots, N \qquad (6.11)$$

and

$$\frac{\partial x_k}{\partial x_{k-1}} = 1 + \frac{1}{2} h \frac{\partial y_k}{\partial x_{k-1}},$$

$$\frac{\partial x_k}{\partial y_{k-1}} = \frac{1}{2} h \left( 1 + \frac{\partial y_k}{\partial x_{k-1}} \right),$$

$$\frac{\partial y_k}{\partial x_{k-1}} = -2 \left( 1 + \frac{1}{2} \delta h + \frac{1}{2} \Delta h \right)^{-1} \Delta, \qquad (6.12)$$

$$\frac{\partial y_k}{\partial y_{k-1}} = \left( 1 + \frac{1}{2} \delta h + \frac{1}{2} \Delta h \right)^{-1} \left( 1 - \frac{1}{2} \delta h - \frac{1}{2} \Delta h \right);$$

with

$$\Delta = \frac{1}{8} h \left[ -4\alpha + 3\beta (x_{k-1} + x_k)^2 \right]. \qquad (6.13)$$

To measure the stability and bifurcation of period-1 motion, the eigenvalues are computed by

$$|DP - \lambda \mathbf{I}| = 0 \qquad (6.14)$$

where

$$DP = \left[ \frac{\partial \mathbf{x}_N}{\partial \mathbf{x}_0} \right]_{(\mathbf{x}_N^*, \mathbf{x}_{N-1}^*, \dots, \mathbf{x}_0^*)} = DP_N \cdot \dots \cdot DP_2 \cdot DP_1 = \prod_{k=N}^{1} \left[ \frac{\partial \mathbf{x}_k}{\partial \mathbf{x}_{k-1}} \right]_{(\mathbf{x}_k^*, \mathbf{x}_{k-1}^*)}. \qquad (6.15)$$

Owing to the two-dimensional mapping, there are two eigenvalues. From Chap. 2, the stability of period-1 motions can be given as follows:

(i) If the magnitudes of two eigenvalues are less than one (i.e., $|\lambda_i| < 1$, $i = 1, 2$), the period-1 motion is stable.

(ii) If one of two eigenvalue magnitudes are greater than one (i.e., $|\lambda_i| > 1$, $i \in \{1, 2\}$), the period-1 motion is unstable.

For the bifurcation conditions, we have the following statements.

(i) If $\lambda_i = 1$, $i \in \{1, 2\}$ and $|\lambda_j| < 1$, $j \in \{1, 2\}$ but $j \neq i$, the saddle-node bifurcation of period-1 motion occurs.

(ii) If $\lambda_i = -1$, $i \in \{1, 2\}$ and $|\lambda_j| < 1$, $j \in \{1, 2\}$ but $j \neq i$, the period-doubling bifurcation of period-1 motion occurs. For the stable period-doubling bifurcation, the period-doubling periodic motion will be observed.

(iii) If $|\lambda_{1,2}| = 1$ with $\lambda_{1,2} = \alpha \pm i\beta$, the Neimark bifurcation of period-1 motion occurs. For the stable Neimark bifurcation, the quasiperiodic motion relative to the period-1 motions will be observed.

To measure the variation characteristics of node point $\mathbf{x}_k$ with the initial condition $\mathbf{x}_0$, we have

$$|DP_{k(k-1)\cdots 1} - \lambda^{(k)}\mathbf{I}| = 0 \tag{6.16}$$

where

$$DP_{k(k-1)\cdots 1} = \left[\frac{\partial \mathbf{x}_k}{\partial \mathbf{x}_0}\right]_{(\mathbf{x}_k^*, \mathbf{x}_{k-1}^*, \ldots, \mathbf{x}_0^*)} = DP_k \cdot \ldots \cdot DP_2 \cdot DP_1 = \prod_{l=k}^{1}\left[\frac{\partial \mathbf{x}_l}{\partial \mathbf{x}_{l-1}}\right]_{(\mathbf{x}_l^*, \mathbf{x}_{l-1}^*)}. \tag{6.17}$$

The dynamics characteristics of $\mathbf{x}_k$ in the neighborhood of $\mathbf{x}_k^*$ varying with the initial condition of $\mathbf{x}_0$ in the neighborhood of $\mathbf{x}_0^*$ can be discussed as follows:

(i) If the magnitudes of two eigenvalues are less than one (i.e., $|\lambda_i^{(k)}| < 1$, $i = 1, 2$), the node point $\mathbf{x}_k$ in the neighborhood of $\mathbf{x}_k^*$ with variation of $\mathbf{x}_0$ will approach to $\mathbf{x}_k^*$ for the period-1 motion.

(ii) If one of two eigenvalue magnitudes are greater than one (i.e., $|\lambda_i^{(k)}| > 1$, $i \in \{1, 2\}$), the node point $\mathbf{x}_k$ in the neighborhood of $\mathbf{x}_k^*$ with variation of $\mathbf{x}_0$ will move away from $\mathbf{x}_k^*$ for the period-1 motion.

## 6.2 Period-m Motions

Once the period-doubling bifurcation of the period-1 motions occurs, the period-2 motions will appear. If the period-doubling bifurcation of the period-2 motion occurs, the period-4 motions will appear and so on. In addition, other periodic motions will exist. In general, to predict the period-m motions in such a Duffing oscillator analytically, consider a mapping structure as follows

$$P = \underbrace{P_{mN} \circ P_{mN-1} \circ \ldots \circ P_2 \circ P_1}_{mN-\text{actions}} : (x_0^{(m)}, y_0^{(m)}) \rightarrow (x_{mN}^{(m)}, y_{mN}^{(m)}) \qquad (6.18)$$

with

$$P_k : (x_{k-1}^{(m)}, y_{k-1}^{(m)}) \rightarrow (x_k^{(m)}, y_k^{(m)}) \Rightarrow (x_k^{(m)}, y_k^{(m)}) = P_k(x_{k-1}^{(m)}, y_{k-1}^{(m)}) \qquad (6.19)$$
$$(k = 1, 2, \ldots, mN).$$

From Eq. (6.4), the corresponding algebraic equations are

$$\left. \begin{array}{l} x_k^{(m)} = x_{k-1}^{(m)} + \frac{1}{2}h(y_{k-1}^{(m)} + y_k^{(m)}), \\ y_k^{(m)} = y_{k-1}^{(m)} + h[Q_0 \cos \Omega(t_{k-1} + \frac{1}{2}h) - \frac{1}{2}\delta(y_{k-1}^{(m)} + y_k^{(m)}) \\ \quad + \frac{1}{2}\alpha(x_{k-1}^{(m)} + x_k^{(m)}) - \frac{1}{8}\beta(x_{k-1}^{(m)} + x_k^{(m)})^3] \end{array} \right\} \text{ for } P_k. \qquad (6.20)$$
$$(k = 1, 2, \ldots, mN)$$

The corresponding periodicity conditions are

$$(x_{mN}^{(m)}, y_{mN}^{(m)}) = (x_0^{(m)}, y_0^{(m)}) \qquad (6.21)$$

From Eqs. (6.20) and (6.21), values of nodes at the discretized Duffing oscillator can be determined by $2(mN + 1)$ equations. Once the node points $\mathbf{x}_k^{(m)*}$ $(k = 1, 2, \ldots, mN)$ of the period-m motion are obtained, the stability of period-m motion can be discussed by the corresponding Jacobian matrix. For a small perturbation in vicinity of $\mathbf{x}_k^{(m)*}$, $\mathbf{x}_k^{(m)} = \mathbf{x}_k^{(m)*} + \Delta\mathbf{x}_k^{(m)}$, $(k = 0, 1, 2, \cdots, mN)$, we have

$$\Delta\mathbf{x}_{mN} = DP\Delta\mathbf{x}_0^{(m)} = \underbrace{DP_{mN} \cdot DP_{mN-1} \cdot \ldots \cdot DP_2 \cdot DP_1}_{mN-\text{muplication}} \Delta\mathbf{x}_0^{(m)} \qquad (6.22)$$

with

$$\Delta\mathbf{x}_k^{(m)} = DP_k\Delta\mathbf{x}_{k-1}^{(m)} \equiv \left[\frac{\partial\mathbf{x}_k^{(m)}}{\partial\mathbf{x}_{k-1}^{(m)}}\right]_{(\mathbf{x}_{k-1}^{(m)}, \mathbf{x}_k^{(m)*})} \Delta\mathbf{x}_{k-1}^{(m)}, \quad (k = 1, 2, \ldots, mN) \qquad (6.23)$$

where

$$DP_k = \left[\frac{\partial\mathbf{x}_k^{(m)}}{\partial\mathbf{x}_{k-1}^{(m)}}\right]_{(\mathbf{x}_{k-1}^{(m)*}, \mathbf{x}_k^{(m)*})} = \left[\begin{array}{cc} \frac{\partial x_k^{(m)}}{\partial x_{k-1}^{(m)}} & \frac{\partial x_k^{(m)}}{\partial y_{k-1}^{(m)}} \\ \frac{\partial y_k^{(m)}}{\partial x_{k-1}^{(m)}} & \frac{\partial y_k^{(m)}}{\partial y_{k-1}^{(m)}} \end{array}\right]_{(\mathbf{x}_{k-1}^{(m)*}, \mathbf{x}_k^{(m)*})} \text{ for } k = 1, 2, \ldots, mN. \quad (6.24)$$

To measure the stability and bifurcation of period-m motion, the eigenvalues are computed by

$$|DP - \lambda \mathbf{I}| = 0 \tag{6.25}$$

where

$$DP = \left[ \frac{\partial \mathbf{x}_{mN}^{(m)}}{\partial \mathbf{x}_0^{(m)}} \right]_{(\mathbf{x}_{mN}^{(m)*}, \mathbf{x}_{mN-1}^{(m)*}, \dots, \mathbf{x}_0^{(m)*})} \tag{6.26}$$

$$= DP_{mN} \cdot \dots \cdot DP_2 \cdot DP_1 = \prod_{k=mN}^{1} \left[ \frac{\partial \mathbf{x}_k^{(m)}}{\partial \mathbf{x}_k^{(m)}} \right]_{(\mathbf{x}_k^{(m)*}, \mathbf{x}_{k-1}^{(m)*})}.$$

Similarly, the stability and bifurcation conditions are the same as for the period-1 motion.

## 6.3 Bifurcation Trees of Periodic Motions

From the foregoing section, the node points of periodic motions for the Duffing oscillator can be computed, and the set of node points of periodic motions with $(N + 1)$ points per period $T = 2\pi/\Omega$ are defined as

$$\sum = \{(x_k, y_k)|t_k = t_0 + kT/N; t_0 = 0; T = 2\pi/\Omega; k = 0, 1, 2, \dots\}. \tag{6.27}$$

The periodicity of period-$m$ motion is $(x_k, y_k) = (x_{k+mN}, y_{k+mN})$. From all analytical prediction of the node points of periodic motion, the FFT can provide the harmonic amplitudes and phases, which will be presented in this section. To avoid presenting all node points of periodic motions, the node points relative to the initial condition point for each period are collected in the Poincare mapping section for period-$m$ motions ($m = 1, 2, \dots$), as defined by

$$\sum_m = \left\{ (x_{\mathrm{mod}(k,N)}, y_{\mathrm{mod}(k,N)}) \middle| \begin{array}{l} t_k = t_0 + kT/N; t_0 = 0; \\ T = 2\pi/\Omega; k = 0, 1, 2, \dots \end{array} \right\} \tag{6.28}$$

which will be used to present periodic motions.

In this section, analytical predictions of both the bifurcation trees of period-1 motions to chaos and period-3 motions to chaos in the Duffing oscillator will be presented, and the corresponding stability and bifurcation analysis will be completed through the eigenvlaue analysis of discrete mapping structures of periodic motions. Consider system parameters

$$\delta = 1.0, \quad \alpha = 5.5, \quad \beta = 20.0, \quad Q_0 = 10.0. \tag{6.29}$$

For a global of view, analytical predictions of the periodic motion in the Duffing oscillator are illustrated in Fig. 6.2. Analytical predictions provide a complete view of the stable and unstable periodic motions. The eigenvalue analysis gives the bifurcation and stability of the periodic motions in the Duffing oscillator. In Fig. 6.4, the prediction of complete bifurcation trees of the period-1 motion to chaos is presented through the period-1 to period-4 motions. In addition, the bifurcation tree of period-3 motion is included to show the coexisting periodic motions. The solid and dashed curves depict the stable and unstable motions, respectively. The solution pairs of asymmetric motions are presented with black and red colors, respectively. The symbols "SN" and "PD" represent the saddle-node and period-doubling bifurcations, respectively. The prediction of the displacement $x_{\mathrm{mod}(k,N)}$ and velocity $y_{\mathrm{mod}(k,N)}$ of the periodic nodes varying with excitation frequency $\Omega$ is presented in Fig. 6.2a, b, respectively. The symmetric and asymmetric periodic motions are labeled by "S" and "A," respectively. The period-1, period-2, period-4, and period-3 motions are labeled by P-1, P-2, P-4, and P-3, respectively. The asymmetric period-1 motions appear from the saddle-node bifurcations of the symmetric period-1 motions. The period-2 motions appear from the period-doubling bifurcations of the asymmetric period-1 motions, and the period-4 motion appear from the period-doubling bifurcation of the period-2 motion. Such period-2 and period-4 motions are asymmetric. The period-3 motions possess the symmetric and asymmetric motions. The asymmetric period-3 motions appears from the symmetric period-3 motion The real part, imaginary part, and magnitudes of eigenvalues for all periodic motions are also illustrated in Fig. 6.2c–e, respectively. In Fig. 6.2c, the saddle-node bifurcations are given by $\lambda_i = 1$ and $|\lambda_j| < 1$ $(i,j \in \{1,2\}$ but $j \neq i)$, and the period-doubling bifurcations are given by $\lambda_i = -1$ and $|\lambda_j| < 1 (i,j \in \{1,2\}$ but $j \neq i)$. For unstable periodic motions, one of the two eigenvalues experiences $|\lambda_i| > 1$ $(i \in \{1,2\})$. For the bifurcation trees of period-1 to period-4 motion, the frequency range lies in $\Omega \in (0, \infty)$. However, the period-3 motions lie in $\Omega \in (1.5, 1.8)$ and $(4.0, 8.0)$.

To make clear illustrations, the bifurcation trees of the period-1 to period-4 motions are presented in Fig. 6.3a–d. The symmetric period-1 motion exist for $\Omega \in (0, \infty)$. The unstable symmetric period-1 motions are in the range of $\Omega \in (1.016, 1.23), (1.50, 2.63)$, and $(4.528, \infty)$, pertaining to the asymmetric period-1 motions. In addition, there are two segments of unstable symmetric period-1 motions, associated with multiple coexisting solutions with jumping phenomena, and the corresponding frequency ranges are $\Omega \in (1.46, 1.513)$ and $(3.96, 5.98)$. The corresponding saddle-node bifurcations for jumping phenomena in the multiple solutions ranges are $\Omega \approx 1.46, 1.513, 3.96, 5.98$. The asymmetric period-1 motions are generated from the symmetric period-1 motions with saddle-node bifurcations. The bifurcation points are at $\Omega \approx 1.016, 1.23, 1.50, 2.63, 4.528, 6.73$. A pair of two asymmetric period-1 motions will be produced, and the two asymmetric period-1 motions are in $\Omega \in (1.016, 1.23)$ for the first branch, $(1.50, 2.63)$ for the second

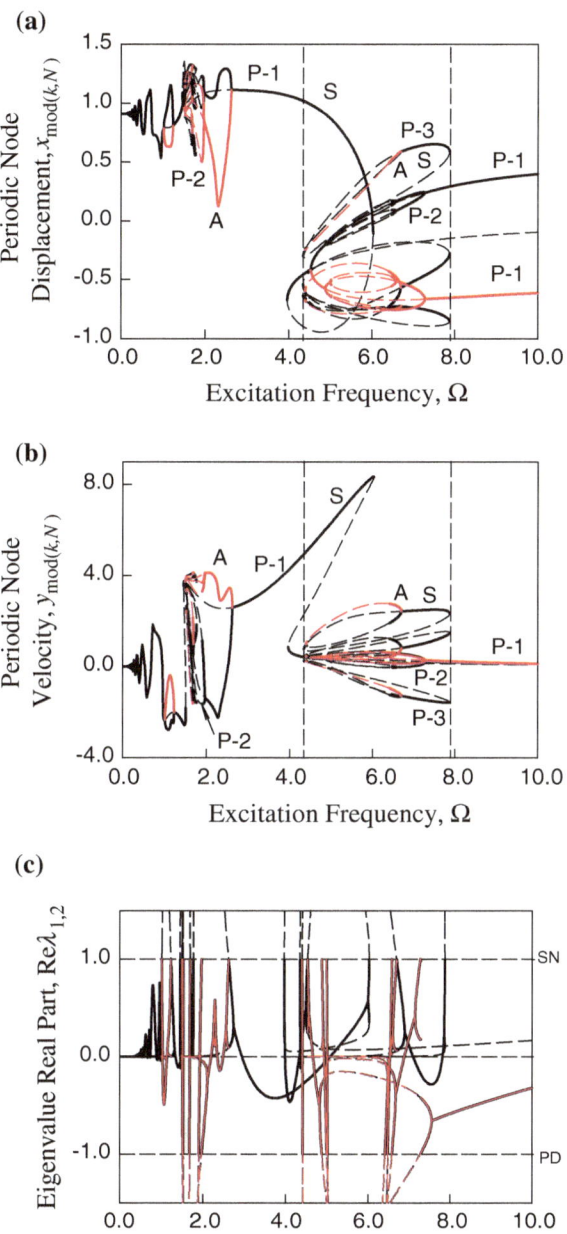

**Fig. 6.2** A global view of the analytical prediction of bifurcation trees of period-1 and period-3 motions to chaos varying with excitation frequency $\Omega$: **a** periodic node displacement $x_{\mathrm{mod}(k,N)}$; **b** periodic node velocity $y_{\mathrm{mod}(k,N)}$; **c** real part of eigenvalues; **d** imaginary part of eigenvalues; and **e** magnitude of eigenvalues ($\alpha = 5.5, \beta = 20.0, \delta = 1.0, Q_0 = 10$) $\mathrm{mod}(k,N) = 0$

**(d)**

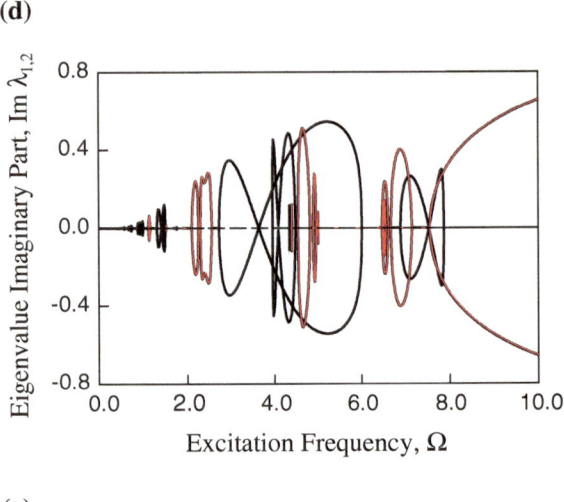

**(e)**

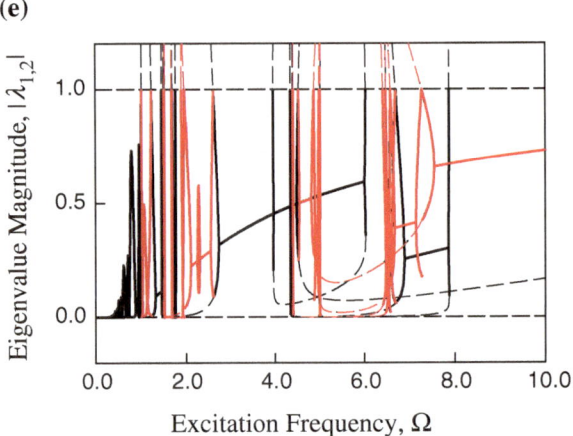

**Fig. 6.2** (continued)

branch, and $(4.528, \infty)$ for the third branch. The stable asymmetric period-1 motions are in the ranges of $\Omega \in (1.016, 1.23)$ for the first branch, $\Omega \in (1.50, 1.517)$ and $(1.97, 2.63)$ for the second branch, and $(4.528, 4.88)$ and $(7.27, \infty)$ for the third branch. The unstable asymmetric period-1 motions are in the range of $\Omega \in (1.517, 1.97)$ for the second branch, and $\Omega \in (4.88, 7.27)$ for the third branch. From the two asymmetric period-1 motions, the period-2 motions will be generated through the period-doubling bifurcation. The period-doubling bifurcation points of the asymmetric period-1 motions are at $\Omega \approx 1.517, 1.97, 4.528, 7.27$ which are also the saddle-node bifurcations of the period-2 motion. The period-2 motions exist in the range of $\Omega \in (1.517, 1.97)$ for the second branch, and $\Omega \in (4.88, 7.27)$ for the third branch. The stable period-2 motions are in $\Omega \in$

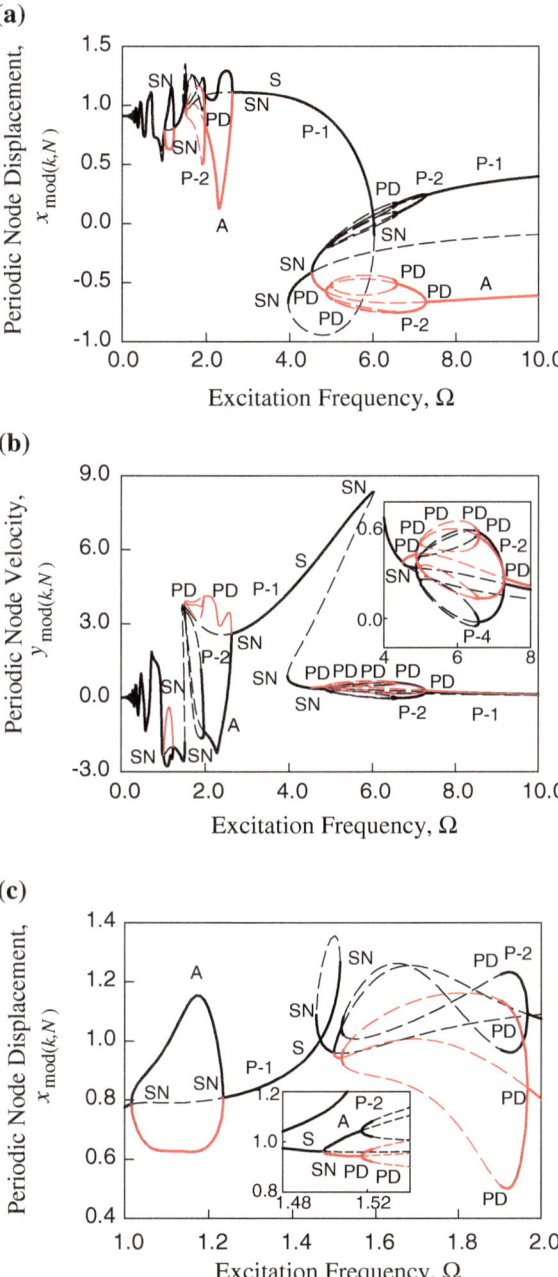

**Fig. 6.3** Analytical prediction of bifurcation trees of period-1 motions to chaos: **a** periodic node displacement $x_{\text{mod}(k,N)}$ and **b** periodic node velocity $y_{\text{mod}(k,N)}$; A zoomed view: **c** periodic node displacement $x_{\text{mod}(k,N)}$ and **d** periodic node velocity $y_{\text{mod}(k,N)}$. ($\alpha = 5.5, \beta = 20.0, \delta = 1.0, Q_0 = 10$). $\text{mod}(k, N) = 0$

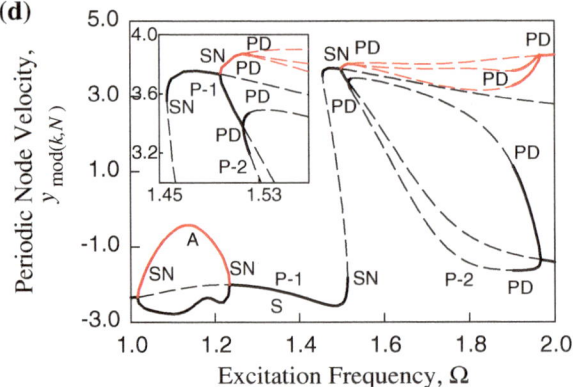

**Fig. 6.3** (continued)

$(1.517, 1.521)$ and $(1.90,\ 1.97)$ for the second branch, and $\Omega \in (4.88, 4.97)$ and $(6.58, 7.27)$ for the third branch. The unstable period-2 motions are in $\Omega \in (1.52, 1.90)$ for the second branch and $\Omega \in (4.97, 6.58)$ for the third branch. The period-doubling bifurcations of period-2 motions are $\Omega \approx 1.52, 1.90$ for the second branch and $\Omega \approx 4.97, 6.58$ for the third branch, and they are the saddle-node bifurcations for the period-4 motions. The period-4 motions are in the range of $\Omega \in (1.52, 1.90)$ for the second branch and $\Omega \in (4.97, 6.58)$ for the third branch. For the third branch, the stable period-4 motions are in $\Omega \in (4.97, 5.03)$ and $\Omega \in (6.49, 6.58)$, and the unstable period-4 motions are in $\Omega \in (5.03, 6.49)$. The period-doubling bifurcations of period-4 motion in the third branch are at $\Omega \approx 5.03, 6.49$, which is the saddle-node bifurcation for period-8 motion. Thus, the period-8 motions exist for $\Omega \in (5.03, 6.49)$. Continuously, we can obtain period-16 motions to chaos. Because the stable motions for period-8 or higher order periodic motions exist for the short range of excitation frequency, the bifurcation tree of period-1 motion to chaos will not be computed anymore further.

To clearly illustrate the bifurcation trees of period-3 motion to chaos, the symmetric and asymmetric period-3 motions are presented in Fig. 6.4. The period-3 motions have two branches. The symmetric period-3 motions are in $\Omega \in (1.523, 1.772)$ for the first branch and $\Omega \in (4.30, 7.89)$ for the second branch. The stable symmetric period-3 motions are in the ranges of $\Omega \in (1.523, 1.526)$ and $(1.695,\ 1.772)$ for the first branch, and $\Omega \in (4.30, 4.39)$ and $\Omega \in (6.69, 7.89)$ for the second branch. The unstable symmetric period-3 motions are in $\Omega \in (1.526, 1.695)$ for the first branch and $\Omega \in (4.39, 6.69)$ for the second branch, which are also for the asymmetric period-3 motions. The four saddle-node bifurcations are at $\Omega \approx 1.523, 1.772, 4.30, 7.89$ for the stable and unstable symmetric period-3 motions, which will not generate the asymmetric period-3 motions. The other four saddle-node bifurcations of the symmetric period-3 motions at $\Omega \approx 1.526, 1.695, 4.39, 6.69$ are not only for the stable and unstable symmetric period-3 motions but also for appearance of the asymmetric period-3 motions. The stable

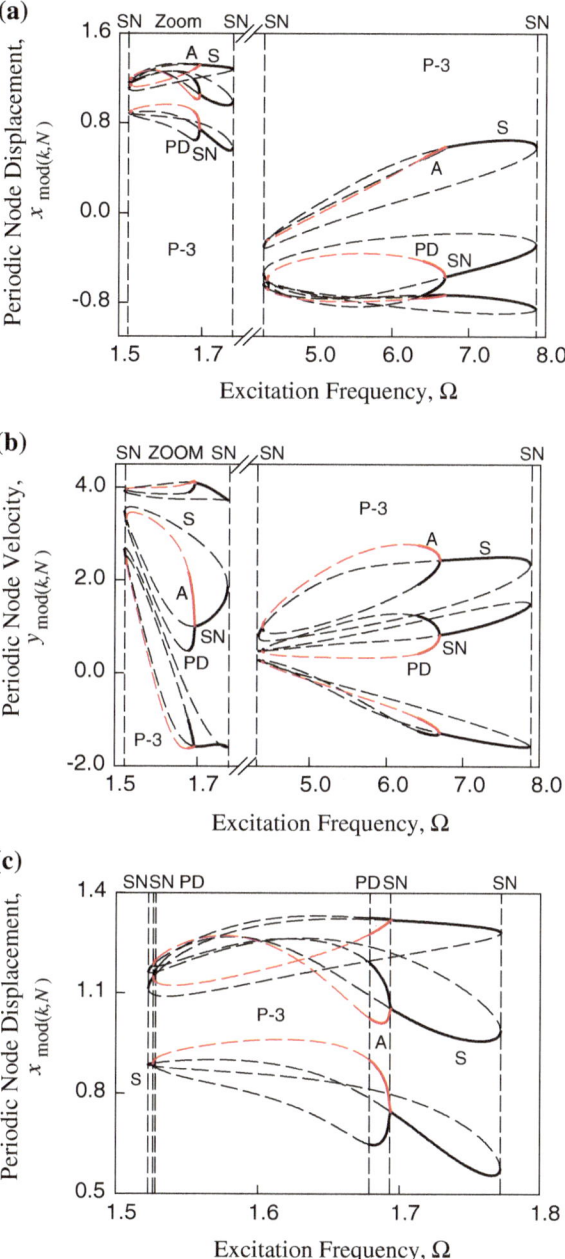

**Fig. 6.4** Analytical prediction of bifurcation trees of period-3 motions to chaos: **a** periodic node displacement $x_{\mathrm{mod}(k,N)}$ and **b** periodic node velocity $y_{\mathrm{mod}(k,N)}$; a zoomed view: **c** periodic node displacement $x_{\mathrm{mod}(k,N)}$ and **d** periodic node velocity $y_{\mathrm{mod}(k,N)}$. ($\alpha = 5.5, \beta = 20.0, \delta = 1.0,$ $Q_0 = 10$). $\mathrm{mod}(k, N) = 0$

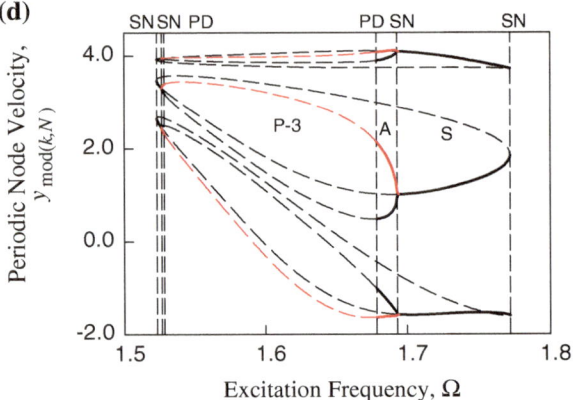

**Fig. 6.4** (continued)

asymmetric period-3 motions are in the ranges of $\Omega \in (1.526, 1.528)$ and $(1.678, 1.695)$ for the first branch, and $\Omega \in (4.39, 4.417)$ and $\Omega \in (6.414, 6.69)$ for the second branch. The unstable asymmetric period-3 motions are in $\Omega \in (1.528, 1.678)$ for the first branch and $\Omega \in (4.417, 6.414)$ for the second branch, which are also for the asymmetric period-3 motions, which are also for the period-6 motions. The period-doubling bifurcations of asymmetric period-3 motions are at $\Omega \approx 1.528, 1.678$ for the first branch and $\Omega \approx 4.417, 6.414$ for the second branch, which are also the saddle-node bifurcation for period-6 motions.

## 6.4 Frequency–Amplitude Characteristics

From discrete mapping structures, the node points of periodic motions are computed. Consider the node points of period-$m$ motions as $\mathbf{x}_k^{(m)} = (x_k^{(m)}, y_k^{(m)})^{\mathrm{T}}$ for $k = 0, 1, 2, \ldots, mN$ in the Duffing oscillator. The approximate expression for period-$m$ motion is determined by the discrete Fourier series as

$$\mathbf{x}^{(m)}(t) \approx \mathbf{a}_0^{(m)} + \sum_{j=1}^{M} \mathbf{b}_{j/m} \cos\left(\frac{j}{m}\Omega t\right) + \mathbf{c}_{j/m} \sin\left(\frac{j}{m}\Omega t\right). \tag{6.30}$$

There are $(2M + 1)$ unknown vector coefficients of $\mathbf{a}_0^{(m)}, \mathbf{b}_{j/m}, \mathbf{c}_{j/m}$. To determine such unknowns, at least we have the given nodes $\mathbf{x}_k^{(m)}$ ($k = 0, 1, 2, \ldots, mN$) with $mN + 1 \geq 2M + 1$. In other words, we have $M \leq mN/2$. The node points $\mathbf{x}_k^{(m)}$ on the period-$m$ motion can be expressed by the discrete Fourier series, as for $t_k \in [0, mT]$

$$\mathbf{x}^{(m)}(t_k) \equiv \mathbf{x}_k^{(m)} = \mathbf{a}_0^{(m)} + \sum_{j=1}^{mN/2} \mathbf{b}_{j/m} \cos\left(\frac{j}{m}\Omega t_k\right) + \mathbf{c}_{j/m} \sin\left(\frac{j}{m}\Omega t_k\right)$$

$$= \mathbf{a}_0^{(m)} + \sum_{j=1}^{mN/2} \mathbf{b}_{j/m} \cos\left(\frac{j}{m}\frac{2k\pi}{N}\right) + \mathbf{c}_{j/m} \sin\left(\frac{j}{m}\frac{2k\pi}{N}\right) \quad (6.31)$$

$$(k = 0, 1, \ldots, mN - 1)$$

where

$$T = \frac{2\pi}{\Omega} = N\Delta t; \quad \Omega t_k = \Omega k \Delta t = \frac{2k\pi}{N};$$

$$\mathbf{a}_0^{(m)} = \frac{1}{N} \sum_{k=0}^{mN} \mathbf{x}_k^{(m)},$$

$$\left. \begin{array}{l} \mathbf{b}_{j/m} = \dfrac{2}{mN} \displaystyle\sum_{k=1}^{mN} \mathbf{x}_k^{(m)} \cos(k\dfrac{2j\pi}{mN}), \\[3mm] \mathbf{c}_{j/m} = \dfrac{2}{mN} \displaystyle\sum_{k=1}^{mN} \mathbf{x}_k^{(m)} \sin(k\dfrac{2j\pi}{mN}) \end{array} \right\} \quad (j = 1, 2, \ldots, mN/2) \quad (6.32)$$

and

$$\mathbf{a}_0^{(m)} = \left(a_{01}^{(m)}, a_{02}^{(m)}\right)^{\mathrm{T}}, \quad \mathbf{b}_{j/m} = \left(b_{j/m1}, b_{j/m2}\right)^{\mathrm{T}}, \quad \mathbf{c}_{j/m} = \left(c_{j/m1}, c_{j/m2}\right)^{\mathrm{T}}. \quad (6.33)$$

The harmonic amplitudes and harmonic phases for period-$m$ motion are

$$\begin{array}{ll} A_{j/m1} = \sqrt{b_{j/m1}^2 + c_{j/m1}^2}, & \varphi_{j/m1} = \arctan\dfrac{c_{j/m1}}{b_{j/m1}}, \\[3mm] A_{j/m2} = \sqrt{b_{j/m2}^2 + c_{j/m2}^2}, & \varphi_{j/m2} = \arctan\dfrac{c_{j/m2}}{b_{j/m2}}. \end{array} \quad (6.34)$$

Thus, the approximate expression for period-m motion in Eq. (6.30) is determined by

$$\mathbf{x}^{(m)}(t) \approx \mathbf{a}_0^{(m)} + \sum_{j=1}^{mN/2} \mathbf{b}_{j/m} \cos\left(\frac{j}{m}\Omega t\right) + \mathbf{c}_{j/m} \sin\left(\frac{j}{m}\Omega t\right). \quad (6.35)$$

The foregoing equation can be expressed as

$$\begin{Bmatrix} x^{(m)}(t) \\ y^{(m)}(t) \end{Bmatrix} \equiv \begin{Bmatrix} x_1^{(m)}(t) \\ x_2^{(m)}(t) \end{Bmatrix} \approx \begin{Bmatrix} a_{01}^{(m)} \\ a_{02}^{(m)} \end{Bmatrix} + \sum_{j=1}^{mN/2} \begin{Bmatrix} A_{j/m1} \cos(\frac{j}{m}\Omega t - \varphi_{j/m1}) \\ A_{j/m1} \cos(\frac{j}{m}\Omega t - \varphi_{j/m2}) \end{Bmatrix}. \quad (6.36)$$

For simplicity, only the excitation frequency–amplitude curves for the displacement $x^{(m)}(t)$ are presented. Similarly, the frequency–amplitudes for velocity $y^{(m)}(t)$ can also be determined. Thus, the displacement can be expressed as

$$x^{(m)}(t) \approx a_0^{(m)} + \sum_{j=1}^{mN/2} b_{j/m} \cos\left(\frac{j}{m}\Omega t\right) + c_{j/m} \sin\left(\frac{j}{m}\Omega t\right) \tag{6.37}$$

and

$$x^{(m)}(t) \approx a_0^{(m)} + \sum_{j=1}^{mN/2} A_{j/m} \cos(\frac{j}{m}\Omega t - \varphi_{j/m}) \tag{6.38}$$

where

$$A_{j/m} = \sqrt{b_{j/m}^2 + c_{j/m}^2}, \quad \varphi_{j/m} = \arctan\frac{c_{j/m}}{b_{j/m}}. \tag{6.39}$$

To discuss nonlinear behaviors of period-$m$ motion for the Duffing oscillator, the frequency–amplitude for displacement will be presented as follows. The acronyms SN and PD are the saddle-node and period-doubling bifurcations for period-m motions, respectively. In all plots, the unstable and stable solutions of period-m motions are represented by the dashed and solid curves, respectively.


## 6.4.1  Period-1 Motions to Chaos

The bifurcation trees of period-1 motion to chaos will be presented through the period-1 to period-4 motions, as shown in Fig. 6.5. The given parameters are listed in Eq. (6.29). The constant term $a_0^{(m)}$ ($m = 1, 2, 4$) is presented in Fig. 6.5i for the solution center on the right side of the y-axis. The bifurcation tree is clearly observed. For the solution center on the left side of the y-axis, we have $a_0^{(m)L} = -a_0^{(m)R}$. For the symmetric period-$m$ motion, we have $a_0^{(m)} = 0$, labeled by "S." However, for asymmetric period-$m$ motion, we have $a_0^{(m)} \neq 0$, labeled by "A." For the symmetric period-1 motion to an asymmetric period-1 motion, the saddle-node bifurcation will occur. The saddle-node bifurcations are at $\Omega \approx 1.016, 1.23, 1.50, 2.63, 4.528$. For such saddle-node bifurcations, the asymmetric periodic motions appear, and the symmetric motions are from the stable to unstable solution or from the unstable to stable solution. The saddle-node bifurcations for symmetric motion jumping points are at $\Omega \approx 1.46, 1.513, 3.96, 5.98$. The symmetric period-1 motion is only from the stable to unstable solution or from the unstable to stable solution. When the asymmetric period-1 motion experiences a period-doubling bifurcations, the period-2 motions will appear and the asymmetric period-1 motion are from the stable to unstable solution. The frequencies of $\Omega \approx 1.517, 1.97, 4.528, 7.27$ are not only for the period-doubling bifurcations of the asymmetric period-1 motions but also for the saddle-node bifurcations of the period-2 motion. When the period-2 motion possesses a period-doubling bifurcation, the period-4 motion appears and the period-2 motion is from the stable to unstable solution. The frequencies of $\Omega \approx 1.52, 1.90, 4.97, 6.58$ are for the period-doubling bifurcations of period-2 motions and for the saddle-node bifurcation for the period-4 motions. The frequencies of $\Omega \approx 5.03, 6.49$ are for the period-doubling bifurcations of period-4 motions and for the saddle-node bifurcation for the period-8 motions. All period-2 and period-4 motions are on the branches of asymmetric period-1 motions, and the centers of the periodic motions are on the right side of the y-axis. In Fig. 6.5ii, the harmonic amplitude $A_{1/4}$ is presented. For period-1 and period-2 motions, $A_{1/4} = 0$. The saddle-node bifurcations are at $\Omega \approx 4.97, 6.58$ for period-1 motion, and the period-doubling bifurcations are at $\Omega \approx 5.03, 6.49$. The bifurcation points are clearly observed, and the quantity level of the harmonic amplitude for period-4 motion is $A_{1/4} \sim 7 \times 10^{-2}$. In Fig. 6.5iii, the harmonic amplitude $A_{1/2}$ for period-4 and period-2 motions is presented. For the second branch, only the period-2 motion are presented because the stability range of period-4 motion is very small and more discrete nodes are needed to obtain such a period-4 motion. For the third branch, the bifurcation trees for period-2 to period-4 motions are clearly illustrated. The period-doubling bifurcations are at $\Omega \approx 5.03, 6.49$ for the third branch. The saddle-node bifurcations of the period-2 motion are at $\Omega \approx 1.517, 1.97, 4.528, 7.27$ for the second and third branches. The quantity level of the harmonic amplitude $A_{1/2}$ is $A_{1/2} \sim 1.5 \times 10^{-1}$. In Fig. 6.5iv, the harmonic amplitude $A_{3/4}$ is presented, which is similar to the harmonic amplitude $A_{1/4}$. The quantity level of such a harmonic amplitude is $A_{3/4} \sim 1.5 \times 10^{-2}$. The other harmonic amplitudes $A_{k/4}$ ($k = 4l + 1, 4l + 3, l = 1, 2, \ldots$) will not be presented herein for reduction of abundant illustrations. In Fig. 6.5v, the primary harmonic amplitudes $A_1$ versus excitation frequency $\Omega$ are presented for the period-1 to period-4 motion. The bifurcation trees are clearly observed. The entire skeleton of the frequency–amplitude for the symmetric period-1 motion is presented, and the asymmetric period-1 motions and the corresponding period-2 and period-4 motions are attached to the symmetric period-1 motion. The quantity level of the primary amplitude is $A_1 \sim 1.8$ for all period-1 to period-4 motions. The bifurcation points are presented as before. In Fig. 6.5vi, the harmonic amplitude $A_{3/2}$ is presented. The bifurcation trees are similar to the harmonic amplitude $A_{1/2}$. The quantity levels of $A_{3/2}$ and $A_{1/2}$ are almost same. That is, $A_{3/2} \sim 0.1$ and $A_{1/2} \sim 0.15$. To reduce abundant illustrations, $A_{k/2}$ ($k = 2l + 1, l = 2, 3, \ldots$) will not be presented anymore. In Fig. 6.5vii, the harmonic amplitude $A_2$ is presented, which is similar to constant term $a_0^{(m)}$. The bifurcation trees have the similar structures for the different harmonic amplitudes, but the corresponding quantity levels of harmonic amplitudes are different. That is, $A_2 \sim 0.6$ are for the first and second branches. However, for the third branch, we have $A_2 \sim 0.01$. In Fig. 6.5viii, the harmonic amplitude $A_3$ is presented, similar to the primary harmonic amplitude $A_1$. The bifurcation trees are different for the different harmonic

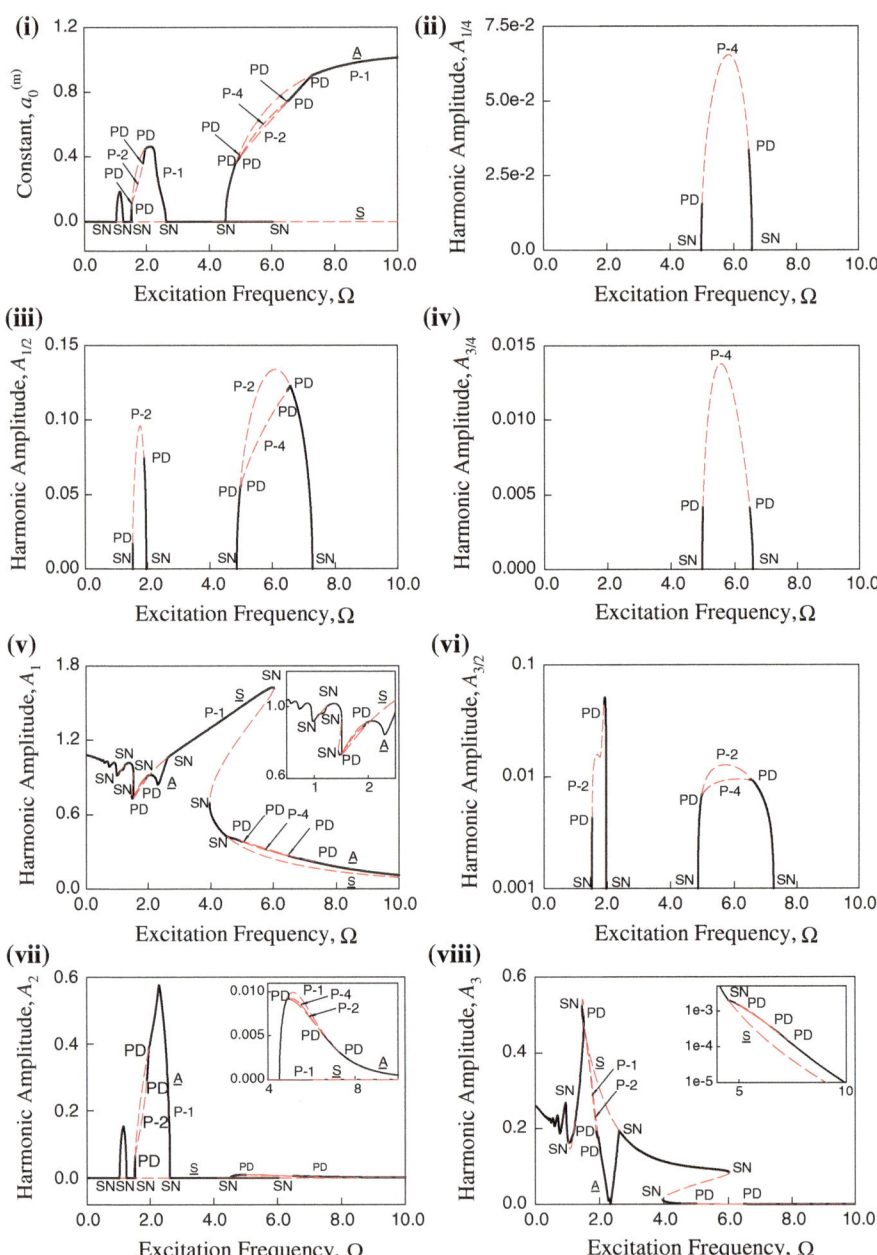

**Fig. 6.5** Frequency–amplitude characteristics for bifurcation trees of period-1 to period-4 motions: $i$  $a_0^{(m)}$  $(m = 1, 2, 4)$  and  $ii$–$x$  $A_{k/m}$  $(m = 4, k = 1, 2, 3, 4; 6, 8, 12, 84, 244)$;  $(\alpha = 5.5, \beta = 20.0,$  $\delta = 1.0, Q_0 = 10)$

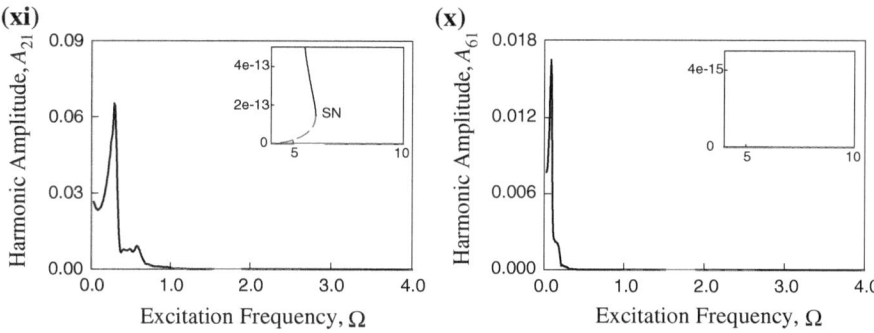

**Fig. 6.5** (continued)

amplitudes, and the corresponding quantity levels of harmonic amplitudes are different. That is, we obtain $A_3 \sim 0.6$, for $\Omega < 5$. However, for $\Omega \geq 5$, we have $A_3 \sim 10^{-3}$. To avoid abundant illustrations, the harmonic amplitudes of $A_{21}$ and $A_{61}$ are presented in Fig. 6.5ix, x. For $\Omega > 1$, $A_{21} < 10^{-5}$ and $A_{61} < 10^{-10}$. For $\Omega < 1$, $A_{21} \sim 0.1$ and $A_{61} \sim 10^{-2}$. From the above discussion on the periodic motion, for $\Omega > 1$, we can use about 80 harmonic terms to approximate period-1, period-2, and period-4 motions. For $\Omega < 1$ but not close to zero, we can use 250 harmonic terms to approximate period-1, period-2, and period-4 motions. For $\Omega \approx 0$, the infinite harmonic terms should be adopted to approximate the periodic motions.

## 6.4.2 Period-3 Motions

The bifurcation trees of period-3 motion to chaos will be presented through the period-3 motions, as shown in Fig. 6.6. Since the period-6 motion has a short stable solution, the bifurcation tree may not be very nice. The given parameters are still listed in Eq. (6.29). The constant term $a_0^{(3)}$ is presented in Fig. 6.6i for the solution center on the right side of the $y$-axis. For the solution center on the left side of the $y$-axis, $a_0^{(3)L} = -a_0^{(3)R}$. For the symmetric period-3 motion, $a_0^{(3)} = 0$, also labeled by "S." However, for the asymmetric period-3 motion, we have $a_0^{(3)} \neq 0$, also labeled by "A". For the symmetric period-3 motion to an asymmetric period-3 motion, the saddle-node bifurcation will occur. The two closed branches of period-3 bifurcations are in range of $\Omega \in (1.5, 1.8)$ and $\Omega \in (4.0, 8.0)$. The four saddle-node bifurcations at $\Omega \approx 1.523, 1.772, 4.30, 7.89$ are for the stable and unstable symmetric period-3 motions only. However, the other four saddle-node bifurcations of the symmetric period-3 motions at $\Omega \approx 1.526, 1.695, 4.39, 6.69$ are not only for the stable and unstable symmetric period-3 motions but also for appearance of the asymmetric period-3 motions. The period-doubling bifurcations of asymmetric period-3 motions are at $\Omega \approx 1.528, 1.678$ for the first branch and $\Omega \approx 4.417, 6.414$

for the second branch, which are also the saddle-node bifurcation for period-6
motions. In Fig. 6.6ii, the harmonic amplitude $A_{1/3}$ is presented. The bifurcation
trees for two branches of period-3 motions are clearly observed. The quantity levels
of such harmonic amplitudes are $A_{1/3} \sim 0.3$ for the first branch and $A_{1/3} \sim 0.9$ for
the second branch of period-3 motions. In Fig. 6.6iii, the harmonic amplitude $A_{2/3}$
is presented. The bifurcation trees of $A_{2/3}$ for two branches of period-3 motions are
similar to $a_0^{(3)}$. The quantity levels of the harmonic amplitudes are $A_{2/3} \sim 0.06$ for
the first branch and $A_{2/3} \sim 0.18$ for the second branch of period-3 motions. To
avoid abundant illustrations, harmonic amplitudes $A_{j/3}$ ($\mathrm{mod}(j, 3) \neq 0$) will not be
presented. The harmonic amplitudes $A_1$ and $A_3$ are presented in Fig. 6.6iv, vi. The
bifurcation trees of these harmonic amplitudes are similar to the harmonic amplitude $A_{1/3}$. The quantity levels of $A_1$ are $A_1 \in (0.75, 0.88)$ for the first branch and
$A_1 \in (0.15, 0.55)$ for the second branch. For the harmonic amplitude $A_3$, we have
$A_3 \in (0.15, 0.5)$ for the first branch and $A_3 \in (0.0, 0.005)$ for the second branch.
The harmonic amplitudes $A_2$ are presented in Fig. 6.6v. The bifurcation trees of
these harmonic amplitudes are similar to constant term $a_0^{(3)}$ and the harmonic
amplitude $A_{2/3}$. The quantity levels of the harmonic amplitude $A_2$ are $A_2 < 0.024$ for
the first branch and $A_2 < 0.012$ for the second branch. To look into the effects of the
higher order harmonic amplitudes, the harmonic amplitudes $A_{60}, A_{181/3}$ are presented in Fig. 6.6vii–viii, respectively. The bifurcation tree of the harmonic
amplitude $A_{60}$ is similar to the constant term $a_0^{(3)}$ and the harmonic amplitude $A_{2/3}$.
For the first branch, $A_{60} < 2.5 \times 10^{-14}$, but for the second branch, $A_{60} < 10^{-16}$. In
fact, for the second branch, the quantity level should be much smaller because our
computational algorithm cannot achieve such more accurate results. The bifurcation
trees of the harmonic amplitude $A_{181/3}$ are similar to the harmonic amplitude $A_{1/3}$.
$A_{181/3} < 3.5 \times 10^{-14}$ for the first branch, but for the second branch, $A_{181/3} < 10^{-16}$.
Once again, for the second branch, the quantity level should be much smaller owing
to the computational accuracy of the discrete algorithm and time step.

## 6.5  Numerical Simulations

In this section, numerical illustrations are given from the semi-analytical solutions
and numerical integration schemes. The initial conditions in numerical simulation
are obtained from analytical prediction of periodic solutions. In all plots for illustration, circular symbols give analytical predictions, and solid curves give numerical
simulation results. Acronym "IC" represents initial conditions. The initial points and
the corresponding periodic points are depicted by the large circular symbols.

   In Fig. 6.7, consider excitation frequency $\Omega = 1.05$ to demonstrate period-1
motion. Other parameters are presented in Eq. (6.29). The analytical prediction
gives the initial condition $(x_0, y_0) \approx (0.888313, -2.694745)$. The displacement,
velocity, trajectory, and harmonic spectrum are presented in Fig. 6.7a–d

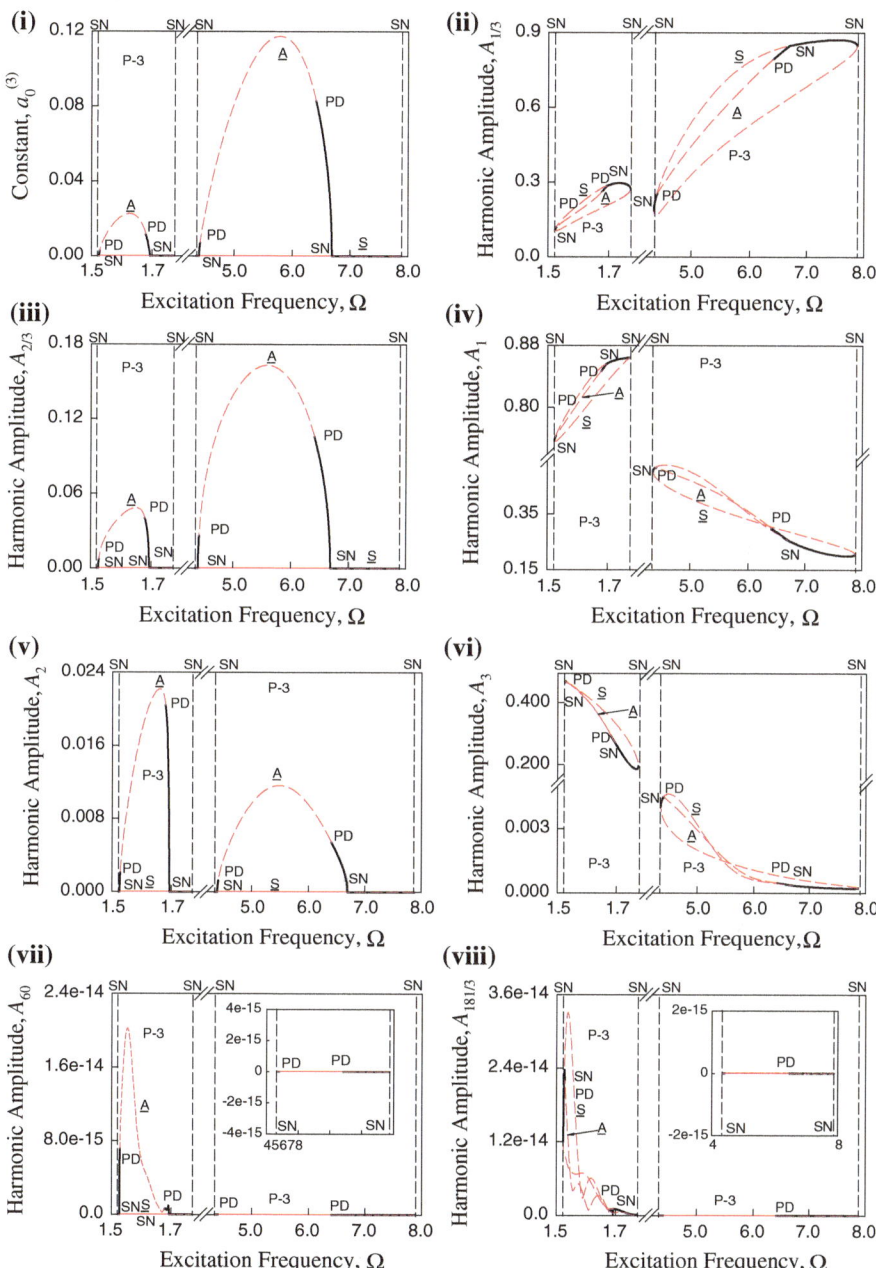

**Fig. 6.6** Frequency–amplitude characteristics for stable and unstable period-3 motions: $i$ $a_0^{(m)}$ ($m = 3$). $ii$–$vii$ $A_{k/m}$ ($m = 3, k = 1, 2, 3; 6, 9, 180, 181$); ($\alpha = 5.5, \beta = 20.0, \delta = 1.0, Q_0 = 10$)

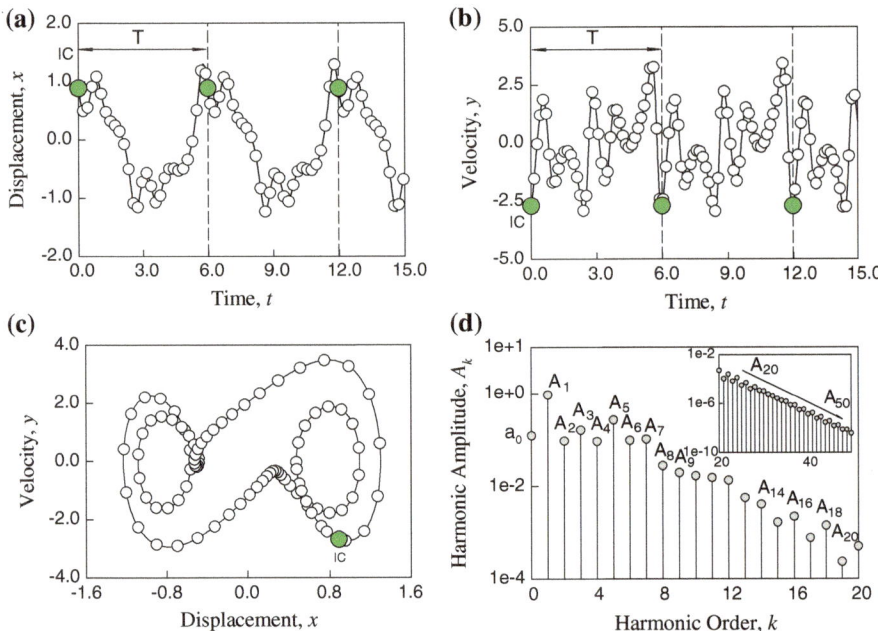

**Fig. 6.7** Period-1 motions ($\Omega = 1.05$): **a** displacement, **b** velocity, **c** trajectory, and **d** harmonic amplitude. Initial condition $(x_0, y_0) \approx (0.888313, -2.694745)$. Parameters ($\alpha = 5.5$, $\beta = 20.0$, $\delta = 1.0, Q_0 = 10$)

respectively. In Fig. 6.7a, b, the time histories of displacement and velocity are not simple sinusoidal alike periodic motion. One period ($1T$) is labeled for the period-1 motion. The trajectory of period-1 motions is presented in Fig. 6.7c. The period-1 motion is like a period-3 motion. The analytical prediction results match very well with numerical simulation results. The harmonic amplitude spectrum is presented in Fig. 6.7d. The constant term is $a_0 \approx 0.1235$. The main harmonic amplitudes are $A_1 \approx 0.9413$, $A_2 \approx 0.0949$, $A_3 \approx 0.1620$, $A_4 \approx 0.0924$, $A_5 \approx 0.2699$, $A_6 \approx 0.0978$, $A_7 \approx 0.1035$, $A_8 \approx 0.0277$, $A_9 \approx 0.0195$, $A_{10} \approx 0.0166$, $A_{11} \approx 0.0153$,. The other harmonic amplitudes are $A_j \in (10^{-9}, 10^{-2})$ ($j = 13, 14, \ldots, 50$) and $A_{50} \approx 3.6100 \times 10^{-9}$. The harmonic amplitudes decrease very slowly with harmonic order. For this period-1 motion, we cannot use one harmonic term to approximate the periodic solutions. From the harmonic amplitudes, at least 12 harmonic terms plus constant term should be included to obtain the rough estimate of periodic motion. The other harmonic amplitudes still can be presented. However, the quantity level is very small, and they will not be presented.

In Fig. 6.8, consider excitation frequency $\Omega = 1.686$ to demonstrate a complex period-3 motion. Other parameters are also presented in Eq. (4.29). The initial condition is $(x_0, y_0) \approx (0.651260, 3.947260)$. The displacement, velocity, trajectory, and harmonic spectrum for the period-3 motion are presented in Fig. 6.8a–d, respectively. The time histories of displacement and velocity are presented in

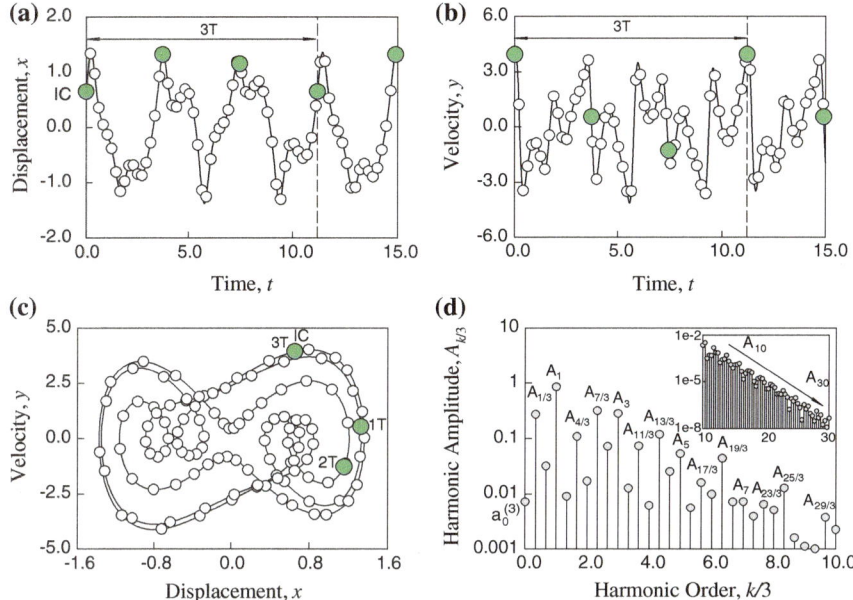

**Fig. 6.8** Period-3 motions ($\Omega = 1.686$): **a** displacement, **b** velocity, **c** trajectory, and **d** harmonic amplitude. Initial condition $(x_0, y_0) \approx (0.651260, 3.947260)$. Parameters ($\alpha = 5.5, \beta = 20.0$, $\delta = 1.0, Q_0 = 10$)

Fig. 6.8a, b, and three periods (3$T$) is labeled for the period-3 motion. The trajectory of period-3 motions is presented in Fig. 6.8c. The period-3 motion is very complex. The initial points with the corresponding periodic points are depicted through the large circular symbols. After three periods, the period-3 motion returns back to the initial condition. The harmonic amplitude spectrum is presented in Fig. 6.8d. The constant term is $a_0^{(3)} \approx 7.2200 \times 10^{-3}$. The main harmonic amplitudes are $A_{1/3} \approx 0.2725$, $A_{2/3} \approx 0.0318$, $A_1 \approx 0.8515$, $A_{4/3} \approx 9.0457 \times 10^{-3}$, $A_{5/3} \approx 0.1088$, $A_2 \approx 0.0170$, $A_{7/3} \approx 0.3172$, $A_{8/3} \approx 0.0721$, $A_3 \approx 0.2823$, $A_{10/3} \approx 0.0127$, $A_{11/3} \approx 0.0733$, $A_4 \approx 6.1598 \times 10^{-3}$, $A_{13/3} \approx 0.1184$, $A_{14/3} \approx 0.0253$, $A_5 \approx 0.0530$, $A_{16/3} \approx 5.5773 \times 10^{-3}$, $A_{17/3} \approx 0.0160$, $A_6 \approx 9.8557 \times 10^{-3}$, $A_{19/3} \approx 0.0440$, $A_{20/3} \approx 7.1797 \times 10^{-3}$, $A_7 \approx 7.2147 \times 10^{-3}$, $A_{22/3} \approx 3.9140 \times 10^{-3}$, $A_{23/3} \approx 6.4677 \times 10^{-3}$, $A_8 \approx 5.0847 \times 10^{-3}$, and $A_{25/3} \approx 0.0128$. The other harmonic amplitudes are $A_{j/3} \in (10^{-9}, 10^{-2})$ ($j = 26, 27, \ldots, 90$) and $A_{30} \approx 4.6072 \times 10^{-3}$. The harmonic amplitudes decrease very slowly with harmonic order. For the period-3 motion, at least 25 harmonic terms plus constant term should be included to obtain the rough estimate of periodic motion. The harmonic amplitudes decrease non-uniformly. For the harmonic amplitudes, the primary harmonic terms of $A_1 \approx 0.8515$ play an important role in the period-3 motion. In traditional analysis, such a period-3 motion cannot be called the superharmonic or subharmonic motion.

To avoid too many illustrations, only trajectories and harmonic amplitude spectrums are presented for periodic motions on the bifurcation tree of period-1 to period-4 motion. Consider an excitation frequency of $\Omega = 8.0$ for period-1 motion. Because the Duffing oscillator possesses the twin-potential well, the two asymmetric solutions will be associated with the twin-potential well. Thus, two initial conditions for the two asymmetric period-1 motions are $(x_0, y_0) \approx$ $(0.301097, 0.195246)$, and $(-0.655147, 0.215682)$. The trajectories of the two asymmetric period-1 motion are presented in Fig. 6.9a. The two asymmetric

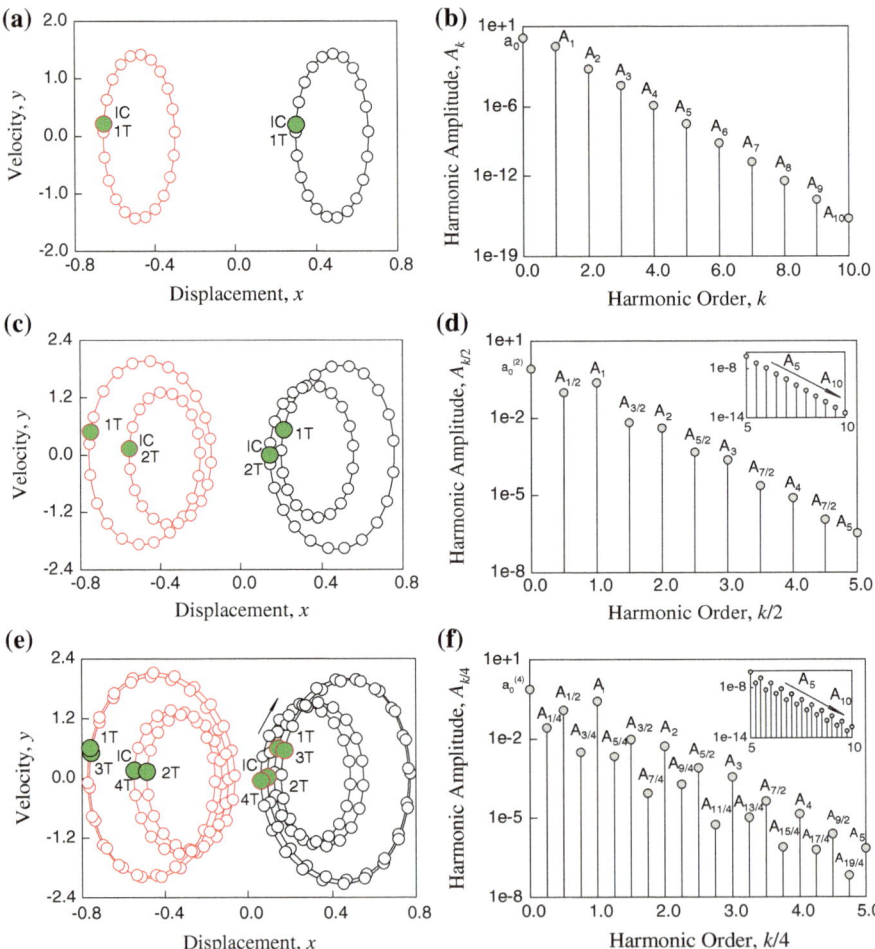

**Fig. 6.9** Period-1 motions ($\Omega = 8.0$): **a** trajectory and **b** harmonic amplitude. Initial condition $(x_0, y_0) \approx (0.301097, 0.195246)$, $(-0.655147, 0.215682)$; period-2 motions ($\Omega = 6.9$): **c** trajectory and **d** harmonic amplitude. Initial condition $(x_0, y_0) \approx (0.214375, 0.524608)$, $(-0.555016, 0.138223)$; period-4 motions ($\Omega = 6.52$): **e** trajectory and **f** harmonic amplitude. Initial condition $(x_0, y_0) \approx (0.062918, -0.051147)$, $(-0.485633, 0.129886)$. Parameters ($\alpha = 5.5$, $\beta = 20.0$, $\delta = 1.0, Q_0 = 10$)

motions are skew symmetric. After one period, the periodic motion returns back to the initial point. The corresponding harmonic amplitudes are presented in Fig. 6.9b. The constant terms are $a_0^L = -a_0^R$, and the other harmonic amplitudes are same. However, the harmonic phases are different with $\varphi_j^L = \mathrm{mod}(\varphi_j^R + (j+1)\pi, 2\pi)$ $(j = 1, 2, \ldots)$. The constant term is $a_0 = a_0^R \approx 0.9528$. The main harmonic amplitudes are $A_1 \approx 0.1788$, and $A_2 \approx 1.8480 \times 10^{-3}$. The other harmonic amplitudes are $A_j \in (10^{-9}, 10^{-3})$ $(j = 3, 4, \ldots, 10)$ and $A_{10} \approx 1.9170 \times 10^{-16}$. For this case, only one harmonic term plus the constant term can provide a good approximation of the period-1 motion. Consider the excitation frequency of $\Omega = 6.9$ for two period-2 motions which are near the asymmetric period-1 motion of $\Omega = 8.0$. The initial conditions for the two period-2 motions are $(x_0, y_0) \approx (0.214375, 0.524608)$, and $(-0.555016, 0.138223)$. The trajectories and harmonic amplitudes for such period-2 motions are presented in Fig. 6.9c, d, respectively. The trajectory of the period-2 motion is more complex than the period-1 motion. After two periods, the period-2 motion returns back to the initial point, and the point at one period is depicted with a large circular symbol. The constant terms are $a_0^{(2)L} = -a_0^{(2)R}$, and the other harmonic amplitudes are the same. However, the harmonic phases are different with $\varphi_{j/2}^L = \mathrm{mod}(\varphi_{j/2}^R + (j/2 + 1)\pi, 2\pi)$ $(j = 1, 2, \ldots)$. The constant term for the period-2 motion on the right-hand side is $a_0^{(2)} = a_0^{(2)R} \approx 0.8246$. The main harmonic amplitudes are $A_{1/2} \approx 0.0985$, $A_1 \approx 0.2393$, $A_{3/2} \approx 6.6685 \times 10^{-3}$, and $A_2 \approx 4.1152 \times 10^{-3}$. The other harmonic amplitudes are $A_{j/2} \in (10^{-9}, 10^{-3})$ $(j = 5, 6, \ldots, 20)$ and $A_{10} \approx 3.5305 \times 10^{-14}$. For this period-2 motion, two harmonic terms plus the constant term can provide a good approximation. Consider the excitation frequency of $\Omega = 6.52$ for period-4 motions. The corresponding trajectories and harmonic amplitudes are presented in Fig. 6.9e, f, respectively. The trajectory of the period-4 motion is much more complex than the period-1 motion. After four periods, the period-4 motion returns back to the initial point, and the points at one, two, and three periods are depicted with large circular symbols, The constant terms are also $a_0^{(4)L} = -a_0^{(4)R}$, and the other harmonic amplitudes are still the same. However, the harmonic phases are different with $\varphi_{j/4}^L = \mathrm{mod}(\varphi_{j/4}^R + (j/4 + 1)\pi, 2\pi)$ $(j = 1, 2, \ldots)$. The constant term for the period-4 motion on the right-hand side is $a_0^{(4)} = a_0^{(4)R} \approx 0.7471$. The main harmonic amplitudes are $A_{1/4} \approx 0.0262$, $A_{1/2} \approx 0.1210$, $A_{3/4} \approx 3.0916 \times 10^{-3}$, $A_1 \approx 0.2632$, $A_{5/4} \approx 2.1204 \times 10^{-3}$, $A_{3/2} \approx 9.3470 \times 10^{-3}$, $A_{7/4} \approx 8.5130 \times 10^{-5}$, and $A_2 \approx 5.2201 \times 10^{-3}$. The other harmonic amplitudes are $A_{j/4} \in (10^{-9}, 10^{-3})$ $(j = 9, 10, \ldots, 40)$ and $A_{10} \approx 1.8765 \times 10^{-13}$. Eight harmonic terms plus constant term can give a good approximation for period-4 motions.

Periodic motions in the bifurcation tree of period-3 motion will be illustrated for demonstration of motion complexity. Consider an excitation frequency of $\Omega = 6.52$ for period-3 motion. The period-3 motion crosses the separatrix of the non-damped, twin-well, Duffing oscillator. The initial condition for such an asymmetric period-3

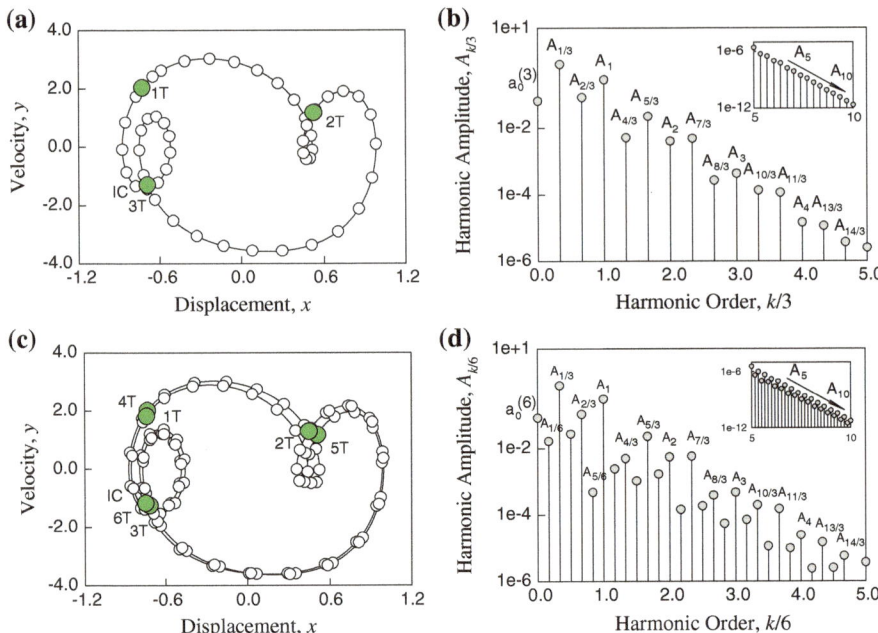

**Fig. 6.10** Period-3 motions ($\Omega = 6.52$): **a** trajectory and **b** harmonic amplitude. Initial condition $(x_0, y_0) \approx (-0.728926, 2.046610)$; period-6 motions ($\Omega = 6.37$): **c** trajectory and **d** harmonic amplitude. Initial condition $(x_0, y_0) \approx (0.448070, 1.286008)$. Parameters ($\alpha = 5.5$, $\beta = 20.0$, $\delta = 1.0, Q_0 = 10$)

motions is $(x_0, y_0) \approx (-0.728926, 2.046610)$. The trajectories of the asymmetric period-3 motion are presented in Fig. 6.10a. After three periods, the period-3 motion returns back to the initial point. The corresponding harmonic amplitudes are presented in Fig. 6.10b. The constant term is $a_0^{(3)} \approx 0.0660$. The main harmonic amplitudes are $A_{1/3} \approx 0.8156$, $A_{2/3} \approx 0.0843$, $A_1 \approx 0.2827$, $A_{4/3} \approx 5.0215 \times 10^{-3}$, $A_{5/3} \approx 0.0222$, $A_2 \approx 4.0123 \times 10^{-3}$, and $A_{7/3} \approx 4.8011 \times 10^{-3}$. The other harmonic amplitudes are $A_{j/3} \in (10^{-9}, 10^{-3})$ ($j = 8, 9, \ldots, 30$) and $A_{10} \approx 2.1566 \times 10^{-12}$. For the period-3 motion, only seven harmonic terms plus the constant term can provide a good approximation of the period-3 motion. For the harmonic amplitudes, $A_{1/3} \approx 0.8156$ plays an important role in the period-3 motion. In traditional analysis, such a period-3 motion is called the subharmonic motion. The center of such a period-3 motion is on the right-hand side. However, there is another asymmetric period-3 motion possessing the center on the left-hand side. The constant terms are $a_0^{(3)L} = -a_0^{(3)R}$, and the other harmonic amplitudes are same. However, the harmonic phases are different with $\varphi_{j/2^l m}^L = \mathrm{mod}(\varphi_{j/2^l m}^R + (j/2^l + 1)\pi, 2\pi)$ ($m = 3$, $j = 1, 2, \ldots$, $l = 0, 1, 2, \ldots$; ). The constant term is $a_0^{(3)} = a_0^{(3)R} \approx 0.0660$. Consider the excitation frequency of $\Omega = 6.37$ for period-6 motions which is given by the analytical prediction. The initial condition for the period-6 motions are $(x_0, y_0) \approx (-0.732997, 2.006567)$. The trajectory and harmonic amplitudes for such a period-6 motion are presented in Fig. 6.10c, d, respectively. After six periods, the period-6 motion returns back to the initial point, and the periodic points for six periods are presented by the large circular symbols, The constant term for the period-6 motion on the right-hand side is $a_0^{(6)} = a_0^{(6)R} \approx 0.0832$. The main harmonic amplitudes are $A_{1/6} \approx 0.0166$, $A_{1/3} \approx 0.7886$, $A_{1/2} \approx 0.0272$, $A_{2/3} \approx 0.1059$, $A_{5/6} \approx 4.7814 \times 10^{-4}$, $A_1 \approx 0.3043$, $A_{7/6} \approx 2.4425 \times 10^{-3}$, $A_{4/3} \approx 4.9541 \times 10^{-3}$, $A_{3/2} \approx 1.0541 \times 10^{-3}$, $A_{5/3} \approx 0.0228$, $A_{11/6} \approx 1.6912 \times 10^{-3}$, $A_2 \approx 5.5408 \times 10^{-3}$, $A_{13/6} \approx 1.4457 \times 10^{-4}$, and $A_{7/3} \approx 5.7796 \times 10^{-3}$. The other harmonic amplitudes are $A_{j/6} \in (10^{-12}, 10^{-3})$ ($j = 15, 16, \ldots, 60$) and $A_{10} \approx 4.9640 \times 10^{-12}$. For this period-6 motion, 14 harmonic terms plus the constant term can provide a good approximation. The constant terms are $a_0^{(6)L} = -a_0^{(6)R}$, and the other harmonic amplitudes are the same. However, the phases are different with $\varphi_{j/2^l m}^L = \text{mod}(\varphi_{j/2^l m}^R + (j/2^l + 1)\pi, 2\pi)(m = 3, j = 1, 2, \ldots, l = 1)$.

## Reference

Luo, A. C. J., & Guo, Y. (2015). A semi-analytical prediction of periodic motions in the Duffing oscillator through mapping structure. In *Discontinuity, Nonlinearity and Complexity, 4*(2), 121–150.

# Index

© Higher Education Press, Beijing and Springer-Verlag Berlin Heidelberg 2015
A.C.J. Luo, *Discretization and Implicit Mapping Dynamics*,
Nonlinear Physical Science, DOI 10.1007/978-3-662-47275-0